Handbook of Experimental Pharmacology

Volume 216

For further volumes:
http://www.springer.com/series/164

Erich Gulbins • Irina Petrache
Editors

Sphingolipids in Disease

Springer

Editors
Erich Gulbins
Department of Molecular Biology
University of Duisburg-Essen
Essen
Germany

Irina Petrache
Division of Pulmonary, Allergy,
Critical Care and Occupational Medicine
Department of Medicine
Indiana University School of Medicine
Indianapolis, IN, USA

ISSN 0171-2004 ISSN 1865-0325 (electronic)
ISBN 978-3-7091-1510-7 ISBN 978-3-7091-1511-4 (eBook)
DOI 10.1007/978-3-7091-1511-4
Springer Wien Heidelberg New York Dordrecht London

Library of Congress Control Number: 2013935589

Printed on acid-free paper

Springer is part of Springer Science+Business Media (www.springer.com)

Preface

Until the late 1980s, sphingolipids were believed to represent structural components of the plasma membrane, whose function was to provide a protective barrier to the cell. This picture dramatically changed within the last years. Sphingolipids are now recognized signals for fundamental cellular processes such as proliferation, survival, cell death, adhesion, migration, angiogenesis, and embryogenesis. The explosion of knowledge regarding sphingolipids was facilitated by biochemical studies of their signaling properties, the cloning of enzymes of the sphingolipid metabolism, development of genetic models for determining their physiologic roles, and the establishment of biochemical, biophysical, and optical methods for their detection and quantitation. The next step in the evolution of sphingolipids will be the transfer of basic insights into the biochemistry and cell biology of human disease. The recent success of the sphingolipid drug, Fingolimod, a sphingosine 1-phosphate agonist, which rapidly became a therapy for multiple sclerosis, exemplifies the potential of targeting sphingolipids for the treatment of human disorders. The aim of our two volumes in this series—*Sphingolipids*: *Basic Science and Drug Development* and *Sphingolipids in Disease*—is to define the state of the art of sphingolipid biology and to present preclinical developments and early clinical applications of this fascinating class of lipids.

Essen, Germany — Erich Gulbins
Indianapolis, IN, USA — Irina Petrache

Contents

Part III Sphingolipids in Inflammation, Infection and Lung Diseases

Part IV Sphingolipids in Neuro-psychiatry and Muscle Diseases

Part I
Sphingolipids in Cancer

Sphingosine Kinase/Sphingosine 1-Phosphate Signaling in Cancer Therapeutics and Drug Resistance

Shanmugam Panneer Selvam and Besim Ogretmen

Contents

Abstract In this chapter, roles of bioactive sphingolipids, specifically sphingosine kinase 1 (SK1) and 2 (SK2) and their product—sphingosine 1-phosphate (S1P)—will be reviewed with respect to regulation of cancer growth, metastasis, chemotherapeutics, and drug resistance. Sphingolipids are known to be key bioeffector molecules that regulate cancer proliferation, angiogenesis, and cell death. Sphingolipid molecules such as ceramide and S1P have been shown to control cancer cell death and proliferation, respectively. Roles of S1P have been described with respect to their intracellular and extracellular pro-survival and drug resistance

S.P. Selvam • B. Ogretmen (✉)
Department of Biochemistry and Molecular Biology, Medical University of South Carolina, 86, Jonathan Lucas Street, Room 512A, Charleston, SC 29425, USA

Hollings Cancer Center, Medical University of South Carolina, 86, Jonathan Lucas Street, Room 512A, Charleston, SC 29425, USA
e-mail: ogretmen@musc.edu

E. Gulbins and I. Petrache (eds.), *Sphingolipids in Disease*,
Handbook of Experimental Pharmacology 216,
DOI 10.1007/978-3-7091-1511-4_1,

functions mostly through S1P receptor (S1PR1-5) engagement. Identification of novel intracellular SK/S1P targets has broadened the existing complex regulatory roles of bioactive sphingolipids in cancer pathogenesis and therapeutics. Thus, deciphering the biochemical and molecular regulation of SK/S1P/S1PR signaling could permit development of novel therapeutic interventions to improve cancer therapy and/or overcome drug resistance.

Keywords SK1 • SK2 • S1P • S1PR • Ceramide • Cancer and Drug resistance • Anti-Cancer therapeutics

1 Introduction

Sphingolipids are structural and functional components of biological membranes (Ogretmen and Hannun 2004; Ponnusamy et al. 2010), which contribute to maintenance of membrane fluidity and subdomain structure. They are also implicated in bioeffector roles in cancer pathogenesis (Hannun and Obeid 2008; Ogretmen and Hannun 2004). Bioactive sphingolipids such as ceramide, sphingosine, and S1P are important in cell death pathways (apoptosis, necrosis, autophagy, anoikis), cancer proliferation, migration, inflammation, and drug resistance (Hannun and Obeid 2008; Ogretmen and Hannun 2004; Ponnusamy et al. 2010; Saddoughi et al. 2008). This chapter will focus on the roles of SK/S1P/S1PR signaling in cancer cell growth, therapeutics, drug resistance, and metastasis.

2 Sphingolipid Metabolism

The de novo sphingolipid synthesis pathway (Fig. 1) begins with the condensation of serine and palmitoyl-coA catalyzed by serine palmitoyl transferase (SPT) (Dolgachev et al. 2004; Reynolds et al. 2004) leading to 3-ketosphinganine generation, which is rapidly reduced to dihydrosphingosine. Dihydrosphingosine is then *N*-acylated by a family of six dihydroceramide synthases (CerS1-6, also known as longevity associated gene, LAG, homologues, LASS1-6), which show preference for varying fatty acyl chain length specificity to synthesize dihydroceramide (Futerman and Hannun 2004). Then, a double bond is introduced between carbons 4 and 5 of the sphingosine backbone to generate ceramide (Kraveka et al. 2007; Michel et al. 1997). Ceramide is at the center of the sphingolipid metabolism, displaying mainly antiproliferative and pro-apoptotic roles (Ogretmen and Hannun 2004; Ponnusamy et al. 2010). Ceramide can be deacylated by ceramidases to generate sphingosine, which is rapidly phosphorylated by SK1 and SK2 to generate S1P, a pleiotropic lipid elucidating pro-survival, anti-apoptotic, metastatic, and/or chemoresistance functions in various cancers (Spiegel and Milstien 2003). S1P can be dephosphorylated by two S1P phosphatases (S1PP1 and S1PP2) in a reversible reaction to generate sphingosine (Mandala 2001; Mao et al. 1999; Spiegel and Milstien 2003), or S1P can be irreversibly cleaved by S1P lyase to form ethanolamine-1-phosphate and hexadecenal (Ikeda et al. 2004). Recently, S1P and

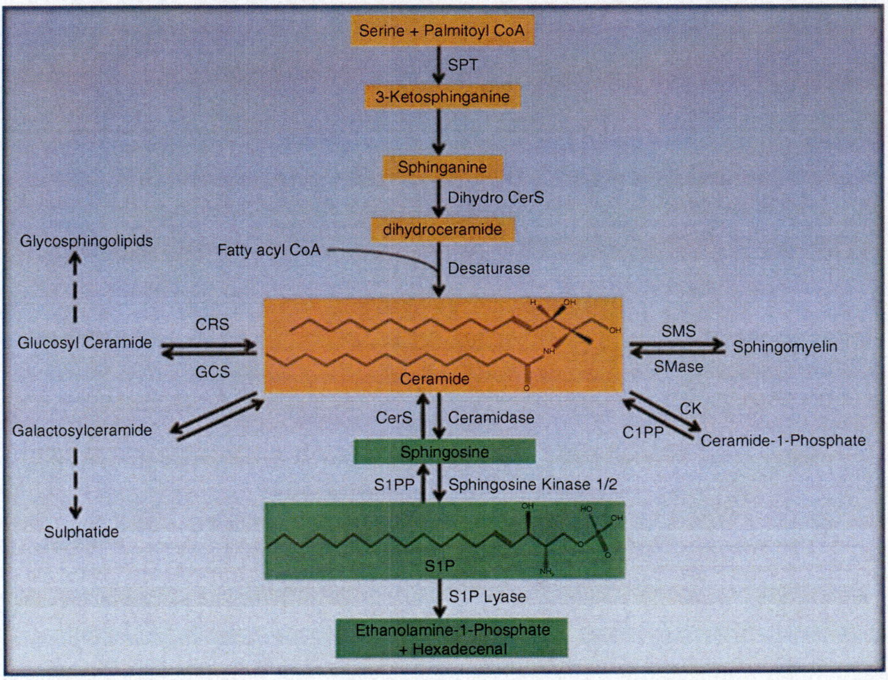

Fig. 1 *De novo sphingolipid synthetic pathway*. The initial step is the condensation of serine and palmitoyl-coA by serine palmitoyl transferase (SPT) followed by the action of ceramide synthases (CerS) and desaturase (DES) to generate ceramide. Ceramide is also generated by the degradation of sphingomyelin (SM) by sphingomyelinase (SMase) or by the action of cerebrosidase (CRS) on glycosphingolipids also by the action of ceramide 1-phosphate phosphatase (C1PP). Ceramide is further metabolized by ceramidase (CDase) to yield sphingosine, which is used as a substrate by SK1 and SK2 to generate S1P. S1P can be dephosphorylated by S1P phosphatases (S1PP) to generate sphingosine, or it can be irreversibly cleaved by S1P lyase into ethanolamine 1-phosphate and C_{18} fatty aldehyde (hexadecenal). Ceramide is further metabolized to generate complex glucosyl and galactosyl-ceramide or glycolipids. Ceramide can be acted upon by sphingomyelin synthase to generate sphingomyelin or by ceramide kinase to generate ceramide 1-phosphate (C1P). *C1P* ceramide 1-phosphate, *C1PP* ceramide 1-phosphate phosphatase, *CDase* ceramidase, *CerS* ceramide synthase, *CRS* cerebrosidase, *DES* desaturase, *GCS* glucosyl ceramide synthase, *S1P* sphingosine 1-phosphate, *S1PP* S1P phosphatase, *SK1* sphingosine kinase 1, *SK2* sphingosine kinase 2, *SM* sphingomyelin, *SMase* sphingomyelinase, *SPT* serine palmitoyl transferase

hexadecenal have been shown to promote MOMP (mitochondrial outer membrane permeabilization). Interestingly, S1P and hexadecenal have been shown to cooperate with BAK and BAX, respectively, to regulate MOMP and cellular responses to apoptosis (Chipuk et al. 2012).

3 Ceramide/S1P Rheostat in Cancer

The fate of a cell is determined by the balance between ceramide and S1P signaling (not necessarily a quantitative ratio for the amount of lipids, but a biological/metabolic balance between these two signaling arms of sphingolipids with

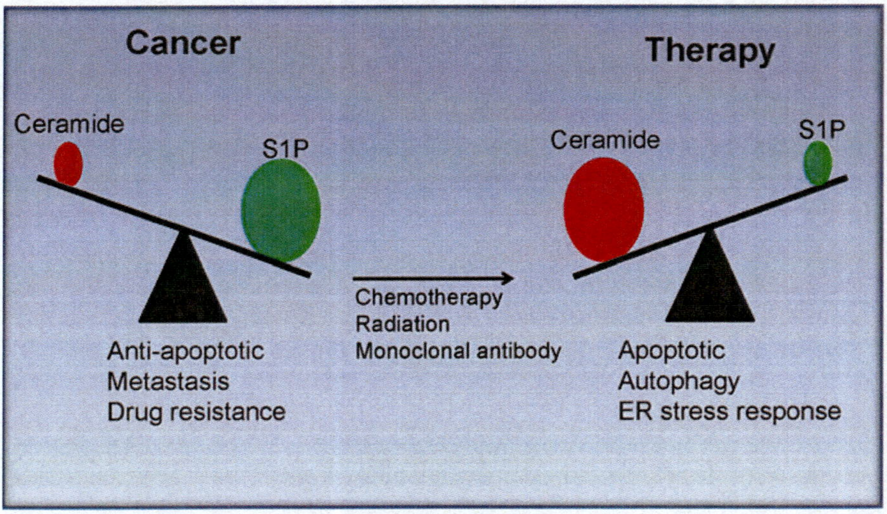

Fig. 2 *Ceramide-S1P rheostat in cancer and therapy*. There exists a balance or rheostat in ceramide to S1P signaling in cancer. A shift towards ceramide accumulation leads to pro-apoptotic, autophagic, ER stress response, and anti-survival effects, whereas a dynamic shift towards S1P accumulation leads to pro-survival, anti-apoptotic, metastatic, and drug-resistant phenotypes. Potential therapeutic approaches will be to increase ceramide and decrease S1P in cancer cells by chemotherapy, radiation, monoclonal antibody, and other molecular approaches. *ER* endoplasmic reticulum, *S1P* sphingosine 1-phosphate

opposing functions), which is often referred to as the ceramide/S1P rheostat (Fig. 2). There exists a dynamic balance between ceramide and S1P signaling, and when a shift towards ceramide is achieved by stress signaling such as radiation, heat, and chemotherapy treatment, this drives cells to undergo cell death and antiproliferation (Hannun and Obeid 2008). On the other hand, when the balance shifts towards S1P accumulation, cells exert pro-survival, anti-apoptosis, and/or chemoresistance (Ponnusamy et al. 2010). Increases in endogenous ceramide by chemotherapeutic agents, TNF-α, CD95, hypoxia, DNA damage, and heat stress can activate cell death pathways. Also, increases in ceramide via inhibiting ceramide metabolism or overexpressing CerS lead to cell death, in general. Moreover, overexpression of bacterial SMase, which generates ceramide by degradation of sphingomyelin, was shown to induce apoptosis (Meyers-Needham et al. 2012a). In contrast, inhibition of de novo ceramide generation by fumonisin B1 blocks ceramide-mediated cell death by chemotherapeutic drugs.

Ceramide has various established intracellular targets such as PP1, PP2A, I2PP2A, cathepsin D, and protein kinase Cζ, which mediate its apoptotic/cell death functions (Fox et al. 2007; Heinrich et al. 2000; Ogretmen and Hannun 2004; Wang et al. 2005). Our laboratory identified telomerase to be a nuclear target of ceramide (exogenous C_6-ceramide or C_{18}-ceramide generated by CerS1) which decreased c-Myc-mediated activation of hTERT promoter in lung cancer cells (Ogretmen et al. 2001). In fact, ceramide deacetylates Sp3 transcription factor and deacetylated Sp3 recruits HDAC1 to the hTERT promoter, thereby decreasing hTERT expression/activity (Wooten-Blanks et al. 2007; Wooten and Ogretmen 2005). We also showed that ceramide binds I2PP2A and relieves PP2A-mediated tumor suppression functions in A549 lung cancer cells. Ceramide mediates the degradation of c-Myc by PP2A activation via prevention of I2PP2A-dependent inhibition of PP2A (Mukhopadhyay et al. 2009).

Recently, ceramides with different fatty acid chain lengths were suggested to have distinct functions. For example, in head and neck cancer cells, CerS1-generated C_{18}-ceramide suppressed tumor growth, whereas CerS6-generated C_{16}-ceramide increased tumor growth/proliferation (Senkal et al. 2011). These distinct and unexpected functions of endogenous ceramides might be due to their downstream targets regulated via ceramide/protein interactions and/or their subcellular localization at distinct biological membranes or membrane microdomains. Although mechanisms that regulate CerS expression and function are largely unknown for de novo ceramide generation, new insights about the modulation of CerS expression suggested that epigenetic and posttranscriptional control by concerted functions of HDAC1 and microRNA-574-5-mediated targeting of CerS1 mRNA is involved in its repression in HNSCC (Meyers-Needham et al. 2012b). In addition, dimerization of CerS proteins has been shown to regulate their function for the generation of ceramide (Laviad et al. 2012).

In contrast to ceramide, S1P plays a pro-survival function. Sphingosine kinases and S1P-phosphatases/lyase are important players in the regulation of S1P generation/metabolism. SiRNA-mediated knockdown of SK1 inhibits cell proliferation and increases ceramide/S1P rheostat in pancreatic, prostate, and leukemia cancer cells (Akao et al. 2006; Baran et al. 2007; Guillermet-Guibert et al. 2009; Pchejetski et al. 2005). In contrast, overexpression of SK1 leads to increased cell proliferation by inducing G1/S phase transition and DNA synthesis (Olivera et al. 1999). Exogenous S1P addition was found to significantly inhibit apoptosis via increased Bcl-2 (Sauer et al. 2005) and Mcl-1 expression (Li et al. 2008) or decreased BAD and BAX expression (Avery et al. 2008). In particular, exogenous S1P prevents BAD/BAX translocation to mitochondria, thereby inhibiting the intrinsic cell death (mitochondrial) pathway (Betito and Cuvillier 2006). Also, SK1 overexpression has been shown to inhibit cytochrome c release and caspase-3 activation by the regulation of BCL-X_L, MCL1, and BIM in chronic myelogenous leukemia (CML) cells (Bonhoure et al. 2008).

Moreover, important in vivo findings indicate the involvement of S1P in cancer progression. For example, breast and prostate cancer cells that overexpressed SK1 formed tumors in mice, and ceramide/S1P was altered in these tumors (Nava et al. 2002; Pchejetski et al. 2005), which were resistant to doxorubicin and had increased neovascularization and decreased ceramide/S1P (Nava et al. 2002; Pchejetski et al. 2005). Interestingly, CML and leukemia cells, which are sensitive to imatinib and daunorubicin, respectively, had higher ceramide/S1P compared to resistant cells. Furthermore, increased SK1/S1P in response to camptothecin treatment in prostate cancer cells suggests a role for S1P signaling in chemoresistance (Fig. 3).

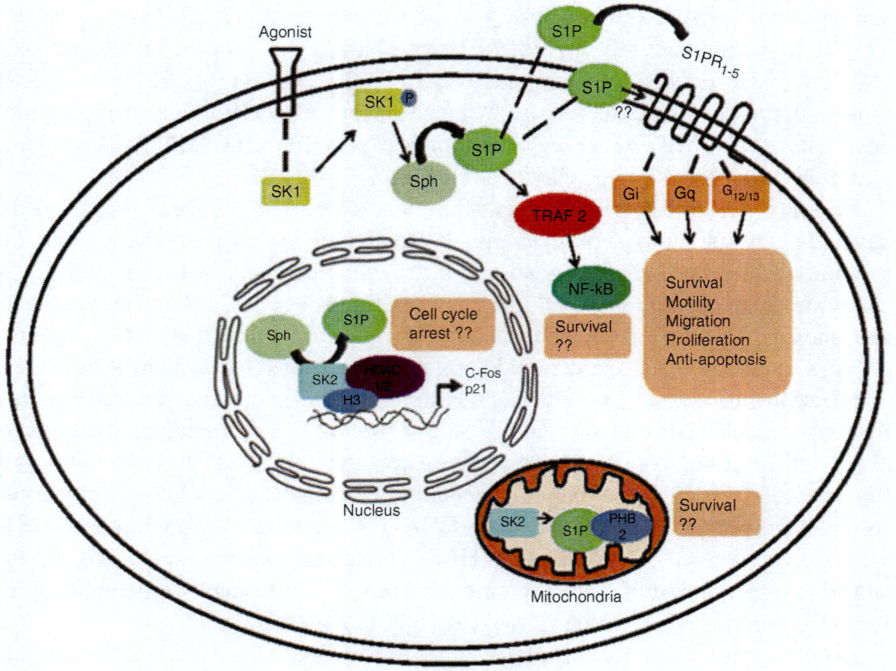

Fig. 3 *SK1/S1P signaling and intracellular targets of S1P*. SK/S1P signaling pathway involves the activation of SK1 by agonist-mediated receptor action. The activated SK1 translocates to the inner leaflet of the plasma membrane to utilize the substrate sphingosine to generate S1P. The S1P generated inside the cells are transported outside the cell by S1P transporters and furthermore engage in an autocrine or paracrine fashion to five G protein-coupled receptors specific for S1P ($S1PR_{1-5}$) to induce an array of downstream mechanisms involved in cell motility, survival, migration, and/or proliferation. S1P generated by SK1 has been reported to play intracellular function by binding with TRAF2 protein to regulate NF-κB function. S1P generated by SK2 has been shown to interact with HDAC1/2 in the nucleus to regulate transcription of p21 and c-Fos genes. Moreover, SK2-generated S1P also binds to prohibitin 2 (PHB2), mitochondrial protein which in turn regulates cytochrome c oxidase (Cox-2) in respiration complex assembly and function. *HDAC* histone deacetylase, *NF-κB* nuclear factor kappa light chain enhance of activated B cells, *PHB2* prohibitin 2; *TRAF2* TNF receptor associated factor 2

4 S1P Signaling

S1P generated by SK1 has been shown to be secreted and engaged with S1P receptors ($S1PR_{1-5}$) to elicit various downstream responses involved in inflammation, cell migration, angiogenesis, and/or lymphocyte trafficking (Spiegel and Milstien 2003; Strub et al. 2010). S1PR1 is important in vascular maturation, immune cell trafficking, endothelial barrier function, and angiogenesis as displayed in S1PR1-null mice. Also the S1P–S1PR1 interaction reveals receptor tyrosine kinase (RTK) activation such as EGF- and PDGF-mediated pathways that dictate cellular growth and migration. Importantly, ATP-dependent ABC transport proteins such as ABCC1, ABCA1, and ABCG2 are involved in S1P transport (Mitra et al. 2006; Sato et al. 2007; Takabe et al. 2010). Also, ABCC1 is involved in S1P export from mast cells independent of degranulation (Mitra et al. 2006). Recently, TOH/SPNS2 (two of hearts protein) has been identified as an S1P transporter in zebra fish, and interestingly TOH/SPNS2 acts upstream of MIL, an orthologue of S1PR2 involved in heart development (Kawahara et al. 2009; Osborne et al. 2008). SPNS2 was shown to be important for the transport of phosphorylated form of FTY-720, an immunomodulatory drug used for lymphocyte egress in multiple sclerosis. Importantly SPNS2-null mice had increased accumulation of mature T cells and a decreased T cell population in blood and secondary lymphoid organs (Fukuhara et al. 2012; Hisano et al. 2011). Thus, S1P-mediated signaling (Fig. 3) is important in various cellular processes, and how this mechanism is dysregulated in cancer needs further investigation. Recently, the structure of ligand-bound S1PR1 was solved, and data suggest that S1P might engage with the receptor within the plasma membrane (Hanson et al. 2012), indicating that lateral movements of S1P within the membrane or ligand swapping between S1PR1 and other receptors might be involved in this process.

5 Roles of SK in Cancer Pathogenesis

Sphingosine kinases (SK1 and SK2) are novel lipid kinases, which are evolutionarily conserved as diverse as in humans, mice, yeast, and plants. SK1 and SK2 are encoded by *SPHK1* and *SPHK2* genes in humans (Pitson 2011; Strub et al. 2010). SphK1 and SphK2 belong to the diacylglycerol kinase family and have five conserved domains C1–C5 (Spiegel and Milstien 2003). The unique catalytic domain is present between domains C1 and C3. SphK1 and SphK2 differ by the presence of a TM (transmembrane domain) present only in SphK2 and not in SphK1. Both SK1 and SK2 enzymes show substrate specificity towards D-e-sphingosine and D-e-dihydrosphingosine. Interestingly, SK1 and SK2 show different tissue distribution with SK1 expression higher in lung and spleen, whereas SK2 levels were found to be higher in the liver and the heart. SK1 is predominantly cytosolic in nature, whereas SK2 has been shown to localize both in the cytoplasm and nucleus (Kohama et al. 1998; Liu et al. 2000, 2002; Pitson 2011). Moreover, SK1 expression was detected at embryonic day 7 (E7) of mice and SK2 at E11 indicating the distinct functions of both of these isoenzymes. The functions of SK1 and SK2 seem redundant during development; SK1 and SK2 null mice survive, whereas the SK double knockout is embryonically lethal (Mizugishi et al. 2005; Pitson 2011; Spiegel and Milstien 2003).

5.1 SK1 and Cancer

SK1 has well-established pro-survival functions in various cancers. Indeed, by virtue of its transformation potential, SK1 is considered to be a bona fide oncogene

(Vadas et al. 2008). Upon transfection with SK1, untransformed NIH3T3 cells undergo transformation with the ability to form colonies in soft agar and tumors in nude mice (Xia et al. 2000). Activation of SK1 by agonists of GPCRs, protein kinases, proinflammatory cytokines, and small GTPases leads to its translocation to plasma membrane where it catalyzes the conversion of sphingosine to S1P. Importantly, phosphorylation of SK1 at Ser 225 by ERK1 and ERK2 is important for its activation and translocation to the plasma membrane (Pitson et al. 2003). This finding was further confirmed by S225A mutant of SK1 that lacks the phosphorylation site (Pitson et al. 2005). When SK1 S225A mutant was targeted to the plasma membrane by tagging with a myristoyl or palmitoyl moiety, the cells became transformed, via targeting to plasma membrane (Pitson et al. 2005). The membrane affinity and plasma membrane selectivity are determined by Thr54 and Asn89 residues of human SK1, wherein these residues interact specifically with phosphatidyl serine in the plasma membrane, thus making sphingosine available to generate S1P, which can be secreted outside the cell and engage with S1PRs to induce prosurvival functions (Stahelin et al. 2005). Interestingly, SK1 expression was found also to be important for Ras-mediated transformation. In cancers, such as breast, lung, ovary, stomach, and kidney, SK1 mRNA increased approximately twofold compared to paired normal tissues (French et al. 2003; Johnson et al. 2005; Kawamori et al. 2006). Also immunohistochemical studies with lung, colon, and breast cancer tissues were positive for SK1 expression in tumor tissues and/or carcinoma cells (Johnson et al. 2005; Kawamori et al. 2006; Kohno et al. 2006; Ruckhaberle et al. 2008). Moreover, microarray data show elevated SK1 in squamous cell carcinoma (Nindl et al. 2006), melanoma cancer (Talantov et al. 2005), *N*-methyl-*N*-nitrosourea-induced rat breast cancer (Chan et al. 2005), recurrent breast cancer after tamoxifen treatment (Ma et al. 2004), cervical cancer (Wong et al. 2003), head and neck cancer, and leukemia (Andersson et al. 2007; Ginos et al. 2004; Pyeon et al. 2007). Interestingly, siRNA-mediated downregulation of SK1 showed decreased migration of EGF, prolactin, and E2-induced migration of MCF-7 breast cancer cells (Doll et al. 2007; Sarkar et al. 2005), whereas downregulation of both SK1 and SK2 inhibited migration of MDA-MB-453 cells (Hait et al. 2005), indicating a possible redundancy/overlap in functions between these isoenzymes in cancer cell migration.

Novel intracellular targets of SK1 have been recently established: SK1-generated S1P has been shown to bind to TRAF2 protein, an E3 ubiquitin ligase, modulating TNF-α-induced activation of NF-κB signaling (Alvarez et al. 2010). SK1 has been shown to bind TRAF2 to induce K63-mediated polyubiquitination of RIP1 leading to IκB degradation and subsequent stimulation of the NF-κB pathway (Fig. 3). These findings suggest SK1/S1P to be important driver of cancer progression.

Remarkably, a recent finding from our laboratory suggests that serum S1P generated by SK1, and not tumor S1P, was important for tumor metastasis to the lungs (Fig. 4). In this study, genetic ablation of systemic SK1 leads to increased breast cancer metastasis suppressor 1 (Brms1) expression via alterations of S1PR2 signaling in tumor cells, leading to suppression of metastasis. Furthermore, treatment with anti-S1P monoclonal antibody (Sphingomab) decreased lung metastasis

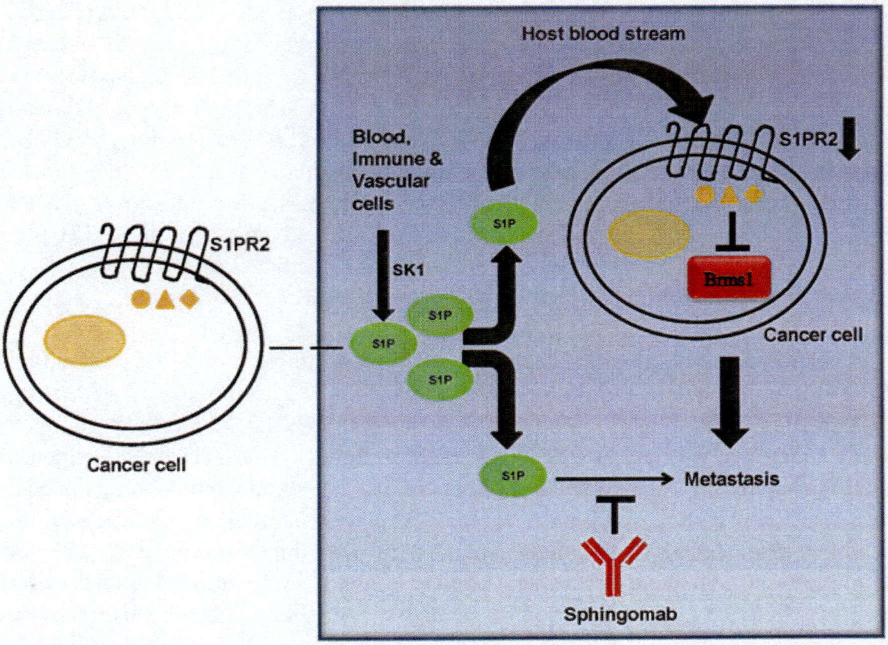

Fig. 4 *Role of systemic SK1/S1P in the modulation of S1PR2/Brms1 expression in lung tumor metastasis.* S1P generated in response to cancer cell exposure to bloodstream elevates systemic S1P levels thereby leading to the induction of S1PR2 expression and further leading to the suppression of Brms1, a master metastatic suppressor gene in cancer cells, inducing their metastatic potential. Use of pharmacologic inhibitors or antibody-based therapeutic tools (Sphingomab, Lpath Inc.) effectively modulates serum S1P via inhibition of cancer cell S1PR2 signaling to elevate tumor Brms1 levels and to suppress lung metastasis. *Brms1* breast cancer metastasis suppressor 1, *SK1* sphingosine kinase 1, *S1P* sphingosine 1 phosphate, *S1PR2* S1P receptor 2

in this model by neutralizing circulating/systemic S1P, inducing cancer Brms1 expression, and thus further proving the importance of SK1-generated systemic S1P in regulating tumor metastasis (Ponnusamy et al., 2012). Recently, the role of SK1 in the regulation of tumor metastasis has been also shown in a breast cancer model (Nagahashi et al. 2012). These data suggest that inhibition of systemic SK1/S1P and/or cancer S1PR2 signaling might inhibit tumor metastasis.

5.2 SK2 and Cancer

SK2 has been shown to predominantly localize to the nucleus, although cytoplasmic and ER localizations were also reported (Igarashi et al. 2003; Maceyka et al. 2005). SK2 localizes to the nucleoplasm and causes cell cycle arrest. Importantly, SK2 retains a nuclear export signal, and activation with phorbol myristate acetate (PMA) leads to protein kinase D-mediated phosphorylation and export from the nucleus. SK2, unlike SK1, possesses a BH-3 domain in its sequence that sequesters

Bcl-XL to diminish its anti-apoptotic functions (Liu et al. 2003). Interestingly, when SK2 is overexpressed, it induces apoptosis, cell cycle arrest, and/or caspase-3 activation (Liu et al. 2003). In contrast to the apoptotic roles, SK2 also displays anti-apoptotic functions. Knockdown of SK2 in HCT116 colon cancer cells and MCF7 breast cancer cells prevents doxorubicin-induced p21 expression and G2/M cell cycle arrest in a p53 independent manner (Sankala et al. 2007). SK2 knockdown decreased glioblastoma cell proliferation, whereas SK1 knockdown did not have any effects (Van Brocklyn et al. 2005). EGF activates SK2 at Ser351 and Thr578 (Hait et al. 2007) residues leading to EGF-mediated migration of MDA-MB-453 breast cancer cells. These studies elucidate SK2's anti-apoptotic and pro-proliferative effects.

Most importantly, the first identified intracellular function of S1P generated by SK2 was in the nucleus to inhibit HDAC1/2 enzymatic activity thereby preventing deacetylation of histone H3 (Hait et al. 2009). The SK2/HDAC repressor complex was found to be associated with promoter regions of P21 (CDKN1) and c-Fos genes thereby modulating their expression. Interestingly, another intracellular target of S1P has been identified as prohibitin 2 (PHB2), a protein regulating mitochondrial function (Strub et al. 2011). S1P generated by SK2 binds prohibitin 2 and not prohibitin 1 to regulate cytochrome c oxidase complex IV in mitochondrial respiration and function. Also, mitochondria isolated from SK2 knockout mice showed decreased association of PHB2 and cytochrome c oxidase (Strub et al. 2011) (Fig. 3).

Collectively, SK1- and SK2-generated S1P have distinct roles in the context of their subcellular localization and function. Also, SK1 and SK2 might also have overlapping functions in different cancer models for inducing pro-survival and anti-apoptosis, possibly via distinct mechanisms of action at different subcellular compartments.

6 Role of S1P in Autophagy

Autophagy is a cellular process to degrade long-lived proteins and organelles through lysosomes. The autophagic process can be protective or lethal; the protective autophagy pathway directs degradation of damaged organelles and recycling of amino acid to overcome nutrient deprivation, whereas lethal autophagy leads to caspase independent cell death involving Atg proteins (Kondo and Kondo 2006; Levine and Kroemer 2008). Use of SK2 inhibitor ABC294640 induces autophagic cell death in A498 cells (Beljanski et al. 2010). Mechanistically, Beljanski and colleagues reported that increased ceramide and sphingosine in response to ABC294640 treatment resulted in autophagy in vivo. Moreover, protective autophagy was activated in response to SK1 upregulation by lack of beclin 1 and mTOR inhibition (Lavieu et al. 2006). Recently, it was shown that S1P metabolism plays an important role in switching from protective autophagy to apoptosis through the involvement of SPP1 (S1P phosphohydrolase). It was found that doxorubicin-mediated autophagy was greatly decreased by SPP1 ablation with a concomitant increase in apoptosis by ceramide accumulation via de novo synthesis.

Ceramide then activates calpain to cleave Atg5 in SPP1-depleted cells to shift from protective autophagy to apoptosis (Lepine et al. 2011).

7 SK1/S1P Signaling in Drug Resistance

SK/S1P protects cancer cells from drug-induced cell death; therefore, levels of bioactive sphingolipids have been shown to modulate drug resistance. For example, PC3 prostate cancer cells, which are androgen insensitive, are resistant to camptothecin treatment by upregulation of SK1 and S1P1/S1P3 signaling, and these cells became sensitive upon inhibition of SK and S1P formation (Bektas et al. 2005). Radiation-resistant prostate cancer cell line LNCaP showed sensitivity upon coadministration of TNF-α and γ-irradiation to increase the generation of sphingosine and decreased S1P formation (Nava et al. 2000). Moreover, treatment of LNCaP cells with SK inhibitor (*N,N*-dimethyl sphingosine) sensitized them to radiation-induced apoptosis. In addition, SK1 induced resistance of melanoma cells to ceramide and Fas-induced cell death. The role of SK1/S1P in breast cancer progression was also reported to be important. A microarray study conducted with 1,269 breast cancer tumor samples showed a significant increase in SK1 gene expression and showed a strong correlation between SK1 expression and poor prognosis (Ruckhaberle et al. 2008). Additionally SK1 expression was responsible for resistance to tamoxifen-induced cell death in MCF7 cells. Tamoxifen-resistant MCF7 cells showed elevated SK1 levels and inhibition of SK1 by DMS/SKI-II restored sensitivity of the cells to tamoxifen-induced apoptosis (Sukocheva et al. 2009). In erythroleukemia progression, overexpression of SK1 in non tumorigenic proerythroblasts showed increased tumorigenicity and resistance to cell death. Of note, a microarray-based transcriptome profile showed transcriptional upregulation of SK1 in tumorigenic proerythroblasts. These data suggest SK1 expression to be important in erythroleukemic progression (Le Scolan et al. 2005). Moreover, SK1 is overexpressed in glioblastoma multiforme (GBM) and targeted inhibition of SK1 with an isoform-specific drug; SK1-I significantly decreased tumor proliferation and markedly suppressed pro-survival AKT and its downstream targets (p70S6K and GSK3-α) (Kapitonov et al. 2009; Knapp et al. 2010). SK1-I not only inhibited S1P generation but also increased ceramide to induce apoptosis (Kapitonov et al. 2009). In endometrial cancer, compared with healthy endometrial tissues, the cancer tissues had increased SK1 activity (2.6-fold) and elevated S1P (1.6-fold) in serum, indicating a significant change in sphingolipid metabolism in driving cancer progression (Knapp et al. 2010).

In CML, ceramide/S1P rheostat plays a crucial role in conferring drug resistance. K562 CML cells generate endogenous C_{18}-ceramide in response to imatinib treatment, and interestingly K562 cells which show resistance to imatinib treatment have increased SK1 expression. Importantly, siRNA-mediated knockdown of SK1 sensitized K562-imatinib-resistant cells to apoptosis. These studies showed that overexpression of SK1 induces drug resistance in CML cells by altering the ceramide/S1P rheostat towards S1P accumulation (Baran et al. 2007). Recently,

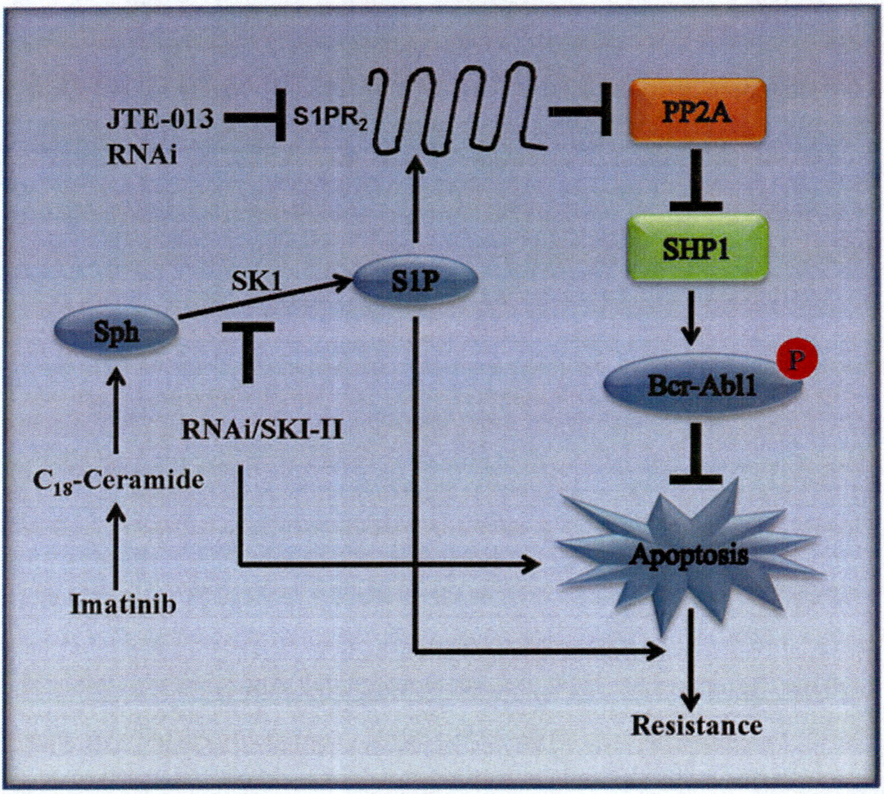

Fig. 5 *Role of SK1/S1P/S1PR-mediated drug resistance in chronic myelogenous leukemia (CML).* SK1/S1P-mediated drug resistance in CML is mediated by S1PR2-mediated PP2A modulation which abolishes proteasomal degradation of Bcr-Abl1 by enhancing its stability, resulting in drug resistance. Pharmacologic inhibitor (SKI-II) or use of molecular approaches (RNAi) by inhibiting SK1 can lead to PP2A-mediated dephosphorylation and degradation of Bcr-Abl1, thereby overcoming drug resistance. *CML* chronic myelogenous leukemia, *PP2A* protein phosphatase 2A, *SK1* sphingosine kinase 1, *S1P* sphingosine 1-phosphate, *S1PR2* S1P2 receptor 2, *SHP1* src homology region 2 domain containing phosphatase 1

we have shown that SK1/S1P and S1PR2 signaling is important for Bcr-Abl1 stability in CML (chronic myeloid leukemia), leading to imatinib resistance. This study showed that SK1/S1P/S1PR2 prevents Bcr-Abl1 dephosphorylation and subsequent degradation by inhibiting PP2A. Inhibition of SK1/S1P/S1PR2 signaling either by pharmacological or molecular approaches restored PP2A-mediated dephosphorylation of Bcr-Abl1 and also enhanced imatinib or nilotinib (drugs used against CML) mediated growth inhibition. Furthermore, allograft tumors derived from 32D cells expressing either wild-type or mutant Bcr-Abl1 genes were more sensitive to nilotinib upon inhibition of SK1/S1PR2 signaling (Fig. 5). These findings suggest that inhibition of SK1/S1P/S1PR2 signaling overcomes drug resistance in CML (Salas et al. 2011). In a separate study, LAMA84 cells which are resistant to imatinib treatment became sensitized when SK1 is inhibited by F-12509a, a SK1 inhibitor, leading to apoptotic cell death (Bonhoure et al. 2008).

8 SK2 and Drug Resistance

Although SK1/S1P signaling has been well established in drug resistance, only recently SK2/S1P has been shown to confer chemoresistance in cancers. Targeted inhibition of SK2 by molecular/pharmacological agents sensitized cancer cells to chemotherapy. Recently, Antoon's laboratory reported that the SK2 inhibitor ABC294640 decreased estrogen receptor (ER)-positive breast cancer tumor growth by 68.4 % compared to vehicle-treated tumors. Mechanistically, SK2 inhibition decreased estrogen-mediated transcription of ER-regulated genes such as SDF1 and the progesterone receptor (Antoon et al. 2010). Subsequently, inhibition of SK2 by ABC294640 at submicromolar concentration was shown to block proliferation of endocrine therapy-resistant MDA-MB-231 and chemoresistant MCF-7TN-R cells (Antoon et al. 2011). Also ABC294640 diminished NF-κB pro-survival signaling by decreasing activation of Ser536 phosphorylation of the p65 subunit (Antoon et al. 2011). Moreover, ABC294640, which is orally bioavailable, resulted in growth inhibition of xenograft-derived tumors of MCF-7TN-R cells at 50 mg/kg in SCID mice.

Schnitzer et al. (2009) elucidated the role of SK2-mediated chemoresistance in A549 lung cancer cells. In this study, it was demonstrated that hypoxia induces SK2 protein/activity leading to S1P secretion via S1PR1/S1PR3 signaling to activate P42/44 MAPK signaling and confer resistance to etoposide-induced apoptosis (Schnitzer et al. 2009). Recently, Xiao et al. (2012) showed that SK2 confers resistance to sodium butyrate-induced apoptosis in HCT116 colon cancer cells. In fact, sodium butyrate treatment induced the phosphorylation of SK2 by PKD leading to nuclear export, leading to chemoresistance (Xiao et al. 2012).

9 S1P/S1PR2 Signaling in Cancer and Drug Resistance

S1P exerts its signaling function in an autocrine or paracrine manner through five G protein-coupled receptors, termed $S1PR_{1-5}$ (formerly referred as the EDG family of receptors). These receptors involve heterotrimeric G proteins such as G_i, G_q, and $G_{12/13}$ to elicit various cellular functions. Importantly, roles of S1P receptors in modulating cancer growth have been studied (Spiegel and Milstien 2003). For instance, S1PR2 has been shown to be important for retinal angiogenesis. When a neonatal mouse was subjected to ischemia-driven retinopathy, the neovascularization event was downregulated in S1P2-null mice compared to wild-type mice. Moreover, the retina of the S1P2-null mouse had significantly reduced proinflammatory Cox-2 mRNA and increased eNOS expression. These findings indicate the involvement of S1PR2 in retinal angiogenesis and prompt the use of receptor antagonists to counteract ocular neovascularization. Additionally, S1P2 plays an important role in Wilm's tumor, a malignant renal cancer condition. In Wilm's tumor, S1PR2 mRNA was higher in ten cancer specimens compared to noncancerous tissues (Li et al. 2009). Additionally S1PR2 overexpression was shown to induce Cox-2 mRNA and increased prostaglandin E2 synthesis. SiRNA-mediated knockdown or pharmacological inhibition of S1PR2 in WiT49 cells decreased Cox-2 mRNA proving an important role of S1PR2 signaling in driving renal cancer progression (Li et al. 2009). In another study, the migration of WiT49 Wilm's tumor cells was attributed to S1PR1 signaling wherein S1PR1/G_i coupling was shown to induce a promigratory phenotype which depended upon the ratio of S1PR1/S1PR2 and further identified PI3K and Rac1 as downstream regulators of cell motility. S1PR2 was also shown important in esophageal cancer motility and migration, in which TGFβ enhanced S1P-mediated activation of ERK1/2 through S1PR2 through non-Smad signaling (Miller et al. 2008).

In a separate study using a cohort of 304 ER-positive, tamoxifen-treated breast cancer samples, increased SK1 and S1P1 and S1P3 receptor expression was observed. Interestingly cytoplasmic expression of S1P1 and S1P3 was found to be associated with reduced disease-specific survival in patients with ER positive breast cancer (Watson et al. 2010). SK1 also confers resistance towards synthetic retinoid *N*-(4-hydroxyphenyl) retinamide in A2780 ovarian cancer cells; the HPR-resistant cells express more SK1 both at the mRNA and the protein level and became sensitized when treated with SK inhibitor (Illuzzi et al. 2010). Recently, SK1 has been shown to protect androgen-independent LNCaP prostate cancer cells; it does so by increasing S1P signaling through $S1P_{2/3}$ receptors, and the resistance was attenuated by treating with FTY-720, a sphingosine analog that inhibits S1PR signaling (except S1PR2), by inducing proteasome-mediated degradation of SK1 (Tonelli et al. 2010).

These findings suggest that the SK/S1P/S1PR signaling pathway is crucial in sphingolipid-mediated drug resistance in various cancers, and hence targeting this pathway would be an effective anticancer therapeutic strategy to overcome drug resistance.

10 SK/S1P-Mediated Anticancer Therapeutics

Targeting S1P produced by SK1 and SK2 can sensitize cancer cells to therapeutic intervention. Also targeting SK/S1P is attractive because indirect generation of ceramide and sphingosine can have antiproliferative and pro-apoptotic functions. An array of sphingolipid and non-sphingolipid agents has been developed to target SK/S1P and/or S1P receptors (Table 1).

Pan Sphingosine Kinase Inhibitors. N,N-dimethyl sphingosine (DMS), a pan SK inhibitor, was shown to be effective against a panel of cancer cell lines and exhibited antitumor growth properties in nude mice. Also, L-threo dihydrosphingosine

Table 1 Sphingosine kinase/S1P/S1PR based anti cancer therapeutics

Compound	Target	Cancer type
N,N-dimethyl sphingosine (DMS)	SK1, SK2	Leukemia, colon and breast
Safingol (L-threo dihydrosphingosine)	SK1, SK2	Solid tumors
SKI-II	SK1, SK2	Breast
Phenoxodiol	SK1, SK2	Ovarian and prostate
SK1-I	SK1	Glioblastoma, breast and AML
ABC294640	SK2	Breast, prostate and kidney
FTY-720	S1PR	Bladder, prostate, breast and lymphoma
Anti-S1P-mAb	S1P	Lung, ovarian, breast and melanoma

SK1 sphingosine kinase 1, *SK2* sphingosine kinase 2, *S1P* sphingosine 1-phosphate, *S1PR* S1P receptor, AML acute myeloid leukemia

(Safingol) inhibits SK and is currently in phase I clinical trials (Schwartz et al. 1997). Both DMS and Safingol inhibit SK competitively and displayed inhibitory effects towards ceramide kinase and PKC and activate sphingosine-mediated targets such as PI3K and casein kinase 2 (Igarashi et al. 1989; Kedderis et al. 1995; King et al. 2000; McDonald et al. 1991; Megidish et al. 1998; Sugiura et al. 2002). Some of the reported off-target effects of DMS include hemolysis and hepatotoxicity (Kedderis et al. 1995). Interestingly, phenoxodiol, a synthetic analog of plant isoflavone genistein, exhibits anticancer and antiangiogenenic functions by inhibiting SK. Phenoxodiol exerts its anticancer function by inhibiting endothelial cell function both in vitro and in an in vivo model of angiogenesis, and currently phenoxodiol is in clinical trials against ovarian and prostate cancers (De Luca et al. 2005; Gamble et al. 2006). French et al. (2003) have demonstrated the use of non-lipid selective inhibitors of SK and showed inhibition against a panel of cancer cell lines, with greater selectivity to SK than other protein kinases; additionally these SK inhibitors were noncompetitive and displayed antiproliferative functions (French et al. 2003). Furthermore, French et al. (2006) showed SKI-II (4-[4-(4-chloro-phenyl)-thiazol-2-ylamino]-phenol), a noncompetitive inhibitor at nanomolar to submicromolar concentration in vitro, and displayed excellent oral bioavailability with antitumor functions in vivo (French et al. 2006). Additionally, other compounds such as F12509a, a sesquiterpene quinone-based competitive inhibitor of SK, and B5354c, a noncompetitive inhibitor isolated from a marine bacterium, were found to inhibit SK in some cancers, but their efficacy and specificity are still under investigation (Kono et al. 2000a, b, 2002).

SK1 Selective Inhibitors. SK1-I (2R, 3S, 4E)-*N*-methyl-5-(4′-pentylphenyl)-2-aminopent-4-ene-1, 3-diol is a SK1 selective inhibitor and has shown efficacy against orthotopic as well as xenograft glioblastoma or AML xenograft tumors (Kapitonov et al. 2009; Paugh et al. 2008). Recently, Nagahashi et al. (2012) demonstrated decreased breast cancer progression in an 4T1-Luc orthotopic mice and also showed decreased lymph node and lung metastasis upon SK1-I treatment. Interestingly SK1-I decreased serum S1P levels and caused mammary tumor cells to undergo apoptosis (Nagahashi et al. 2012). Moreover, novel SK1-specific inhibitors such as 6ag, 9ab, and12aa were synthesized and tested in vitro, but the in vivo efficacy still needs to be validated (Xiang et al. 2009). There is a need for the development of SK1-specific small molecule inhibitors to be utilized as effective anticancer agents.

SK2 Selective Inhibitors. ABC294640 is a novel SK2-specific inhibitor with excellent oral bioavailability and toxicology properties. ABC294640 displays antiproliferative effects against a variety of cancer cell lines at nanomolar up to submicromolar concentrations. This SK2 selective compound induced autophagic cell death in breast, prostate, and kidney xenograft tumors in vivo (Antoon et al. 2010; Beljanski et al. 2010, 2011; French et al. 2010). ABC294640 is currently in phase I clinical trials against solid tumors. Moreover, sphingoid analogs such as SG12 and SG14 displayed specificity towards SK2, and in particular SG14 did not inhibit PKC (Kim et al. 2005).

S1P Receptor Selective Compounds. FTY-720, a sphingosine analog, phosphorylated by SK2, acts through S1P receptors (S1PR1, 3, 4, 5) to cause receptor endocytosis and rapid lymphocyte egress in multiple sclerosis (Brinkmann et al. 2010, 2002). FTY-720 has been efficacious against refractory multiple sclerosis and approved by FDA as an anti-MS drug (Brinkmann et al. 2010). FTY-720 also inhibited SK1, ceramide synthase, phospolipase A2, and SGPL1, and it was shown to activate tumor suppressor PP2A (Bandhuvula et al. 2005; Berdyshev et al. 2009; Lahiri et al. 2009; Matsuoka et al. 2003; Neviani et al. 2007; Payne et al. 2007; Vessey et al. 2007). FTY-720 has been reported to inhibit colon and breast cancer cell lines in situ through S1P receptor independent effects. In addition, phosphorylated-FTY720 (P = FTY-720) was found to be important for antiangiogenesis in an in vitro spheroid model, indicating a receptor-dependent function, which can also be mimicked by SEW2871, a S1PR1-specific antagonist (Nagaoka et al. 2008; Schmid et al. 2007). Further investigation is required to delineate the receptor-dependent and receptor-independent functions of FTY-720 against cancer growth and/or proliferation. Recently, (R)-FTY-720-OMe, a stereospecific analog of FTY-720, showed SK2-specific inhibition and caused actin rearrangement in MCF-7 cells (Lim et al. 2011), suggesting that this novel enantiomer can be used against breast cancer. Other receptor antagonists such as VPC4416, VPC2309, VPC25239, and W146 against S1PR1/3 and S1PR1, respectively, have shown promising results in situ (Davis et al. 2005; Sanna et al. 2006).

Antibody-Based Therapeutics. An anti-S1P-mAb that specifically targets and neutralizes S1P has been shown to be effective against lung A549, ovarian SKOV3, breast MDA-MB-231, and melanoma F16/B10 cancer models in situ and in vivo. Anti-S1P-mAb functions as a molecular sponge to neutralize S1P signaling and cause tumor regression in both xenograft and allograft models or to inhibit lung metastasis (Ponnusamy et al. 2012). Sphingomab (LT1002) and its humanized form (LT1009) neutralize bFGF- and VEGF-induced angiogenesis and block S1P-induced endothelial cell tube formation and migration in numerous in vitro assays (Visentin et al. 2006). Interestingly, LT1009 (sonepcizumab, Lpath Inc.) is currently in phase I/II clinical trials as an anticancer drug (Sabbadini 2011).

11 Conclusion and Future Perspectives

Recent developments in sphingolipid biology, especially the discovery of the roles of SK/S1P/S1PR2 signaling in the regulation of cell proliferation, angiogenesis, drug resistance, and metastasis, have added to our understanding of these processes in various cancers. Importantly emerging evidence suggests that sphingolipids have various functions based on their subcellular localization, relative distribution in tissues, fatty acyl chain-length specificities, and their direct downstream targets via lipid-protein binding. Importantly, chain length-specific ceramides were found to be altered in cancers; thus—from a therapeutic perspective—reconstitution of ceramide generation by use of small molecule inhibitors or ceramide analogs/ mimetics could be effective anticancer techniques. In contrast, SK and S1P were elevated in various cancers and inhibitors; antagonists and monoclonal antibodies against SK/S1P/S1PR signaling might hold promise for decreasing cancer growth, proliferation, and metastasis. Recent identification of SPNS2 as a novel S1P transporter (Kawahara et al. 2009) will help delineate the mechanism of S1P transport from various cell types, altering systemic S1P accumulation, which then plays a role in inducing tumor metastasis (Ponnusamy et al., 2012). Moreover, identification of the S1PR1 crystal structure can point towards more potent antagonists of S1PR-specific compounds to mediate anticancer functions (Hanson et al. 2012). Although recent research in sphingolipid metabolism and biology has yielded significant mechanistic information towards cancer pathogenesis and therapeutics, limitations exist regarding our understanding of their roles in varying cancer subtypes, subcellular compartmentalization, and direct downstream targets, which often yield context-dependent effects in response to changes in their generation/ accumulation. Moreover, the inherent obstacle in studying membrane-bound enzymes and their lipid products involved in metabolism must be overcome to identify molecular and structural details of their functions in the regulation of cancer growth, proliferation, metastasis, and for the development of novel anticancer therapeutics.

Acknowledgments We thank the members of the Ogretmen laboratory for their helpful discussions and apologize to those investigators whose important work was not included in this chapter because of space limitations. BO is funded by research grants from the National Institutes of Health. We thank Dr. Jennifer Schnellmann for her editorial review.

References

Akao Y, Banno Y, Nakagawa Y, Hasegawa N, Kim TJ, Murate T, Igarashi Y, Nozawa Y (2006) High expression of sphingosine kinase 1 and S1P receptors in chemotherapy-resistant prostate cancer PC3 cells and their camptothecin-induced up-regulation. Biochem Biophys Res Commun 342:1284–1290

Alvarez SE, Harikumar KB, Hait NC, Allegood J, Strub GM, Kim EY, Maceyka M, Jiang H, Luo C, Kordula T, Milstien S, Spiegel S (2010) Sphingosine-1-phosphate is a missing cofactor for the E3 ubiquitin ligase TRAF2. Nature 465:1084–1088

Andersson A, Ritz C, Lindgren D, Eden P, Lassen C, Heldrup J, Olofsson T, Rade J, Fontes M, Porwit-Macdonald A, Behrendtz M, Hoglund M, Johansson B, Fioretos T (2007) Microarray-based classification of a consecutive series of 121 childhood acute leukemias: prediction of leukemic and genetic subtype as well as of minimal residual disease status. Leukemia 21:1198–1203

Antoon JW, White MD, Meacham WD, Slaughter EM, Muir SE, Elliott S, Rhodes LV, Ashe HB, Wiese TE, Smith CD, Burow ME, Beckman BS (2010) Antiestrogenic effects of the novel sphingosine kinase-2 inhibitor ABC294640. Endocrinology 151:5124–5135

Antoon JW, White MD, Slaughter EM, Driver JL, Khalili HS, Elliott S, Smith CD, Burow ME, Beckman BS (2011) Targeting NFkB mediated breast cancer chemoresistance through selective inhibition of sphingosine kinase-2. Cancer Biol Ther 11:678–689

Avery K, Avery S, Shepherd J, Heath PR, Moore H (2008) Targeting NFkB mediated breast cancer chemoresistance through selective inhibition of sphingosine kinase-2. Stem Cells Dev 17:1195–1205

Bandhuvula P, Tam YY, Oskouian B, Saba JD (2005) The immune modulator FTY720 inhibits sphingosine-1-phosphate lyase activity. J Biol Chem 280:33697–33700

Baran Y, Salas A, Senkal CE, Gunduz U, Bielawski J, Obeid LM, Ogretmen B (2007) Alterations of ceramide/sphingosine 1-phosphate rheostat involved in the regulation of resistance to imatinib-induced apoptosis in K562 human chronic myeloid leukemia cells. J Biol Chem 282:10922–10934

Bektas M, Jolly PS, Muller C, Eberle J, Spiegel S, Geilen CC (2005) Sphingosine kinase activity counteracts ceramide-mediated cell death in human melanoma cells: role of Bcl-2 expression. Oncogene 24:178–187

Beljanski V, Knaak C, Smith CD (2010) A novel sphingosine kinase inhibitor induces autophagy in tumor cells. J Pharmacol Exp Ther 333:454–464

Beljanski V, Lewis CS, Smith CD (2011) Antitumor activity of sphingosine kinase 2 inhibitor ABC294640 and sorafenib in hepatocellular carcinoma xenografts. Cancer Biol Ther 11:524–534

Berdyshev EV, Gorshkova I, Skobeleva A, Bittman R, Lu X, Dudek SM, Mirzapoiazova T, Garcia JG, Natarajan V (2009) FTY720 inhibits ceramide synthases and up-regulates dihydrosphingosine 1-phosphate formation in human lung endothelial cells. J Biol Chem 284:5467–5477

Betito S, Cuvillier O (2006) Regulation by sphingosine 1-phosphate of Bax and Bad activities during apoptosis in a MEK-dependent manner. Biochem Biophys Res Commun 340:1273–1277

Bonhoure E, Lauret A, Barnes DJ, Martin C, Malavaud B, Kohama T, Melo JV, Cuvillier O (2008) Sphingosine kinase-1 is a downstream regulator of imatinib-induced apoptosis in chronic myeloid leukemia cells. Leukemia 22:971–979

Brinkmann V, Davis MD, Heise CE, Albert R, Cottens S, Hof R, Bruns C, Prieschl E, Baumruker T, Hiestand P, Foster CA, Zollinger M, Lynch KR (2002) The immune modulator FTY720 targets sphingosine 1-phosphate receptors. J Biol Chem 277:21453–21457

Brinkmann V, Billich A, Baumruker T, Heining P, Schmouder R, Francis G, Aradhye S, Burtin P (2010) Fingolimod (FTY720): discovery and development of an oral drug to treat multiple sclerosis. Nat Rev Drug Discov 9:883–897

Chan MM, Lu X, Merchant FM, Iglehart JD, Miron PL (2005) Gene expression profiling of NMU-induced rat mammary tumors: cross species comparison with human breast cancer. Carcinogenesis 26:1343–1353

Chipuk JE, McStay GP, Bharti A, Kuwana T, Clarke CJ, Siskind LJ, Obeid LM, Green DR (2012) Sphingolipid metabolism cooperates with BAK and BAX to promote the mitochondrial pathway of apoptosis. Cell 148:988–1000

Davis MD, Clemens JJ, Macdonald TL, Lynch KR (2005) Sphingosine 1-phosphate analogs as receptor antagonists. J Biol Chem 280:9833–9841

De Luca T, Morre DM, Zhao H, Morre DJ (2005) NAD+/NADH and/or CoQ/CoQH2 ratios from plasma membrane electron transport may determine ceramide and sphingosine-1-phosphate levels accompanying G1 arrest and apoptosis. Biofactors 25:43–60

Dolgachev V, Farooqui MS, Kulaeva OI, Tainsky MA, Nagy B, Hanada K, Separovic D (2004) De novo ceramide accumulation due to inhibition of its conversion to complex sphingolipids in apoptotic photosensitized cells. J Biol Chem 279:23238–23249

Doll F, Pfeilschifter J, Huwiler A (2007) Prolactin upregulates sphingosine kinase-1 expression and activity in the human breast cancer cell line MCF7 and triggers enhanced proliferation and migration. Endocr Relat Cancer 14:325–335

Fox TE, Houck KL, O'Neill SM, Nagarajan M, Stover TC, Pomianowski PT, Unal O, Yun JK, Naides SJ, Kester M (2007) Ceramide recruits and activates protein kinase C zeta (PKC zeta) within structured membrane microdomains. J Biol Chem 282:12450–12457

French KJ, Schrecengost RS, Lee BD, Zhuang Y, Smith SN, Eberly JL, Yun JK, Smith CD (2003) Discovery and evaluation of inhibitors of human sphingosine kinase. Cancer Res 63:5962–5969

French KJ, Upson JJ, Keller SN, Zhuang Y, Yun JK, Smith CD (2006) Antitumor activity of sphingosine kinase inhibitors. J Pharmacol Exp Ther 318:596–603

French KJ, Zhuang Y, Maines LW, Gao P, Wang W, Beljanski V, Upson JJ, Green CL, Keller SN, Smith CD (2010) Pharmacology and antitumor activity of ABC294640, a selective inhibitor of sphingosine kinase-2. J Pharmacol Exp Ther 333:129–139

Fukuhara S, Simmons S, Kawamura S, Inoue A, Orba Y, Tokudome T, Sunden Y, Arai Y, Moriwaki K, Ishida J, Uemura A, Kiyonari H, Abe T, Fukamizu A, Hirashima M, Sawa H, Aoki J, Ishii M, Mochizuki N (2012) The sphingosine-1-phosphate transporter Spns2 expressed on endothelial cells regulates lymphocyte trafficking in mice. J Clin Invest 122:1416–1426

Futerman AH, Hannun YA (2004) The complex life of simple sphingolipids. EMBO Rep 5:777–782

Gamble JR, Xia P, Hahn CN, Drew JJ, Drogemuller CJ, Brown D, Vadas MA (2006) Phenoxodiol, an experimental anticancer drug, shows potent antiangiogenic properties in addition to its antitumour effects. Int J Cancer 118:2412–2420

Ginos MA, Page GP, Michalowicz BS, Patel KJ, Volker SE, Pambuccian SE, Ondrey FG, Adams GL, Gaffney PM (2004) Identification of a gene expression signature associated with recurrent disease in squamous cell carcinoma of the head and neck. Cancer Res 64:55–63

Guillermet-Guibert J, Davenne L, Pchejetski D, Saint-Laurent N, Brizuela L, Guilbeau-Frugier C, Delisle MB, Cuvillier O, Susini C, Bousquet C (2009) Targeting the sphingolipid metabolism to defeat pancreatic cancer cell resistance to the chemotherapeutic gemcitabine drug. Mol Cancer Ther 8:809–820

Hait NC, Sarkar S, Le Stunff H, Mikami A, Maceyka M, Milstien S, Spiegel S (2005) Role of sphingosine kinase 2 in cell migration toward epidermal growth factor. J Biol Chem 280:29462–29469

Hait NC, Bellamy A, Milstien S, Kordula T, Spiegel S (2007) Sphingosine kinase type 2 activation by ERK-mediated phosphorylation. J Biol Chem 282:12058–12065

Hait NC, Allegood J, Maceyka M, Strub GM, Harikumar KB, Singh SK, Luo C, Marmorstein R, Kordula T, Milstien S, Spiegel S (2009) Regulation of histone acetylation in the nucleus by sphingosine-1-phosphate. Science 325:1254–1257

Hannun YA, Obeid LM (2008) Principles of bioactive lipid signalling: lessons from sphingolipids. Nat Rev Mol Cell Biol 9:139–150

Hanson MA, Roth CB, Jo E, Griffith MT, Scott FL, Reinhart G, Desale H, Clemons B, Cahalan SM, Schuerer SC, Sanna MG, Han GW, Kuhn P, Rosen H, Stevens RC (2012) Crystal structure of a lipid G protein-coupled receptor. Science 335:851–855

Heinrich M, Wickel M, Winoto-Morbach S, Schneider-Brachert W, Weber T, Brunner J, Saftig P, Peters C, Kronke M, Schutze S (2000) Ceramide as an activator lipid of cathepsin D. Adv Exp Med Biol 477:305–315

Hisano Y, Kobayashi N, Kawahara A, Yamaguchi A, Nishi T (2011) The sphingosine 1-phosphate transporter, SPNS2, functions as a transporter of the phosphorylated form of the immunomodulating agent FTY720. J Biol Chem 286:1758–1766

Igarashi Y, Hakomori S, Toyokuni T, Dean B, Fujita S, Sugimoto M, Ogawa T, el-Ghendy K, Racker E (1989) Effect of chemically well-defined sphingosine and its N-methyl derivatives on protein kinase C and src kinase activities. Biochemistry 28:6796–6800

Igarashi N, Okada T, Hayashi S, Fujita T, Jahangeer S, Nakamura S (2003) Sphingosine kinase 2 is a nuclear protein and inhibits DNA synthesis. J Biol Chem 278:46832–46839

Ikeda M, Kihara A, Igarashi Y (2004) Sphingosine-1-phosphate lyase SPL is an endoplasmic reticulum-resident, integral membrane protein with the pyridoxal 5′-phosphate binding domain exposed to the cytosol. Biochem Biophys Res Commun 325:338–343

Illuzzi G, Bernacchioni C, Aureli M, Prioni S, Frera G, Donati C, Valsecchi M, Chigorno V, Bruni P, Sonnino S, Prinetti A (2010) Sphingosine kinase mediates resistance to the synthetic retinoid N-(4-hydroxyphenyl)retinamide in human ovarian cancer cells. J Biol Chem 285: 18594–18602

Johnson KR, Johnson KY, Crellin HG, Ogretmen B, Boylan AM, Harley RA, Obeid LM (2005) Immunohistochemical distribution of sphingosine kinase 1 in normal and tumor lung tissue. J Histochem Cytochem 53:1159–1166

Kapitonov D, Allegood JC, Mitchell C, Hait NC, Almenara JA, Adams JK, Zipkin RE, Dent P, Kordula T, Milstien S, Spiegel S (2009) Targeting sphingosine kinase 1 inhibits Akt signaling, induces apoptosis, and suppresses growth of human glioblastoma cells and xenografts. Cancer Res 69:6915–6923

Kawahara A, Nishi T, Hisano Y, Fukui H, Yamaguchi A, Mochizuki N (2009) The sphingolipid transporter spns2 functions in migration of zebrafish myocardial precursors. Science 323:524–527

Kawamori T, Osta W, Johnson KR, Pettus BJ, Bielawski J, Tanaka T, Wargovich MJ, Reddy BS, Hannun YA, Obeid LM, Zhou D (2006) Sphingosine kinase 1 is up-regulated in colon carcinogenesis. FASEB J 20:386–388

Kedderis LB, Bozigian HP, Kleeman JM, Hall RL, Palmer TE, Harrison SD Jr, Susick RL Jr (1995) Toxicity of the protein kinase C inhibitor safingol administered alone and in combination with chemotherapeutic agents. Fundam Appl Toxicol 25:201–217

Kim JW, Kim YW, Inagaki Y, Hwang YA, Mitsutake S, Ryu YW, Lee WK, Ha HJ, Park CS, Igarashi Y (2005) Synthesis and evaluation of sphingoid analogs as inhibitors of sphingosine kinases. Bioorg Med Chem 13:3475–3485

King CC, Zenke FT, Dawson PE, Dutil EM, Newton AC, Hemmings BA, Bokoch GM (2000) Sphingosine is a novel activator of 3-phosphoinositide-dependent kinase 1. J Biol Chem 275:18108–18113

Knapp P, Baranowski M, Knapp M, Zabielski P, Blachnio-Zabielska AU, Gorski J (2010) Altered sphingolipid metabolism in human endometrial cancer. Prostaglandins Other Lipid Mediat 92:62–66

Kohama T, Olivera A, Edsall L, Nagiec MM, Dickson R, Spiegel S (1998) Molecular cloning and functional characterization of murine sphingosine kinase. J Biol Chem 273:23722–23728

Kohno M, Momoi M, Oo ML, Paik JH, Lee YM, Venkataraman K, Ai Y, Ristimaki AP, Fyrst H, Sano H, Rosenberg D, Saba JD, Proia RL, Hla T (2006) Intracellular role for sphingosine kinase 1 in intestinal adenoma cell proliferation. Mol Cell Biol 26:7211–7223

Kondo Y, Kondo S (2006) Autophagy and cancer therapy. Autophagy 2:85–90

Kono K, Tanaka M, Mizuno T, Kodama K, Ogita T, Kohama T (2000a) B-535a, b and c, new sphingosine kinase inhibitors, produced by a marine bacterium; taxonomy, fermentation, isolation, physico-chemical properties and structure determination. J Antibiot (Tokyo) 53:753–758

Kono K, Tanaka M, Ogita T, Kohama T (2000b) Characterization of B-5354c, a new sphingosine kinase inhibitor, produced by a marine bacterium. J Antibiot (Tokyo) 53:759–764

Kono K, Sugiura M, Kohama T (2002) Inhibition of recombinant sphingosine kinases by novel inhibitors of microbial origin, F-12509A and B-5354c. J Antibiot (Tokyo) 55:99–103

Kraveka JM, Li L, Szulc ZM, Bielawski J, Ogretmen B, Hannun YA, Obeid LM, Bielawska A (2007) Involvement of dihydroceramide desaturase in cell cycle progression in human neuroblastoma cells. J Biol Chem 282:16718–16728

Lahiri S, Park H, Laviad EL, Lu X, Bittman R, Futerman AH (2009) Ceramide synthesis is modulated by the sphingosine analog FTY720 via a mixture of uncompetitive and noncompetitive inhibition in an Acyl-CoA chain length-dependent manner. J Biol Chem 284:16090–16098

Laviad EL, Kelly S, Merrill AH Jr, Futerman AH (2012) Modulation of ceramide synthase activity via dimerization. J Biol Chem 287(25):21025–21033

Lavieu G, Scarlatti F, Sala G, Carpentier S, Levade T, Ghidoni R, Botti J, Codogno P (2006) Regulation of autophagy by sphingosine kinase 1 and its role in cell survival during nutrient starvation. J Biol Chem 281:8518–8527

Le Scolan E, Pchejetski D, Banno Y, Denis N, Mayeux P, Vainchenker W, Levade T, Moreau-Gachelin F (2005) Overexpression of sphingosine kinase 1 is an oncogenic event in erythroleukemic progression. Blood 106:1808–1816

Lepine S, Allegood JC, Edmonds Y, Milstien S, Spiegel S (2011) Autophagy induced by deficiency of sphingosine-1-phosphate phosphohydrolase 1 is switched to apoptosis by calpain-mediated autophagy-related gene 5 (Atg5) cleavage. J Biol Chem 286:44380–44390

Levine B, Kroemer G (2008) Autophagy in the pathogenesis of disease. Cell 132:27–42

Li QF, Wu CT, Guo Q, Wang H, Wang LS (2008) Sphingosine 1-phosphate induces Mcl-1 upregulation and protects multiple myeloma cells against apoptosis. Biochem Biophys Res Commun 371:159–162

Li MH, Sanchez T, Milne GL, Morrow JD, Hla T, Ferrer F (2009) S1P/S1P2 signaling induces cyclooxygenase-2 expression in Wilms tumor. J Urol 181:1347–1352

Lim KG, Sun C, Bittman R, Pyne NJ, Pyne S (2011) (R)-FTY720 methyl ether is a specific sphingosine kinase 2 inhibitor: effect on sphingosine kinase 2 expression in HEK 293 cells and actin rearrangement and survival of MCF-7 breast cancer cells. Cell Signal 23:1590–1595

Liu H, Sugiura M, Nava VE, Edsall LC, Kono K, Poulton S, Milstien S, Kohama T, Spiegel S (2000) Molecular cloning and functional characterization of a novel mammalian sphingosine kinase type 2 isoform. J Biol Chem 275:19513–19520

Liu H, Chakravarty D, Maceyka M, Milstien S, Spiegel S (2002) Sphingosine kinases: a novel family of lipid kinases. Prog Nucleic Acid Res Mol Biol 71:493–511

Liu H, Toman RE, Goparaju SK, Maceyka M, Nava VE, Sankala H, Payne SG, Bektas M, Ishii I, Chun J, Milstien S, Spiegel S (2003) Sphingosine kinase type 2 is a putative BH3-only protein that induces apoptosis. J Biol Chem 278:40330–40336

Ma XJ, Wang Z, Ryan PD, Isakoff SJ, Barmettler A, Fuller A, Muir B, Mohapatra G, Salunga R, Tuggle JT, Tran Y, Tran D, Tassin A, Amon P, Wang W, Enright E, Stecker K, Estepa-Sabal E, Smith B, Younger J, Balis U, Michaelson J, Bhan A, Habin K, Baer TM, Brugge J, Haber DA, Erlander MG, Sgroi DC (2004) A two-gene expression ratio predicts clinical outcome in breast cancer patients treated with tamoxifen. Cancer Cell 5:607–616

Maceyka M, Sankala H, Hait NC, Le Stunff H, Liu H, Toman R, Collier C, Zhang M, Satin LS, Merrill AH Jr, Milstien S, Spiegel S (2005) SphK1 and SphK2, sphingosine kinase isoenzymes with opposing functions in sphingolipid metabolism. J Biol Chem 280:37118–37129

Mandala SM (2001) Sphingosine-1-phosphate phosphatases. Prostaglandins Other Lipid Mediat 64:143–156

Mao C, Saba JD, Obeid LM (1999) The dihydrosphingosine-1-phosphate phosphatases of Saccharomyces cerevisiae are important regulators of cell proliferation and heat stress responses. Biochem J 342(Pt 3):667–675

Matsuoka Y, Nagahara Y, Ikekita M, Shinomiya T (2003) A novel immunosuppressive agent FTY720 induced Akt dephosphorylation in leukemia cells. Br J Pharmacol 138:1303–1312

McDonald OB, Hannun YA, Reynolds CH, Sahyoun N (1991) Activation of casein kinase II by sphingosine. J Biol Chem 266:21773–21776

Megidish T, Cooper J, Zhang L, Fu H, Hakomori S (1998) A novel sphingosine-dependent protein kinase (SDK1) specifically phosphorylates certain isoforms of 14-3-3 protein. J Biol Chem 273:21834–21845

Meyers-Needham M, Lewis JA, Gencer S, Sentelle RD, Saddoughi SA, Clarke CJ, Hannun YA, Norell H, da Palma TM, Nishimura M, Kraveka JM, Khavandgar Z, Murshed M, Cevik MO, Ogretmen B (2012a) Off-target function of the Sonic hedgehog inhibitor cyclopamine in mediating apoptosis via nitric oxide-dependent neutral sphingomyelinase 2/ceramide induction. Mol Cancer Ther 11:1092–1102

Meyers-Needham M, Ponnusamy S, Gencer S, Jiang W, Thomas RJ, Senkal CE, Ogretmen B (2012b) Concerted functions of HDAC1 and microRNA-574-5p repress alternatively spliced ceramide synthase 1 expression in human cancer cells. EMBO Mol Med 4:78–92

Michel C, van Echten-Deckert G, Rother J, Sandhoff K, Wang E, Merrill AH Jr (1997) Characterization of ceramide synthesis. A dihydroceramide desaturase introduces the 4,5-trans-double bond of sphingosine at the level of dihydroceramide. J Biol Chem 272:22432–22437

Miller AV, Alvarez SE, Spiegel S, Lebman DA (2008) Sphingosine kinases and sphingosine-1-phosphate are critical for transforming growth factor beta-induced extracellular signal-regulated kinase 1 and 2 activation and promotion of migration and invasion of esophageal cancer cells. Mol Cell Biol 28:4142–4151

Mitra P, Oskeritzian CA, Payne SG, Beaven MA, Milstien S, Spiegel S (2006) Role of ABCC1 in export of sphingosine-1-phosphate from mast cells. Proc Natl Acad Sci USA 103:16394–16399

Mizugishi K, Yamashita T, Olivera A, Miller GF, Spiegel S, Proia RL (2005) Essential role for sphingosine kinases in neural and vascular development. Mol Cell Biol 25:11113–11121

Mukhopadhyay A, Saddoughi SA, Song P, Sultan I, Ponnusamy S, Senkal CE, Snook CF, Arnold HK, Sears RC, Hannun YA, Ogretmen B (2009) Direct interaction between the inhibitor 2 and ceramide via sphingolipid-protein binding is involved in the regulation of protein phosphatase 2A activity and signaling. FASEB J 23:751–763

Nagahashi M, Ramachandran S, Kim EY, Allegood JC, Rashid OM, Yamada A, Zhao R, Milstien S, Zhou H, Spiegel S, Takabe K (2012) Sphingosine-1-phosphate produced by sphingosine kinase 1 promotes breast cancer progression by stimulating angiogenesis and lymphangiogenesis. Cancer Res 72:726–735

Nagaoka Y, Otsuki K, Fujita T, Uesato S (2008) Effects of phosphorylation of immunomodulatory agent FTY720 (fingolimod) on antiproliferative activity against breast and colon cancer cells. Biol Pharm Bull 31:1177–1181

Nava VE, Cuvillier O, Edsall LC, Kimura K, Milstien S, Gelmann EP, Spiegel S (2000) Sphingosine enhances apoptosis of radiation-resistant prostate cancer cells. Cancer Res 60:4468–4474

Nava VE, Hobson JP, Murthy S, Milstien S, Spiegel S (2002) Sphingosine kinase type 1 promotes estrogen-dependent tumorigenesis of breast cancer MCF-7 cells. Exp Cell Res 281:115–127

Neviani P, Santhanam R, Oaks JJ, Eiring AM, Notari M, Blaser BW, Liu S, Trotta R, Muthusamy N, Gambacorti-Passerini C, Druker BJ, Cortes J, Marcucci G, Chen CS, Verrills NM, Roy DC, Caligiuri MA, Bloomfield CD, Byrd JC, Perrotti D (2007) FTY720, a new alternative for treating blast crisis chronic myelogenous leukemia and Philadelphia chromosome-positive acute lymphocytic leukemia. J Clin Invest 117:2408–2421

Nindl I, Dang C, Forschner T, Kuban RJ, Meyer T, Sterry W, Stockfleth E (2006) Identification of differentially expressed genes in cutaneous squamous cell carcinoma by microarray expression profiling. Mol Cancer 5:30

Ogretmen B, Hannun YA (2004) Biologically active sphingolipids in cancer pathogenesis and treatment. Nat Rev Cancer 4:604–616

Ogretmen B, Kraveka JM, Schady D, Usta J, Hannun YA, Obeid LM (2001) Molecular mechanisms of ceramide-mediated telomerase inhibition in the A549 human lung adenocarcinoma cell line. J Biol Chem 276:32506–32514

Olivera A, Kohama T, Edsall L, Nava V, Cuvillier O, Poulton S, Spiegel S (1999) Sphingosine kinase expression increases intracellular sphingosine-1-phosphate and promotes cell growth and survival. J Cell Biol 147:545–558

Osborne N, Brand-Arzamendi K, Ober EA, Jin SW, Verkade H, Holtzman NG, Yelon D, Stainier DY (2008) The spinster homolog, two of hearts, is required for sphingosine 1-phosphate signaling in zebrafish. Curr Biol 18:1882–1888

Paugh SW, Paugh BS, Rahmani M, Kapitonov D, Almenara JA, Kordula T, Milstien S, Adams JK, Zipkin RE, Grant S, Spiegel S (2008) A selective sphingosine kinase 1 inhibitor integrates multiple molecular therapeutic targets in human leukemia. Blood 112:1382–1391

Payne SG, Oskeritzian CA, Griffiths R, Subramanian P, Barbour SE, Chalfant CE, Milstien S, Spiegel S (2007) The immunosuppressant drug FTY720 inhibits cytosolic phospholipase A2 independently of sphingosine-1-phosphate receptors. Blood 109:1077–1085

Pchejetski D, Golzio M, Bonhoure E, Calvet C, Doumerc N, Garcia V, Mazerolles C, Rischmann P, Teissie J, Malavaud B, Cuvillier O (2005) Sphingosine kinase-1 as a chemotherapy sensor in prostate adenocarcinoma cell and mouse models. Cancer Res 65:11667–11675

Pitson SM (2011) Regulation of sphingosine kinase and sphingolipid signaling. Trends Biochem Sci 36:97–107

Pitson SM, Moretti PA, Zebol JR, Lynn HE, Xia P, Vadas MA, Wattenberg BW (2003) Activation of sphingosine kinase 1 by ERK1/2-mediated phosphorylation. EMBO J 22:5491–5500

Pitson SM, Xia P, Leclercq TM, Moretti PA, Zebol JR, Lynn HE, Wattenberg BW, Vadas MA (2005) Phosphorylation-dependent translocation of sphingosine kinase to the plasma membrane drives its oncogenic signalling. J Exp Med 201:49–54

Ponnusamy S, Meyers-Needham M, Senkal CE, Saddoughi SA, Sentelle D, Selvam SP, Salas A, Ogretmen B (2010) Sphingolipids and cancer: ceramide and sphingosine-1-phosphate in the regulation of cell death and drug resistance. Future Oncol 6:1603–1624

Ponnusamy S, Selvam SP, Mehrotra S, Kawamori T, Snider AJ, Obeid LM, Shao Y, Sabbadini R, Ogretmen B (2012) Communication between host organism and cancer cells is transduced by systemic sphingosine kinase 1/sphingosine 1-phosphate signalling to regulate tumour metastasis. EMBO Mol Med 4(8):761–775

Pyeon D, Newton MA, Lambert PF, den Boon JA, Sengupta S, Marsit CJ, Woodworth CD, Connor JP, Haugen TH, Smith EM, Kelsey KT, Turek LP, Ahlquist P (2007) Fundamental differences in cell cycle deregulation in human papillomavirus-positive and human papillomavirus-negative head/neck and cervical cancers. Cancer Res 67:4605–4619

Reynolds CP, Maurer BJ, Kolesnick RN (2004) Ceramide synthesis and metabolism as a target for cancer therapy. Cancer Lett 206:169–180

Ruckhaberle E, Rody A, Engels K, Gaetje R, von Minckwitz G, Schiffmann S, Grosch S, Geisslinger G, Holtrich U, Karn T, Kaufmann M (2008) Microarray analysis of altered sphingolipid metabolism reveals prognostic significance of sphingosine kinase 1 in breast cancer. Breast Cancer Res Treat 112:41–52

Sabbadini RA (2011) Sphingosine-1-phosphate antibodies as potential agents in the treatment of cancer and age-related macular degeneration. Br J Pharmacol 162:1225–1238

Saddoughi SA, Song P, Ogretmen B (2008) Roles of bioactive sphingolipids in cancer biology and therapeutics. Subcell Biochem 49:413–440

Salas A, Ponnusamy S, Senkal CE, Meyers-Needham M, Selvam SP, Saddoughi SA, Apohan E, Sentelle RD, Smith C, Gault CR, Obeid LM, El-Shewy HM, Oaks J, Santhanam R, Marcucci G, Baran Y, Mahajan S, Fernandes D, Stuart R, Perrotti D, Ogretmen B (2011) Sphingosine kinase-1 and sphingosine 1-phosphate receptor 2 mediate Bcr-Abl1 stability and drug resistance by modulation of protein phosphatase 2A. Blood 117:5941–5952

Sankala HM, Hait NC, Paugh SW, Shida D, Lepine S, Elmore LW, Dent P, Milstien S, Spiegel S (2007) Involvement of sphingosine kinase 2 in p53-independent induction of p21 by the chemotherapeutic drug doxorubicin. Cancer Res 67:10466–10474

Sanna MG, Wang SK, Gonzalez-Cabrera PJ, Don A, Marsolais D, Matheu MP, Wei SH, Parker I, Jo E, Cheng WC, Cahalan MD, Wong CH, Rosen H (2006) Enhancement of capillary leakage and restoration of lymphocyte egress by a chiral S1P1 antagonist in vivo. Nat Chem Biol 2:434–441

Sarkar S, Maceyka M, Hait NC, Paugh SW, Sankala H, Milstien S, Spiegel S (2005) Sphingosine kinase 1 is required for migration, proliferation and survival of MCF-7 human breast cancer cells. FEBS Lett 579:5313–5317

Sato K, Malchinkhuu E, Horiuchi Y, Mogi C, Tomura H, Tosaka M, Yoshimoto Y, Kuwabara A, Okajima F (2007) Critical role of ABCA1 transporter in sphingosine 1-phosphate release from astrocytes. J Neurochem 103:2610–2619

Sauer B, Gonska H, Manggau M, Kim DS, Schraut C, Schafer-Korting M, Kleuser B (2005) Sphingosine 1-phosphate is involved in cytoprotective actions of calcitriol in human fibroblasts and enhances the intracellular Bcl-2/Bax rheostat. Pharmazie 60:298–304

Schmid G, Guba M, Ischenko I, Papyan A, Joka M, Schrepfer S, Bruns CJ, Jauch KW, Heeschen C, Graeb C (2007) The immunosuppressant FTY720 inhibits tumor angiogenesis via the sphingosine 1-phosphate receptor 1. J Cell Biochem 101:259–270

Schnitzer SE, Weigert A, Zhou J, Brune B (2009) Hypoxia enhances sphingosine kinase 2 activity and provokes sphingosine-1-phosphate-mediated chemoresistance in A549 lung cancer cells. Mol Cancer Res 7:393–401

Schwartz GK, Ward D, Saltz L, Casper ES, Spiess T, Mullen E, Woodworth J, Venuti R, Zervos P, Storniolo AM, Kelsen DP (1997) A pilot clinical/pharmacological study of the protein kinase C-specific inhibitor safingol alone and in combination with doxorubicin. Clin Cancer Res 3:537–543

Senkal CE, Ponnusamy S, Manevich Y, Meyers-Needham M, Saddoughi SA, Mukhopadyay A, Dent P, Bielawski J, Ogretmen B (2011) Alteration of ceramide synthase 6/C16-ceramide induces activating transcription factor 6-mediated endoplasmic reticulum (ER) stress and apoptosis via perturbation of cellular Ca2+ and ER/Golgi membrane network. J Biol Chem 286(49):42446–42458

Spiegel S, Milstien S (2003) Sphingosine-1-phosphate: an enigmatic signalling lipid. Nat Rev Mol Cell Biol 4:397–407

Stahelin RV, Hwang JH, Kim JH, Park ZY, Johnson KR, Obeid LM, Cho W (2005) The mechanism of membrane targeting of human sphingosine kinase 1. J Biol Chem 280:43030–43038

Strub GM, Maceyka M, Hait NC, Milstien S, Spiegel S (2010) Extracellular and intracellular actions of sphingosine-1-phosphate. Adv Exp Med Biol 688:141–155

Strub GM, Paillard M, Liang J, Gomez L, Allegood JC, Hait NC, Maceyka M, Price MM, Chen Q, Simpson DC, Kordula T, Milstien S, Lesnefsky EJ, Spiegel S (2011) Sphingosine-1-phosphate produced by sphingosine kinase 2 in mitochondria interacts with prohibitin 2 to regulate complex IV assembly and respiration. FASEB J 25:600–612

Sugiura M, Kono K, Liu H, Shimizugawa T, Minekura H, Spiegel S, Kohama T (2002) Ceramide kinase, a novel lipid kinase. Molecular cloning and functional characterization. J Biol Chem 277:23294–23300

Sukocheva O, Wang L, Verrier E, Vadas MA, Xia P (2009) Restoring endocrine response in breast cancer cells by inhibition of the sphingosine kinase-1 signaling pathway. Endocrinology 150:4484–4492

Takabe K, Kim RH, Allegood JC, Mitra P, Ramachandran S, Nagahashi M, Harikumar KB, Hait NC, Milstien S, Spiegel S (2010) Estradiol induces export of sphingosine 1-phosphate from breast cancer cells via ABCC1 and ABCG2. J Biol Chem 285:10477–10486

Talantov D, Mazumder A, Yu JX, Briggs T, Jiang Y, Backus J, Atkins D, Wang Y (2005) Novel genes associated with malignant melanoma but not benign melanocytic lesions. Clin Cancer Res 11:7234–7242

Tonelli F, Lim KG, Loveridge C, Long J, Pitson SM, Tigyi G, Bittman R, Pyne S, Pyne NJ (2010) FTY720 and (S)-FTY720 vinylphosphonate inhibit sphingosine kinase 1 and promote its proteasomal degradation in human pulmonary artery smooth muscle, breast cancer and androgen-independent prostate cancer cells. Cell Signal 22:1536–1542

Vadas M, Xia P, McCaughan G, Gamble J (2008) The role of sphingosine kinase 1 in cancer: oncogene or non-oncogene addiction? Biochim Biophys Acta 1781:442–447

Van Brocklyn JR, Jackson CA, Pearl DK, Kotur MS, Snyder PJ, Prior TW (2005) Sphingosine kinase-1 expression correlates with poor survival of patients with glioblastoma multiforme: roles of sphingosine kinase isoforms in growth of glioblastoma cell lines. J Neuropathol Exp Neurol 64:695–705

Vessey DA, Kelley M, Zhang J, Li L, Tao R, Karliner JS (2007) Dimethylsphingosine and FTY720 inhibit the SK1 form but activate the SK2 form of sphingosine kinase from rat heart. J Biochem Mol Toxicol 21:273–279

Visentin B, Vekich JA, Sibbald BJ, Cavalli AL, Moreno KM, Matteo RG, Garland WA, Lu Y, Yu S, Hall HS, Kundra V, Mills GB, Sabbadini RA (2006) Validation of an anti-sphingosine-1-phosphate antibody as a potential therapeutic in reducing growth, invasion, and angiogenesis in multiple tumor lineages. Cancer Cell 9:225–238

Wang G, Silva J, Krishnamurthy K, Tran E, Condie BG, Bieberich E (2005) Direct binding to ceramide activates protein kinase Czeta before the formation of a pro-apoptotic complex with PAR-4 in differentiating stem cells. J Biol Chem 280:26415–26424

Watson C, Long JS, Orange C, Tannahill CL, Mallon E, McGlynn LM, Pyne S, Pyne NJ, Edwards J (2010) High expression of sphingosine 1-phosphate receptors, S1P1 and S1P3, sphingosine kinase 1, and extracellular signal-regulated kinase-1/2 is associated with development of tamoxifen resistance in estrogen receptor-positive breast cancer patients. Am J Pathol 177:2205–2215

Wong YF, Selvanayagam ZE, Wei N, Porter J, Vittal R, Hu R, Lin Y, Liao J, Shih JW, Cheung TH, Lo KW, Yim SF, Yip SK, Ngong DT, Siu N, Chan LK, Chan CS, Kong T, Kutlina E, McKinnon RD, Denhardt DT, Chin KV, Chung TK (2003) Expression genomics of cervical cancer: molecular classification and prediction of radiotherapy response by DNA microarray. Clin Cancer Res 9:5486–5492

Wooten LG, Ogretmen B (2005) Sp1/Sp3-dependent regulation of human telomerase reverse transcriptase promoter activity by the bioactive sphingolipid ceramide. J Biol Chem 280:28867–28876

Wooten-Blanks LG, Song P, Senkal CE, Ogretmen B (2007) Mechanisms of ceramide-mediated repression of the human telomerase reverse transcriptase promoter via deacetylation of Sp3 by histone deacetylase 1. FASEB J 21:3386–3397

Xia P, Gamble JR, Wang L, Pitson SM, Moretti PA, Wattenberg BW, D'Andrea RJ, Vadas MA (2000) An oncogenic role of sphingosine kinase. Curr Biol 10:1527–1530

Xiang Y, Asmussen G, Booker M, Hirth B, Kane JL Jr, Liao J, Noson KD, Yee C (2009) Discovery of novel sphingosine kinase 1 inhibitors. Bioorg Med Chem Lett 19:6119–6121

Xiao M, Liu Y, Zou F (2012) Sensitization of human colon cancer cells to sodium butyrate-induced apoptosis by modulation of sphingosine kinase 2 and protein kinase D. Exp Cell Res 318:43–52

Using ASMase Knockout Mice to Model Human Diseases

Guoqiang Hua and Richard Kolesnick

Contents

Abstract Acid sphingomyelinase (ASMase) is a key initiator of sphingomyelin/ceramide signal transduction activated by many stress stimuli. Over the past two decades, much progress has been made in defining the clinical relevance of sphingomyelin/ceramide signaling in numerous diseases using ASMase knockout mice. Organs that operate this pathway are numerous and the disease states regulated are diverse, with ceramide generation governing injury in tumor, gut, ovary, brain, lung, heart, liver, and during infection. This chapter emphasizes evolutionary conservation of sphingolipid stress signaling and mammalian adaptations that permit transduction of organotypic responses. Recognition that

---

G. Hua • R. Kolesnick (✉)
Laboratory of Signal Transduction, Memorial Sloan-Kettering Cancer Center,
1275 York Avenue, New York, NY 10065, USA
e-mail: r-kolesnick@ski.mskcc.org

E. Gulbins and I. Petrache (eds.), *Sphingolipids in Disease*,
Handbook of Experimental Pharmacology 216,
DOI 10.1007/978-3-7091-1511-4_2,

the sphingomyelin/ceramide transducer calibrates extent of tissue injury, ultimately acting as a molecular switch that determines organ fate, is driving development of new pharmacologic concepts and tools to intervene therapeutically.

Keywords ASMase • Ceramide-rich platforms (CRPs) • ASMase Knockout • Single-dose radiotherapy (SDRT) • Cancer • Niemann-Pick Disease (NPD) • Stroke • Blood-brain barrier (BBB) • Acute respiratory distress syndrome (ARDS) • Cystic Fibrosis (CF) • Diabetes • Infections

1 Introduction

1.1 ASMase

The sphingomyelin pathway is a ubiquitous, evolutionarily conserved signaling system initiated by hydrolysis of sphingomyelin to generate the second messenger ceramide. Sphingomyelinase (SMase) is a specialized form of phospholipase C, which cleaves the phosphodiester bond of sphingomyelin, generating ceramide. Several SMase isoforms can be distinguished by their pH optima, cofactor dependence, and subcellular location. SMases are classified into three groups—acid SMase (ASMase), neutral SMase (NSMase), and alkaline SMase (Alk-SMase) (Kolesnick 2002). While ASMase was originally considered a strictly lysosomal enzyme because of its pH optimum at 4.5–5.0, an ASMase isoform was found within secretory vesicles at the plasma membrane (Liu and Anderson 1995; Schissel et al. 1998a). Subsequent studies showed that ASMase exists in two forms, termed lysosomal ASMase (L-ASMase) and secretory ASMase (S-ASMase), differing in glycosylation pattern and NH_2-terminal processing, and consequently in subcellular targeting. ASMase is present in all types of cells, preferentially endowed in endothelium of blood vessels, and cells of the reticuloendothelial system (RES) of liver (Kupffer cells), spleen, bone marrow, lung, as well as macrophages (Otterbach and Stoffel 1995). Activation of ASMase has been demonstrated in response to various unrelated stress stimuli. ASMase is also a key enzyme responsible for ceramide homeostasis.

1.2 Ceramide

Ceramide is a sphingosine-based lipid capable of initiating signaling for numerous stress stimuli, including, but not limited to TNF-α, Fas ligand, ionizing radiation (IR), heat shock, ultraviolet light, and oxidative stress (Fuks et al. 1995; Grassme et al. 2003a; Gulbins 2003; Gulbins et al. 1995; Haimovitz-Friedman et al. 1994; Jarvis et al. 1994; Obeid et al. 1993; Verheij et al. 1996). Structurally, ceramide contains a long-chain sphingoid base backbone, an amide-linked long-chain fatty acid, and a hydroxyl head group. Monounsaturated or saturated fatty acids of various lengths from 2 to 28 carbon atoms are usually found in natural ceramides (Kolesnick et al. 2000). Physical properties of ceramide are determined by length of the fatty acyl chain. Ceramides with long fatty acyl chain (12 carbons or longer) belong to the category of "non-swelling amphiphiles," as they cannot give rise to micelles or other aggregates in aqueous suspension and hence cannot serve as detergents (Small 1970). In contrast, short-chain ceramides swell in water, a property that limits their utility in research (Stancevic and Kolesnick 2010). In mammalian cells, ceramides usually contain fatty acyl chains of 16–24 carbon atoms. Depending on stimulus and cell type, ceramide can be generated either through SMase-dependent hydrolysis of sphingomyelin or a de novo synthetic pathway. SMase-mediated ceramide generation is usually a rapid event localized in most cells at the plasma membrane, while de novo synthesis of ceramide occurs in a more prolonged fashion, exclusively intracellularly within endoplasmic reticulum or mitochondrial-associated membrane (MAM) (Shimeno et al. 1995; Bionda et al. 2004; Kirschnek et al. 2000; Schissel et al. 1998b). As a second messenger, ceramide plays a ubiquitous role in diverse biologic processes (Kolesnick and Hannun 1999). Accumulating evidence suggests that changes in local membrane structure induced by ceramide elevation are essential for its biological function. Once generated, ceramide may amass or be converted into a variety of metabolites that serve as bioeffector molecules, often inducing proliferation, and except for sphingosine and GD3 usually do not signal cell death. Further, some cells manifest enhanced conversion of ceramide to S1P. Balance between pro-apoptotic properties of ceramide and anti-apoptotic properties of S1P has been referred to as the "S1P Rheostat" (Cuvillier et al. 1996) and may under some circumstances play a crucial role in determining outcome of stress damage in tissues.

1.3 Mechanism of Sphingomyelin/Ceramide Signaling

In 1987, Paley et al. reported that 1,2-diacylglycerols induce SMase activation in GH3 pituitary cells and suggested for the first time that a sphingomyelin-based signaling pathway may be active in this response (Kolesnick and Paley 1987). Subsequently, Okazaki et al. (1989, 1990) confirmed this notion, demonstrating that receptor binding of vitamin D3 can activate a sphingomyelin/ceramide signaling pathway in HL-60 cells. Over the past two decades, extensive literature supports the sphingomyelin/ceramide signaling pathway as an evolutionarily conserved response system for stress (UV, heat, mechanical stress, etc.) (Grassme et al. 2007; Hannun and Obeid 2008). This system is usually in the "off" state under baseline conditions, activated upon contacting stress, and calibrates extent of cellular stress, evoking an adaptive or apoptotic cellular response depending on strength of signal. The most widely studied ceramide-generating mechanism involves translocation of ASMase to the outer plasma membrane, where it hydrolyses its substrate sphingomyelin, generating ceramide therein (Grassme et al. 2001a). This event usually occurs within seconds to minutes of encountering

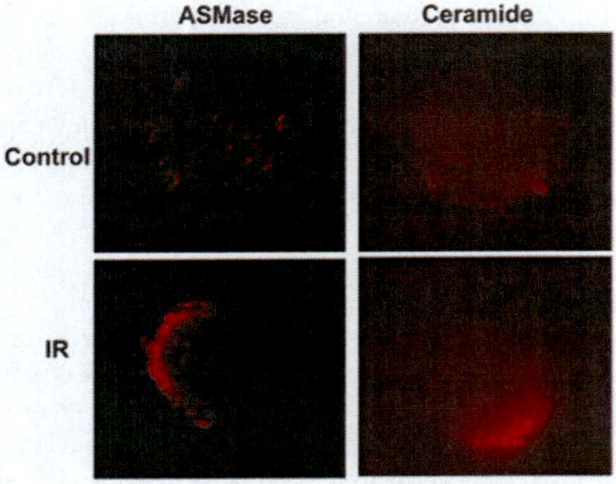

Fig. 1 Ionizing radiation triggers ceramide-rich platform (CRP) formation in BAEC. Clustering of ceramide and ASMase into macrodomains on the outer leaflet of the plasma membrane of BAEC at 1 min post 15 Gy. Cells were fixed 1 min post-irradiation and platforms identified by standard fluorescent microscopy after staining with anti-ceramide IgM MID 15B4 (1:50 dilution, Alexis Biochemicals) or rabbit anti-ASMase Ab 1598 (1:100 dilution), followed by Cy3-conjugated anti-mouse or anti-rabbit IgM (1:500 dilution, Roche Molecular Biochemicals), respectively [Adapted from Rotolo et al. (2012)]

stress, generating ceramide rapidly at the plasma membrane (Rotolo et al. 2012) (Fig. 1). While the mechanism by which diverse stresses are registered at the plasma membrane is uncertain, stress-induced translocation of ASMase to the exoplasmic leaflet of the plasma membrane to initiate stress signaling appears to involve fusion of a subset of S-ASMase-containing acidic vesicles (Kolesnick, Paris, and Rao, unpublished) that abut the inner plasma membrane and upon contacting stress fuse with the bilayer to expose ASMase on the surface (Grassme et al. 2001a). Ceramide, once generated, has a unique biophysical property, that of self-association, forming ceramide-rich platforms (CRPs) on the cell surface, 1–5 μm in diameter, which serve as sites for protein oligomerization and trans-membrane signal transduction (Grassme et al. 2001a, 2003a; Rotolo et al. 2005). These CRPs are sites of protein–protein interactions that lead to downstream signaling, and perturbation of CRP formation has been implicated in pathogenesis of a variety of human diseases (Smith and Schuchman 2008).

Evidence indicates that ceramide also acts as a classic second messenger in proliferative and stress responses, stoichiometrically activating protein targets (Kolesnick 2002; Hannun and Obeid 2002). Direct targets for ceramide include protein kinase C (PKC) δ, ε and ζ (Kashiwagi et al. 2002; Bourbon et al. 2000) kinase suppressor of Ras (KSR1) (Grassme et al. 2001b), c-Raf-1 (Yu et al. 2000), protein phosphatase 1 and 2a (Chalfant et al. 1999), phospholipase A2 (Huwiler et al. 2001), and cathepsin D (Heinrich et al. 1999). Ceramide binds the C1B lipid-binding domain of KSR1 and c-Raf-1 and via a C1B domain induces PKCε translocation to an intracellular site distinct from the site phorbol esters regulate.

A KSR1 C1B-domain polypeptide has been used as a tool to detect surface ceramide on JY B lymphocytes (Grassme et al. 2001b) or as a pharmacologic reagent to inhibit CD95 clustering, hepatocyte apoptosis, and death of mice after intravenous anti-CD95 antibody. This diversity of targets for ceramide interaction is consistent with its pleiotropic involvement in cellular and tissue responses.

1.4 Generation of ASMase Knockout Mice

An inherited deficiency of ASMase activity had been identified as the cause of the lysosomal storage disorder Niemann–Pick disease (NPD) in 1966 by the pioneering work of Brady and colleagues (Brady et al. 1966). To better understand the importance of ASMase in pathogenesis of this disease, two ASMase knockout mouse models were independently developed by Ed Schuchman and Wilhelm Stoffel in 1995 (Otterbach and Stoffel 1995; Horinouchi et al. 1995). These ASMase knockout mice display a phenotype essentially identical to type A NPD (type A NPD is a severe neurodegenerative disease of infancy usually fatal by ages 2–3; see below), including failure to thrive, neurodegeneration, and a shortened life span (Horinouchi et al. 1995). ASMase knockout mice exhibit progressive lipid storage in the RES of liver, lung, and bone marrow and in macrophages of lung, as well as in brain (Otterbach and Stoffel 1995; Horinouchi et al. 1995). The Schuchman ASMase knockout mice are normal at birth and develop routinely until about 12 weeks of age when ataxia and mild tremors became noticeable. At time of death, usually between 6 and 8 months of age, ASMase knockout mice are less than half the weight of wild-type littermates and display an obvious hunched appearance (Fig. 2) (Horinouchi et al. 1995). While the principal lipid accumulating is sphingomyelin, cholesterol and ganglioside storage have also been found. Total blood cholesterol levels are elevated nearly 80 % in ASMase knockout mice compared with wild-type mice. Furthermore, remarkable loss of Purkinje cells in the cerebellum leads to severe impairment of neuromotor coordination. Pulmonary inflammatory disease has also been reported in ASMase knockout mice (Dhami et al. 2001). It should, however, be pointed out that the phenotype of Schuchman's ASMase knockout mouse line is different from that of Stoffel's, despite the fact that the technology used to generate these mice was similar. The life expectancy of the Stoffel ASMase-deficient mice was around 4 months, with mice succumbing to advanced NPD (Otterbach and Stoffel 1995). In contrast, mice from the Schuchman group survive to 8 months of age and display a delay in accumulation of sphingomyelin until 12 weeks or so and delay in significant NPD symptomatology until 16 weeks (Horinouchi et al. 1995). The reason for the differences is not clear. It should be emphasized that all studies demonstrating abnormalities in the apoptotic response to various stress should be carried out in ASMase knockout mice before biochemical, histologic, or clinical manifestations of NPD are apparent.

Availability of ASMase knockout mice and development of pharmacologic modulators of SMase function have provided new insight into involvement of the sphingomyelin/ceramide transducer in animal models of human disease. Tissues

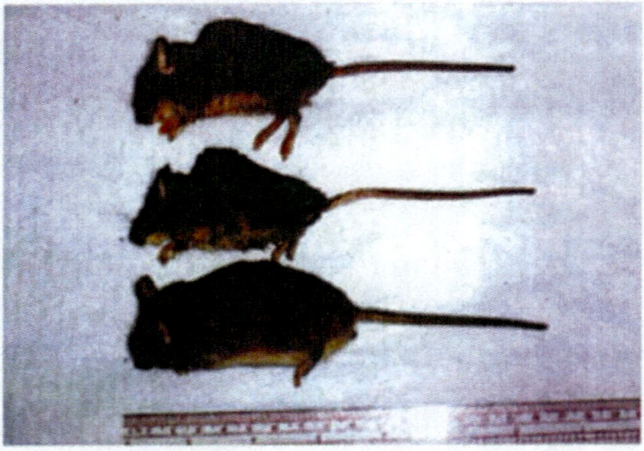

Fig. 2 Phenotype of the ASMase knockout mouse. The dramatic size difference and "hunched" appearance of ASMase-deficient mice (*top*) compared to a control littermate (*bottom*) at 4 months of age [Adapted from Horinouchi et al. (1995)]

possess specific cells that represent primary sensors of environmental stress. These sensor cells operate distinct intracellular pathways that detect and calibrate magnitude of stress by converting stress into biochemical signals (Ch'ang et al. 2005). The sphingomyelin/ceramide pathway represents a sensor cell-transduction system operative in select tissues. For instance, IR induces apoptosis of thymocytes via p53 leading to thymus involution, whereas endothelium in the irradiated GI tract uses ASMase, not p53, to initiate apoptosis, conferring GI damage (Paris et al. 2001). Further, while the sensor cell type that utilizes the sphingomyelin/ceramide transducer varies between organs, for instance, germ cells in ovaries and hepatocytes in liver, microvascular endothelium appears as the most common primary sensor in the ASMase knockout mouse model. Endothelial cells are 20-fold enriched in S-ASMase relative to other mammalian cells (Marathe et al. 1998) and are particularly sensitive to stress-induced apoptosis in vitro and in vivo (Kolesnick and Fuks 2003). Endothelium in lung and throughout the central nervous system (CNS) of ASMase knockout mice is almost completely resistant to apoptosis induced by irradiation (Santana et al. 1996; Pena et al. 2000). Furthermore, ASMase knockout mice display defects in hepatocyte apoptosis, liver failure, and animal lethality upon intravenous injection of anti-CD95 antibodies (Lin et al. 2000). ASMase knockout mice manifest a marked defect in the ovarian developmental program (Perez et al. 1997). A failure to normally delete oocytes in ASMase knockout females during embryogenesis leads to ovarian hyperplasia at birth (Morita et al. 2000). Furthermore, the primary cellular sensor may vary between stresses even within the same organ (Ch'ang et al. 2005). ASMase knockout mice provide an invaluable tool to evaluate defective apoptotic signaling and explore the role of the sphingomyelin/ceramide signaling pathway in pathophysiology of various human diseases. The following section defines how this pathway is being actively explored in a number of experimental models of human disease using ASMase knockout mice.

2 ASMase Knockout Mice and Human Disease

2.1 ASMase Knockout Mice for Cancer Radiation Therapy

As stress is known to increase ceramide levels in select mammalian cells, it is not surprising that increased ceramide content has been observed in response to a number of cancer treatments. Interestingly, many tumors seem to have developed strategies to reduce baseline ceramide levels, associated with resistance to chemotherapy or radiotherapy. Several studies showed that ceramide levels were dramatically decreased in some types of human cancer, including colon cancers (Selzner et al. 2001), gliomas (Riboni et al. 2002), and ovarian cancers (Rylova et al. 1998). There is growing evidence that sphingomyelin/ceramide signaling pathway is activated during various cancer treatments, especially during high single-dose radiotherapy (SDRT).

While the classical theory on the cellular effect of IR identifies DNA as the major target for initiating death pathways, a growing body of data now demonstrates that ASMase-mediated ceramide generation regulates apoptosis in response to IR. In 1994, Haimovitz-Friedman et al. reported that irradiation of bovine aortic endothelial cells (BAEC) induced sphingomyelin conversion to ceramide and apoptosis, suggesting that radiation damage to plasma membrane might initiate apoptotic signaling through ceramide. Activation of ASMase was also observed in Rat-1 Myc-ER cells and primary skin fibroblasts (Zundel and Giaccia 1998; Bohler et al. 1997). Definitive evidence for a role of ASMase in radiation-induced apoptosis was derived from studies with ASMase-deficient models. B lymphoblastoid cells immortalized from an NPD type A patient fail to hydrolyze sphingomyelin in response to radiation, generate ceramide, or undergo apoptosis (Santana et al. 1996). Retroviral transfer of normal ASMase cDNA into these cells restored ceramide generation, leading to restoration of apoptotic cell death upon radiation treatment. Mouse embryonic fibroblasts (MEFs) from ASMase knockout mice also failed to generate ceramide and were completely resistant to radiation-induced apoptosis, yet remained sensitive to staurosporine-induced apoptosis, which is not mediated via ceramide signaling (Lozano et al. 2001). Additional studies showed that sensitivity to radiation-induced apoptosis could be restored in MEFs by exogenous addition of nanomolar concentrations of natural ceramide. Rescue of the radiation apoptotic phenotype by adding ceramide provides strong evidence that ceramide is obligate for this form of apoptosis induced by radiation.

IR, delivered as a small daily fraction of 1.8–3.0 Gy, cures approximately 60 % of localized, nonmetastatic cancer. Fractioned radiotherapy has been favored because normal tissue stem cell clonogens repair DNA double-strand breaks (DSBs) more proficiently than their tumor counterparts during interfraction intervals, enabling dose escalation to tumor cure levels with acceptable normal tissue damage. However, in many instances, curative tumor doses cannot be reached due to high radiosensitivity of adjacent normal organs. The recent

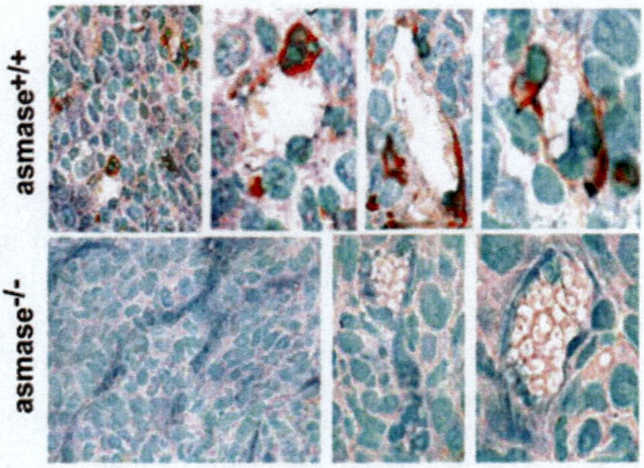

Fig. 3 MCA/129 fibrosarcomas implanted into *ASMase*$^{-/-}$ mice display reduced radiation-induced endothelial cell apoptosis. MCA/129 fibrosarcomas, grown to 150–200 mm^3, were irradiated with 15 Gy. Tumor specimens were obtained 4 h post-irradiation, fixed in 4 % fresh formaldehyde, embedded in paraffin, and 5-μm sections were evaluated for apoptosis by terminal deoxytransferase-mediated deoxyuridine triphosphate nick end labeling (TUNEL). Thereafter, endothelial cells were identified using an antibody specific for the endothelial cell surface marker CD-34. Apoptotic endothelial cells, identified as a *red-brown* TUNEL-positive nuclear signal surrounded by *dark-blue* plasma membrane signal indicative of CD-34 staining, are lacking in sections of tumor grown in *ASMase*$^{-/-}$ mice. Similar results were obtained using B16F1 melanomas [Adapted from Garcia-Barros et al. (2003)]

emergence of image-guided, high-precision targeting of human tumors that effectively excludes normal tissues from the treatment field may obviate this issue, allowing for high SDRT (20–24 Gy), a modality whose success has exceeded all expectations (Song et al. 2004).

The mechanism of tumor response after SDRT may differ from that of conventional fractioned radiotherapy (Moeller et al. 2005; Fuks and Kolesnick 2005). The central role of ASMase-mediated apoptosis in endothelial cells for optimal SDRT response was detailed in two tumor models (melanoma and fibrosarcoma). Studies by Garcia-Barros et al. (2003) indicated that exposure to single radiation doses >8 Gy engages an apoptotic response in tumor microvessels by activating the sphingomyelin/ceramide transducer in endothelium (Fig. 3). MCA129 fibrosarcomas and B16 melanomas transplanted into ASMase or BAX knockout mice, which provide tumors with host-derived apoptosis-resistant microvascular networks, were completely resistant to potentially curative doses of 15–20 Gy SDRT. Hence, high-dose radiation exposure appears to induce primarily sublethal lesions in tumor stem cell clonogens that at these doses are converted into lethal damage upon induction of apoptotic microvascular dysfunction. Discovery of the critical role of the sphingomyelin/ceramide transducer in curative SDRT defines endothelium as a valid pharmacologic target, consistent with early data suggesting that antagonism of VEGF or the VEGFR2 receptor radiosensitizes tumors by derepressing endothelial ASMase (Truman et al. 2010).

2.2 ASMase Knockout Mice for Study of GI Damage

In addition to the potential importance of ASMase in cancer therapy, researchers have been investigating impact of ASMase on normal tissue damage. The radiation GI syndrome is a major toxicity that may occur in a Fukushima-like nuclear accident or upon deliberate radiation exposure. The GI syndrome involves destruction of crypt/villus units, loss of mucosal integrity, and infection by resident enterobacterial flora, characterized clinically as anorexia, vomiting, diarrhea, dehydration, systemic infection, and in extreme cases, septic shock and death. Studies of Paris et al. first reported that microvascular endothelium serves as a primary target for radiation in induction of tissue damage (Paris et al. 2001). Vascular dysfunction in the GI mucosa occurs following a rapid wave of radiation-induced endothelial apoptosis (2–6 h post-IR) regulated by ASMase-mediated ceramide generation within the villus/crypt microvascular network. Genetic inactivation of ASMase inhibits this ceramide-driven mode of endothelial apoptosis, preventing IR-induced animal lethality (Paris et al. 2001). Cho et al. (2004) supported this hypothesis, reporting intravenous injection of the angiopoietin-1 variant COMP-Ang1, which specifically targeted Tie-2 receptors of intestinal microvascular endothelial but not epithelial cells, attenuated radiation-induced endothelial apoptosis, and protected against GI syndrome lethality.

Graft-versus-host disease (GVHD) represents a different process in which ASMase-mediated apoptosis plays a role in morbidity and mortality. GVHD is a frequent complication in leukemia patients receiving allogeneic bone marrow transplantation. In GVHD, donor cytotoxic T lymphocytes (CTLs) attack host tissues leading to organ damage. Recently, Rotolo et al. (2009) showed that CTLs generate CRPs on critical cells within GVHD-responsive target organs (small intestines, skin, liver), signaling apoptosis to initiate GVHD. Evidence indicated cytokines of the TNF superfamily as critical to CTL-induced pathophysiology. Using clinically relevant mouse models of acute GVHD in which allogeneic bone marrow and T cells were transplanted into ASMase$^{+/+}$ and ASMase$^{-/-}$ mice, host ASMase was identified as critical for full-blown GVHD. Lack of host ASMase reduced apoptosis of relevant GVH target cells, including hepatocytes and intestinal and skin cells, as target cells failed to form CRPs upon contacting activated CTLs and consequently resisted CTL-mediated cell death. Hence, the sphingomyelinase/ceramide signaling pathway may provide targets for pharmacologic modulation of GI and perhaps other damage in GVHD, the major dose-limiting toxicity in leukemia patients receiving allogeneic marrow transplantation (Waselenko et al. 2004).

2.3 ASMase Knockout Mice for Study of Ovarian Damage

The ovary represents another organ whose chemosensitivity/radiosensitivity is regulated by ASMase. Almost 90 % of oocytes die by apoptosis prior to birth, a process that continues throughout adulthood resulting in menopause. Tilly and coworkers showed that both fetal and postpartum oocyte deletion in mice require sphingomyelin/ceramide signaling, with ASMase knockout mice displaying 1.6-fold increased oocytes at birth, maintained throughout adulthood (Morita et al. 2000; Casper and Jurisicova 2000). This group defined a mechanism for ceramide-mediated deletion in aging oocytes (Perez et al. 2005). Young wild-type oocytes isolated with their granulosa cell layer display elevated ceramide in both cell populations and were resistant to spontaneous or ceramide-induced apoptosis. In contrast, aged oocytes manifested high granulosa cell ceramide but low oocyte ceramide. Over 24 h in culture, ceramide was transferred from granulosa cells to oocytes by a process requiring gap junctions and intact rafts, and apoptosis, requiring Bax downstream, ensued. S1P treatment did not prevent ceramide transfer but prevented apoptosis. Similarly, genetic ASMase deletion or binding intracellular ceramide by injecting anti-ceramide antibody lowered available ceramide in young oocytes rendering them susceptible to exogenously added ceramide. These studies indicate that the sphingomyelin/ceramide transducer regulates normal ovarian physiology.

The sphingomyelin/ceramide transducer also mediates lethal effects of drugs and IR on oocytes. Morita et al. showed that apoptosis of isolated murine oocytes ex vivo to 200 nM doxorubicin was prevented by S1P pretreatment. Ceramide generation in response to doxorubicin appeared to occur through sphingomyelin hydrolysis rather than de novo ceramide synthesis as there was a robust apoptotic response in ASMase wild-type oocytes cultured with doxorubicin for 24 h, whereas ASMase knockout oocytes were almost completely resistant to doxorubicin destruction. Further, S1P injection into the bursa surrounding the mouse ovary dose-dependently prevented IR-induced oocyte apoptosis and sterility, preserving oocyte function, enabling successful in vitro and in vivo fertilization. In contrast to unprotected mice, irradiated S1P-protected mice maintained normal fertility for over 1 year, and their F1 and F2 offspring did not display behavioral, hematologic, histologic, biochemical, or genetic abnormalities (Paris et al. 2002). Moreover, Tilly and coworkers recently reported a successful monkey trial to evaluate the effect of S1P and its long-acting mimetic FTY720 on ovarian function and fertility after irradiation. In these studies, S1P and FTY720 protected ovaries of adult female rhesus monkeys from damage caused by 15 Gy targeted radiotherapy, allowing for long-term fertility (Zelinski et al. 2011). These studies indicate that chemotherapy-induced stress and IR-induced stress engage the sphingomyelin/ceramide physiologic mechanism for oocyte deletion, accelerating organ damage. Further, these studies suggest a small molecule approach to this intractable side effect of cancer therapy.

2.4 ASMase Knockout Mice for Study of Brain Diseases

2.4.1 Niemann–Pick Disease

Types A and B NPD result from inherited ASMase deficiency. The first type A NPD patient was described by Albert Niemann in 1914. Type A NPD is a severe neurodegenerative disease of infancy usually fatal by ages 2–3. In contrast, type B NPD patients have minimal or no neurologic involvement and often survive into adulthood but may have severe and progressive visceral organ abnormalities, including hepatosplenomegaly, and pulmonary and cardiovascular disease (Schuchman 2010). Differences between types appear due to the level of residual ASMase activity. ASMase knockout mice show a similar phenotype to type A NPD patients, owing to complete absence of ASMase activity. It has been reported that calcium homeostasis is altered in the cerebellum of ASMase knockout mice, suggesting that calcium dysfunction may cause Purkinje cell degeneration (Ginzburg and Futerman 2005).

Several groups have used these mice for testing of novel therapeutics for NPD, including recombinant enzyme replacement therapy (ERT), gene therapy, and stem cell transplantation. In 1999, large-scale purification of recombinant human ASMase was achieved in Chinese hamster ovary cells (He et al. 1999), allowing for human ERT. Initial studies that evaluated the effect of ERT in ASMase knockout mice were performed by Miranda and coworkers, who demonstrated that the sphingomyelin levels were significantly reduced in RES organs when recombinant ASMase (rASMase) was administered intravenously into young mice (Miranda et al. 2000). However, there was no impact on progression of neurologic disease, and mouse life span was not extended due to failure of rASMase to pass the blood–brain barrier into the CNS (Sly and Vogler 2002). Recent studies showed that intraparenchymal injection of rASMase results in regional reduction in sphingomyelin and cholesterol levels, but re-accumulation is observed at 2 weeks post-injection (Yang et al. 2007). Follow-up studies from the same lab demonstrated that intracerebroventricular infusion of rASMase led to widespread ASMase distribution and significant reduction in lysosomal sphingomyelin accumulation (Dodge et al. 2009).

The effect of stem cell transplantation on progression of neurologic disease also has been extensively evaluated in ASMase knockout mice. Intravenous transplantation of ASMase-expressing bone marrow cells into ASMase knockout mice showed positive effects on RES organs, but effects on neurologic disease were modest (Miranda et al. 1998). Further studies showed that intracerebral transplantation of ASMase-expressing mesenchymal stem cells into ASMase knockout mice delayed onset of neurologic abnormalities and extended life span, but progressive neurologic disease was not prevented (Jin et al. 2002; Jin and Schuchman 2003). Overall, these findings suggest stem cell transplantation as therapeutically promising.

In addition to the therapeutic approaches mentioned above, gene therapy has been intensively investigated during the last decade. Adeno-associated viral (AAV) vectors have been widely used. Several studies demonstrated that intracranial injections of AAV encoding human ASMase are effective in decreasing the burden of sphingolipid storage in the brains of ASMase knockout mice. ASMase activity was found not only within deep cerebellar nuclei but also throughout the CNS. Moreover, progressive ataxia was prevented and life span normalized (Dodge et al. 2005; Passini et al. 2005, 2007). These data suggest AAV-based therapy as a promising therapeutic modality for treating NPD.

2.4.2 Stroke

Stroke, the third leading cause of death in the industrialized world, in the large majority of cases results from occlusion of arterial blood flow into the brain. While pathogenesis of the ischemic lesion is complex, excess excitotoxicity, peri-infarct depolarization, inflammation, and apoptosis are factors contributing to evolution of tissue damage. These events do not impact the ischemic region homogenously. Rather, a perfusion-occluded anoxic core, manifesting necrosis, is surrounded by a rim of restricted blood flow, termed the penumbra. Although penumbra ischemic cells, and especially neurons, are at high risk for apoptotic death, this region is potentially salvageable. Hence, a recent emphasis of pharmacologic intervention in stroke is in apoptosis.

Nakane and coworkers reported in gerbils (Nakane et al. 2000) and rats (Kubota et al. 2000) that lethal forebrain ischemia (5 min) induced by bilateral carotid occlusion induced rapid sphingomyelin hydrolysis to ceramide (30 min), preceding neuronal apoptotic death. Debatin and colleagues (Herr et al. 1999) subsequently showed that rat middle cerebral artery (MCA) occlusion leads to twofold increased ceramide levels compared to the ipsilateral non-occluded brain hemisphere at 6 h after reperfusion. ASMase-generated ceramide purportedly initiated neuronal apoptosis in this model by upregulating death-inducing ligands (Fas ligand, TRAIL, and TNF-α), an effect prevented by the neuroprotector FK506 (Herr et al. 1999). Genetic evidence supporting this notion was provided by Mattson and coworkers (Yu et al. 2000) reporting that ASMase knockout mice neither generated ceramide nor upregulated inflammatory cytokines upon MCA occlusion and displayed markedly reduced neuronal apoptosis in the penumbra. Infarct size was reduced 30 %, and neurologic (primarily motor) deficiencies improved dramatically. Primary cultures of ASMase$^{-/-}$ cortical neurons displayed markedly reduced excitotoxicity upon glutamate treatment and chemical hypoxia induced by cyanide, accompanied by protection from intracellular Ca^{2+} elevation, and generation of reactive oxygen species (note: ischemic and excitotoxic neuronal death is considered mediated by calcium overload and oxyradical production). These studies indicate cell autonomous utilization of the sphingomyelin/ceramide transducer by cortical neurons to discriminate death signals.

2.4.3 Blood–Brain Barrier Dysfunction

The blood–brain barrier (BBB) is a highly specialized microvascular network characterized by tight cell–cell junctions lacking fenestrations, which restricts transcapillary flux of water-soluble compounds into the CNS (Neuwelt 2004). The dysfunctions of BBB is observed during CNS bacterial and viral infections, inflammatory and degenerative CNS disorders, cerebrovascular disease, trauma, primary and metastatic brain tumors, and after IR. BBB disruption often results in vasogenic edema that contributes significantly to disease-associated symptoms. Concomitantly, however, it facilitates transport of cytokines, antibodies, inflammatory and immune cells, as well as drugs, to disease-stricken areas.

Recent studies demonstrate a role for sphingomyelin/ceramide transduction in IR-induced BBB dysfunction. Pena et al. (2000) showed that irradiated murine CNS endothelium undergoes dose-dependent apoptosis at 5–100 Gy SDRT, peaking after 12 h. After 50 Gy, which induces subacute and chronic damage, approximately 20 % of endothelium underwent apoptosis at 12 h, inhibitable by genetic ASMase inactivation or pretreatment with the endothelial survival factor bFGF. Li et al. (2003) showed that this level of endothelial apoptosis resulted in a 40–60 % reduction in endothelial cell density 24 h after 50 Gy, an event abrogated in ASMase knockout mice, but not in p53 knockout mice. Further, BBB dysfunction, assessed by leakage of albumin or Evans blue dye into the CNS, was abrogated in ASMase$^{-/-}$ mice. These observations suggest that pharmacologic activation of the sphingomyelin/ceramide transducer might be used therapeutically to produce transient BBB disruption, designed to facilitate drug delivery to disease sanctuary regions within the CNS (van Vulpen et al. 2002).

2.5 *ASMase Knockout Mice for Study of Lung Diseases*

ASMase deficiency in both type B NPD patients and in ASMase knockout mice alters pulmonary cellular and organ structure secondary to accumulation of sphingomyelin, leading to lung abnormalities and decreased pulmonary function. Type B NPD is linked to progressive pulmonary function decline and frequent respiratory infection. X-ray and CT examination in type B NPD patients revealed that over 90 % displayed evidence of interstitial lung disease (Mendelson et al. 2006). ASMase knockout mice begin at 10 weeks of age to have a significantly higher number of cells in their pulmonary airspaces than normal mice, consisting primarily of enlarged and often multinucleated macrophages. In mice and humans, the alveolar macrophage serves as the first line of host defense to clear extracellular bacteria from the lung, implying an important role of ASMase in lung host defense against pathogens. Further, mechanical research indicates that ASMase is required for normal surfactant catabolism by alveolar macrophages. In this regard, ASMase knockout mice exhibit elevated levels of total surfactant lipid and protein (Buccoliero et al. 2004) and alterations in surfactant composition, including increased sphingomyelin content, that appear to contribute to abnormal surfactant function observed in ASMase knockout mice (Buccoliero et al. 2004; Tuder et al. 2003). In principle, these studies suggest that ASMase might regulate fundamental aspects of lung disease pathologies.

2.5.1 Acute Respiratory Distress Syndrome

Acute respiratory distress syndrome (ARDS) is defined by severe noncardiac respiratory distress, impaired arterial oxygenation (hypoxemia), and bilateral pulmonary infiltrates. It has diverse origins, including sepsis, pneumonia, aspiration, trauma, smoke or toxic gas inhalation, and some drugs (Piantadosi and Schwartz 2004). Pathogenesis involves primary endothelial dysfunction, leakage of fluid across the alveolar-capillary barrier, and fluid accumulation in the alveolar space (edema), causing refractory hypoxemia. Cytokines and chemokines may be crucial in endothelial cell dysfunction. Recent attention has focused on platelet-activating factor (PAF) that mediates acute lung injury (ALI) that progresses to ARDS in part through prostaglandin E2 production.

Goggel et al. (2004) demonstrated a role for ceramide in PAF-induced pulmonary edema in mouse and rat models (Barnes 2004). PAF injection into mice or perfusion of isolated intact rat lungs with PAF resulted in rapid serum ASMase and pulmonary ceramide elevation. PAF-induced pulmonary edema was reduced in ASMase knockout mice. Rat pulmonary edema was mimicked by perfusion of short-chain C2-ceramide, but not C2-dihydroceramide, which differs only in the trans double bond at position 4–5. Further, injection of anti-ceramide antibodies antagonized mouse pulmonary edema by about 50 %. Other less specific ASMase inhibitors (desipramine and D609) acted similarly and, when combined with a cyclooxygenase inhibitor to prevent prostaglandin E2 production, abolished edema. These studies suggest that concomitant inhibition of two intracellular lipid signaling pathways might serve as a new approach to ARDS therapy (Barnes 2004).

2.5.2 Pulmonary Fibrosis

Pulmonary fibrosis is another common form of interstitial lung disease associated with inflammation and apoptosis. Recently, the Schuchman group found that intratracheal instillation of bleomycin which can cause pulmonary fibrosis in normal mice leads to increase in lung ASMase activity, while bleomycin instillation in ASMase knockout mice does not cause fibrosis. This suggests that inhibition of ASMase transiently might be a useful way to prevent toxicity of some cancer drugs (Dhami et al. 2010).

Since ASMase plays an important role in surfactant clearance in lung where pathogens interact closely with target cell membranes, there has been an expanding literature regarding ASMase regulation of pathogen infection using ASMase knockout mice. Cystic fibrosis (CF) is a genetic disorder caused by the mutation of cystic fibrosis transmembrane conductance regulator (cftr) gene. Growing evidence shows that sphingolipids, in particular ceramide, play an important role in CF and the bacterial infections associated with this disease (Teichgraber et al. 2008; Becker et al. 2010).

Furthermore, TNF-α administration to mice recapitulates acute hepatic injury syndromes (Ding and Yin 2004). Fernandez-Checa and colleagues showed that wild-type hepatocytes respond to TNF-α with rapid ASMase activation, ceramide generation, and massive apoptosis (Garcia-Ruiz et al. 2003), while ASMase-deficient hepatocytes were TNF-resistant. Consistent with these observations, ASMase knockout mice were resistant to TNF-α-induced hepatitis and death from liver failure. Similarly, ASMase knockout mice were resistant to anti-Fas-induced ceramide elevation and hepatic injury (Lin et al. 2000), and attenuation of ceramide elevation via adenoviral overexpression of neutral ceramidase prevented TNF-induced hepatitis (Osawa et al. 2005). Furthermore, knockout of the mouse adaptor protein FAN (Malagarie-Cazenave et al. 2004), which binds TNFR1 to transduce NSMase activation, displayed reduced IL-6 production and TNF-induced hepatic injury and improved survival. Evidence suggests that ASMase-generated ceramide may act by inducing downregulation of methionine adenosyltransferase 1A (MAT1A) (Mari et al. 2004), an enzyme required for maintaining levels of the one-carbon donor *S*-adenosyl-methionine (SAM). MAT1A impairment lowers SAM, contributing to injury in experimental models of liver damage and in patients with alcohol- and viral-mediated cirrhosis. Intraperitoneal SAM injection into wild-type mice mimicked the ASMase$^{-/-}$ phenotype, preventing TNF- and lipopolysaccharide-induced hepatitis, improving survival. Altogether, these studies indicate that elevating ceramide enhances liver damage, while preventing elevation delivers protection, legitimizing this target for pharmacologic intervention in hepatic injury.

2.8 ASMase Knockout Mice for Study of Infection

Several studies support ASMase as having a central role in infection of mammalian cells with diverse microbial pathogens. CRPs are co-opted by a wide range of pathogens including bacteria, viruses, and parasites to infect mammalian cells. Initial studies with *N. gonorrhoeae* showed ASMase activation and release of ceramide upon infection of human epithelial cells and macrophages. Moreover, pharmacologic inhibition or genetic knockout of ASMase prevented infection of human epithelial cells with *N. gonorrhoeae* (Grassme et al. 1997; Hauck et al. 2000). These data suggested that ASMase is required for internalization of *N. gonorrhoeae*, confirmed in subsequent studies (Grassme et al. 2003b). *P. aeruginosa* also activates ASMase and triggers ceramide release within minutes of contacting target cells. *Pseudomonas*-induced CRP formation on the surface of the target cell appears required for acute infection (Fig. 4), as ASMase knockout prevented CRP formation, internalization of *P. aeruginosa* by lung epithelial cells, and subsequent epithelial cell apoptosis (Grassme et al. 2003b). Evidence also indicates ASMase involvement in infection by *Staphylococcus aureus*, *Salmonella typhimurium*, *Escherichia coli*, and *Mycobacterium* (for further details please see Grassmé and Becker 2013).

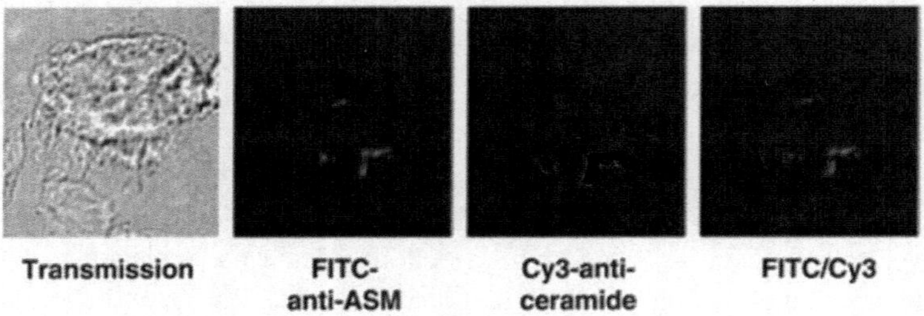

Fig. 4 *P. aeruginosa* triggers platform formation by ASMase-mediated release of ceramide. Infection (20 min) of murine tracheal epithelial cells with *P. aeruginosa* strain ATCC 27853 induced formation of CRPs that contain ASMase. Cells were infected for 20 min with *P. aeruginosa*, fixed in PFA, and isolated and stained with FITC-labeled anti-ASMase and Cy3-coupled anti-ceramide antibodies. The cells were analyzed by fluorescence microscopy [Adapted from Grassme et al. (2003b)]

In addition to bacteria, virus–host cell interaction may require ASMase activity. Griffin and coworkers reported ASMase knockout mice more susceptible to sindbis virus (SV) infection than wild-type or heterozygous (Het) littermates (Ng and Griffin 2006). In SV infection, sphingomyelin is required for virus entry, an event impacted by altered sphingomyelin storage in ASMase-deficient mice. Accumulation of sphingomyelin in brains of ASMase knockout mice allows more rapid entry and spread of SV, leading to higher viral titers in the nervous system, and subsequent increased neuronal death compared with wild-type mice. Several cell-based studies also confirm a fundamental role of ASMase in virus infection. For instance, infection of human epithelial cells with rhinovirus induces ASMase activation, release of ceramide, and CRP formation, events blocked in human fibroblasts genetically deficient for ASMase or by pharmacologic inhibitors of ASMase (Grassme et al. 2005). Inhibition of SMase by *N*-palmitoyl-DL-dihydrosphingosine similarly decreased yield of infectious herpes virus, suggesting ASMase involvement in herpes infection or replication (Steinhart et al. 1984). Moreover, several studies suggest that the sphingomyelin/ceramide signaling pathway may influence the course of HIV-1 infection. HIV-cell infection involves interaction of HIV receptor molecule gp120 with CD4 on the T cell surface. CD4 ligation initiates T cell depletion by clustering of endogenous Fas/Fas ligand (Wang et al. 1994), an event requiring CRPs. Altogether, these data suggest that manipulation of sphingolipids might be strategic for treatment of virus infection.

In addition to a role of ASMase in virus uptake, a recent study showed that ASMase activity might be required by host immune system to eliminate virus-infected cells. $CD8^+$ CTLs and natural killer cells lyse virus-infected target cells through exocytosis of cytotoxic granules. CTLs from ASMase knockout mice are defective in exocytosis of cytolytic effector molecules, leading to attenuated cytotoxic activity of CTLs and delayed elimination of lymphocytic choriomeningitis virus (Herz et al. 2009). Moreover, macrophages from ASMase knockout mice are defective in killing of the bacterium *L. monocytogenes*, resulting in unrestricted spread of infection (Utermohlen et al. 2003). Given the complexity of the impact of

Table 1 Human diseases/therapies modeled via ASMase knockout

Organs	Human diseases
Small intestines	GI syndrome and GVHD
Ovary	Normal ovarian physiology
	Cancer therapy-induced infertility
Brain	Niemann–Pick disease
	Stroke
	Blood–brain barrier dysfunction (BBB)
Lung	Acute respiratory distress syndrome (ARDS)
	Pulmonary fibrosis
	Cystic fibrosis
Heart	Atherosclerosis
	Chronic heart failure
	Diabetes
Liver	T cell-mediated auto-aggressive liver disease
	Wilson's disease
	Chronic liver disease
Infections	*Pseudomonas aeruginosa*
	N. gonorrhoeae
	L. monocytogenes
	Sindbis virus
	P. falciparum
Cancer	Single-dose radiotherapy (SDRT)

ASMase deficiency on the course of infection, full molecular characterization of sphingolipid involvement in specific infectious states would appear warranted before clinical pharmacologic manipulation is attempted.

Finally, at least one parasitic infection is influenced by SMase-induced ceramide generation in a unique way. *Plasmodium* seems to have developed a strategy to circumvent a requirement for mammalian target cell ASMase. In fact, *Plasmodium* expresses its own SMase for generation of ceramide *in trans* in target cells, an effect required for erythrocyte infection with *P. falciparum or P. berghei*. Therefore, only a slightly different course of malarial infection was observed between ASMase wild-type and knockout mice (Brand et al. 2008). Thus, in addition to host ASMase, pathogen SMase might represent a novel target for malaria, defined using the ASMase mouse knockout.

2.9 Perspective

ASMase knockout mice provide an invaluable tool to explore the role of the sphingomyelin/ceramide signaling pathway in pathophysiology of various human diseases (Table 1). Accumulating evidence indicates a crucial involvement of sphingomyelin/ceramide in a number of disease initiation states, as well as disease treatments, making the pathway an attractive target for therapy. In particular, manipulation of CRPs might be strategic for treatment of various human diseases.

Further studies, however, are needed to more fully molecularly characterize the enzymes involved in generation and metabolism of ceramide in these disease states to develop a more comprehensive understanding of the signaling network that regulates their individual sphingomyelin/ceramide transduction. Nonetheless, human diseases regulated through sphingomyelin/ceramide signal transduction represent a newly defined class of clinical entities that provides unique targets of opportunity for pharmacologic intervention. This developmental process undoubtedly will continue to rely on use of the original ASMase knockout mice strain, and on newer versions currently in development, in the process of disease discovery and therapeutics.

References

Barnes P (2004) Ceramide lances the lungs. Nat Med 10:130–131

Becker K, Grassme H, Zhang Y, Gulbins E (2010) Ceramide in Pseudomonas aeruginosa infections and cystic fibrosis. Cell Physiol Biochem 26:57–66

Bionda C, Portoukalian J, Schmitt D, Rodriguez-Lafrasse C, Ardail D (2004) Subcellular compartmentalization of ceramide metabolism: MAM (mitochondria-associated membrane) and/or mitochondria? Biochem J 382:527–533

Bohler T, Baumler C, Herr I, Groll A, Kurz M, Debatin K (1997) Activation of the CD95 system increases with disease progression in human immunodeficiency virus type 1-infected children and adolescents. Pediatr Infect Dis J 16:754–759

Bourbon N, Yun J, Kester M (2000) Ceramide directly activates protein kinase C zeta to regulate a stress-activated protein kinase signaling complex. J Biol Chem 275:35617–35623

Brady R, Kanfer J, Mock M, Fredrickson D (1966) The metabolism of sphingomyelin. II. Evidence of an enzymatic deficiency in Niemann-Pick disease. Proc Natl Acad Sci USA 55:366–369

Brand V, Koka S, Lang C, Jendrossek V, Huber S, Gulbins E, Lang F (2008) Influence of amitriptyline on eryptosis, parasitemia and survival of Plasmodium berghei-infected mice. Cell Physiol Biochem 22:405–412

Buccoliero R, Ginzburg L, Futerman A (2004) Elevation of lung surfactant phosphatidylcholine in mouse models of Sandhoff and of Niemann-Pick A disease. J Inherit Metab Dis 27:641–648

Casper R, Jurisicova A (2000) Protecting the female germ line from cancer therapy. Nat Med 6:1100–1101

Ch'ang H, Maj J, Paris F, Xing H, Zhang J, Truman J, Cardon-Cardo C, Haimovitz-Friedman A, Kolesnick R, Fuks Z (2005) ATM regulates target switching to escalating doses of radiation in the intestines. Nat Med 11:484–490

Chalfant C, Kishikawa K, Mumby M, Kamibayashi C, Bielawska A, Hannun Y (1999) Long chain ceramides activate protein phosphatase-1 and protein phosphatase-2A. Activation is stereospecific and regulated by phosphatidic acid. J Biol Chem 274:20313–20317

Cho C, Kammerer R, Lee H, Yasunaga K, Kim K, Choi H, Kim W, Kim S, Park S, Lee G, Koh G (2004) Designed angiopoietin-1 variant, COMP-Ang1, protects against radiation-induced endothelial cell apoptosis. Proc Natl Acad Sci USA 101:5553–5558

Cuvillier O, Pirianov G, Kleuser B, Vanek P, Coso O, Gutkind S, Spiegel S (1996) Suppression of ceramide-mediated programmed cell death by sphingosine-1-phosphate. Nature 381: 800–803

Deevska G, Rozenova K, Giltiay N, Chambers M, White J, Boyanovsky B, Wei J, Daugherty A, Smart E, Reid M, Merrill A, Nikolova-Karakashian M (2009) Acid sphingomyelinase

deficiency prevents diet-induced hepatic triacylglycerol accumulation and hyperglycemia in mice. J Biol Chem 284:8359–8368
Devlin C, Leventhal A, Kuriakose G, Schuchman E, Williams K, Tabas I (2008) Acid sphingomyelinase promotes lipoprotein retention within early atheromata and accelerates lesion progression. Arterioscler Thromb Vasc Biol 28:1723–1730
Dhami R, He X, Gordon R, Schuchman E (2001) Analysis of the lung pathology and alveolar macrophage function in the acid sphingomyelinase–deficient mouse model of Niemann-Pick disease. Lab Invest 81:987–999
Dhami R, He X, Schuchman E (2010) Acid sphingomyelinase deficiency attenuates bleomycin-induced lung inflammation and fibrosis in mice. Cell Physiol Biochem 26:749–760
Ding W, Yin X (2004) Dissection of the multiple mechanisms of TNF-alpha-induced apoptosis in liver injury. J Cell Mol Med 8:445–454
Dodge J, Clarke J, Song A, Bu J, Yang W, Taksir T, Griffiths D, Zhao M, Schuchman E, Cheng S, O'Riordan C, Shihabuddin L, Passini M, Stewart G (2005) Gene transfer of human acid sphingomyelinase corrects neuropathology and motor deficits in a mouse model of Niemann-Pick type A disease. Proc Natl Acad Sci USA 102:17822–17827
Dodge J, Clarke J, Treleaven C, Taksir T, Griffiths D, Yang W, Fidler J, Passini M, Karey K, Schuchman E, Cheng S, Shihabuddin L (2009) Intracerebroventricular infusion of acid sphingomyelinase corrects CNS manifestations in a mouse model of Niemann-Pick A disease. Exp Neurol 215:349–357
Doehner W, Bunck A, Rauchhaus M, von Haehling S, Brunkhorst F, Cicoira M, Tschope C, Ponikowski P, Claus R, Anker S (2007) Secretory sphingomyelinase is upregulated in chronic heart failure: a second messenger system of immune activation relates to body composition, muscular functional capacity, and peripheral blood flow. Eur Heart J 28:821–828
Fuks Z, Kolesnick R (2005) Engaging the vascular component of the tumor response. Cancer Cell 8:89–91
Fuks Z, Alfieri A, Haimovitz-Friedman A, Seddon A, Cordon-Cardo C (1995) Intravenous basic fibroblast growth factor protects the lung but not mediastinal organs against radiation-induced apoptosis in vivo. Cancer J Sci Am 1:62–72
Garcia-Barros M, Paris F, Cordon-Cardo C, Lyden D, Rafii S, Haimovitz-Friedman A, Fuks Z, Kolesnick R (2003) Tumor response to radiotherapy regulated by endothelial cell apoptosis. Science 300:1155–1159
Garcia-Ruiz C, Colell A, Mari M, Morales A, Calvo M, Enrich C, Fernandez-Checa J (2003) Defective TNF-alpha-mediated hepatocellular apoptosis and liver damage in acidic sphingomyelinase knockout mice. J Clin Invest 111:197–208
Ginzburg L, Futerman A (2005) Defective calcium homeostasis in the cerebellum in a mouse model of Niemann-Pick A disease. J Neurochem 95:1619–1628
Goggel R, Winoto-Morbach S, Vielhaber G, Imai Y, Lindner K, Brade L, Brade H, Ehlers S, Slutsky A, Schutze S, Gulbins E, Uhlig S (2004) PAF-mediated pulmonary edema: a new role for acid sphingomyelinase and ceramide. Nat Med 10:155–160
Grassmé H, Becker KA (2013) Bacterial infections and ceramide. In: Gulbins E, Petrache I (eds) Sphingolipids in disease, vol 216, Handbook of experimental pharmacology. Springer, Heidelberg
Grassme H, Gulbins E, Brenner B, Ferlinz K, Sandhoff K, Harzer K, Lang F, Meyer T (1997) Acidic sphingomyelinase mediates entry of N. gonorrhoeae into nonphagocytic cells. Cell 91:605–615
Grassme H, Jekle A, Riehle A, Schwarz H, Berger J, Sandhoff K, Kolesnick R, Gulbins E (2001a) CD95 signaling via ceramide-rich membrane rafts. J Biol Chem 276:20589–20596
Grassme H, Schwarz H, Gulbins E (2001b) Molecular mechanisms of ceramide-mediated CD95 clustering. Biochem Biophys Res Commun 284:1016–1030
Grassme H, Cremesti A, Kolesnick R, Gulbins E (2003a) Ceramide-mediated clustering is required for CD95-DISC formation. Oncogene 22:5457–5470

Grassme H, Jendrossek V, Riehle A, von Kurthy G, Berger J, Schwarz H, Weller M, Kolesnick R, Gulbins E (2003b) Host defense against Pseudomonas aeruginosa requires ceramide-rich membrane rafts. Nat Med 9:322–330

Grassme H, Riehle A, Wilker B, Gulbins E (2005) Rhinoviruses infect human epithelial cells via ceramide-enriched membrane platforms. J Biol Chem 280:26256–26262

Grassme H, Riethmuller J, Gulbins E (2007) Biological aspects of ceramide-enriched membrane domains. Prog Lipid Res 46:161–170

Guicciardi M, Gores G (2005) Apoptosis: a mechanism of acute and chronic liver injury. Gut 54:1024–1033

Gulbins E (2003) Regulation of death receptor signaling and apoptosis by ceramide. Pharmacol Res 47:393–399

Gulbins E, Bissonnette R, Mahboubi A, Martin S, Nishioka W, Brunner T, Baier G, Baier-Bitterlich G, Byrd C, Lang F (1995) FAS-induced apoptosis is mediated via a ceramide-initiated RAS signaling pathway. Immunity 2:341–351

Haimovitz-Friedman A, Kan C, Ehleiter D, Persaud R, McLoughlin M, Fuks Z, Kolesnick R (1994) Ionizing radiation acts on cellular membranes to generate ceramide and initiate apoptosis. J Exp Med 180:525–535

Hannun Y, Obeid L (2002) The ceramide-centric universe of lipid-mediated cell regulation: stress encounters of the lipid kind. J Biol Chem 277:25847–25850

Hannun Y, Obeid L (2008) Principles of bioactive lipid signalling: lessons from sphingolipids. Nat Rev Mol Cell Biol 9:139–150

Hauck C, Grassme H, Bock J, Jendrossek V, Ferlinz K, Meyer T, Gulbins E (2000) Acid sphingomyelinase is involved in CEACAM receptor-mediated phagocytosis of Neisseria gonorrhoeae. FEBS Lett 478:260–266

He X, Miranda S, Xiong X, Dagan A, Gatt S, Schuchman E (1999) Characterization of human acid sphingomyelinase purified from the media of overexpressing Chinese hamster ovary cells. Biochim Biophys Acta 1432:251–264

Heinrich M, Wickel M, Schneider-Brachert W, Sandberg C, Gahr J, Schwandner R, Weber T, Saftig P, Peters C, Brunner J, Kronke M, Schutze S (1999) Cathepsin D targeted by acid sphingomyelinase-derived ceramide. EMBO J 18:5252–5263

Herr I, Martin-Villalba A, Kurz E, Roncaioli P, Schenkel J, Cifone M, Debatin K (1999) FK506 prevents stroke-induced generation of ceramide and apoptosis signaling. Brain Res 826:210–219

Herz J, Pardo J, Kashkar H, Schramm M, Kuzmenkina E, Bos E, Wiegmann K, Wallich R, Peters P, Herzig S, Schmelzer E, Kronke M, Simon MM, Utermohlen O (2009) Acid sphingomyelinase is a key regulator of cytotoxic granule secretion by primary T lymphocytes. Nat Immunol 10:761–768

Horinouchi K, Erlich S, Perl D, Ferlinz K, Bisgaier C, Sandhoff K, Desnick R, Stewart C, Schuchman E (1995) Acid sphingomyelinase deficient mice: a model of types A and B Niemann-Pick disease. Nat Genet 10:288–293

Huwiler A, Johansen B, Skarstad A, Pfeilschifter J (2001) Ceramide binds to the CaLB domain of cytosolic phospholipase A2 and facilitates its membrane docking and arachidonic acid release. FASEB J 15:7–9

Jarvis W, Kolesnick R, Fornari F, Traylor R, Gewirtz D, Grant S (1994) Induction of apoptotic DNA damage and cell death by activation of the sphingomyelin pathway. Proc Natl Acad Sci USA 91:73–77

Jin H, Schuchman E (2003) Ex vivo gene therapy using bone marrow-derived cells: combined effects of intracerebral and intravenous transplantation in a mouse model of Niemann-Pick disease. Mol Ther 8:876–885

Jin H, Carter J, Huntley G, Schuchman E (2002) Intracerebral transplantation of mesenchymal stem cells into acid sphingomyelinase-deficient mice delays the onset of neurological abnormalities and extends their life span. J Clin Invest 109:1183–1191

Kashiwagi K, Shirai Y, Kuriyama M, Sakai N, Saito N (2002) Importance of C1B domain for lipid messenger-induced targeting of protein kinase C. J Biol Chem 277:18037–18045

Kirschnek S, Paris F, Weller M, Grassme H, Ferlinz K, Riehle A, Fuks Z, Kolesnick R, Gulbins E (2000) CD95-mediated apoptosis in vivo involves acid sphingomyelinase. J Biol Chem 275:27316–27323

Kolesnick R (2002) The therapeutic potential of modulating the ceramide/sphingomyelin pathway. J Clin Invest 110:3–8

Kolesnick R, Fuks Z (2003) Radiation and ceramide-induced apoptosis. Oncogene 22:5897–5906

Kolesnick R, Hannun Y (1999) Ceramide and apoptosis. Trends Biochem Sci 24:224–225

Kolesnick R, Paley A (1987) 1,2-Diacylglycerols and phorbol esters stimulate phosphatidylcholine metabolism in GH3 pituitary cells. Evidence for separate mechanisms of action. J Biol Chem 262:9204–9210

Kolesnick R, Goni F, Alonso A (2000) Compartmentalization of ceramide signaling: physical foundations and biological effects. J Cell Physiol 184:285–300

Kubota M, Nakane M, Nakagomi T, Tamura A, Hisaki H, Ueta N, Inokuchi J, Hirayama A (2000) Sphingolipid biosynthesis by L-PDMP after rat MCA occlusion. Acta Neurochir Suppl 76:339–341

Lang P, Schenck M, Nicolay J, Becker J, Kempe D, Lupescu A, Koka S, Eisele K, Klarl B, Rubben H, Schmid K, Mann K, Hildenbrand S, Hefter H, Huber S, Wieder T, Erhardt A, Haussinger D, Gulbins E, Lang F (2007) Liver cell death and anemia in Wilson disease involve acid sphingomyelinase and ceramide. Nat Med 13:164–170

Leventhal A, Chen W, Tall A, Tabas I (2001) Acid sphingomyelinase-deficient macrophages have defective cholesterol trafficking and efflux. J Biol Chem 276:44976–44983

Li Y, Chen P, Haimovitz-Friedman A, Reilly R, Wong C (2003) Endothelial apoptosis initiates acute blood–brain barrier disruption after ionizing radiation. Cancer Res 63:5950–5956

Lin T, Genestier L, Pinkoski M, Castro A, Nicholas S, Mogil R, Paris F, Fuks Z, Schuchman E, Kolesnick R, Green D (2000) Role of acidic sphingomyelinase in Fas/CD95-mediated cell death. J Biol Chem 275:8657–8663

Liu P, Anderson RG (1995) Compartmentalized production of ceramide at the cell surface. J Biol Chem 270:27179–27185

Lozano J, Menendez S, Morales A, Ehleiter D, Liao W, Wagman R, Haimovitz-Friedman A, Fuks Z, Kolesnick R (2001) Cell autonomous apoptosis defects in acid sphingomyelinase knockout fibroblasts. J Biol Chem 276:442–448

Malagarie-Cazenave S, Segui B, Leveque S, Garcia V, Carpentier S, Altie M, Brouchet A, Gouaze V, Andrieu-Abadie N, Barreira Y, Benoist H, Levade T (2004) Role of FAN in tumor necrosis factor-alpha and lipopolysaccharide-induced interleukin-6 secretion and lethality in D-galactosamine-sensitized mice. J Biol Chem 279:18648–18655

Marathe S, Schissel S, Yellin M, Beatini N, Mintzer R, Williams K, Tabas I (1998) Human vascular endothelial cells are a rich and regulatable source of secretory sphingomyelinase. Implications for early atherogenesis and ceramide-mediated cell signaling. J Biol Chem 273:4081–4088

Mari M, Colell A, Morales A, Paneda C, Varela-Nieto I, Garcia-Ruiz C, Fernandez-Checa J (2004) Acidic sphingomyelinase downregulates the liver-specific methionine adenosyltransferase 1A, contributing to tumor necrosis factor-induced lethal hepatitis. J Clin Invest 113:895–904

McGovern M, Pohl-Worgall T, Deckelbaum R, Simpson W, Mendelson D, Desnick R, Schuchman E, Wasserstein M (2004) Lipid abnormalities in children with types A and B Niemann Pick disease. J Pediatr 145:77–81

Mendelson D, Wasserstein M, Desnick R, Glass R, Simpson W, Skloot G, Vanier M, Bembi B, Giugliani R, Mengel E, Cox G, McGovern M (2006) Type B Niemann-Pick disease: findings at chest radiography, thin-section CT, and pulmonary function testing. Radiology 238:339–345

Miranda S, Erlich S, Friedrich V, Haskins M, Gatt S, Schuchman E (1998) Biochemical, pathological, and clinical response to transplantation of normal bone marrow cells into acid sphingomyelinase-deficient mice. Transplantation 65:884–892

Miranda S, He X, Simonaro C, Gatt S, Dagan A, Desnick R, Schuchman E (2000) Infusion of recombinant human acid sphingomyelinase into niemann-pick disease mice leads to visceral, but not neurological, correction of the pathophysiology. FASEB J 14:1988–1995

Moeller B, Dreher M, Rabbani Z, Schroeder T, Cao Y, Li C, Dewhirst M (2005) Pleiotropic effects of HIF-1 blockade on tumor radiosensitivity. Cancer Cell 8:99–110

Morita Y, Perez G, Paris F, Miranda S, Ehleiter D, Haimovitz-Friedman A, Fuks Z, Xie Z, Reed J, Schuchman E, Kolesnick R, Tilly J (2000) Oocyte apoptosis is suppressed by disruption of the acid sphingomyelinase gene or by sphingosine-1-phosphate therapy. Nat Med 6:1109–1114

Nakane M, Kubota M, Nakagomi T, Tamura A, Hisaki H, Shimasaki H, Ueta N (2000) Lethal forebrain ischemia stimulates sphingomyelin hydrolysis and ceramide generation in the gerbil hippocampus. Neurosci Lett 296:89–92

Neuwelt E (2004) Mechanisms of disease: the blood–brain barrier. Neurosurgery 54:131–140

Ng C, Griffin D (2006) Acid sphingomyelinase deficiency increases susceptibility to fatal alphavirus encephalomyelitis. J Virol 80:10989–10999

Obeid L, Linardic C, Karolak L, Hannun Y (1993) Programmed cell death induced by ceramide. Science 259:1769–1771

Okazaki T, Bell R, Hannun Y (1989) Sphingomyelin turnover induced by vitamin D3 in HL-60 cells. Role in cell differentiation. J Biol Chem 264:19076–19080

Okazaki T, Bielawska A, Bell R, Hannun Y (1990) Role of ceramide as a lipid mediator of 1 alpha,25-dihydroxyvitamin D3-induced HL-60 cell differentiation. J Biol Chem 265:15823–15831

Osawa Y, Uchinami H, Bielawski J, Schwabe R, Hannun Y, Brenner D (2005) Roles for C16-ceramide and sphingosine 1-phosphate in regulating hepatocyte apoptosis in response to tumor necrosis factor-alpha. J Biol Chem 280:27879–27887

Otterbach B, Stoffel W (1995) Acid sphingomyelinase-deficient mice mimic the neurovisceral form of human lysosomal storage disease (Niemann-Pick disease). Cell 81:1053–1061

Paris F, Fuks Z, Kang A, Capodieci P, Juan G, Ehleiter D, Haimovitz-Friedman A, Cordon-Cardo C, Kolesnick R (2001) Endothelial apoptosis as the primary lesion initiating intestinal radiation damage in mice. Science 293:293–297

Paris F, Perez G, Fuks Z, Haimovitz-Friedman A, Nguyen H, Bose M, Ilagan A, Hunt P, Morgan W, Tilly J, Kolesnick R (2002) Sphingosine 1-phosphate preserves fertility in irradiated female mice without propagating genomic damage in offspring. Nat Med 8:901–902

Passini M, Macauley S, Huff M, Taksir T, Bu J, Wu I, Piepenhagen P, Dodge J, Shihabuddin L, O'Riordan C, Schuchman E, Stewart G (2005) AAV vector-mediated correction of brain pathology in a mouse model of Niemann-Pick A disease. Mol Ther 11:754–762

Passini M, Bu J, Fidler J, Ziegler R, Foley J, Dodge J, Yang W, Clarke J, Taksir T, Griffiths D, Zhao M, O'Riordan C, Schuchman E, Shihabuddin L, Cheng S (2007) Combination brain and systemic injections of AAV provide maximal functional and survival benefits in the Niemann-Pick mouse. Proc Natl Acad Sci USA 104:9505–9510

Pena L, Fuks Z, Kolesnick R (2000) Radiation-induced apoptosis of endothelial cells in the murine central nervous system: protection by fibroblast growth factor and sphingomyelinase deficiency. Cancer Res 60:321–327

Perez G, Knudson C, Leykin L, Korsmeyer S, Tilly J (1997) Apoptosis-associated signaling pathways are required for chemotherapy-mediated female germ cell destruction. Nat Med 3:1228–1232

Perez G, Jurisicova A, Matikainen T, Moriyama T, Kim M, Takai Y, Pru J, Kolesnick R, Tilly J (2005) A central role for ceramide in the age-related acceleration of apoptosis in the female germline. FASEB J 19:860–862

Piantadosi C, Schwartz D (2004) The acute respiratory distress syndrome. Ann Intern Med 141:460–470

Pier G (2002) CFTR mutations and host susceptibility to Pseudomonas aeruginosa lung infection. Curr Opin Microbiol 5:81–86

Riboni L, Campanella R, Bassi R, Villani R, Gaini S, Martinelli-Boneschi F, Viani P, Tettamanti G (2002) Ceramide levels are inversely associated with malignant progression of human glial tumors. Glia 39:105–113

Riethmuller J, Anthonysamy J, Serra E, Schwab M, Doring G, Gulbins E (2009) Therapeutic efficacy and safety of amitriptyline in patients with cystic fibrosis. Cell Physiol Biochem 24:65–72

Rotolo J, Zhang J, Donepudi M, Lee H, Fuks Z, Kolesnick R (2005) Caspase-dependent and -independent activation of acid sphingomyelinase signaling. J Biol Chem 280:26425–26434

Rotolo J, Stancevic B, Lu S, Zhang J, Suh D, King C, Kappel L, Murphy G, Liu C, Fuks Z, van den Brink M, Kolesnick R (2009) Cytolytic T cells induce ceramide-rich platforms in target cell membranes to initiate graft-versus-host disease. Blood 114:3693–3706

Rotolo J, Stancevic B, Zhang J, Hua G, Fuller J, Yin X, Haimovitz-Friedman A, Kim K, Qian M, Cardó-Vila M, Fuks Z, Pasqualini R, Arap W, Kolesnick R (2012) Anti-ceramide antibody prevents the radiation GI syndrome in mice. J Clin Invest 122:1786–1790

Rylova S, Somova O, Dyatlovitskaya E (1998) Comparative investigation of sphingoid bases and fatty acids in ceramides and sphingomyelins from human ovarian malignant tumors and normal ovary. Biochemistry 63:1057–1060

Santana P, Pena L, Haimovitz-Friedman A, Martin S, Green D, McLoughlin M, Cordon-Cardo C, Schuchman E, Fuks Z, Kolesnick R (1996) Acid sphingomyelinase-deficient human lymphoblasts and mice are defective in radiation-induced apoptosis. Cell 86:189–199

Schissel SL, Keesle GA, Schuchman EH, Williams KJ, Tabas I (1998a) The cellular trafficking and zinc dependence of secretory and lysosomal sphingomyelinase, two products of the acid sphingomyelinase gene. J Biol Chem 273:18250–18259

Schissel S, Jiang X, Tweedie-Hardman J, Jeong T, Camejo E, Najib J, Rapp J, Williams K, Tabas I (1998b) Secretory sphingomyelinase, a product of the acid sphingomyelinase gene, can hydrolyze atherogenic lipoproteins at neutral pH. Implications for atherosclerotic lesion development. J Biol Chem 273:2738–2746

Schuchman E (2010) Acid sphingomyelinase, cell membranes and human disease: lessons from Niemann-Pick disease. FEBS Lett 584:1895–1900

Seino K, Kayagaki N, Takeda K, Fukao K, Okumura K, Yagita H (1997) Contribution of Fas ligand to T cell-mediated hepatic injury in mice. Gastroenterology 113:1315–1322

Selzner M, Bielawska A, Morse M, Rudiger H, Sindram D, Hannun Y, Clavien P (2001) Induction of apoptotic cell death and prevention of tumor growth by ceramide analogues in metastatic human colon cancer. Cancer Res 61:1233–1240

Shimeno H, Soeda S, Yasukouchi M, Okamura N, Nagamatsu A (1995) Fatty acyl-Co A: sphingosine acyltransferase in bovine brain mitochondria: its solubilization and reconstitution onto the membrane lipid liposomes. Biol Pharm Bull 18:1335–1339

Sly W, Vogler C (2002) Brain-directed gene therapy for lysosomal storage disease: going well beyond the blood-brain barrier. Proc Natl Acad Sci USA 99:5760–5762

Small D (1970) Surface and bulk interactions of lipids and water with a classification of biologically active lipids based on these interactions. Fed Proc 29:1320–1326

Smith E, Schuchman E (2008) The unexpected role of acid sphingomyelinase in cell death and the pathophysiology of common diseases. FASEB J 22:3419–3431

Song D, Kavanagh B, Benedict S, Schefter T (2004) Stereotactic body radiation therapy. Rationale, techniques, applications, and optimization. Oncology 18:1419–1430

Stancevic B, Kolesnick R (2010) Ceramide-rich platforms in transmembrane signaling. FEBS Lett 584:1728–1740

Steinhart W, Busch J, Oettgen J, Howland J (1984) Sphingolipid metabolism during infection of human fibroblasts by herpes simplex virus type 1. Intervirology 21:70–76

Teichgraber V, Ulrich M, Endlich N, Riethmuller J, Wilker B, De Oliveira-Munding C, van Heeckeren A, Barr M, von Kurthy G, Schmid K, Weller M, Tummler B, Lang F, Grassme H, Doring G, Gulbins E (2008) Ceramide accumulation mediates inflammation, cell death and infection susceptibility in cystic fibrosis. Nat Med 14:382–391

Truman J, Garcia-Barros M, Kaag M, Hambardzumyan D, Stancevic B, Chan M, Fuks Z, Kolesnick R, Haimovitz-Friedman A (2010) Endothelial membrane remodeling is obligate for anti-angiogenic radiosensitization during tumor radiosurgery. PLoS One 5(8):e12310

Tuder R, Petrache I, Elias J, Voelkel N, Henson P (2003) Apoptosis and emphysema: the missing link. Am J Respir Cell Mol Biol 28:551–554

Utermohlen O, Karow U, Lohler J, Kronke M (2003) Severe impairment in early host defense against Listeria monocytogenes in mice deficient in acid sphingomyelinase. J Immunol 170:2621–2628

van Vulpen M, Kal H, Taphoorn M, El-Sharouni S (2002) Changes in blood–brain barrier permeability induced by radiotherapy: implications for timing of chemotherapy? (Review). Oncol Rep 9:683–688

Verheij M, Bose R, Lin X, Yao B, Jarvis W, Grant S, Birrer M, Szabo E, Zon L, Kyriakis J, Haimovitz-Friedman A, Fuks Z, Kolesnick R (1996) Requirement for ceramide-initiated SAPK/JNK signalling in stress-induced apoptosis. Nature 380:75–79

Wang Z, Dudhane A, Orlikowsky T, Clarke K, Li X, Darzynkiewicz Z, Hoffmann M (1994) CD4 engagement induces Fas antigen-dependent apoptosis of T cells in vivo. Eur J Immunol 24:1549–1552

Waselenko J, MacVittie T, Blakely W, Pesik N, Wiley A, Dickerson W, Tsu H, Confer D, Coleman C, Seed T, Lowry P, Armitage J, Dainiak N (2004) Medical management of the acute radiation syndrome: recommendations of the Strategic National Stockpile Radiation Working Group. Ann Intern Med 140:1037–1051

Yang W, Dodge J, Passini M, Taksir T, Griffiths D, Schuchman E, Cheng S, Shihabuddin L (2007) Intraparenchymal injections of acid sphingomyelinase results in regional correction of lysosomal storage pathology in the Niemann-Pick A mouse. Exp Neurol 207:258–266

Yu Z, Nikolova-Karakashian M, Zhou D, Cheng G, Schuchman E, Mattson M (2000) Pivotal role for acidic sphingomyelinase in cerebral ischemia-induced ceramide and cytokine production, and neuronal apoptosis. J Mol Neurosci 15:85–97

Zelinski M, Murphy M, Lawson M, Jurisicova A, Pau K, Toscano N, Jacob D, Fanton J, Casper R, Dertinger S, Tilly J (2011) In vivo delivery of FTY720 prevents radiation-induced ovarian failure and infertility in adult female nonhuman primates. Fertil Steril 95(1440–1445): e1441–e1447

Zundel W, Giaccia A (1998) Inhibition of the anti-apoptotic PI(3)K/Akt/Bad pathway by stress. Genes Dev 12:1941–1946

New Perspectives on the Role of Sphingosine 1-Phosphate in Cancer

Susan Pyne and Nigel J. Pyne

Contents

Abstract In this chapter, we review the latest developments concerning the role of sphingosine 1-phosphate (S1P) in cancer. Particular focus is paid to the role of sphingosine kinases 1 and 2, S1P lyase and S1P-dependent signalling networks in both solid tumours and haematological cancer. The potential of this S1P-dependent pathophysiology as a therapeutic target for the treatment of cancer is also discussed.

Keywords sphingosine 1-phosphate • sphingosine kinase • cancer

1 Introduction

There are six major hallmarks of cancer. These are (1) enhanced proliferation, (2) evasion of growth suppression, (3) enhanced cell survival/reduced apoptosis, (4) acquisition of replicative immortality/reduced senescence, (5) angiogenesis and

S. Pyne (✉) • N.J. Pyne (✉)
Cell Biology Group, SIPBS, University of Strathclyde, 161 Cathedral St, Glasgow, UK
e-mail: susan.pyne@strath.ac.uk; n.j.pyne@strath.ac.uk

E. Gulbins and I. Petrache (eds.), *Sphingolipids in Disease*,
Handbook of Experimental Pharmacology 216,
DOI 10.1007/978-3-7091-1511-4_3,

creation of a tumour microenvironment and (6) increased invasiveness and metastasis. These hallmarks can be initiated by genetic instability including activating or inactivating mutations of oncogenes and tumour suppressors, respectively. Sphingosine 1-phosphate (S1P) is a bioactive lipid that has been implicated in cancer. S1P is formed by the phosphorylation of sphingosine, catalysed by sphingosine kinase (two isoforms termed SK1 and SK2) and is degraded by S1P lyase or dephosphorylated by S1P phosphatase. Extracellular S1P activates a family of G-protein-coupled S1P receptors that couple to various cellular signalling pathways (Pyne and Pyne 2000), whereas intracellular S1P directly binds to intracellular protein targets such as the E3 ligase, TRAF2. The role of S1P in cancer has recently been extensively reviewed (Pyne and Pyne 2010), and therefore, the purpose of this chapter is to update progress in this area in the last 2 years. We describe recent advances which demonstrate that S1P, S1P receptors, sphingosine kinase and S1P lyase are involved in aberrant cellular processes that contribute to the hallmarks of cancer. Strategies targeting SK1 are described which provide impetus for the development of novel SK1 inhibitors that might be used for therapeutic treatment of cancer in the future and that might offer better disease management options.

2 Sphingosine Kinase and Cancer

There is substantial evidence of a role for SK1 in numerous cancers (Pyne and Pyne 2010). For instance, there is increased SK1 expression in stomach, lung, brain, colon, kidney and breast cancers and non-Hodgkins lymphoma (Pyne and Pyne 2010). Many factors influence SK1 expression, such as hypoxia, growth factors and cytokines, and this has a clear indication concerning prognosis. Thus, high expression of SK1 in the tumours of ER^+ breast cancer patients is associated with reduced patient survival and increased disease recurrence (Long et al. 2010a; Watson et al. 2010). One of the reasons for poor prognosis might be because SK1 is a sensor for S1P binding to S1P receptors that enables tuning of the migratory response to S1P in ER^+ breast cancer cells (Long et al. 2010a). In this regard, the binding of S1P to $S1P_3$ receptors promotes the translocation of SK1 from the cytoplasm to the plasma membrane of ER^+ MCF-7 breast cancer cells, and this is a key step in promoting migration. This is exemplified by the finding that siRNA knockdown of SK1 reduces the ability of S1P to induce activation of ERK-1/2, which involves EGF receptor transactivation (Sukocheva et al. 2006; Long et al. 2010a, b) and is required for migration (Long et al. 2010a). Moreover, siRNA knockdown of SK1 reduces the expression of $S1P_3$ in these cells (Long et al. 2010a). Thus, SK1 appears to modulate responsiveness of these cancer cells to S1P by regulating the expression of the $S1P_3$ receptor (Fig. 1). A poor prognosis of grade 4 astrocytoma patients also correlates with high expression of SK1 in tumours (Van Brocklyn et al. 2005). Additional evidence highlighting a key role for SK1 in cancer progression is the finding that the genetic deletion of SK1 in mice significantly reduced 4-nitroquinoline-1-oxide (4-NQO)-induced head and neck squamous cell

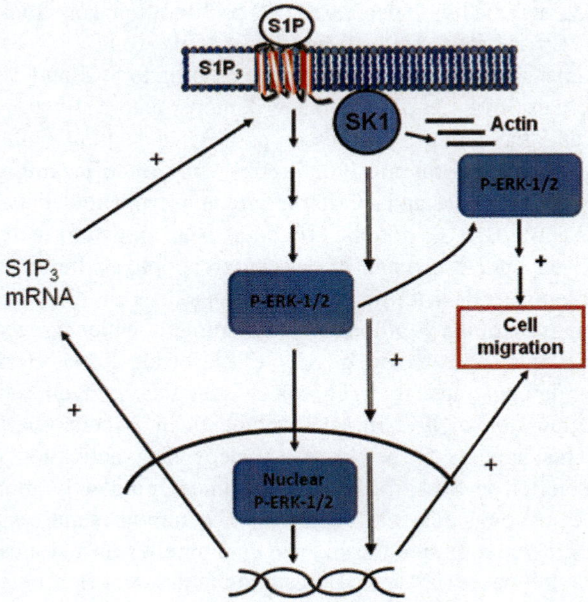

Fig. 1 Schematic showing how SK1 acts as a sensor that governs responsiveness of oestrogen receptor-positive breast cancer cells to S1P in terms of regulating the formation of a migratory phenotype. SK1 regulates the expression of $S1P_3$, which binds S1P to activate the ERK-1/2 pathway. Phosphorylated ERK-1/2 accumulates in actin-enriched lamellipodia and the nucleus where it can regulate metalloproteinase expression, both of which are required for allowing these cells to migrate in response to S1P

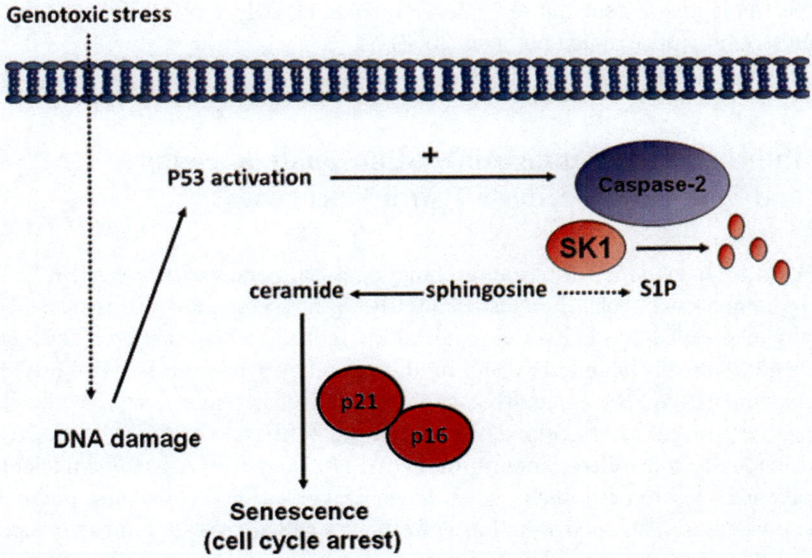

Fig. 2 Schematic showing how genotoxic stress activates p53 which induces a possible caspase-2 catalysed proteolysis of SK1. The downregulation of SK1 perturbs the ceramide-sphingosine-S1P rheostat leading to ceramide-dependent induction of p21 and p16 which are involved in promoting tumour senescence

carcinogenesis, associated with decreased cell proliferation, activation of caspase-3 and ablated AKT pro-survival signalling (Shirai et al. 2011).

One of the most significant recent advances relating to SK1 and cancer has been made by Heffernan-Stroud et al. who have demonstrated a functional interaction between SK1 and the tumour suppressor p53. Mice lacking both p53 tumour-suppressor alleles develop thymic lymphoma, while mice lacking only one p53 allele develop osteosarcomas and soft tissue sarcoma (Donehower et al. 1992; Jacks et al. 1994; Lozano 2010). Genotoxic stress has been shown to increase ceramide levels via a p53-dependent mechanism leading to apoptosis. Indeed, the treatment of Molt-4 leukaemia cells with actinomycin D induces the degradation of SK1, while over-expression of the papilloma virus E6 protein, which targets p53 itself for degradation, prevents the removal of SK1 (Taha et al. 2004). Heffernan-Stroud et al. have now demonstrated that genotoxic stress (actinomycin, etoposide, UV) induces the degradation of SK1 in MCF-7 breast cancer cells (Heffernan-Stroud et al. 2011). This appears to be via a p53-dependent activation of caspase-2. Significantly, deletion of SK1 in p53 deficient mice completely abrogates thymic lymphomas and prolongs survival by 30 %. The tumour suppression by p53 is linked with the elevation of sphingosine and ceramide levels and increased expression of cell cycle inhibitors p21 and p16 resulting in tumour cell senescence (Fig. 2). Therefore, certain cancers might be driven by p53-inactivating mutations that increase SK1 stability and expression. The role of SK1 in cancer is also exemplified by its regulation by other tumour suppressors. In this regard, various extracellular stimuli can activate SK1 via its phosphorylation by ERK-1/2 at Ser225 which increases its activity and promotes its translocation to the plasma membrane (Pitson et al. 2003). Deactivation of SK1 occurs via its dephosphorylation by protein phosphatase 2A (PP2A) (Barr et al. 2008), facilitated by the B'α (B56α/PR61α/PPP2R5A) regulatory subunit of PP2A (Pitman et al. 2011), a well-defined tumour-suppressor protein (Arnold and Sears 2008).

3 Interaction of Sphingosine 1-Phosphate Receptors and Sphingosine Kinase 1 with Oncogenes

There is now evidence for several functional partnerships between SK1, S1P receptors and oncogenes. For instance, H-RAS has been shown to increase SK1 activity in NIH3T3 fibroblasts (Xia et al. 2000). Indeed, transformation of these cells into fibrosarcoma is inhibited by SK1 inhibitors and by over-expression of the G82D dominant negative SK1 mutant (Xia et al. 2000). SK1 is also activated by the GDP bound form of eukaryotic elongation factor 1A (eEF1A) (Leclercq et al. 2011). Translationally controlled tumour protein (TCTP) is an eEF1A guanine nucleotide dissociation inhibitor which when over-expressed in cells also activates SK1. Intriguingly, an oncogenic form of eEF1A1 called prostate tumour inducer-1 (PTI-1) which lacks the GDP–GTP-binding domain activates SK1, and this is required for PTI-1-induced neoplastic transformation (Leclercq et al. 2011). HER2/ErbB2 also functionally interacts with $S1P_4$ in ER^- MDA-MB-453 breast cancer cells to promote enhanced ERK-1/2 signalling (Long et al. 2010b), and S1P stimulates the tyrosine phosphorylation of HER2 in MKN28 and MKN74 gastric cancer cells (Shida et al. 2005). Paradoxically, SK1 can negate the function of HER2 in ER^+ breast cancer cells (Long et al. 2010a). This involves a process in which HER2 increases SK1 expression, which results in a negative feedback reduction in HER2 expression and ablated migration of these cells in response to S1P. High levels of SK1 in a HER2/ER^+ background induce a deregulation of p21-activated protein kinase 1 which normally functions to promote migration of these cells in response to S1P. Therefore, SK1 can induce tolerance to HER2 in ER^+ breast cancer cells, and this might have clinical significance as high cytoplasmic SK1 expression in $HER2^+$/ER^+ breast tumours is associated with *increased* patient survival and *reduced* disease recurrence in patients treated with tamoxifen (Long et al. 2010a).

4 Advances in Sphingosine Kinase Inhibitors

In the past 10 years, a number of SK1 inhibitors have been synthesised. SK1-I (BML-258; (2R,3S,4E)-*N*-methyl-5-(4′-pentylphenyl)-2-aminopent-4-ene-1,3-diol) is a water-soluble sphingosine analogue that is specific for SK1 and enhances survival of mice in an orthotopic intracranial tumour model (Kapitonov et al. 2009). SKi (or SKI-II, 2-(*p*-hydroxyanilino)-4-(*p*-chlorophenyl) thiazole) inhibits both SK1 and SK2 and reduces S1P levels, inhibits proliferation and induces apoptosis in various cancer cell lines (French et al. 2003). Several natural products, B-5354c (Kono et al. 2000a), F-12509A (Kono et al. 2000b) and S-15183a,b (Kono et al. 2001), have been isolated, which inhibit SK activity in vitro. More recently, FTY720 (2-amino-2-[2-(4-octylphenyl)ethyl]-1,3-propanediol hydrochloride) has been identified as a SK1 inhibitor (Tonelli et al. 2010; Lim et al. 2011a). Indeed, the inhibition of SK1 activity by FTY720 induces prostate cancer cell apoptosis in a manner independent of S1P receptors (Pchejetski et al. 2010). Moreover, γ irradiation, which does not affect SK1 activity alone, synergised with FTY720 in inhibiting SK1 activity and inducing apoptosis. This can be reverted by over-expression of SK1, which protects the cancer cells from FTY720-induced apoptosis. These findings have translational significance because FTY720 also increases the sensitivity of prostate tumours to radiation in vivo without toxic side effects (Pchejetski et al. 2010).

There are also other examples expanding the utility of FTY720 as an anticancer agent. For example, FTY720 has been shown to reduce gastric cancer cell proliferation, and this is associated with G1 phase cell cycle arrest and apoptosis (Zheng et al. 2010). This involves an increase in the expression of PTEN, which subsequently ablates the phosphorylation of AKT and MDM2 (Zheng et al. 2010). MDM2 is an E3 ubiquitin ligase that catalyses the ubiquitination of p53, which enables its proteasomal degradation and which maintains low levels of p53 in cells. Therefore, FTY720 induces a post-translational increase in p53 stability and expression, which is associated with the induction of p53-regulated genes involved in growth arrest and apoptosis, such as Cip1/p21, p27 and BH3-only proteins.

FTY720 is also cytotoxic in human ovarian cancer cells (Zhang et al. 2010). This occurs via a mechanism that is independent of caspase-3 activity and involves cellular swelling and cytoplasmic vacuolisation, typical of necrotic cell death (Zhang et al. 2010). FTY720 also promotes autophagy in these cells as evidenced by formation of autophagosomes and the accumulation of LC3-II. Interestingly, the siRNA knockdown of Beclin 1 or LC3 leads to enhanced necrotic cell death in response to FTY720, suggesting that stimulation of autophagy is a rebound survival response that impedes FTY720-induced necrotic death. Thus, targeting autophagy might augment the necrotic death promoting activity of FTY720 in ovarian cancer.

FTY720 also has marked inhibitory effects on hepatocellular carcinoma (Omar et al. 2011). In this case, FTY720 induces apoptosis by activating reactive oxygen species and PKCδ. In addition, the compound OSU-2S has been synthesised from FTY720. This compound lacks $S1P_1$ receptor activity, thereby excluding the potential side effect of lymphopenia (as seen with FTY720, once phosphorylated) that might compromise anticancer activity. OSU-2S exhibits higher potency than FTY720 and is also active in vivo, reducing tumour growth in both ectopic and orthotopic models of hepatocellular carcinoma without significant toxicity (Omar et al. 2011).

Structural modification of the FTY720 scaffold has produced (*S*)-FTY720 vinylphosphonate, which is also an SK1 inhibitor (Tonelli et al. 2010). The treatment of MCF-7 breast cancer cells with FTY720, (*S*)-FTY720 vinylphosphonate or SKi prevents the formation of actin-enriched lamellipodia in response to S1P (Lim et al. 2011a), suggesting that these inhibitors block formation of the migratory phenotype, and which might have application for preventing metastasis. Indeed, treatment of MCF-7 cells with SKi reduces their migration to S1P (Long et al. 2010a). Furthermore, siRNA knockdown of SK1 similarly prevents actin rearrangement into lamellipodia (Long et al. 2010a) (Fig. 1). This is reminiscent of the action of FTY720 in vivo where it decreases metastasis in a mouse model of breast cancer and which is associated with impaired formation of filopodia (Azuma et al. 2002). Remarkably, the treatment of MCF-7 breast cancer and LNCaP prostate cancer cells with FTY720, (*S*)-FTY720 vinylphosphonate or SKi also induces a novel ubiquitin-proteasomal degradation of SK1 (Tonelli et al. 2010; Loveridge et al. 2010; Lim et al. 2011a), and this results in apoptosis, thereby offering additional enhanced efficacy options for the treatment of cancer. Recently, Macdonald et al. have synthesised amidine-based SK1 inhibitors which exhibit nanomolar potency and reduce intracellular S1P levels in human leukaemia U937 cells (Kennedy et al. 2011), thereby providing evidence for the first time that it is possible to achieve high, clinically relevant potency.

There is also new emerging evidence for an important role of SK2 in cancer. This is exemplified by the finding that the siRNA knockdown of SK2 enhances doxorubicin-induced apoptosis of breast or colon cancer cells (Sankala et al. 2007). Surprisingly, siRNA knockdown of SK2 elevates SK1 expression and increases intracellular S1P in A498, Caki-1 or MDA-MB-231 cells (Gao and Smith 2011). However, knockdown of SK2 in these cells reduces cell proliferation and migration/invasion, and this is actually more effective than knockdown of SK1, which results in reduced intracellular S1P and has no effect on SK2 expression. The knockdown of SK1 or SK2 also have differential effects on p53, p21, ERK-1, ERK-2, FAK and VCAM1 indicating that SK1 and SK2 have non-overlapping functions. These studies provide evidence that loss of SK2 has a stronger anticancer effect in these cell lines compared with loss of SK1 (Gao and Smith 2011) although this is unexpected given the widely documented pro-cancerous role of SK1-derived S1P. The increase in intracellular S1P upon loss of SK2 together with reduced proliferation suggests that SK2 regulates a discrete intracellular pool of S1P that is functionally compartmentalised from S1P formed by SK1 in the cancer cell types studied (see Pyne et al. 2009 for a review of spatial aspects of S1P signalling). The reduction in SK2-derived S1P upon knockdown of SK2 might then be masked by the net increase in intracellular S1P levels caused by the compensatory elevation in SK1 expression. Alternatively, it is possible that unidentified non-catalytic functions of SK2 regulate some of the effectors. This is not without foundation as SK2 interacts with Bcl2 via its BH3 domain, although this interaction promotes apoptosis (Liu et al. 2003). Nevertheless, these studies strongly support the idea that the synthesis of SK2 inhibitors might be a strategy for the development of anticancer agents. In this regard, two new SK2-selective inhibitors, namely (*R*)-FTY720 methyl ether ((*R*)-FTY720-OMe) and ABC294640, have been synthesised and characterised (Lim et al. 2011b; French et al. 2010).

(*R*)-FTY720-OMe has one of the prochiral hydroxyl groups of FTY720 replaced by a methoxy group, thereby blocking the site of phosphorylation by SK2. (*R*)-FTY720-OMe is a selective, competitive (with sphingosine) inhibitor of SK2. The treatment of MCF-7 cells with (*R*)-FTY720-OMe prevents actin enrichment into lamellipodia in response to S1P suggesting application to inhibit metastasis (Lim et al. 2011b). The aryladamantane compound, ABC294640 [3-(4-chlorophenyl)-adamantane-1-carboxylic acid (pyridin-4-ylmethyl)amide] is a competitive inhibitor (with sphingosine) of SK2 activity and reduces intracellular S1P formation in cancer cells (French et al. 2010). The K_i values for inhibition of SK2 by both (*R*)-FTY720-OMe and ABC294640 are very similar (16.5 vs. 10 μM, respectively). ABC294640 suppresses the proliferation of several tumour cell lines and inhibits tumour cell migration associated with the loss of actin microfilaments. In addition, ABC294640 has significant in vivo activity and inhibits tumour progression in mice with mammary adenocarcinoma xenografts (French et al. 2010).

More recent studies have demonstrated that ABC294640 induces autophagy in A-498 kidney carcinoma, PC-3 prostate and MDA-MB-231 breast cancer cells (Beljanski et al. 2010). Interestingly, combined treatment of A-498 cells with ABC294640 and an autophagy inhibitor (3-methyladenine) induces apoptosis, demonstrating a close balance between autophagic and apoptotic processes. In this regard, Smith et al. proposed a model for the action of ABC294640 (Beljanski et al. 2010): in normal conditions, tumour cells operate under conditions of a low autophagic flux. This enables balancing the energy demand for growth, which is high and is driven by the ERK pathway. In contrast, the apoptotic rate is low due to high AKT pro-survival signalling. When the cells are treated with ABC294640, there is a marked reduction in ERK and AKT signalling leading to the inhibition of proliferation and pro-survival signalling. However, the inhibition of SK2 activity by ABC294640 results in the accumulation of ceramide which, in turn, stimulates autophagic death and which predominates over apoptotic death. However, autophagy induced by ABC294640 is blocked in the presence of 3-methyladenine, and therefore, ABC294640 induces apoptosis as the predominant mechanism. In this respect, it is interesting to note that (*R*)-FTY720-OMe induces apoptosis in HEK 293 cells (Lim et al. 2011b). Smith et al. have also found that ABC294640 delays tumour growth and increases autophagy markers in severe combined immunodeficient mice with A-498 xenografts, albeit the number of apoptotic cells did not increase (Beljanski et al. 2010).

ABC294640 has also been shown to decrease ER-regulated gene expression, and this is accounted for by the ability of ABC294640 to also bind to the antagonist ligand-binding domain of the oestrogen receptor, ERα. In this case, ABC294640 functions as a partial antagonist similar to tamoxifen (Antoon et al. 2010). This property of ABC294640 might be physiologically relevant as ABC294640 inhibits ER^+ breast cancer tumour formation in vivo. Similarly, SKi has also been reported to function as an ER antagonist (Antoon et al. 2011). These additional activities of ABC294640 and SKi enable a multipronged attack on breast cancer. In this regard, (*S*)-FTY720 vinylphosphonate might also be exploited to launch a multipronged attack on cancer as it can affect multiple targets. With regard to S1P-related targets, (*S*)-FTY720 vinylphosphonate inhibits both SK1 and SK2 activities (Lim et al. 2011a). It is also a full antagonist of $S1P_{1,3,4}$ (K_i 208, 15 and 1,190 nM, respectively) and a partial antagonist of $S1P_2$ and $S1P_5$ (Valentine et al. 2010). Additionally, (*S*)-FTY720 vinylphosphonate reduces the expression of the androgen receptor in androgen-independent LNCaP-AI cells (Tonelli et al. 2010), which might provide additional anti-proliferative activity.

The importance of S1P receptors in cancer is highlighted by several studies. For instance, S1P stimulates the migration of fibrosarcoma cells through $S1P_1$ (Fisher et al. 2006) and gastric cancer cells through $S1P_3$ (Yamashita et al. 2006). High membrane $S1P_1$ expression in the ER^+ breast cancer tumours is associated with shorter time to disease recurrence on tamoxifen, and high cytoplasmic $S1P_1$ and $S1P_3$ tumour expression is linked with reduced disease specific survival times (Watson et al. 2010). In addition, IL-6/Jak2 signalling is linked with $S1P_1$, and this is required for the sustained activation of Stat3, which promotes cancer progression (Lee et al. 2010). A positive feedback loop exists, whereby Stat3 increases the expression of $S1P_1$ resulting in further activation of Stat3 and up-regulation of IL-6 expression, which enhances tumour growth and metastasis.

Taking all these findings together, it is logical to propose that the combined inhibition of SK1 and/or SK2 and/or the antagonism of S1P receptors in cancer might provide better efficacy options compared with agents that inhibit each of these targets alone.

5 Advances on the Role of S1P Lyase in Cancer

The human SGPL1 gene, which encodes S1P lyase (SPL), maps to chromosomal regions that are commonly mutated in cancer, and one can consider SPL as having tumour-suppressor activity. Expression of SPL is regulated principally by Sp1 and a Sp1/GATA-4 complex which bind to the SGPL1 gene promoter (Ito et al. 2011). However, SPL expression is not directly correlated with cellular S1P levels in lung cancer cell lines (Ito et al. 2011). Nevertheless, deficiency in SPL induces resistance to etoposide and doxorubicin. Moreover, the reduction in doxorubicin-induced apoptosis in $Sgpl^{-/-}$ cells is associated with up-regulation of the anti-apoptotic proteins, Bcl2 and Bcl-xl. The deficiency of SPL is also linked with increased proliferation, anchorage-independent growth and formation of tumours in mice. Importantly, SPL expression is decreased in human melanoma cell lines, thereby suggesting reduced tumour-suppressor activity that might contribute to the progression of this cancer (Colié et al. 2009).

6 Advances in Other Specific Cancers

Several recent significant advances have been made concerning the role of S1P in haematological cancers:

6.1 Myeloid Leukaemia

The *BCR/ABL1* oncogene is produced by the Philadelphia chromosome translocation t(9;22)(q34;q11) and is a linked to poor prognosis in chronic myelogenous leukaemia (CML). It is also present in 20 % of adult acute lymphoblastic leukaemias (ALL). *BCR/ABL1* encodes a 210 kDa constitutively active tyrosine kinase that is present in CML patients, where it enhances survival of myeloid progenitors leading to myeloproliferative disease. The p190-encoded Bcr/Abl has similar transforming ability but stimulates proliferation of lymphoid progenitors. The Bcr/Abl oncoprotein activates signalling pathways that result in increased survival and ablated differentiation of hematopoietic progenitors.

The Bcr/Abl tyrosine kinase has been shown to increase SK1 expression, and this appears necessary for regulating the level of Mcl-1 and other anti-apoptotic Bcl-2 family members that enhance cell survival (Li et al. 2007; Bonhoure et al. 2008). In CML cells, Bcr/Abl increases SK1 expression through its activation of the ERK-1/2, PI3K and Janus kinase 2 pathways (Li et al. 2007) and AKT2/mTOR in imatinib-resistant CML cells (Marfe et al. 2011). The major treatment option for CML involves the use of the Bcr/Abl tyrosine kinase inhibitor, imatinib, which reduces SK1 expression, and this appears to contribute to the therapeutic efficacy of

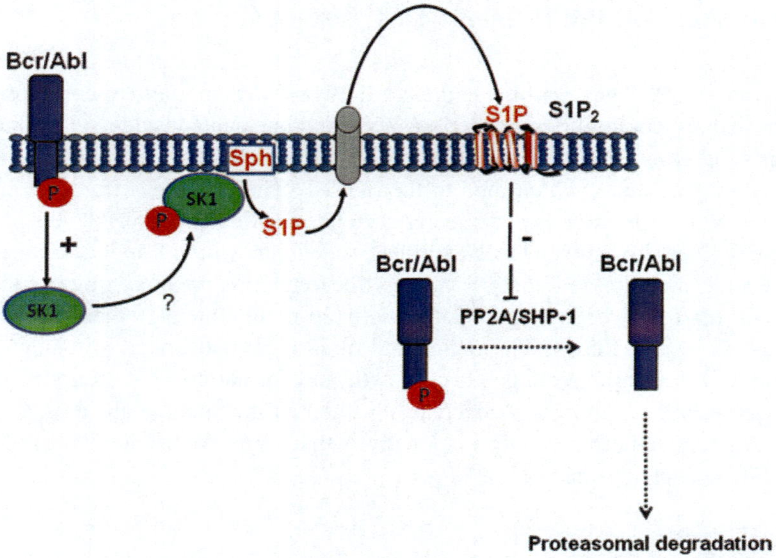

Fig. 3 Schematic showing how SK1 catalyses formation of S1P which is released to act on $S1P_2$. S1P bound to $S1P_2$ inhibits PP2A/SHP-1-induced degradation of Bcr/Abl, thereby increasing its stability and conferring resistance of CML to imatinib

imatinib (Bonhoure et al. 2008). However, CML and Ph1 ALL patients develop resistance dependent on the over-expression or mutation of Bcr/Abl. Moreover, SK1 confers resistance of CML cells to imatinib (Baran et al. 2007). Indeed, the knockdown of SK1 expression by small interference RNA increases the sensitivity of resistant cells to imatinib in terms of the induction of apoptosis (Marfe et al. 2011) and enforced expression of SK1 in K562 cells increases the S1P/C18-ceramide ratio and prevents apoptosis to imatinib (Baran et al. 2007). These findings have been replicated in other CML cells. For instance, the ceramide:S1P ratio is increased in response to imatinib in imatinib-sensitive LAMA84 cells, while the ratio is unaltered in imatinib-resistant cells. Additionally, over-expression of SK1 in imatinib-sensitive cells impairs apoptosis by inhibiting caspase-3 activation and cytochrome C/Smac release. This involves the SK1-dependent modulation of Bim, Bcl-xL and Mcl-1 expression (Bonhoure et al. 2008). Furthermore, daunorubicin-sensitive but not insensitive leukaemia cells (CML, AML and ALL) exhibit an elevated ceramide:S1P ratio when treated with daunorubicin and sensitivity to daunorubicin is restored by inhibiting SK1 activity (Sobue et al. 2008).

Taken together, these findings identify SK1 as having a very important role in regulating the sensitivity of leukaemic cells to chemotherapeutic agents. Indeed, Ogretman et al. have recently provided a molecular explanation for the role of SK1 in the acquisition of resistance to imatinib in CML (Salas et al. 2011). They demonstrated that increased SK1/S1P enhances Bcr/Abl1 protein stability. This involves S1P formed by SK1 being released to act on $S1P_2$ receptor which inactivates PP2A and thereby prevents Bcr/Abl1 degradation (Fig. 3). This is consistent with the fact that the proteasomal degradation of Bcr/Abl1 requires its dephosphorylation by PP2A/SHP-1. Moreover, pharmacological inhibition or siRNA knockdown of SK1 restores dephosphorylation and proteasomal degradation of Bcr/Abl1 and enhances the inhibition of growth by imatinib or nilotinib in primary $CD34^+$ mononuclear cells obtained from chronic phase and blast crisis CML patients, imatinib-insensitive CML cells and murine progenitor cells expressing the wild-type or mutant (Y253H or T315I) Bcr/Abl1. SK1 inhibition also enhances the sensitivity of 32D/T315I-Bcr/Abl1-derived mouse allografts to growth inhibition by nilotinib (Salas et al. 2011).

Interestingly, Bcr/Abl1 inactivates the tumour suppressor PP2A by enhancing the expression of the PP2A inhibitor SET (Neviani et al. 2005). The effect of the functional loss of PP2A in chronic phase and blast crisis CML patient cells has been assessed using FTY720, which is an activator of PP2A (Neviani et al. 2007). The recent findings made by Salas et al. (2011) and Tonelli et al. (2010) can now be used to elaborate the mechanism of action of FTY720. Neviani et al. (2007) showed that whilst normal $CD34^+$ or $CD34^+/CD19^+$ bone marrow cells were unaffected, FTY720 induced apoptosis of $CML\text{-}BC^{CD34+}$ and Ph1 $ALL^{CD34+/CD19+}$ progenitors and impaired clonogenicity of imatinib/dasatinib-sensitive and -resistant Bcr/Abl myeloid and lymphoid cell lines. These workers proposed that the restoration of PP2A activity by FTY720 reduces Bcr/Abl expression by inducing its proteasomal degradation. In addition, exogenous S1P counteracts the effect of FTY720 on Bcr/Abl stability, which is consistent with a role for extracellular S1P in preventing Bcr/Abl degradation, as reported by Salas et al. (2011). Moreover, SK inhibitors reduce the phosphorylation and stability of Bcr/Abl suggesting that inhibition of SK1 is sufficient to prevent the SK1-$S1P_2$ autocrine loop that functions to inactivate PP2A. As mentioned previously, we have shown that FTY720 is a competitive inhibitor of SK1 (Lim et al. 2011a). In addition, FTY720 promotes the ubiquitin-proteasomal degradation of SK1 to create SK1 null cancer cells (Tonelli et al. 2010). It is therefore possible that the effect of FTY720 on Bcr/Abl stability is mediated not only by the direct activation of PP2A but also by inhibition/downregulation of SK1, which would theoretically restore PP2A activity by interrupting the S1P/$S1P_2$ functional loop (Fig. 4). Alternatively, the activation of PP2A activity by FTY720 might function in a negative feedback loop to inactivate SK1 (Barr et al. 2008) that has been phosphorylated by ERK-1/2 on S225 (Pitson et al. 2003), thereby providing an additional route for reducing SK1 activity. FTY720 also suppresses Bcr/Abl-dependent (including mutated p210/p190Bcr/Abl (T315I)) leukaemogenesis without exerting any toxicity. Therefore, the action of FTY720 on SK1 inhibition/downregulation and restoration of PP2A activity might represent a clear example of where modulation of ceramide-sphingosine-S1P rheostat has a well-defined clinical translational potential.

ALL is the most common form of childhood cancer. Notably, FTY720 reduces survival of Ph(+) and Ph(−) ALL cell lines. However, this occurs by a mechanism that is independent of PP2A and caspase-3 (Wallington-Beddoe et al. 2011). FTY720 also induces autophagy as evidenced by increased LC3II expression and autophagic flux. Interestingly, FTY720 stimulates reactive oxygen species (ROS),

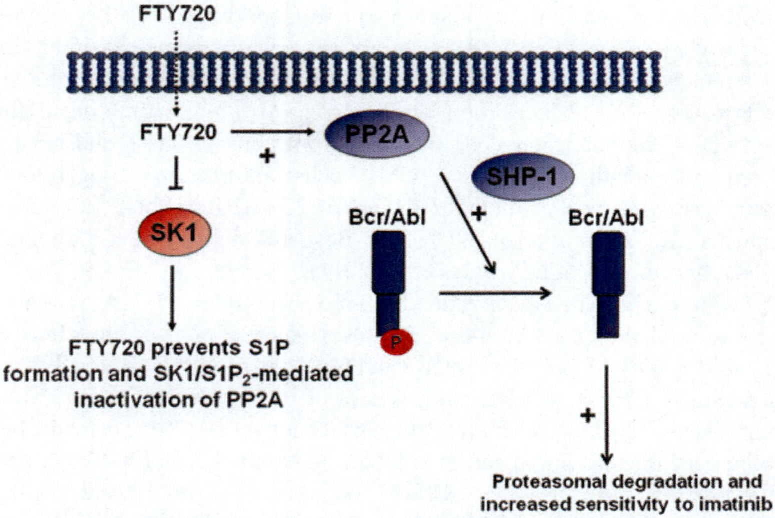

Fig. 4 Schematic showing how FTY720-dependent inhibition/degradation of SK1 might perturb the ceramide-sphingosine-S1P rheostat to prevent the inhibitory effect of S1P/S1P$_2$ on PP2A/SHP-1, thereby enhancing Bcr/Abl degradation and improving sensitivity of CML to imatinib. This might represent an additional mechanism to direct activation of PP2A by FTY720

and therefore, the antioxidant N-acetyl-cysteine (NAC) partially reverses the cytotoxic effects of this compound. In this respect, it is noteworthy that shRNA knockdown of SK1 or an SK1 inhibitor induces ROS formation in carcinoma (Huwlier et al. 2011), suggesting close coupling between SK1 and ROS formation.

6.2 *Multiple Myeloma*

Multiple myeloma cell lines and cells isolated from patients that are unresponsive to conventional therapeutics undergo apoptosis in response to FTY720. This is mediated by activation of caspase-8, -9 and -2 and altered BAX cleavage and mitochondrial potential (Yasui et al. 2005). However, FTY720 also reduced IL-6-stimulated Akt phosphorylation, Stat3 and ERK-1/2 activation, IGF-I-stimulated Akt activation and TNFα-stimulated IKB and NFκB phosphorylation in these cells. A Mcl-1-dependent pathway is involved in the FTY720-induced apoptosis of B cell malignancies and primary B cells from chronic lymphocytic leukaemia (CLL) patients (Liu et al. 2008). FTY720 can also induce prolonged survival of mice with disseminated B cell lymphoma/leukaemia (Liu et al. 2008).

Lymphoma: The oncogenic activity of Runx (Runx1, 2 and 3) genes has been demonstrated in transgenic mice that develop lymphoma and which display marked synergism with over-expressed Myc. Interestingly, three genes involved in sphingolipid metabolism have been identified as direct targets for Runx-dependent transcriptional regulation (Kilbey et al. 2010). *Sgpp1* (which encodes S1P phosphatase-1) is decreased, whereas *Ugcg* (which encodes UDP–glucose ceramide galactosyltransferase) and *St3gal5/Siat9* (which encodes GM3 synthase) are both increased. This is in line with the finding that ectopic over-expression of Runx reduces intracellular long-chain ceramides and elevates extracellular S1P in NIH3T3 fibroblasts. Runx expression also reduces the activation JNK and p38 MAPK, which are essential proteins involved in ceramide-induced death.

7 Summary and Future Directions

There is now ample evidence that sphingosine kinase has tumour-promoting activity via direct or functional interaction with oncogenes. Cancer progression might occur because positive selection and clonal expansion are consequences of SK1 conferring a significant survival and growth advantage to the cancer cells. S1P lyase is pro-apoptotic and therefore exhibits tumour-suppressing activity. This highlights the key role for regulating the position of the ceramide-sphingosine-S1P rheostat in cancer. The release of S1P from cells or its partitioning into the plasma membrane and close proximity association with S1P receptors in addition to its binding to intracellular targets such as HDAC1/2 (Hait et al. 2009) and TRAF2 (Alvarez et al. 2010) might influence ubiquitination, phosphorylation, methylation, glycosylation and redox pathways conducive to enhanced survival and invasiveness of cancer cells. The functional interaction between S1P receptors and oncogenes is an additional level of complexity that might contribute to cancer and is worthy of further investigation. These processes are therefore viable targets for therapeutic intervention as exemplified by the effectiveness of SK inhibitors in in vivo cancer models (Pyne and Pyne 2010). Indeed, the SK inhibitor L-*threo*-dihydrosphingosine (safingol) in combination with the DNA cross-linking agent, cisplatin is effective in phase I clinical trials for the treatment of advanced solid phase tumours (Dickson et al. 2011). More successful translation to the clinic requires the development of compounds with nanomolar potency, enhanced efficacy (e.g. exemplified by inhibitor-induced proteasomal degradation of SK1) and improved bioavailability. In addition, we believe that a major focus of future research will be in the identification and characterisation of new intracellular targets that bind S1P. Indeed, it is possible that the activation/inhibition of subsets of intracellular targets might, in concert with oncogenes, confer a particular phenotype to the cancer. Identification of these S1P-dependent signalling pathways will open up new translational possibilities for therapeutically targeting cancer using personalised medicines so that this disease can be treated more effectively than is currently the case.

References

Alvarez SE, Harikumar KB, Hait NC, Allegood J, Strub GM, Kim EY, Maceyka M, Jiang H, Luo C, Kordula T, Milstien S, Spiegel S (2010) Sphingosine 1-phosphate is a missing co-factor for the E3 ligase TRAF2. Nature 465:1084–1088

Antoon JW, White MD, Meacham WD, Slaughter EM, Muir SE, Elliott S, Rhodes LV, Ashe HB, Wiese TE, Smith CD, Burow ME, Beckman BS (2010) Antiestrogenic effects of the novel sphingosine kinase-2 inhibitor ABC294640. Endocrinology 151:5124–5135

Antoon JW, Meacham WD, Bratton MR, Slaughter EM, Rhodes LV, Ashe HB, Wiese TE, Burow ME, Beckman BS (2011) Pharmacological inhibition of sphingosine kinase isoforms alters estrogen receptor signalling in human breast cancer. J Mol Endocrinol 46:205–216

Arnold HK, Sears RC (2008) A tumor suppressor role for PP2A–alpha through negative regulation of c-Myc and other key oncoproteins. Cancer Metastasis Rev 27:147–158

Azuma H, Takahara S, Ichimaru N, Wang JD, Itoh Y, Otsuki Y, Morimoto J, Fukui R, Hoshiga M, Ishihara T, Nonomura N, Suzuki S, Okuyama A, Katsuoka Y (2002) Marked prevention of tumor growth and metastasis by a novel immunosuppressive agent, FTY720, in mouse breast cancer models. Cancer Res 62:1410–1419

Baran Y, Salas A, Senkal CE, Gunduz U, Bielawski J, Obeid LM, Ogretmen B (2007) Alterations of ceramide/sphingosine 1-phosphate rheostat involved in the regulation of resistance to imatinib induced apoptosis in K562 human chronic myeloid lekaemia cells. J Biol Chem 282:10922–10934

Barr RK, Lynn HE, Moretti PA, Khew-Goodall Y, Pitson SM (2008) Deactivation of sphingosine kinase 1 by protein phosphatase 2A. J Biol Chem 283:34994–35002

Beljanski V, Knaak C, Smith CD (2010) A novel sphingosine kinase inhibitor induces autophagy in tumour cells. J Pharmacol Exp Ther 333:454–464

Bonhoure E, Lauret A, Barnes DJ, Martin C, Malavaud B, Kohama T, Melo JB, Cuvillier O (2008) Sphingosine kinase-1 is a downstream regulator of imatinib-induced apoptosis in chronic myeloid leukaemia cells. Leukemia 22:971–979

Colié S, Van Veldhoven PP, Kedjouar B, Bedia C, Albinet V, Sorli SC, Garcia V, Djavaheri-Mergny M, Bauvy C, Codogno P, Levade T, Andrieu-Abadie N (2009) Disruption of sphingosine 1-phosphate lyase confers resistance to chemotherapy and promotes oncogenesis through Bcl-2/Bcl-xL upregulation. Cancer Res 69:9346–9353

Dickson MA, Carvajal RD, Merrill AH Jr, Gonen M, Cane LM, Schwartz GK (2011) A phase I clinical trial of safingol in combination with Cisplatin in advanced solid tumours. Clin Cancer Res 17:2484–2492

Donehower LA, Harvey M, Slagle BL, McArthur MJ, Montgomery CA Jr, Butel JS, Bradley A (1992) Mice deficient for p53 are developmentally normal but susceptible to spontaneous tumours. Nature 356:215–221

Fisher KE, Pop A, Koh W, Anthis NJ, Saunders WB, Davis GE (2006) Tumour cell invasion of collagen matrices requires coordinate lipid agonist-induced G protein and membrane-type matrix metalloproteinase-1-dependent signalling. Mol Cancer 5:69

French KJ, Schrecengost RS, Lee BD, Zhuang Y, Smith SN, Eberly JLM, Yun JK, Smith CD (2003) Discovery and evaluation of inhibitors of human sphingosine kinase. Cancer Res 63:5962–5969

French KJ, Zhuang Y, Maines LW, Gao P, Wang W, Beljanski V, Upson JJ, Green CL, Keller SN, Smith CD (2010) Pharmacology and antitumour activity of ABC294640, a selective inhibitor of sphingosine kinase-2. J Pharmacol Exp Ther 333:129–139

Gao P, Smith CD (2011) Ablation of sphingosine kinase-2 inhibits tumour cell proliferation and migration. Mol Cancer Res 9(11):1509–19

Hait NC, Allegood J, Maceyka M, Strub GM, Harikumar KB, Singh SK, Luo C, Marmorstein R, Kordula T, Milstien S, Spiegel S (2009) Regulation of histone acetylation in the nucleus by sphingosine 1-phosphate. Science 325:1254–1257

Heffernan-Stroud LA, Helke KL, Jenkins RW, De Costa AM, Hannum YA, Obeid LM (2011) Defining a role for sphingosine kinase 1 in p53-dependent tumours. Oncogene 31(9):1166–75

Huwlier A, Kotelevets N, Xin C, Pastukhov O, Pfeilschiffter J, Zangemeister-Wittke U (2011) Loss of sphingosine kinase 1 in carcinoma cells increases formation of reactive oxygen species and sensitivity to doxorubicin-induced DNA damage. Br J Pharmacol 162:532–543

Ito H, Yoshida K, Murakami M, Hagiwara K, Sasaki N, Kobayashi M, Takagi A, Kojima T, Sobue S, Suzuki M, Tamiya-Koizumi K, Nakamura M, Banno Y, Nozawa Y, Murate T (2011) Heterogeneous sphingosine-1-phosphate lyase gene expression and its regulatory mechanism in human lung cancer cell lines. Biochim Biophys Acta 1811:119–128

Jacks T, Remington L, Williams BO, Schmitt EM, Halachmi S, Bronson RT, Weinberg RA (1994) Tumour spectrum analysis in p53-mutant mice. Curr Biol 4:1–7

Kapitonov D, Allegood JC, Mitchell C, Hait NC, Almenara JA, Adams JK, Zipkin RE, Dent P, Kordula T, Milstien S, Spiegel S (2009) Targeting sphingosine kinase 1 inhibits AKT signalling, induces apoptosis, and suppresses growth of human glioblastoma cells and xenografts. Cancer Res 69:6915–6923

Kennedy AJ, Mathews TP, Kharel Y, Field SD, Moyer ML, East JE, Houck JD, Lynch KR, Macdonlad TL (2011) Development of amidine-based sphingosine kinase 1 nanomolar inhibitors and reduction of sphingosine 1-phosphate in human leukemia cells. J Med Chem 54:3524–3548

Kilbey A, Terry A, Jenkins A, Borland G, Zhang Q, Wakelam MJ, Cameron ER, Neil JC (2010) Runx regulation of sphingolipid metabolism and survival signalling. Cancer Res 70:5860–5869

Kono K, Tanaka M, Ogita T, Kohama T (2000a) Characterisation of B-5354c, a new sphingosine kinase inhibitor, produced by a marine bacterium. J Antibiot (Tokyo) 53:759–764

Kono K, Tanaka M, Ogita T, Hosoya T, Koyama T (2000b) F-12509A, a new sphingosine kinase inhibitor, produced by a discomycete. J Antibiot (Tokyo) 53:459–466

Kono K, Tanaka M, Ono Y, Hosoya T, Ogita T, Kohama T (2001) S-15183a and b, new sphingosine kinase inhibitors, produced by a fungus. J Antibiot (Tokyo) 54:415–420

Leclercq TM, Moretti PA, Pitson SM (2011) Guanine nucleotides regulate sphingosine kinase 1 activation by eukaryotic elongation factor 1A and provide a mechanism for eEF1A-associated oncogenesis. Oncogene 30:372–378

Lee H, Deng J, Kujawski M, Yang C, Liu Y, Herrmann A, Kortylewski M, Horne D, Somlo G, Forman S, Jove R, Yu H (2010) STAT3-induced S1PR1 expression is crucial for persistent STAT3 activation in tumors. Nat Med 16:1421–1428

Li QF, Huang WR, Duan HF, Wang H, Wu CT, Wang LS (2007) Sphingosine kinase-1 mediates BCR/ABL-induced upregulation of Mcl-1 in chronic myeloid leukemia cells. Oncogene 26:7904–7908

Lim KG, Tonelli F, Li Z, Lu X, Bittman R, Pyne S, Pyne NJ (2011a) FTY720 analogues as sphingosine kinase 1 inhibitors: Enzyme inhibition kinetics, allosterism, proteasomal degradation and actin rearrangement in MCF-7 breast cancer cells. J Biol Chem 286:18633–18640

Lim KG, Sun C, Bittman R, Pyne NJ, Pyne S (2011b) (*R*)-FTY720 methyl ether is a specific sphingosine kinase 2 inhibitor: effect on sphingosine kinase 2 expression in HEK 293 cells and actin rearrangement and survival of MCF-7 breast cancer cells. Cell Signal 23:1590–1595

Liu H, Toman RE, Goparaju SK, Maceyka M, Nava VE, Sankala H, Payne SG, Bektas M, Ishii I, Chun J, Milstien S, Spiegel S (2003) Sphingosine kinase type 2 is a putative BH3-only protein that induces apoptosis. J Biol Chem 278:40330–40336

Liu Q, Zhao X, Frissora F, Ma Y, Santhanam R, Jarjoura D, Lehman A, Perrotti D, Chen CS, Dalton JT, Muthusamy N, Byrd JC (2008) FTY720 demonstrates promising pre-clinical activity for chronic lymphocytic leukemia and lymphoblastic leukemia/lymphoma. Blood 111:275–284

Long JS, Edwards J, Watson C, Tovey S, Mair K, Schiff R, Natarajan V, Pyne NJ, Pyne S (2010a) Sphingosine kinase 1 induces tolerance to human epidermal growth factor receptor 2 and prevents formation of a migratory phenotype in response to sphingosine 1-phosphate in estrogen receptor positive breast cancer cells. Mol Cell Biol 30:3827–3841

Long JS, Fujiwara Y, Edwards J, Tannahill C, Tigyi G, Pyne S, Pyne NJ (2010b) Sphingosine 1-phosphate 4 uses HER2 (ErbB2) to regulate extracellular signal regulated kinase-1/2 in MDA-MB-453 breast cancer cells. J Biol Chem 285:35957–35966

Loveridge C, Tonelli F, Leclecq T, Lim KG, Long S, Berdyshev E, Tate RJ, Natarajan V, Pitson SM, Pyne NJ, Pyne S (2010) The sphingosine kinase 1 inhibitor 2-(p-hydroxyanilino)-4-(p-chlorophenyl)thiazole induces proteasomal degradation of sphingosine kinase 1 in mammalian cells. J Biol Chem 285:38841–38852

Lozano G (2010) Mouse models of p53 functions. Cold Spring Harb Perspect Biol 2(4):a001115

Marfe G, Di Stefano C, Gambacurta A, Ottone T, Martini V, Abruzzese E, Mologni L, Sinibaldi-Salimei P, de Fabritis P, Gambacorti-Passerini C, Amadori S, Birge RB (2011) Sphingosine kinase 1 overexpression is regulated by signalling through PI3K, AKT2, and mTOR in imatinib-resistant chronic myeloid leukemia cells. Exp Hematol 39:653–665.e6

Neviani P, Santhanam R, Trotta R, Notari M, Blaser BW, Liu S, Mao H, Chang JS, Galietta A, Uttam A, Roy DC, Valtieri M, Bruner-Klisovic R, Caligiuri MA, Bloomfield CD, Marcucci G, Perrotti D (2005) The tumor suppressor PP2A is functionally inactivated in blast crisis CML through the inhibitory activity of the BCR/ABL-regulated SET protein. Cancer Cell 8:355–368

Neviani P, Santhanam R, Oaks JJ, Eiring AM, Notari M, Blaser BW, Liu S, Trotta R, Muthusamy N, Gambacorti-Passerini C, Druker BJ, Cortes J, Marcucci G, Chen C-S, Verrills NM, Roy DC, Caligiuri MA, Bloomfield CD, Byrd JC, Perrotti D (2007) FTY720, a new alternative for treating blast crisis chronic myelogenous leukemia and Philadelphia chromosome–positive acute lymphocytic leukemia. J Clin Invest 117:2408–2421

Omar HA, Chou CC, Berman-Booty LD, Ma Y, Hung JH, Wang D, Kogure T, Patel T, Terracciano L, Muthusamy N, Byrd JC, Kulp SK, Chen CS (2011) Antitumor effects of OSU-2S, a nonimmunosuppressive analogue of FTY720, in hepatocellular carcinoma. Hepatology 53:1943–1958

Pchejetski D, Bohler T, Brizuela L, Sauer L, Doumerc N, Golzio M, Salunkhe V, Teissié J, Malavaud B, Waxman J, Cuvillier O (2010) FTY720 (fingolimod) sensitizes prostate cancer cells to radiotherapy by inhibition of sphingosine kinase-1. Cancer Res 70:8651–8661

Pitman MR, Barr RK, Gliddon BL, Magarey AM, Moretti PA, Pitson SM (2011) A critical role for the protein phosphatase 2A B'alpha regulatory subunit in dephosphorylation of sphingosine kinase 1. Int J Biochem Cell Biol 43:342–347

Pitson SM, Moretti PA, Zebol JR, Lynn HE, Xia P, Vadas MA, Wattenberg BW (2003) Activation of sphingosine kinase 1 by ERK1/2-mediated phosphorylation. EMBO J 22:5491–5500

Pyne S, Pyne NJ (2000) Sphingosine 1-phosphate signalling in mammalian cells. Biochem J 349:385–402

Pyne NJ, Pyne S (2010) Sphingosine 1-phosphate and cancer. Nat Rev Cancer 10:489–503

Pyne S, Lee SC, Long J, Pyne NJ (2009) Role of sphingosine kinases and lipid phosphate phosphatases in regulating spatial sphingosine 1-phosphate signalling in health and disease. Cell Signal 21:14–21

Salas A, Ponnusamy S, Senkal CE, Meyers-Needham M, Selvam SP, Saddoughi SA, Apohan E, Sentelle RD, Smith C, Gault CR, Obeid LM, El-Shewy HM, Oaks J, Santhanam R, Marcucci G, Baran Y, Mahajan S, Fernandes D, Stuart R, Perrotti D, Ogretmen B (2011) Sphingosine kinase-1 and sphingosine 1-phosphate receptor 2 mediate Bcr/Abl1 stability and drug resistance by modulation of protein phosphatase 2A. Blood 117:5941–5952

Sankala HM, Hait NC, Paugh SW, Shida D, Lépine S, Elmore LW, Dent P, Milstien S, Spiegel S (2007) Involvement of sphingosine kinase 2 in p53-independent induction of p21 by the chemotherapeutic drug doxorubicin. Cancer Res 67:10466–10474

Shida D, Kitayama J, Yamaguchi H, Yamashita H, Mori K, Watanabe T, Nagawa H (2005) Lysophospholipids transactivate HER2/neu (erbB-2) in human gastric cancer cells. Biochem Biophys Res Commun 327:907–914

Shirai K, Kaneshiro T, Wada M, Furuya H, Bielawski J, Hannun YA, Obeid LM, Ogretmen B, Kawamori T (2011) A role of sphingosine kinase 1 in head and neck carcinogenesis. Cancer Prev Res 4:454–462

Sobue S, Nemoto S, Murakami M, Ito H, Kimura A, Gao S, Furuhata A, Takagi A, Kojima T, Nakamura M, Ito Y, Suzuki M, Banno Y, Nozawa Y, Murati T (2008) Implications of sphingosine kinase 1 expression level for the cellular sphingolipid rheostat: relevance as a marker for daunorubicin sensitivity of leukaemia cells. Int J Haematol 87:266–275

Sukocheva O, Wadham C, Holmes A, Albanese N, Verrier E, Feng F, Bernal A, Derian CK, Ullrich A, Vadas MA, Xia P (2006) Estrogen transactivates EGFR via the sphingosine 1-phosphate receptor Edg-3: the role of sphingosine kinase-1. J Cell Biol 173:301–310

Taha TA, Osta W, Kozhaya L, Bielawski J, Johnson KR, Gillanders WE, Dbaibo GS, Hannun YA, Obeid LM (2004) Downregulation of sphingosine kinase-1 by DNA damage: dependence on proteases and p53. J Biol Chem 279:20546–20554

Tonelli F, Lim KG, Loveridge C, Long J, Pitson SM, Tigyi G, Bittman R, Pyne S, Pyne NJ (2010) FTY720 and (S)-FTY720 vinylphosphonate inhibit sphingosine kinase 1 and promote its proteasomal degradation in human pulmonary artery smooth muscle, breast cancer and androgen-independent prostate cancer cells. Cell Signal 22:1536–1542

Valentine WJ, Kiss GN, Liu J, Gotoh M, Murakami-Murofushi K, Pham TC, Baker DL, Parrill AL, Lu X, Sun C, Bittman R, Pyne NJ, Tigyi G (2010) (*S*)-FTY720-vinylphosphonate, an analogue of the immunosuppressive agent FTY720, signals as a pan-antagonist of sphingosine 1-phosphate GPCR and inhibits autotaxin activity. Cell Signal 22:1543–1553

Van Brocklyn JR, Jackson CA, Pearl DK, Kotur MS, Snyder PJ, Prior TW (2005) Sphingosine kinase-1 expression correlates with poor survival of patients with glioblastoma multiforme: role of sphingosine kinase isoforms in growth of glioblastoma cell lines. J Neuropathol Exp Neurol 64:695–705

Wallington-Beddoe CT, Hewson J, Bradstock KF, Bendall LJ (2011) FTY720 produces caspase-independent cell death of acute lymphoblastic leukemia cells. Autophagy 7:707–715

Watson C, Long JS, Orange C, Tannahill CL, Mallon E, McGlynn LM, Pyne S, Pyne NJ, Edwards J (2010) High expression of sphingosine 1-phosphate receptors, $S1P_1$ and $S1P_3$, sphingosine kinase 1, and extracellular signal-regulated kinase-1/2 is associated with development of tamoxifen resistance in estrogen receptor-positive breast cancer patients. Am J Pathol 177:2205–2215

Xia P, Gamble JR, Wang L, Pitson SM, Moretti PA, Wattenberg BW, D'Andrea RJ, Vadas MA (2000) An oncogenic role of sphingosine kinase. Curr Biol 10:1527–1530

Yamashita H, Kitayama J, Shida D, Yamaguchi H, Mori K, Osada M, Aoki S, Yatomi Y, Takuwa Y, Nagawa H (2006) Sphingosine 1-phosphate receptor expression profile in human gastric cancer cells: differential regulation on the migration and proliferation. J Surg Res 130:80–87

Yasui H, Hideshima T, Raje N, Roccaro AM, Shiraishi N, Kumar S, Hamasaki M, Ishitsuka K, Tai YT, Podar K, Catley L, Mitsiades CS, Richardson PG, Albert R, Brinkmann V, Chauhan D, Anderson KC (2005) FTY720 induces apoptosis in multiple myeloma cells and overcomes drug resistance. Cancer Res 65:7478–7484

Zhang N, Qi Y, Wadham C, Wang L, Warren A, Di W, Xia P (2010) FTY720 induces necrotic cell death and autophagy in ovarian cancer cells: a protective role of autophagy. Autophagy 6:1157–1167

Zheng T, Meng X, Wang JB, Chen X, Yin D, Liang Y, Song X, Pan S, Jiang H, Liu L (2010) PTEN- and p53-mediated apoptosis and cell cycle arrest by FTY720 in gastric cancer cells and nude mice. J Cell Biochem 111:218–228

Sphingolipids and Response to Chemotherapy

Marie-Thérèse Dimanche-Boitrel and Amélie Rebillard

Contents

Abstract Chemotherapy is frequently used to treat primary or metastatic cancers, but intrinsic or acquired drug resistance limits its efficiency. Sphingolipids are important regulators of various cellular processes including proliferation, apoptosis, differentiation, angiogenesis, stress, and inflammatory responses which are linked to various aspects of cancer, like tumor growth, neoangiogenesis, and response to chemotherapy. Ceramide, the central molecule of sphingolipid metabolism, generally mediates antiproliferative and proapoptotic functions, whereas sphingosine-1-phosphate and other derivatives have opposing effects. Among the variety of enzymes that control ceramide generation, acid or neutral

M.-T. Dimanche-Boitrel (✉)
Université de Rennes 1, Institut de Recherche Santé Environnement et Travail (IRSET), 35043 Rennes, France

Institut National de la Santé et de la Recherche Médicale (INSERM), U1085, Team "Stress, Membrane and Signaling", 35043 Rennes, France
e-mail: marie-therese.boitrel@univ-rennes1.fr

A. Rebillard
EA 1274, Laboratoire "Mouvement, Sport, Santé", University of Rennes 2, 35044 Rennes, France

E. Gulbins and I. Petrache (eds.), *Sphingolipids in Disease*,
Handbook of Experimental Pharmacology 216,
DOI 10.1007/978-3-7091-1511-4_4,

sphingomyelinases and ceramide synthases are important targets to allow killing of cancer cells by chemotherapeutic drugs. On the contrary, glucosylceramide synthase, ceramidase, and sphingosine kinase are other targets driving cancer cell resistance to chemotherapy. This chapter focuses on ceramide-based mechanisms leading to cancer therapy sensitization or resistance which could have some impacts on the development of novel cancer therapeutic strategies.

Keywords chemotherapy • ceramide • sphingomyelinases • ceramide kinase • sphingosine kinase

1 Introduction

Sphingolipids are ubiquitous components of eukaryotic cell membranes known to be involved in a variety of cellular processes including proliferation, growth, differentiation, apoptosis, and membrane structure. Dysregulation of their metabolism is evident in various pathological conditions including metabolic disorders and cancer. Thus, sphingolipids represent interesting targets for the development of novel chemotherapy. Studies performed over the past decade demonstrate that chemotherapy induces the death of cancer cells by apoptosis via two major death pathways. The first one is called the extrinsic death receptor pathway that begins with ligation of cell surface death receptors like CD95 by CD95L (Suda et al. 1993) or via CD95 clustering at the cell surface independently of its ligand (Micheau et al. 1999; Shao et al. 2001). Activation of CD95 death receptor leads to the recruitment of the protein Fas-associated death domain (FADD) and procaspase-8 to form a death-inducing signaling complex (DISC). Within the DISC, the procaspase-8 is transactivated to release mature caspase-8 initiating directly the apoptotic cascade (Muzio et al. 1996) or indirectly via the cleavage of Bid in t-Bid (truncated Bid) that consequently activates the mitochondrial death pathway (Li et al. 1998; Luo et al. 1998). The second one is named the intrinsic mitochondrial death pathway regulated by members of the Bcl-2 family (Green 2000). Most chemotherapeutic drugs induce disruption of the outer mitochondrial membrane and the release of proapoptotic molecules such as cytochrome c, Smac/DIABLO, HtrA2/Omi, apoptosis-inducing factor (AIF), and endonuclease G (Endo G) from the mitochondrial intermembrane space (Ravagnan et al. 2002). In the cytosol, cytochrome c forms a complex with apoptosis protease activating factor-1 (Apaf-1), which induces via caspase-9 activation of caspase-3 leading to cell death. Since exogenous treatment with ceramide (Obeid et al. 1993) or endogenous production of ceramide following many apoptotic stimuli (Hannun 1994; Lin et al. 2006) induces apoptosis, ceramide is considered as an important mediator in both intrinsic and extrinsic death pathway. Ceramide levels are significantly decreased in human colon (Selzner et al. 2001), glial (Riboni et al. 2002), ovarian (Rylova et al. 1998), and head and neck cancers (Koybasi et al. 2004), and malignant cells with low level of ceramide are resistant to apoptosis (Chmura et al. 1997). Moreover, absence of generation of ceramide is associated with chemoresistance (Wang et al. 1999,

2003). In this context, modulation of ceramide content may favor apoptosis and targeting altered sphingolipid metabolism in cancer could contribute to potentiate chemotherapy and counteract chemoresistance. This chapter describes the current knowledge about sphingolipids and their roles in response to chemotherapy.

2 Sphingolipid Metabolism

Sphingolipids are membrane lipids containing a sphingoid base (sphingosine or sphinganine) which is in most cases acylated with a fatty acid. The resulting ceramides can carry hydrophilic headgroups such as phosphorylcholine in sphingomyelin (SM), carbohydrate residues in glycosphingolipids, and a phosphate moiety in ceramide-1-phosphate. Sphingolipids (like sphingomyelin or glycosphingolipids) with glycerophospholipids and cholesterol are characteristic components of cell membranes. As structural lipids, they are mainly found in the plasma membrane and to a lesser extent in intracellular membranes (Van Meer and Hoetzl 2010). Plasma membrane is characterized by the presence of distinct microdomains enriched in sphingolipids and cholesterol, termed lipid rafts (Brown and London 2000; Simons and Van Meer 1988; Simons and Ikonen 1997). These membrane domains have the properties to be insoluble in nonionic detergents at 4 °C, of weak density on sucrose gradient, and destabilized by cholesterol-depleting agents. Moreover, these structures actively participate to metabolic and signal transduction processes (Verkleij and Post 2000), especially in the CD95 death receptor pathway (Hueber et al. 2002) but also in response to chemotherapy (Bezombes et al. 2003; Dimanche-Boitrel et al. 2005). Then, sphingolipids play a role as mediator lipids and are involved in the regulation of cellular functions. Their biosynthesis and catabolism involve a large number of intermediate metabolites with distinct biological activities (Fig. 1).

Intracellular ceramide can be formed either by de novo synthesis that requires the action of serine palmitoyltransferase (SPT) (Kang et al. 2010) and/or ceramide synthase (Bose et al. 1995) or through the SMase-dependent catabolism of SM, in various separate cellular compartments. In response to chemotherapy, SMase activation is the predominant pathway to generate ceramide (Rizzieri and Hannun 1998; Pettus et al. 2002; Ogretmen and Hannun 2004) besides the activation of de novo synthesis (Bose et al. 1995). SMases are phospholipase C-like enzymes that mediate the hydrolysis of SM to phosphocholine and ceramide. Today, three classes of SMases have been described: the acid, the neutral, and the alkaline form, according to their optimum pH, cation dependency, and subcellular location (Levade and Jaffrézou 1999; Goni and Alonso 2002). Acid sphingomyelinase (ASMase), a soluble glycoprotein, was the first described. It was originally identified in the lysosomes with an optimum pH at 4.5–5, but this isoform could be translocated to the plasma membrane, particularly in lipid rafts after activation by ligand binding to specific receptor (Gulbins and Grassmé 2002) or chemotherapy (Lacour et al. 2004; Dimanche-Boitrel et al. 2005; Carpinteiro et al. 2008),

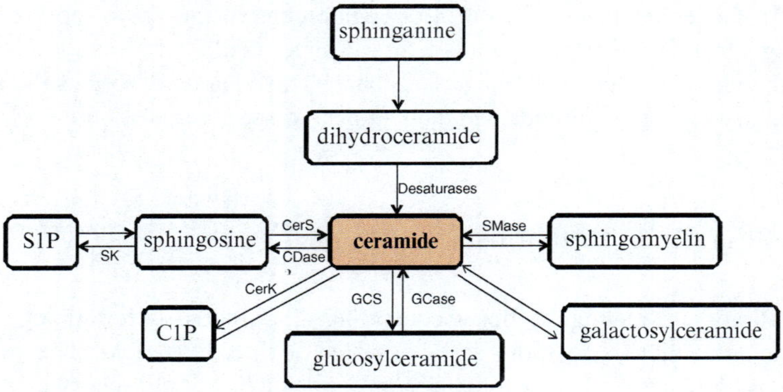

Fig. 1 Overview of sphingolipids metabolism. Ceramide generation can arise from the de novo synthesis pathway and/or the hydrolysis of membrane sphingomyelin by various sphingomyelinases. Ceramide can be transformed in different metabolic intermediates such as ceramide-1-phosphate, glucosylceramide, galactosylceramide, or sphingosine-1-phosphate. *CDase* ceramidase, *CerK* ceramide kinase, *GCS* glucosylceramide synthase, *GCase* glucosylceramidase, *SK* sphingosine kinase, *SMase* sphingomyelinase

promoting subsequently receptor clustering. The secretory sphingomyelinase (SSMase) arises from the ASMase gene through differential protein trafficking of a common precursor localized either in lysosomes or in Golgi (Schissel et al. 1998). This protein is activated by physiological concentrations of Zn^{2+} (Schissel et al. 1996; Spence et al. 1989). Different forms of neutral sphingomyelinases (NSMase) are characterized with a pH optimum at 7.4: a plasma membrane, Mg^{2+}- or Mn^{2+}-dependent form and a cytosolic, Mg^{2+}-independent form as well as nuclear and mitochondrial forms (Goni and Alonso 2002; Birbes et al. 2002; Tomiuk et al. 2000; Wu et al. 2010). Several ceramide synthases (CerS1–6) have been described (Pewzner-Jung et al. 2006). These enzymes locate in microsomes, are integral membrane proteins of the endoplasmic reticulum, and synthesize ceramides with different fatty acid chain lengths (Mizutani et al. 2005). As reported above, many inductors of apoptosis generate ceramide, suggesting a role of this sphingolipid in programmed cell death (Pettus et al. 2002). Additionally, increasing the levels of endogenous ceramide results in apoptosis and growth arrest (Abe et al. 1995; Bielawska et al. 1996). Most of studies suggest that ceramide levels decrease in cancers, notably in ovarian, head and neck, colon, and brain tumors. However, this notion must be seen with caution because it seems that it is the content in specific ceramides which is important rather than their global level. For example, only C_{18}-ceramide and no other ceramide species is significantly lower in tumor tissues of HNSCC patients when compared with controls, which correlates with lymphovascular invasion and nodal metastasis (Koybasi et al. 2004; Karahatay et al. 2007), suggesting that this ceramide inhibits tumor growth. By contrast, recent data suggest that elevated C_{16}-ceramide is associated with a positive lymph node status in breast cancer patients (Schiffmann et al. 2009).

Since a network of specialized and compartmentalized enzymes regulates the levels of ceramide, ceramide metabolites are produced in distinct localization and have different functions. For example, an important metabolite of ceramide is ceramide-1-phosphate (C1P). This metabolite is described as mitogenic and antiapoptotic. It is formed from ceramide by the action of a specific ceramide kinase (CerK), which is distinct from the sphingosine kinases (SK1 and SK2) that synthesize sphingosine-1-phosphate (S1P). CerK is localized in three major compartments: the Golgi complex, the plasma membrane, and cytoplasmic vesicles (Bornancin 2011). C1P blocks apoptosis in bone-marrow-derived macrophages through inhibition of ASMase, thereby reducing ceramide generation (Gómez-Muñoz et al. 2004). CerK mRNA level is upregulated in estrogen receptor (ER)-negative breast cancer tumors in comparison with ER-positive ones (Ruckhäberle et al. 2009). Moreover, CerK is upregulated in several hepatoma cell lines and knockdown of CerK increases susceptibility to UV-induced apoptosis (Hsieh et al. 2009). Specific inhibition of CerK by NVP-231 in combination with tamoxifen increases ceramide levels and reduces cell growth (Graf et al. 2008).

Like many products of sphingomyelin, S1P exhibits a wide range of biological activities. It is a pleiotropic lipid mediator that has been shown to regulate cell growth, cell survival, cell invasion, vascular maturation, and angiogenesis, processes that are important for cancer progression (Olivera and Spiegel 1993; Cuvillier et al. 1996). However, S1P may have opposing roles depending on synthesis via SK1 and SK2 and subcellular localization, since SK1 is localized mainly in the cytosol, whereas SK2 is present in several intracellular compartments (nucleus, mitochondria, and intracellular membranes). In fact, S1P synthesized by SK1 is involved in proliferative signaling (Taha et al. 2006), whereas S1P synthesized by SK2 is described as an antiproliferative and apoptotic mediator (Maceyka et al. 2005).

In summary, ceramide and sphingosine are very often involved in apoptosis, cell-cycle arrest, and cell senescence, whereas sphingosine-1-phosphate and ceramide-1-phosphate promote cell survival, proliferation, and inflammation. The balance between these sphingolipids may affect the fate of the cell.

3 Sphingolipids and Sensitivity to Chemotherapy

Ceramide is composed of sphingosine linked to a fatty acyl chain varying in length from 16 to 26 carbon atoms (Mimeault 2002; Pettus et al. 2002; Ogretmen and Hannun 2004; Zheng et al. 2006). These distinct ceramides have different role in apoptosis, differentiation, and cell growth depending on stimulus and cell context. A number of cytotoxic agents appear to be effective because of their ability to activate ceramide-mediated pathways in cancer cells. Daunorubicin and 1-beta-D arabinofuranosylcytosine (Ara-C) are the first two anticancer agents to induce apoptosis via the generation of ceramide (Bose et al. 1995; Strum et al. 1994; Jaffrézou et al. 1996). Many other chemotherapeutic drugs are shown to produce ceramide: vincristine (Zhang et al. 1996), vinblastine (Cabot et al. 1999), etoposide (Tepper et al. 1999), paclitaxel (Charles et al. 2001), irinotecan (Suzuki et al. 1997), mitoxantrone (Bettaieb et al. 1999), and cisplatin (Lacour et al. 2004). And more recently, it was shown that arsenic trioxide induces accumulation of cytotoxic levels of ceramide in acute promyelocytic leukemia and adult T-cell leukemia/ lymphoma cells (Dbaibo et al. 2007). As reported in many studies, chemotherapy can impact ceramide metabolism by promoting ceramide synthesis de novo, by activating sphingomyelinase, and/or by blocking glucosylceramide formation. In each case, the result is an increase in ceramide-induced cytotoxic response.

4 Ceramide Synthases

The role of ceramide synthases as targets for chemotherapeutic drugs is currently emerging. Ceramide synthases (CerS) differ by their specificity for the generation of endogenous ceramides with distinct fatty acid chain lengths (Spassieva et al. 2006). Particularly, CerS1/4 mainly generates ceramide with a C_{18}-containing fatty acid chain (C_{18}-ceramide) (Venkataraman et al. 2002), CerS2 rather generates very long chain ceramides (C_{24}-ceramide) (Laviad et al. 2008; Mizutani et al. 2005), whereas CerS5/6 preferentially mediates the generation of C_{16}-ceramide and, to a lesser extent, C_{12}- and C_{14}-ceramides (Riebeling et al. 2003). Myeloid leukemia cells treated with daunorubicin exhibit ceramide accumulation via activation of CerS and inhibition of ceramide synthase with fumonisin B1 prevents daunorubicin-induced apoptosis (Bose et al. 1995). Activation of CerS is also observed in response to various cytotoxic agents including lymphotoxin, TNF, camptothecin, doxorubicin, Taxol, oxidative stress, and androgen ablation (Plo et al. 1999; Xu et al. 1998; Rath et al. 2009; Ueda et al. 2001; Eto et al. 2003). Expression of CerS1 sensitizes cancer cells to several chemotherapeutic agents including cisplatin, gemcitabine, doxorubicin, imatinib, and vincristine (Min et al. 2007; Senkal et al. 2007; Baran et al. 2007), and small interfering RNA directed against CerS1 reduces the effects of these drugs. CerS1 mRNA and enzymatic activity is increased in HSNNC cells upon treatment with gemcitabine and doxorubicin leading to C_{18}-ceramide generation and cell death (Senkal et al. 2007). In addition, CerS5 seems to increase the sensitivity of mammalian cells to doxorubicin and vincristine but not to cisplatin and carboplatin (Min et al. 2007). And overexpression of CerS6 in resistant cells resensitized them to TRAIL-induced apoptosis via increased C_{16}-ceramide (White-Gilbertson et al. 2009). Cannabinoids lead to transcriptional induction of CerS3 and CerS6 and generation of C_{16}, C_{18}, C_{24}, and $C_{24:1}$-ceramides inducing cell death in lymphoma cells (Gustafsson et al. 2009), and knockdown of CerS6 expression abolishes activation of CD95 and significantly reduces toxicity of vorinostat combined with sorafenib in hepatoma and pancreatic carcinoma cells (Park et al. 2010). On the other hand, CerS2 and CerS4 levels seem to have no effects on cell response to cytotoxic drugs.

5 Sphingomyelinases

Several studies indicate a role of ASMase in chemotherapy-induced ceramide generation (Pettus et al. 2002; Ogretmen and Hannun 2004), particularly after treatment with gemcitabine, fenretinide, and paclitaxel (Modrak et al. 2004; Lovat et al. 2004). The hydrolysis of SM by ASMase produces ceramide in specific membrane domains termed lipid rafts (Liu and Anderson 1995), resulting in the formation of large ceramide-enriched membrane platforms where membrane receptors are clustered (Grassmé et al. 2001, 2002). Sub-toxic doses of doxorubicin result in ASMase activation, release of ceramide, and formation of ceramide-enriched membrane platforms that facilitate DR5 clustering after treatment with very low doses of TRAIL in BJAB Burkitt lymphoma cell line and murine T splenocytes (Dumitru et al. 2007). In addition, ceramide and ASMase are also important in the induction of apoptosis by other antineoplastic agents such as rituximab (Bezombes et al. 2004) and TRAIL (Dumitru and Gulbins 2006), and overexpression of ASMase sensitizes glioma cells to doxorubicin and gemcitabine (Grammatikos et al. 2007). Similar mechanisms are described after treatment with cisplatin which can induce a redistribution of CD95 death receptor in lipid rafts of human colon cancer cells (Lacour et al. 2004). Few minutes after treatment, the sodium-proton exchanger-1 NHE1 located at the plasma membrane is inhibited, leading to an intracellular acidification which facilitates the activation of ASMase. Then, this enzyme hydrolyzes membrane sphingomyelin into ceramide, allowing the aggregation of lipid rafts into large signaling platforms in which CD95 death receptors are oligomerized, inducing cell death (Rebillard et al. 2007). Concomitantly, an increase in membrane fluidity is measured early after cisplatin treatment by electron paramagnetic resonance which could be related to the ability of ceramide to induce membrane fusion/fission (Cremesti et al. 2002; Dimanche-Boitrel et al. 2005; Rebillard et al. 2008a). A pretreatment with imipramine, an inhibitor of ASMase, or reduced expression of ASMase by RNA interference strongly reduces cisplatin-induced apoptosis in human colon cancer cells, suggesting a main role of ceramide in cell death. In the same way, cisplatin early activates ASMase in breast cancer cells, leading to ceramide production (Zeidan et al. 2008). These data confirm the important involvement of ceramide pathway in apoptosis induction (Gulbins and Grassmé 2002; Gulbins and Kolesnick 2003; Gulbins and Li 2006). Moreover, reactive oxygen species (ROS) seem to be involved in ASMase activation by fenretinide, doxorubicin, and TRAIL (Lovat et al. 2004; Dumitru and Gulbins 2006; Grammatikos et al. 2007), as demonstrated by the use of ROS scavengers. Consistent with this, a mechanism implicating ROS in ASMase activation is characterized in vitro with the oxidation of the cysteine residue 629 in purified ASMase, leading to activation and dimerization of the enzyme (Qiu et al. 2003). However, the role of ASMase in response to chemotherapy has been mainly studied on in vitro cellular models, and the precise mechanisms of its action in vivo remain not well defined. A recent work points out that in a Niemann–Pick disease (NPD; type B) patient who developed a marginal zone lymphoma, rituximab is still acting, suggesting that ASMase is dispensable for rituximab efficacy (Sabourdy et al. 2011). Further studies are needed to confirm the role of ASMase in response to chemotherapy in cancer patients. The role of ASMase is not limited to tumors since this enzyme is also involved in deleterious effects induced by anticancer drugs in normal tissues. Doxorubicin induces ASMase-dependent oocyte lethality, which is responsible for sterility (Morita et al. 2000). In the same way, as irradiation (Paris et al. 2001), cisplatin induces apoptosis of endothelial cells in small intestine leading to the gastrointestinal (GI) syndrome which is not observed in ASMase-knockout mice (Rebillard et al. 2008b). Moreover, cisplatin triggers dendritic cells (DC) apoptosis through increased expression and activation of ASMase, limiting the use of chemoimmunotherapy in cancer treatment. However, the ex vivo nitric oxide (NO) donors treatment protects DC from cisplatin toxicity and enhances tumor regression in B16 mouse melanoma model and animal survival following cisplatin treatment (Perrotta et al. 2007). As previously described, daunorubicin triggers the release of ceramide in leukemic cells through the activation of NSMases (Mansat et al. 1997; Mansat-de Mas et al. 1999) and this activation is mediated by both serine proteases, protein kinase C, and ROS, resulting in the consecutive activation of Jun-N-terminal kinases (JNK). Another chemotherapeutic agent, 1-beta-D-arabinofuranosyl-cytosine (Ara-C), is shown to rapidly enhance NSMases activation in leukemia cells (Whitman et al. 1997; Strum et al. 1994) by altering the cellular redox status. Ceramide generation activates the Src-like tyrosine kinase Lyn and JNK to mediate apoptosis (Bezombes et al. 2001; Grazide et al. 2002). Consistent with the proposed proapoptotic role of NSMases and NSMase-generated ceramide, NSMase3 expression is induced upon Adriamycin treatment and its overexpression sensitizes cells to Adriamycin (Corcoran et al. 2008). Further, daunorubicin transcriptionally regulates *neutral sphingomyelinase 2* in human breast cancer cell lines, leading to an increased ceramide production and cell death (Ito et al. 2009).

6 Sphingolipid and Chemoresistance

As well known, the main obstacle against cancer therapy is the development of drug resistance resulting in chemotherapy failure. A possible mechanism to overcome this drug resistance is modulation of the sphingolipid metabolism.

Among the sphingolipids, S1P seems to be a key regulator of chemoresistance. As previously described, S1P is generated by the conversion of ceramide to sphingosine by ceramidase and the subsequent rapid phosphorylation of sphingosine to S1P, which is catalyzed by sphingosine kinase.

7 Sphingosine Kinases

High expression of sphingosine kinase-1 SK1 and S1P is observed in many types of cancers such as gastric, lung, colon, breast, uterus, and kidney (Kawamori et al. 2009; Visentin et al. 2006). SK1 activity is increased 2.5-fold in endometrial tumors compared with healthy sections, and S1P levels are 1.6-fold higher in cancer tissues. In breast tumors samples, elevated SK1 expression is correlated with poor prognosis and promotion of metastasis (Ruckhäberle et al. 2008). Moreover, SK1 increases RAS V12-dependent transformation of NIH3T3 fibroblasts to form fibrosarcoma cells, demonstrating for the first time the role of SK1 in cancer transformation (Xia et al. 2000). Targeting SK1 induces apoptosis and suppresses growth of human glioblastoma cells and xenografts (Kapitonov et al. 2009). S1P and SK1 are involved in resistance to apoptosis induced by CD95, ceramide, (Bektas et al. 2005) and a myriad of other stimuli such as camptothecin, gemcitabine, imatinib, and Taxol. Cancer cell lines that are resistant to chemotherapeutic agents have a high expression of SK1 and SP1, such as prostate cancer cells that are resistant to camptothecin (Akao et al. 2006; Pchejetski et al. 2005), pancreatic cancer cells resistant to gemcitabine (Guillermet-Guibert et al. 2009), and chronic myeloid leukemia (CML) cells that are resistant to imatinib (Baran et al. 2007). Imatinib-sensitive CML cells and daunorubicin-sensitive leukemia cells have a higher ceramide/S1P ratio than their chemotherapeutic resistant counterparts (Baran et al. 2007; Sobue et al. 2008). The involvement of SK1 in drug resistance is also established in breast cancer cells. SK1 overexpression causes promotion of cell proliferation and resistance to tamoxifen-induced apoptosis, and the inhibition of SK1 by a specific inhibitor induces the resensitization of breast cancer cells to tamoxifen-induced apoptosis (Sukocheva et al. 2009). In contrast, the role of SK2 is much less known. Whereas, endogenous SK2 also promotes survival by direct involvement of S1P (Hait et al. 2009), its enforced overexpression suppresses cell growth and enhances apoptosis and sensitivity to doxorubicin (Liu et al. 2003; Sankala et al. 2007). However, colon cancer cells with high SK1 and SK2 expression were resistant to oxaliplatin (L-OHP), and inhibition of both SK isoenzymes renders the colon cancer cells sensitive to L-OHP (Nemoto et al. 2009).

8 Ceramidases

Due to their ability to break down ceramide to regulate sphingosine and S1P levels, acid, neutral, and alkaline ceramidases (Canals et al. 2011) are important regulators of cell survival (Mao and Obeid 2008). Human acid ceramidase is overexpressed in prostate cancer (Seelan et al. 2000). Overexpression of acid ceramidase in prostate cancer cell line DU145 or in fibrosarcoma cell line L929 elevates resistance to chemotherapy (Saad et al. 2007) or TNF-α (Strelow et al. 2000), respectively. On the contrary, downregulation of acid ceramidase sensitizes A375 melanoma cells to dacarbazine (Bedia et al. 2011). Moreover, addition of acid ceramidase inhibitors,

B13 or N-oleoyl ethanolamine, induces apoptosis in prostate and colon cancer cell lines and xenografts (Samsel et al. 2004; Holman et al. 2008; Selzner et al. 2001) or overcomes TNF-α resistance (Strelow et al. 2000), respectively. In the same way, overexpression of neutral ceramidase confers resistance of primary hepatocytes to TNF-α and protection against TNF-α-induced liver damage (Osawa et al. 2005).

9 Glucosylceramide Synthase and Gangliosides

Other dysfunctions in ceramide metabolism also contribute to multidrug resistance. Specifically, ceramide glycosylation by the glucosylceramide synthase (GCS), which forms the metabolite glucosylceramide, may be an important pathway for bypassing apoptosis. Tumors from patients who fail to respond to chemotherapy express elevated glucosylceramide levels (Lucci et al. 1998) as well as a human ovarian adenocarcinoma cell line established from a patient resistant to doxorubicin, melphalan, and cisplatin. A number of drug-resistant cancer cell lines accumulate this noncytotoxic metabolite (Lavie et al. 1996). Drug-resistant breast cancer cells and cutaneous cancer cells have higher levels of glucosylceramide than their drug-sensitive counterparts. The level of GCS activity may determine the multidrug resistance phenotype in cancer cells. The introduction of GCS gene into drug-sensitive breast cancer cell lines results in an 11-fold higher level of GCS activity, leading to resistance to doxorubicin, exogenous ceramide (Liu et al. 1999a), and TNF-α-induced cell death (Liu et al. 1999b). This process seems to be related to hyperglycosylation of ceramide and not to changes in the levels of P-glycoprotein, Bcl-2, or TNF receptor-1 expression. However, recent data demonstrated that knockdown of GCS expression significantly inhibits the expression of MDR1, a gene encoding for P-glycoprotein (P-gp), and reverses drug resistance (Gouazé et al. 2005; Gouazé-Andersson et al. 2007). On the contrary, overexpression of GCS increases P-gp expression and resistance acquisition in breast cancer cells (Gouazé et al. 2004; Liu et al. 2010). Interestingly, P-gp is proposed as a specific transporter for glucosylceramide, translocating this molecule across the Golgi to deliver it for the synthesis of glycosphingolipids (De Rosa et al. 2004). Thus, P-gp and GCS appear to function in the same pathway of ceramide/GlcCer metabolism, and this may provide an important link for the function of GCS in drug resistance. These data are consistent with an earlier study demonstrating that the inhibition of P-gp prevents GCS activity and alters glucosylceramide levels (Goulding et al. 2000). Additionally, P-gp overexpressing cells have an increased accumulation of glucosylceramide (Gouazé et al. 2004; Morjani et al. 2001) which is the precursor for the generation of complex glycosphingolipids and gangliosides (Futerman and Hannun 2004). Most gangliosides are known to protect cells from apoptosis (Bektas and Spiegel 2004). For example, GM1 could prevent cell death in growth factor-deprived neuronal cells (Ferrari et al. 1995), could enhance S1P production in rat heart fibroblasts through the activation of sphingosine kinase, and could protect cells from C2-ceramide or staurosporine-induced cell death (Cavallini et al. 1999).

Moreover, gangliosides GM2 and GM3 have been associated with multidrug resistance phenotype in cancer cells (Gouazé-Andersson and Cabot 2006).

10 Conclusion and Future Directions

Experimental evidence suggests that there is an alteration of ceramide contents and of the expression of enzymes involved in sphingolipid metabolism in several cancers that contributes to cancer therapy resistance. In fact, the roles of sphingolipids in the regulation of response to chemotherapy are demonstrated in various cellular models but need to be further studied in cancer patients. As described, chemotherapy induces ceramide generation via activation of several enzymes in different subcellular localizations (plasma membrane, reticulum endoplasmic, nuclear membrane, and mitochondria-associated membranes). Moreover, different ceramide species could be generated with distinct biological properties showing the complexity of cell response to chemotherapy. Therefore, ceramide analogues, modulators of sphingolipids metabolism, inhibitors of SK or CerK, might be exploited for the development of new therapeutic cancer strategies via the increase in ceramide levels. Such therapeutic strategies based on the modulation of ceramide level in tumors have already been the subjects of novel patents (Dimanche-Boitrel et al. 2011). However, further studies are needed to better understand the role of sphingolipid metabolism in drug resistance in patients and to elaborate new therapeutic strategies by comparing their toxicity in malignant and normal tissues using conventional methods of biochemistry and molecular biology and also more complex approaches such as lipidomics and bioinformatics.

Acknowledgements Research in the IRSET INSERM U1085 group was supported by grants from the Ligue Nationale Contre le Cancer (the Côte d'Armor, Ille et Vilaine, Morbihan, Vendée, and Sarthe Comittees), INSERM, University of Rennes 1, and the Region Bretagne.

References

Abe A, Radin NS, Shayman JA, Wotring LL, Zipkin RE, Sivakumar R, Ruggieri JM, Carson KG, Ganem B (1995) Structural and stereochemical studies of potent inhibitors of glucosylceramide synthase and tumor cell growth. J Lipid Res 36:611–621

Akao Y, Banno Y, Nakagawa Y, Hasegawa N, Kim TJ, Murate T, Igarashi Y, Nozawa Y (2006) High expression of sphingosine kinase 1 and S1P receptors in chemotherapy-resistant prostate cancer PC-3 cells and their camptothecin-induced up-regulation. Biochem Biophys Res Commun 342:1284–1290

Baran Y, Salas A, Senkal CE, Gunduz U, Bielawski J, Obeid LM, Ogretmen B (2007) Alterations of ceramide/sphingosine 1-phosphate rheostat involved in the regulation of resistance to imatinib-induced apoptosis in K562 human chronic myeloid leukemia cells. J Biol Chem 282:10922–10934

Bedia C, Casas J, Andrieu-Abadie N, Fabriàs G, Levade T (2011) Acid ceramidase expression modulates the sensitivity of A375 melanoma cells to dacarbazine. J Biol Chem 286:28200–28209

Bektas M, Jolly PS, Muller C, Eberle J, Spiegel S, Geilen CC (2005) Sphingosine kinase activity counteracts ceramide-mediated cell death in human melanoma cells: role of Bcl-2 expression. Oncogene 24:178–187

Bektas M, Spiegel S (2004) Glycosphingolipids and cell death. Glycoconj J 20:39–47

Bettaieb A, Plo I, Mansat V, Quillet-Mary A, Levade T, Laurent G, Jaffrézou JP (1999) Daunorubicin and mitoxantrone-triggered phosphatidylcholine hydrolysis: implication in drug-induced ceramide generation and apoptosis. Mol Pharmacol 55:118–125

Bezombes C, Grazide S, Garret C, Fabre C, Quillet-Mary A, Müller S, Jaffrézou JP, Laurent G (2004) Rituximab antiproliferative effect in B-lymphoma cells is associated with acid-sphingomyelinase activation in raft microdomains. Blood 104:1166–1173

Bezombes C, Laurent G, Jaffrézou JP (2003) Implication of raft microdomains in drug induced apoptosis. Curr Med Chem Anticancer Agents 3:263–270

Bezombes C, Plo I, Mansat-De Mas V, Quillet-Mary A, Nègre-Salvayre A, Laurent G, Jaffrézou JP (2001) Oxidative stress-induced activation of Lyn recruits sphingomyelinase and is requisite for its stimulation by Ara-C. FASEB J 15:1583–1585

Bielawska A, Greenberg MS, Perry D, Jayadev S, Shayman JA, McKay C, Hannun YA (1996) (1S,2R)-D-erythro-2-(N-myristoylamino)-1-phenyl-1-propanol as an inhibitor of ceramidase. J Biol Chem 271:12646–12654

Birbes H, Bawab SE, Obeid LM, Hannun YA (2002) Mitochondria and ceramide: intertwined roles in regulation of apoptosis. Adv Enzyme Regul 42:113–129

Bornancin F (2011) Ceramide kinase: the first decade. Cell Signal 23:999–1008

Bose R, Verheij M, Haimovitz-Friedman A, Scotto K, Fuks Z, Kolesnick R (1995) Ceramide synthase mediates daunorubicin-induced apoptosis: an alternative mechanism for generating death signals. Cell 82:405–414

Brown DA, London E (2000) Structure and function of sphingolipid- and cholesterol-rich membrane rafts. J Biol Chem 275:17221–17224

Cabot MC, Giuliano AE, Han TY, Liu YY (1999) SDZ PSC 833, the cyclosporine A analogue and multidrug resistance modulator, activates ceramide synthesis and increases vinblastine sensitivity in drug-sensitive and drug-resistant cancer cells. Cancer Res 59:880–885

Canals D, Perry DM, Jenkins RW, Hannun YA (2011) Drug targeting of sphingolipid metabolism: sphingomyelinases and ceramidases. Br J Pharmacol 163:694–712

Carpinteiro A, Dumitru C, Schenck M, Gulbins E (2008) Ceramide-induced cell death in malignant cells. Cancer Lett 264:1–10

Cavallini L, Venerando R, Miotto G, Alexandre A (1999) Ganglioside GM1 protection from apoptosis of rat heart fibroblasts. Arch Biochem Biophys 370:156–162

Charles AG, Han TY, Liu YY, Hansen N, Giuliano AE, Cabot MC (2001) Taxol-induced ceramide generation and apoptosis in human breast cancer cells. Cancer Chemother Pharmacol 47:444–450

Chmura SJ, Nodzenski E, Beckett MA, Kufe DW, Quintans J, Weichselbaum RR (1997) Loss of ceramide production confers resistance to radiation-induced apoptosis. Cancer Res 57:1270–1275

Corcoran CA, He Q, Ponnusamy S, Ogretmen B, Huang Y, Sheikh MS (2008) Neutral sphingomyelinase-3 is a DNA damage and nongenotoxic stress-regulated gene that is deregulated in human malignancies. Mol Cancer Res 6:795–807

Cremesti AE, Goni FM, Kolesnick R (2002) Role of sphingomyelinase and ceramide in modulating rafts: do biophysical properties determine biologic outcome? FEBS Lett 531:47–53

Cuvillier O, Pirianov G, Kleuser B, Vanek PG, Coso OA, Gutkind S, Spiegel S (1996) Suppression of ceramide-mediated programmed cell death by sphingosine-1-phosphate. Nature 381:800–803

Dbaibo GS, Kfoury Y, Darwiche N, Panjarian S, Kozhaya L, Nasr R, Abdallah M, Hermine O, El-Sabban M, de Thé H, Bazarbachi A (2007) Arsenic trioxide induces accumulation of cytotoxic levels of ceramide in acute promyelocytic leukemia and adult T-cell leukemia/lymphoma cells through de novo ceramide synthesis and inhibition of glucosylceramide synthase activity. Haematologica 92:753–762

De Rosa MF, Sillence D, Ackerley C, Lingwood C (2004) Role of multiple drug resistance protein 1 in neutral but not acidic glycosphingolipid biosynthesis. J Biol Chem 279:7867–7876

Dimanche-Boitrel MT, Meurette O, Rebillard A, Lacour S (2005) Role of early plasma membrane events in chemotherapy-induced cell death. Drug Resist Updat 8:5–14

Dimanche-Boitrel MT, Rebillard A, Gulbins E (2011) Ceramide in chemotherapy of tumors. Recent Pat Anticancer Drug Discov 6:284–293

Dumitru CA, Carpinteiro A, Trarbach T, Hengge UR, Gulbins E (2007) Doxorubicin enhances TRAIL-induced cell death via ceramide-enriched membrane platforms. Apoptosis 12:1533–1541

Dumitru CA, Gulbins E (2006) TRAIL activates acid sphingomyelinase via a redox mechanism and releases ceramide to trigger apoptosis. Oncogene 25:5612–5625

Eto M, Bennouna J, Hunter OC, Hershberger PA, Kanto T, Johnson CS, Lotze MT, Amoscato AA (2003) C16 ceramide accumulates following androgen ablation in LNCaP prostate cancer cells. Prostate 57:66–79

Ferrari G, Anderson BL, Stephens RM, Kaplan DR, Greene LA (1995) Prevention of apoptotic neuronal death by GM1 ganglioside. Involvement of Trk neurotrophin receptors. J Biol Chem 270:3074–3080

Futerman AH, Hannun YA (2004) The complex life of simple sphingolipids. EMBO Rep 5:777–782

Gómez-Muñoz A, Kong JY, Salh B, Steinbrecher UP (2004) Ceramide-1-phosphate blocks apoptosis through inhibition of acid sphingomyelinase in macrophages. J Lipid Res 45:99–105

Goni FM, Alonso A (2002) Sphingomyelinases: enzymology and membrane activity. FEBS Lett 531:38–46

Gouazé V, Liu YY, Prickett CS, Yu JY, Giuliano AE, Cabot MC (2005) Glucosylceramide synthase blockade down-regulates P-glycoprotein and resensitizes multidrug-resistant breast cancer cells to anticancer drugs. Cancer Res 65:3861–3867

Gouazé V, Yu JY, Bleicher RJ, Han TY, Liu YY, Wang H, Gottesman MM, Bitterman A, Giuliano AE, Cabot MC (2004) Overexpression of glucosylceramide synthase and P-glycoprotein in cancer cells selected for resistance to natural product chemotherapy. Mol Cancer Ther 3:633–639

Gouazé-Andersson V, Yu JY, Kreitenberg AJ, Bielawska A, Giuliano AE, Cabot MC (2007) Ceramide and glucosylceramide upregulate expression of the multidrug resistance gene MDR1 in cancer cells. Biochim Biophys Acta 1771:1407–1417

Gouazé-Andersson V, Cabot MC (2006) Glycosphingolipids and drug resistance. Biochim Biophys Acta 1758:2096–2103

Goulding CW, Giuliano AE, Cabot MC (2000) SDZ PSC 833 the drug resistance modulator activates cellular ceramide formation by a pathway independent of P-glycoprotein. Cancer Lett 149:143–151

Graf C, Klumpp M, Habig M, Rovina P, Billich A, Baumruker T, Oberhauser B, Bornancin F (2008) Targeting ceramide metabolism with a potent and specific ceramide kinase inhibitor. Mol Pharmacol 74:925–932

Grammatikos G, Teichgräber V, Carpinteiro A, Trarbach T, Weller M, Hengge UR, Gulbins E (2007) Overexpression of acid sphingomyelinase sensitizes glioma cells to chemotherapy. Antioxid Redox Signal 9:1449–1456

Grassmé H, Jekle A, Riehle A, Schwarz H, Berger J, Sandhoff K, Kolesnick R, Gulbins E (2001) CD95 signaling via ceramide-rich membrane rafts. J Biol Chem 276:20589–20596

Grassmé H, Jendrossek V, Bock J, Riehle A, Gulbins E (2002) Ceramide-rich membrane rafts mediate CD40 clustering. J Immunol 168:298–307

Grazide S, Maestre N, Veldman RJ, Bezombes C, Maddens S, Levade T, Laurent G, Jaffrézou JP (2002) Ara-C- and daunorubicin-induced recruitment of Lyn in sphingomyelinase-enriched membrane rafts. FASEB J 16:1685–1687

Green DR (2000) Apoptotic pathways: paper wraps stone blunts scissors. Cell 102:1–4

Guillermet-Guibert J, Davenne L, Pchejetski D, Saint-Laurent N, Brizuela L, Guilbeau-Frugier C, Delisle MB, Cuvillier O, Susini C, Bousquet C (2009) Targeting the sphingolipid metabolism to defeat pancreatic cancer resistance to the chemotherapeutic gemcitabine drug. Mol Cancer Res 8:809–820

Gulbins E, Grassmé H (2002) Ceramide and cell death receptor clustering. Biochim Biophys Acta 1585:139–145

Gulbins E, Kolesnick R (2003) Raft ceramide in molecular medicine. Oncogene 22:7070–7077

Gulbins E, Li PL (2006) Physiological and pathophysiological aspects of ceramide. Am J Physiol Regul Integr Comp Physiol 290:R11–26

Gustafsson K, Sander B, Bielawski J, Hannun YA, Flygare J (2009) Potentiation of cannabinoid-induced cytotoxicity in mantle cell lymphoma through modulation of ceramide metabolism. Mol Cancer Res 7:1086–1098

Hait NC, Allegood J, Maceyka M, Strub GM, Harikumar KB, Singh SK, Luo C, Marmorstein R, Kordula T, Milstien S, Spiegel S (2009) Regulation of histone acetylation in the nucleus by sphingosine-1-phosphate. Science 325:1254–1257

Hannun YA (1994) The sphingomyelin cycle and the second messenger function of ceramide. J Biol Chem 269:3125–3128

Holman DH, Turner LS, El-Zawahry A, Elojeimy S, Liu X, Bielawski J, Szulc ZM, Norris K, Zeidan YH, Hannun YA, Bielawska A, Norris JS (2008) Lysosomotropic acid ceramidase inhibitor induces apoptosis in prostate cancer cells. Cancer Chemother Pharmacol 61:231–242

Hsieh SY, Hsu CY, He JR, Liu CL, Lo SJ, Chen YC, Huang HY (2009) Identifying apoptosis-evasion proteins/pathways in human hepatoma cells via induction of cellular hormesis by UV irradiation. J Proteome Res 8:3977–3986

Hueber AO, Bernard AM, Herincs Z, Couzinet A, He HT (2002) An essential role for membrane rafts in the initiation of Fas/CD95-triggered cell death in mouse thymocytes. EMBO Rep 3:190–196

Ito H, Murakami M, Furuhata A, Gao S, Yoshida K, Sobue S, Hagiwara K, Takagi A, Kojima T, Suzuki M, Banno Y, Tanaka K, Tamiya-Koizumi K, Kyogashima M, Nozawa Y, Murate T (2009) Transcriptional regulation of neutral sphingomyelinase 2 gene expression of a human breast cancer cell line, MCF-7, induced by the anti-cancer drug, daunorubicin. Biochim Biophys Acta 1789:681–690

Jaffrézou JP, Levade T, Bettaïeb A, Andrieu N, Bezombes C, Maestre N, Vermeersch S, Rousse A, Laurent G (1996) Daunorubicin-induced apoptosis: triggering of ceramide generation through sphingomyelin hydrolysis. EMBO J 15:2417–2424

Kang MS, Ahn KH, Kim SK, Jeon HJ, Ji JE, Choi JM, Jung KM, Jung SY, Kim DK (2010) Hypoxia-induced neuronal apoptosis is mediated by de novo synthesis of ceramide through activation of serine palmitoyltransferase. Cell Signal 22:610–618

Kapitonov D, Allegood JC, Mitchell C, Hait NC, Almenara JA, Adams JK, Zipkin RE, Dent P, Kordula T, Milstien S, Spiegel S (2009) Targeting sphingosine kinase 1 inhibits Akt signaling, induces apoptosis, and suppresses growth of human glioblastoma cells and xenografts. Cancer Res 69:6915–6923

Karahatay S, Thomas K, Koybasi S, Senkal CE, Elojeimy S, Liu X, Bielawski J, Day TA, Gillespie MB, Sinha D, Norris JS, Hannun YA, Ogretmen B (2007) Clinical relevance of ceramide metabolism in the pathogenesis of human head and neck squamous cell carcinoma (HNSCC): attenuation of C(18)-ceramide in HNSCC tumors correlates with lymphovascular invasion and nodal metastasis. Cancer Lett 256:101–111

Kawamori T, Kaneshiro T, Okumura M, Maalouf S, Uflacker A, Bielawski J, Hannun YA, Obeid LM (2009) Role for sphingosine kinase 1 in colon carcinogenesis. FASEB J 23:405–414

Koybasi S, Senkal CE, Sundararaj K, Spassieva S, Bielawski J, Osta W, Day TA, Jiang JC, Jazwinski SM, Hannun YA, Obeid LM, Ogretmen B (2004) Defects in cell growth regulation by C18:0-ceramide and longevity assurance gene 1 in human head and neck squamous cell carcinomas. J Biol Chem 279:44311–44319

Lacour S, Hammann A, Grazide S, Lagadic-Gossmann D, Athias A, Sergent O, Laurent G, Gambert P, Solary E, Dimanche-Boitrel MT (2004) Cisplatin-induced CD95 redistribution into membrane lipid rafts of HT29 human colon cancer cells. Cancer Res 64:3593–3598

Laviad EL, Albee L, Pankova-Kholmyansky I, Epstein S, Park H, Merrill AH Jr, Futerman AH (2008) Characterization of ceramide synthase 2: tissue distribution, substrate specificity and inhibition by sphingosine 1-phosphate. J Biol Chem 283:5677–5684

Lavie Y, Cao H, Bursten SL, Giuliano AE, Cabot MC (1996) Accumulation of glucosylceramides in multidrug-resistant cancer cells. J Biol Chem 271:19530–19536

Levade T, Jaffrézou JP (1999) Signaling sphingomyelinases: which, where, how and why? Biochim Biophys Acta 1438:1–17

Li H, Zhu H, Xu CJ, Yuan J (1998) Cleavage of BID by caspase 8 mediates the mitochondrial damage in the Fas pathway of apoptosis. Cell 94:491–501

Lin CF, Chen CL, Lin YS (2006) Ceramide in apoptotic signaling and anticancer therapy. Curr Med Chem 13:1609–1616

Liu P, Anderson R (1995) Compartmentalized production of ceramide at the cell surface. J Biol Chem 270:27179–27185

Liu YY, Han TY, Giuliano AE, Cabot MC (1999a) Expression of glucosylceramide synthase, converting ceramide to glucosylceramide, confers adriamycin resistance in human breast cancer cells. J Biol Chem 274:1140–1146

Liu YY, Han TY, Giuliano AE, Ichikawa S, Hirabayashi Y, Cabot MC (1999b) Glycosylation of ceramide potentiates cellular resistance to tumor necrosis factor-alpha-induced apoptosis. Exp Cell Res 252:464–470

Liu H, Toman RE, Goparaju SK, Maceyka M, Nava VE, Sankala H, Payne SG, Bektas M, Ishii I, Chun J, Milstien S, Spiegel S (2003) Sphingosine kinase type 2 is a putative BH3-only protein that induces apoptosis. J Biol Chem 278:40330–40336

Liu YY, Gupta V, Patwardhan GA, Bhinge K, Zhao Y, Bao J, Mehendale H, Cabot MC, Li YT, Jazwinski SM (2010) Glucosylceramide synthase upregulates MDR1 expression in the regulation of cancer drug resistance through cSrc and beta-catenin signaling. Mol Cancer 9:145

Lovat PE, Di Sano F, Corazzari M, Fazi B, Donnorso RP, Pearson AD, Hall AG, Redfern CP, Piacentini M (2004) Gangliosides link the acidic sphingomyelinase-mediated induction of ceramide to 12-lipoxygenase-dependent apoptosis of neuroblastoma in response to fenretinide. J Natl Cancer Inst 96:1288–1299

Lucci A, Cho WI, Han TY, Giuliano AE, Morton DL, Cabot MC (1998) Glucosylceramide: a marker for multiple-drug resistant cancers. Anticancer Res 18:475–480

Luo X, Budihardjo I, Zou H, Slaughter C, Wang X (1998) Bid, a Bcl2 interacting protein, mediates cytochrome c release from mitochondria in response to activation of cell surface death receptors. Cell 94:481–490

Mao C, Obeid LM (2008) Ceramidases: regulators of cellular responses mediated by ceramide, sphingosine, and sphingosine-1-phosphate. Biochim Biophys Acta 1781:424–434

Maceyka M, Sankala H, Hait NC, Le Stunff H, Liu H, Toman R, Collier C, Zhang M, Satin LS, Merrill AH Jr, Milstien S, Spiegel S (2005) SphK1 and SphK2, sphingosine kinase isoenzymes with opposing functions in sphingolipid metabolism. J Biol Chem 280:37118–37129

Mansat V, Laurent G, Levade T, Bettaïeb A, Jaffrézou JP (1997) The protein kinase C activators phorbol esters and phosphatidylserine inhibit neutral sphingomyelinase activation, ceramide generation, and apoptosis triggered by daunorubicin. Cancer Res 57:5300–5304

Mansat-de Mas V, Bezombes C, Quillet-Mary A, Bettaïeb A, D'orgeix AD, Laurent G, Jaffrézou JP (1999) Implication of radical oxygen species in ceramide generation, c-Jun N-terminal kinase activation and apoptosis induced by daunorubicin. Mol Pharmacol 56:867–874

Micheau O, Solary E, Hammann A, Dimanche-Boitrel MT (1999) Fas ligand-independent, FADD-mediated activation of the Fas death pathway by anticancer drugs. J Biol Chem 274:7987–7992

Mimeault M (2002) New advances on structural and biological functions of ceramide in apoptotic/necrotic cell death and cancer. FEBS Lett 530:9–16

Min J, Mesika A, Sivaguru M, Van Veldhoven PP, Alexander H, Futerman AH, Alexander S (2007) (Dihydro)ceramide synthase 1 regulated sensitivity to cisplatin is associated with the activation of p38 mitogen-activated protein kinase and is abrogated by sphingosine kinase 1. Mol Cancer Res 5:801–812

Mizutani Y, Kihara A, Igarashi Y (2005) Mammalian Lass6 and its related family members regulate synthesis of specific ceramides. Biochem J 390:263–271

Modrak DE, Cardillo TM, Newsome GA, Goldenberg DM, Gold DV (2004) Synergistic interaction between sphingomyelin and gemcitabine potentiates ceramide-mediated apoptosis in pancreatic cancer. Cancer Res 64:8405–8410

Morita Y, Perez GI, Paris F, Miranda SR, Ehleiter D, Haimovitz-Friedman A, Fuks Z, Xie Z, Reed JC, Schuchman EH, Kolesnick RN, Tilly JL (2000) Oocyte apoptosis is suppressed by disruption of the acid sphingomyelinase gene or by sphingosine-1-phosphate therapy. Nat Med 6:1109–1114

Morjani H, Aouali N, Belhoussine R, Veldman RJ, Levade T, Manfait M (2001) Elevation of glucosylceramide in multidrug-resistant cancer cells and accumulation in cytoplasmic droplets. Int J Cancer 94:157–165

Muzio M, Chinnaiyan AM, Kischkel FC, O'Rourke K, Shevchenko A, Ni J, Scaffidi C, Bretz JD, Zhang M, Gentz R, Mann M, Krammer PH, Peter ME, Dixit VM (1996) FLICE, a novel FADD-homologous ICE/CED-3-like protease, is recruited to the CD95 (Fas/APO-1) death-inducing signaling complex. Cell 85:817–827

Nemoto S, Nakamura M, Osawa Y, Kono S, Itoh Y, Okano Y, Murate T, Hara A, Ueda H, Nozawa Y, Banno Y (2009) Sphingosine kinase isoforms regulate oxaliplatin sensitivity of human colon cancer cells through ceramide accumulation and Akt activation. J Biol Chem 284:10422–10432

Obeid LM, Linardic CM, Karolak LA, Hannun YA (1993) Programmed cell death induced by ceramide. Science 259:1769–1771

Ogretmen B, Hannun YA (2004) Biologically active sphingolipids in cancer pathogenesis and treatment. Nat Rev Cancer 4:604–616

Olivera A, Spiegel S (1993) Sphingosine-1-phosphate as second messenger in cell proliferation induced by PDGF and FCS mitogens. Nature 365:557–560

Osawa Y, Uchinami H, Bielawski J, Schwabe RF, Hannun YA, Brenner DA (2005) Roles for C16-ceramide and sphingosine 1-phosphate in regulating hepatocyte apoptosis in response to tumor necrosis factor-alpha. J Biol Chem 280:27879–27887

Paris F, Fuks Z, Kang A, Capodieci P, Juan G, Ehleiter D, Haimovitz-Friedman A, Cordon-Cardo C, Kolesnick R (2001) Endothelial apoptosis as the primary lesion initiating intestinal radiation damage in mice. Science 293:293–297

Park MA, Mitchell C, Zhang G, Yacoub A, Allegood J, Häussinger D, Reinehr R, Larner A, Spiegel S, Fisher PB, Voelkel-Johnson C, Ogretmen B, Grant S, Dent P (2010) Vorinostat and sorafenib increase CD95 activation in gastrointestinal tumor cells through a Ca(2+)-de novo ceramide-PP2A-reactive oxygen species-dependent signaling pathway. Cancer Res 70:6313–6324

Pchejetski D, Golzio M, Bonhoure E, Calvet C, Doumerc N, Garcia V, Mazerolles C, Rischmann P, Teissié J, Malavaud B, Cuvillier O (2005) Sphingosine kinase-1 as a chemotherapy sensor in prostate adenocarcinoma cell and mouse models. Cancer Res 65:11667–11675

Perrotta C, Bizzozero L, Falcone S, Rovere-Querini P, Prinetti A, Schuchman EH, Sonnino S, Manfredi AA, Clementi E (2007) Nitric oxide boosts chemoimmunotherapy via inhibition of acid sphingomyelinase in a mouse model of melanoma. Cancer Res 67:7559–7564

Pettus BJ, Chalfant CE, Hannun YA (2002) Ceramide in apoptosis: an overview and current perspectives. Biochim Biophys Acta 1585:114–125

Pewzner-Jung Y, Ben-Dor S, Futerman AH (2006) When do Lasses (longevity assurance genes) become CerS (ceramide synthases)? Insights into the regulation of ceramide synthesis. J Biol Chem 281:25001–25005

Plo I, Ghandour S, Feutz AC, Clanet M, Laurent G, Bettaieb A (1999) Involvement of de novo ceramide biosynthesis in lymphotoxin-induced oligodendrocyte death. Neuroreport 10:2373–2376

Qiu H, Edmunds T, Baker-Malcolm J, Karey KP, Estes S, Schwarz C, Hughes H, Van Patten SM (2003) Activation of human acid sphingomyelinase through modification or deletion of C-terminal cysteine. J Biol Chem 278:32744–32752

Rath G, Schneider C, Langlois B, Sartelet H, Morjani H, Btaouri HE, Dedieu S, Martiny L (2009) De novo ceramide synthesis is responsible for the anti-tumor properties of camptothecin and doxorubicin in follicular thyroid carcinoma. Int J Biochem Cell Biol 41:1165–1172

Ravagnan L, Roumier T, Kroemer G (2002) Mitochondria, the killer organelles and their weapons. J Cell Physiol 2:131–137

Rebillard A, Tekpli X, Meurette O, Sergent O, Le Moigne-Muller G, Vernhet L, Gorria M, Chevanne M, Christmann M, Kaina B, Counillon L, Gulbins E, Lagadic-Gossmann D, Dimanche-Boitrel MT (2007) Cisplatin-induced apoptosis involved membrane fluidification via inhibition of NHE1 in human colon cancer cells. Cancer Res 67:7865–7874

Rebillard A, Lagadic-Gossmann D, Dimanche-Boitrel MT (2008a) Cisplatin cytotoxicity: DNA and plasma membrane targets. Curr Med Chem 15:2656–2663

Rebillard A, Rioux-Leclercq N, Muller C, Bellaud P, Jouan F, Meurette O, Jouan E, Vernhet L, Le Quément C, Carpinteiro A, Schenck M, Lagadic-Gossmann D, Gulbins E, Dimanche-Boitrel MT (2008b) Acid sphingomyelinase deficiency protects from cisplatin-induced gastrointestinal damage. Oncogene 27:6590–6595

Riboni L, Campanella R, Bassi R, Villani R, Gaini SM, Martinelli-Boneschi F, Viani P, Tettamanti G (2002) Ceramide levels are inversely associated with malignant progression of human glial cells. Glia 39:105–113

Riebeling C, Allegood JC, Wang E, Merill AH Jr, Futerman AH (2003) Two mammalian longevity assurance gene (LAG1) family members, trh1 and trh4, regulate dihydroceramide synthesis using different fatty acyl-CoA donors. J Biol Chem 278:43452–43459

Rizzieri KE, Hannun YA (1998) Sphingolipid metabolism, apoptosis and resistance to cytotoxic agents: can we interfere? Drug Resist Updat 1:359–376

Ruckhäberle E, Karn T, Rody A, Hanker L, Gätje R, Metzler D, Holtrich U, Kaufmann M (2009) Gene expression of ceramide kinase, galactosyl ceramide synthase and ganglioside GD3 synthase is associated with prognosis in breast cancer. J Cancer Res Clin Oncol 135:1005–1013

Ruckhäberle E, Rody A, Engels K, Gaetje R, von Minckwitz G, Schiffmann S, Grösch S, Geisslinger G, Holtrich U, Karn T, Kaufmann M (2008) Microarray analysis of altered sphingolipid metabolism reveals prognostic significance of sphingosine kinase 1 in breast cancer. Breast Cancer Res Treat 112:41–52

Rylova SN, Somova OG, Dyatlovitskaya EV (1998) Comparative investigation of sphingoid bases and fatty acids in ceramides and sphingomyelins from human ovarian malignant tumors and normal ovary. Biochemistry 63:1057–1060

Saad AF, Meacham WD, Bai A, Anelli V, Elojeimy S, Mahdy AE, Turner LS, Cheng J, Bielawska A, Bielawski J, Keane TE, Obeid LM, Hannun YA, Norris JS, Liu X (2007) The functional effects of acid ceramidase overexpression in prostate cancer progression and resistance to chemotherapy. Cancer Biol Ther 6:1455–1460

Sabourdy F, Selves J, Astudillo L, Laurent C, Brousset P, Delisle MB, Therville N, Andrieu-Abadie N, Ségui B, Recher C, Levade T (2011) Is active acid sphingomyelinase required for the antiproliferative response to rituximab? Blood 117:3695–3696

Samsel L, Zaidel G, Drumgoole HM, Jelovac D, Drachenberg C, Rhee JG, Brodie AM, Bielawska A, Smyth MJ (2004) The ceramide analog, B13, induces apoptosis in prostate cancer cell lines and inhibits tumor growth in prostate cancer xenografts. Prostate 58:382–393

Sankala HM, Hait NC, Paugh SW, Shida D, Lépine S, Elmore LW, Dent P, Milstien S, Spiegel S (2007) Involvement of sphingosine kinase 2 in p53-independent induction of p21 by the chemotherapeutic drug doxorubicin. Cancer Res 67:10466–10474

Schiffmann S, Sandner J, Birod K, Wobst I, Angioni C, Ruckhäberle E, Kaufmann M, Ackermann H, Lötsch J, Schmidt H, Geisslinger G, Grösch S (2009) Ceramide synthases and ceramide levels are increased in breast cancer tissue. Carcinogenesis 30:745–752

Schissel S, Keesler G, Schuchman E, Williams K, Tabas I (1998) The cellular trafficking and zinc dependence of secretory and lysosomal sphingomyelinase, two products of the acid sphingomyelinase gene. J Biol Chem 273:18250–18259

Schissel S, Schuchman E, Williams K, Tabas I (1996) Zn2+-stimulated sphingomyelinase is secreted by many cell types and is a product of the acid sphingomyelinase gene. J Biol Chem 271:18431–18436

Seelan RS, Qian C, Yokomizo A, Bostwick DG, Smith DI, Liu W (2000) Human acid ceramidase is overexpressed but not mutated in prostate cancer. Genes Chromosomes Cancer 29:137–146

Selzner M, Bielawska A, Morse MA, Rüdiger HA, Sindram D, Hannun YA, Clavien PA (2001) Induction of apoptotic cell death and prevention of tumor growth by ceramide analogues in metastatic human colon cancer. Cancer Res 61:1233–1240

Senkal CE, Ponnusamy S, Rossi MJ, Bialewski J, Sinha D, Jiang JC, Jazwinski SM, Hannun YA, Ogretmen B (2007) Role of human longevity assurance gene 1 and C18-ceramide in chemotherapy-induced cell death in human head and neck squamous cell carcinomas. Mol Cancer Ther 6:712–722

Shao RG, Cao X, Nieves-Neira W, Dimanche-Boitrel MT, Solary E, Pommier Y (2001) Activation of the Fas pathway independently of Fas ligand during apoptosis induced by camptothecin in p53 mutant human colon carcinoma cells. Oncogene 20:1852–1859

Simons K, Ikonen E (1997) Functional rafts in cell membranes. Nature 387:569–572

Simons K, Van Meer G (1988) Lipid sorting in epithelial cells. Biochemistry 27:6197–6202

Sobue S, Nemoto S, Murakami M, Ito H, Kimura A, Gao S, Furuhata A, Takagi A, Kojima T, Nakamura M, Ito Y, Suzuki M, Banno Y, Nozawa Y, Murate T (2008) Implications of sphingosine kinase 1 expression level for the cellular sphingolipid rheostat: relevance as a marker for daunorubicin sensitivity of leukaemia cells. Int J Haematol 87:266–275

Spassieva S, Seo JG, Jiang JC, Bielawski J, Alvarez-Vasquez F, Jazwinski SM, Hannun YA, Obeid LM (2006) Necessary role for the Lag1p motif in (dihydro)ceramide synthase activity. J Biol Chem 281:33931–33938

Spence MW, Byers DM, Palmer FBSC, Cook HW (1989) A new Zn2+-stimulated sphingomyelinase in fetal bovine serum. J Biol Chem 264:5358–53563

Strelow A, Bernardo K, Adam-Klages S, Linke T, Sandhoff K, Krönke M, Adam D (2000) Overexpression of acid ceramidase protects from tumor necrosis factor-induced cell death. J Exp Med 192:601–612

Strum JC, Small GW, Pauig SB, Daniel LW (1994) 1-beta-D-Arabinofuranosylcytosine stimulates ceramide and diglyceride formation in HL-60 cells. J Biol Chem 269:15493–15497

Suda T, Takahashi T, Golstein P, Nagata S (1993) Molecular cloning and expression of the Fas ligand, a novel member of the tumor necrosis factor family. Cell 75:1169–1178

Sukocheva O, Wang L, Verrier E, Vadas MA, Xia P (2009) Restoring endocrine response in breast cancer cells by inhibition of the sphingosine kinase-1 signaling pathway. Endocrinology 150:4484–4492

Suzuki A, Iwasaki M, Kato M, Wagai N (1997) Sequential operation of ceramide synthesis and ICE cascade in CPT-11-initiated apoptotic death signaling. Exp Cell Res 233:41–47

Taha TA, Kitatani K, El-Alwani M, Bielawski J, Hannun YA, Obeid LM (2006) Loss of sphingosine kinase-1 activates the intrinsic pathway of programmed cell death: modulation of sphingolipid levels and the induction of apoptosis. FASEB J 20:482–484

Tepper AD, de Vries E, van Blitterswijk WJ, Borst J (1999) Ordering of ceramide formation, caspase activation, and mitochondrial changes during CD95- and DNA damage-induced apoptosis. J Clin Invest 103:971–978

Tomiuk S, Zumbansen M, Stoffel W (2000) Characterization and subcellular localization of murine and human magnesium-dependent neutral sphingomyelinase. J Biol Chem 275:5710–5717

Ueda N, Camargo SM, Hong X, Basnakian AG, Walker PD, Shah SV (2001) Role of ceramide synthase in oxidant injury to renal tubular epithelial cells. J Am Soc Nephrol 12:2384–2391

Van Meer G, Hoetzl S (2010) Sphingolipid topology and the dynamic organization and function of membrane proteins. FEBS Lett 584:1800–1805

Venkataraman K, Riebeling C, Bodennec J, Riezman H, Allegood JC, Sullards MC, Merrill AH Jr, Futerman AH (2002) Upstream of growth and differentiation factor 1 (uog1), a mammalian homolog of the yeast longevity assurance gene 1 (LAG1), regulates N-stearoyl-sphinganine (C18-(dihydro)ceramide) synthesis in a fumonisin B1-independent manner in mammalian cells. J Biol Chem 277:35642–35649

Verkleij AJ, Post JA (2000) Membrane phospholipid asymmetry and signal transduction. J Membr Biol 178:1–10

Visentin B, Vekich JA, Sibbald BJ, Cavalli AL, Moreno KM, Matteo RG, Garland WA, Lu Y, Yu S, Hall HS, Kundra V, Mills GB, Sabbadini RA (2006) Validation of an anti-sphingosine-1-phosphate antibody as a potential therapeutic in reducing growth, invasion, and angiogenesis in multiple tumor lineages. Cancer Cell 9:225–238

Wang H, Charles AG, Frankel AJ, Cabot MC (2003) Increasing intracellular ceramide: an approach that enhances the cytotoxic response in prostate cancer cells. Urology 61:1047–1052

Wang XZ, Beebe JR, Pwiti L, Bielawska A, Smyth MJ (1999) Aberrant sphingolipid signaling is involved in the resistance of prostate cancer cell lines to chemotherapy. Cancer Res 59:5842–5848

White-Gilbertson S, Mullen T, Senkal C, Lu P, Ogretmen B, Obeid L, Voelkel-Johnson C (2009) Ceramide synthase 6 modulates TRAIL sensitivity and nuclear translocation of active caspase-3 in colon cancer cells. Oncogene 28:1132–1141

Whitman SP, Civoli F, Daniel LW (1997) Protein kinase CbetaII activation by 1-beta-D-arabinofuranosylcytosine is antagonistic to stimulation of apoptosis and Bcl-2alpha down-regulation. J Biol Chem 272:23481–23484

Wu BX, Clarke CJ, Hannun YA (2010) Mammalian neutral sphingomyelinases: regulation and roles in cell signaling responses. Neuromolecular Med 12:320–330

Xia P, Gamble JR, Wang L, Pitson SM, Moretti PA, Wattenberg BW, D'Andrea RJ, Vadas MA (2000) An oncogenic role of sphingosine kinase. Curr Biol 10:1527–1530

Xu J, Yeh CH, Chen S, He L, Sensi SL, Canzoniero LM, Choi DW, Hsu CY (1998) Involvement of de novo ceramide biosynthesis in tumor necrosis factor-cycloheximide-induced cerebral endothelial cell death. J Biol Chem 273:16521–16526

Zeidan YH, Jenkins RW, Hannun YA (2008) Remodeling of cellular cytoskeleton by the acid sphingomyelinase/ceramide pathway. J Cell Biol 181:335–350

Zhang J, Alter N, Reed JC, Borner C, Obeid LM, Hannun YA (1996) Bcl-2 interrupts the ceramide-mediated pathway of cell death. Proc Natl Acad Sci USA 93:5325–538

Zheng W, Kollmeyer J, Symolon H, Momin A, Munter E, Wang E, Kelly S, Allegood JC, Liu Y, Peng Q, Ramaraju H, Sullards MC, Cabot M, Merrill AH Jr (2006) Ceramides and other bioactive sphingolipid backbones in health and disease: lipidomic analysis, metabolism and roles in membrane structure, dynamics, signaling and autophagy. Biochim Biophys Acta 1758:1864–1884

Lung Cancer and Lung Injury: The Dual Role of Ceramide

Tzipora Goldkorn, Samuel Chung, and Simone Filosto

Contents

Abstract Sphingolipids play key roles in cancer, yet our current understanding of sphingolipid function in lung cancer is limited to a few key players. The best characterized of these are sphingosine-1-phoshate and ceramide which are described for their opposing roles in cell fate. However, because sphingolipids as a whole are readily interconverted by a complex enzymatic machinery, no single sphingolipid appears to have exactly one role. Instead, the roles of specific sphingolipids appear to be context specific as demonstrated by findings that ceramide-1-phosphate has both proliferative and apoptotic effects depending on

T. Goldkorn (✉) • S. Chung • S. Filosto
Center for Comparative Respiratory Biology and Medicine, Genome and Biomedical Sciences Facility, University of California School of Medicine, Davis, CA 95616, USA
e-mail: ttgoldkorn@ucdavis.edu

E. Gulbins and I. Petrache (eds.), *Sphingolipids in Disease*,
Handbook of Experimental Pharmacology 216,
DOI 10.1007/978-3-7091-1511-4_5, © Springer-Verlag Wien 2013

its concentration. Therefore, we present herein several years of research on ceramide, a sphingolipid linked to apoptotic signaling, that is emerging in cancer research for its potential roles in proliferation and cell-to-cell communication via exosomes.

Ceramide is a well-studied sphingolipid in both normal and pathological conditions ranging from skin development to lung cancer. Interestingly, several groups have previously reported its increased levels in emphysema patients who are smokers, a patient subpopulation greatly susceptible to lung cancer. However, the molecular mechanisms through which cigarette smoke (CS) and ceramide accumulation lead to lung cancer, non-small cell lung cancer (NSCLC) specifically, are unknown.

Interestingly, recent studies clearly establish that two signaling pathways are activated during CS exposure in the lung airway. One centers on the activation of neutral sphingomyelinase2 (nSMase2), an enzyme that hydrolyzes sphingomyelin to ceramide. The other pathway focuses on the oncogenic EGF receptor (EGFR), which becomes aberrantly activated but not degraded, leading to prolonged proliferative signaling. Recent studies show that these two signaling pathways may actually converge and integrate. Specifically, Goldkorn et al. demonstrated that during CS exposure, EGFR is favorably co-localized in ceramide-enriched regions of the plasma membrane, proposing that nSMase2/ceramide plays a role in the aberrant EGFR activation, leading to augmented tumorigenic signaling. Moreover, new findings indicate that CS exposure may induce resistance to the tyrosine kinase inhibitors (TKIs), used for treatment of NSCLC, merely through posttranslational molecular alterations. Furthermore, structural anomalies of the CS-activated EGFR appear to be supported by the excess ceramide produced by the CS-activated nSMase2 in the plasma membrane of lung epithelial cells.

We present in this chapter the progression of the sphingolipid field in lung cancer using ceramide as an example. However, many crucial questions remain to be answered regarding the role of sphingolipids in lung cancer because of the glut of promising observations.

Keywords Sphingolipids • Lung cancer and injury • Ceramide • EGFR • Neutral sphingomyelinase 2 • Cigarette smoke • Oxidative stress

1 Introduction

It is well established that sphingolipids play crucial roles in a number of biological processes, including cell survival, proliferation, and apoptosis of many cell types. Consistently, it has been demonstrated that several different sphingolipids and sphingolipid-metabolizing enzymes are critical regulators of cancer cell fate. However, as of this moment, very little is known about the role of sphingolipids in lung cancer development; therefore, much more research and better models are required.

Indeed, despite the potential role of sphingolipids in several aspects of lung cancer biology such as cell growth and death, metastasis, and interaction/communication between cancer cells and their microenvironment, only a few fragmented investigations have been carried out to date. We present herein these studies together with our vision of the ceramide-generating machinery as a very promising, challenging, and still mostly undefined field of research that could lead to a paradigm shift and a better understanding of lung cancer. Particular emphasis is given to the potential function of ceramide in the context of smoking-related lung tumorigenesis, as tobacco smoking is the most prominent cause of lung cancer.

2 The Ceramide-Generating Machinery as the Center of Sphingolipids Biology

Among the sphingolipids, ceramide has been one of the most studied compounds because it represents the core of the complex machinery, often referred to as "the ceramide-generating machinery," which allows the conversion of one type of sphingolipid to another. Indeed, ceramide can be generated via at least three different pathways, namely, de novo, salvage, and sphingomyelin hydrolysis pathway (Mullen et al. 2011; Wu et al. 2010). At the same time, ceramide represents the main hub that provides many other sphingolipids, such as ceramide-1-phosphate, sphingosine, sphingosine-1-phosphate (S1P), and sphingomyelin, each of which can be involved in many and different biological processes (Hannun and Obeid 2011; Maceyka et al. 2012; Merrill 2011; Morales et al. 2007; Ponnusamy et al. 2010).

To date, the best-characterized sphingolipids in cancer cell fate are ceramide and S1P. These two lipids have been "canonically" described as harboring opposite roles in driving cell fate: ceramide often described as a pro-apoptotic or "tumor suppressor lipid" and S1P as pro-survival/-proliferation or "oncogenic lipid" (Oskouian and Saba 2010). However, this certainly oversimplifies the complex enzymatic machinery controlling sphingolipids biology, where activation of specific enzymes can rapidly alter the balance between ceramide and S1P as well as trigger their conversion to several other bioactive ones. Moreover, recent evidences suggested that just amongst ceramides, different lengths of the fatty acid constituents of ceramide are associated with opposite effects on cell fate. For instance, it has been shown in various cancer models that C(16)-ceramide increases tumor growth, while C(18)-ceramide suppresses it (Ponnusamy et al. 2010), suggesting that different ceramide synthases may lead to opposite outcomes of cell fate.

Thus, it is important to point out that the effect of sphingolipids, particularly of ceramides, on cell fate (and thus on cancer growth/death) is the combined effect of several interconvertible sphingolipids which are regulated in distinct subcellular compartments, executing distinct cellular functions (Hannun and Obeid 2011).

3 Sphingolipids in Lung Cancer: A Field that Needs Research and Discovery

Our understanding of sphingolipids' role in lung cancer is still very limited and fragmented, perhaps suffering from the lack of suitable in vivo models to assess the long-term effects of certain sphingolipids or sphingolipid-regulating enzymes on tumor growth. Most studies on sphingolipids and lung cancer were performed in vitro, investigating short-term effects of sphingolipids on cell proliferation and apoptosis of lung cancer cell. Mitra et al. (2007) showed in A549 lung adenocarcinoma cells that ceramide-1-phosphate (C1P) can have either proliferative and pro-survival effects during serum starvation or pro-apoptotic effects depending on the concentration: low concentrations (<5 μM C1P) leading to proliferation, whereas high concentration (>5 μM C1P) causing apoptosis. In this study, Mitra et al. claimed that the cytotoxic effects of high C1P levels were due to ceramide accumulation, generated by dephosphorylation of C1P. On the other hand, the proliferative effects of C1P low concentration were attributed to reduced cell ceramide levels, possibly caused by inhibition of a sphingomyelinase (Gomez-Munoz et al. 2004; Mitra et al. 2007). Importantly, this study also showed that ceramide kinase expression directly correlated with enhanced cell growth. This strongly supports the notion that sphingolipids can be rapidly converted amongst themselves and that their overall effect on cell fate was a function of the prevailing sphingolipid-controlling enzymes that become activated (or inhibited) in a certain context.

Consistent with this idea, de Molina et al. (2012) reported that the conversion of ceramide to sphingosine by acid ceramidase is a mechanism regulating non-small cell lung cancer (NSCLC) resistance to targeted therapy (choline kinase-α inhibitors) despite an initial increase in ceramide levels. Importantly, this study highlights how sphingolipids can be critically involved in modulating lung cancer therapy resistance. Indeed, Schnitzer et al. (2009) recently showed that enhanced sphingosine kinase 2 expression, activity, and subsequent upregulation of sphingosine-1-phosphate (S1P) are part of the stress response to hypoxia in A549 cells, which was not only essential for cell viability but also protected cells from chemotherapy (etoposide)-induced apoptosis.

In the past, Inokuchi et al. (1990) proposed a potential role for several glycosphingolipids (glucosylceramide, lactosylceramide, ceramide trihexoside, globoside, and ganglioside GM3) in lung cancer, finding that glucosylceramide synthase contributes to metastatic spreading of Lewis lung carcinoma. Moreover, a recent study by Yamada et al. provides important evidence supporting both the role of sphingolipids in lung cancer growth and metastasis and the potential for effective therapeutic intervention by targeting sphingolipids. Their study showed high expression of the glycosphingolipid ganglioside GM2 (monosialic ganglioside 2) in small-cell lung cancer (SCLC) human specimens and, using an in vivo tumor xenograft mouse model, demonstrated that an anti-GM2 antibody-based therapy could block SCLC metastases and increase cancer cell apoptosis, thereby prolonging the mouse' survival (Yamada et al. 2011). However, as of yet, there are no molecular mechanisms explaining this observation.

Adding a level of complexity, many pathways can be activated downstream of bioactive sphingolipids which appear to be context specific. For instance, Mukhopadhyay et al. (2009) showed that membrane C(18)-ceramide can mediate its pro-apoptotic effects in lung cancer cells via its binding partner protein phosphatase 2A (PP2A), whose function is upstream of c-Myc degradation, thereby presenting antiproliferation effects. These findings are in line with earlier studies from the same group (Ogretmen and colleagues) showing that exogenous administration of C(6)-ceramide to A549 cells could inhibit both c-Myc and telomerase activity which are important for proliferation (Ogretmen et al. 2001). However, there could be many reasons why this PP2A-dependent apoptotic pathway is not undertaken or is not prevalent. For instance, Mukhopadhyay et al. (2009) also reported that okadaic acid treatment could completely counteract c-Myc degradation following exposure of A549 cells to exogenous ("pro-apoptotic") ceramide. Furthermore, this PP2A-dependent pathway relies on both low levels of its endogenous inhibitor I2PP2A and a functional PP2A. Since it has been shown that PP2A is inhibited by oxidative stress (Foley et al. 2007) and given that oxidative stress is a common denominator of many lung diseases and lung cancer, it is easy to envision that this PP2A-dependent apoptotic trigger would fail in lung cancer (Esme et al. 2008), particularly in the context of smoking-related lung cancer. Currently, much more research is needed to understand the role of sphingolipids in lung cancer. Moreover, future research should be oriented toward identifying bioactive sphingolipids in vivo: namely, models of lung cancer that better mimic the conditions where human lung cancer initiates and progresses.

To this extent, our group has been committed to investigate the cell signaling events initiated in lung epithelial cells exposed to cigarette smoke-induced oxidative stress, a model relevant to both smoking-induced lung injury and lung cancer (Filosto et al. 2011a; Goldkorn et al. 2005). Therefore, we will provide our perspective on sphingolipids research, with particular focus on ceramide signaling, in lung cancer.

4 Cigarette Smoke-Induced Lung Cancer and Lung Injury

Despite the well-documented dangers of smoking, millions of smokers continue to puff away every day, each puff containing roughly 5,000 toxic compounds that include 10^{15} free radicals such as hydrogen peroxide (H_2O_2) and hydroxyl and organic radicals in the gaseous phase (Church and Pryor 1985). Toxic oxidants in cigarette smoke have a detrimental effect on epithelial cells lining the airway and contribute to several lung pathological conditions including emphysema and bronchitis [jointly known as chronic obstructive pulmonary disease (COPD)], which are injury-related lung diseases, and lung cancer. Indeed, tobacco smoking is the major risk for developing lung cancer, causing approximately 80–90 % of all cases (Alberg and Samet 2003).

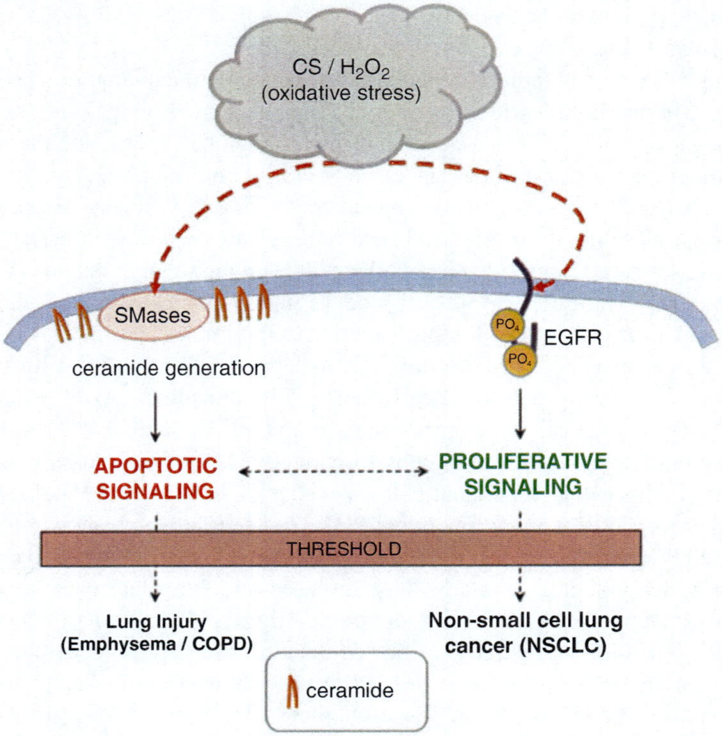

Fig. 1 Cigarette smoke (CS)-induced oxidative stress concomitantly generates ceramide which mediates apoptotic signaling and activates the epidermal growth factor receptor (EGFR) which mediates proliferative signaling. Though CS-induced lung injury is initially dominant, epidemiological studies suggest smoking-related lung cancer becomes promoted in this environment

Interestingly, several prospective clinical studies over the years show that lung cancer incidence becomes significantly increased in COPD patient populations despite that lung injury and lung cancer are apoptosis dominant and proliferation dominant, respectively. For instance, de Torres et al. (2007) reported that in their 5-year prospective cohort of former and current smokers, individuals with emphysema were more likely to develop lung cancer than those without emphysema. Moreover, individuals with both emphysema and airway obstruction were at an even higher risk than those with neither conditions alone (de Torres et al. 2007). Similarly, studies by Caballero et al. (2008) convincingly reported that even the existence of some low-grade emphysema without noticeable airflow obstruction is associated with significantly elevated risk of lung cancer.

Thus far, clinical data demands novel mechanistic insight into how apoptosis-dominant lung injury leads to proliferation-dominant lung cancer. Indeed, both smoking-induced lung injury and lung cancer, though originating from opposing phenotypes at the cellular level, may actually share common molecular mechanisms that underlie both pathologies and explain their clinical association. To respond to this major problem, we discuss herein the potential role for ceramide generation as a signaling event in both lung injury and lung cancer (Fig. 1).

5 Cigarette Smoke-Generated Oxidative Stress and Ceramide Generation in Airway Epithelial Cells

Epithelial cells lining both the lung main airways and the terminal alveoli represent the lung's first line of defense and thus are extensively exposed to reactive oxidants such as those present in the gas phase of CS. Over the last 15 years in our laboratory, we have been investigating the cell signaling events initiated in human airway epithelial (HAE) cells during CS exposure, leading to either apoptosis (hallmark of injury) or proliferation (hallmark of cancer).

Studies by Goldkorn et al. in the direction of oxidative stress-induced lung injury addressed whether HAE cells undergo apoptosis when exposed to micromolar concentrations of H_2O_2 and, subsequently, whether this process is mediated by ceramide generation (Chan and Goldkorn 2000; Goldkorn et al. 1998). Both a dose-dependent (50–250 μM H_2O_2) and a time-dependent relationship were observed between H_2O_2 exposure and increased cellular ceramide levels and subsequent apoptosis. This initial model for H_2O_2-induced cellular ox-stress and ceramide generation consisted of direct exposures of airway epithelial cells to H_2O_2, glutathione (GSH), or both by media supplementation (Khan et al. 2006; Lavrentiadou et al. 2001; Ravid et al. 2003). Even though these early studies undertook a reductionist, simplified approach, both scenarios were relevant to the lung epithelium.

H_2O_2 is a ubiquitous molecule that is freely miscible and able to cross cell membranes readily and is present in several air pollutants, including the vapor phase of tobacco smoke (Khan et al. 2008). It is detected in exhaled air of humans (Williams and Chance 1983), and the amounts of exhaled H_2O_2 appear greater in subjects with pulmonary disease (Sznajder et al. 1989) and in cigarette smokers (Nowak et al. 1996). In addition, it was shown that either administration of exogenous H_2O_2 or enhancement of endogenously generated H_2O_2 was effective in depleting cellular GSH and initiating ceramide generation and apoptosis events (Ravid et al. 2003).

Interestingly, later studies showed that CS exposure regulates both growth and death of airway epithelial cells via its H_2O_2 component (Khan et al. 2008; Lavrentiadou et al. 2001): CS exposure could generate between 100 and 800 μM H_2O_2, as measured in the media of cultured HAE cells, a factor that was dependent on the length of exposure and the amount of combusted tobacco. Furthermore, when the CS-generated H_2O_2 was quenched by antioxidants such as GSH or NAC (*N*-acetylcysteine), no ceramide upregulation was observed (Khan et al. 2008). Since then, it has been shown that exposure to oxidative stress leads to ceramide level increase prior to caspase-3-dependent apoptosis in airway epithelial cells, mimicking the dominant apoptotic process of emphysema pathogenesis (Ravid et al.

2003). Following this discovery, Goldkorn et al. sought to identify the enzymatic regulator(s) controlling this phenomenon.

6 Identifying Neutral Sphingomyelinase 2, a Novel Sphingomyelinase Upregulated in Cigarette Smoke-Induced Lung Injury

Sphingomyelin (SM) hydrolysis by sphingomyelinases (SMases) is considered a major pathway for cell stress-induced ceramide generation and subsequent apoptosis (Clarke et al. 2008; Filosto et al. 2010, 2011a; Goldkorn et al. 1998; Goldkorn and Filosto 2010; Nikolova-Karakashian et al. 2008; Ravid et al. 2003; Rutkute et al. 2007b; Wu et al. 2010). According to current models, SM may be metabolized to ceramide by several SMases, but some SMases may have special pathophysiological significance (Uhlig and Gulbins 2008).

Several years of purifying and cloning proteins from the lungs of nonhuman primates and HAE cells led to the independent identification of a novel sphingomyelinase, neutral sphingomyelinase 2 (nSMase2, SMPD3 gene), as the sole sphingomyelinase responsible for ceramide generation in response to oxidative stress (Levy et al. 2006). Importantly, it was demonstrated that nSMase2 is a redox-sensitive enzyme that is activated in HAE cells in response to ox-stress induced by H_2O_2 or by CS exposure (Levy et al. 2006, 2009). Moreover, silencing of nSMase2 expression using siRNA completely abrogates ceramide production and, more importantly, the downstream apoptotic events in response to oxidative stress. It was also demonstrated that the H_2O_2 component of the CS gas phase was responsible for the nSMase2 activation and ceramide generation (Levy et al. 2009).

Convincingly, both Goldkorn et al. and others have reported that the major SMase enzyme that becomes activated by ox-stress appears to be nSMase2 (Levy et al. 2006, 2009; Rutkute et al. 2007a). Last of all, but most notably, it has been shown that nSMase2 becomes overexpressed in human emphysema patients and rodents chronically exposed to cigarette smoke and that mice heterozygous for nSMase2 (+/−) have dramatically reduced ceramide generation in response to CS (Filosto et al. 2011a).

Filosto et al. have since demonstrated a molecular mechanism that regulates nSMase2 function in response to ox-stress by finding that nSMase2 is controlled via a complex phosphorylation-based machinery. It was proven that nSMase2 phosphorylation on specific serines triggers both its activation and protein stabilization upon exposure of HAE cells to ox-stress (Filosto et al. 2012a). Furthermore, we identified some of the protein regulators of nSMase2 phosphorylation: Calcineurin (CaN) phosphatase binds directly to nSMase2 and dephosphorylates it, but not under ox-stress, when CaN is inhibited/degraded, allowing nSMase2 to be

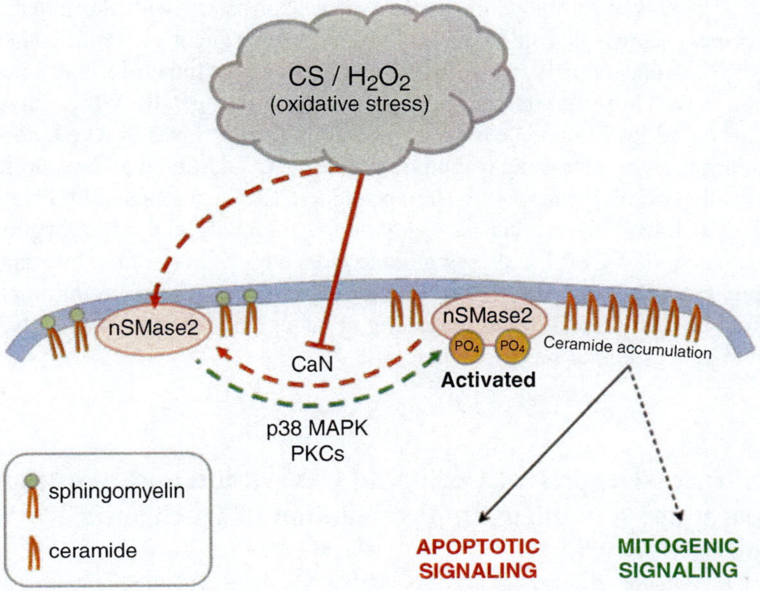

Fig. 2 Cigarette smoke (CS)-induced oxidative stress activates the neutral sphingomyelinase 2 (nSMase2) in human airway epithelial cells via phosphorylation of its serine residues, leading to ceramide accumulation in the inner leaflet of the plasma membrane

phosphorylated downstream of p38 mitogen-activated protein kinase (MAPK) and/or protein kinase C (PKCs) (Filosto et al. 2010) (Fig. 2).

7 Can nSMase2 Activation and Ceramide Generation Support Mitogenic Responses?

The above studies identified nSMase2 as the ceramide-generating enzyme during oxidative stress whose activation is a requisite to trigger apoptosis of HAE cells. However, several considerations should be taken before concluding that apoptosis is the sole downstream effect of ceramide generation. First, while increased ceramide levels could be expected for any cell exposed to ox-stress, only 20–50 % (depending on the exposure conditions) of the exposed cells would actually undergo apoptosis (Khan et al. 2008; Levy et al. 2009). Second, oxidative stress has also been proposed as a mitogenic signal promoting lung cancer growth (Khan et al. 2008; Levy et al. 2009). Third, most studies investigating ceramide generation under stress conditions, including ours, were conducted during acute, short-term exposure to cell stressors, and therefore, the long-term effects of ceramide generation on cancer cell growth have not been evaluated.

Presently, though it is plausible that during cellular stress ceramide acts as a direct stimulus for apoptosis, it is equally possible that ceramide generation has mitogenic effects favoring tumor development. Consistent with the latter, it has been recently shown that nSMase2 activity and subsequent ceramide generation under TNF-α-induced cell stress can be mechanistically linked to both apoptotic initiation as well as proliferative signaling (Devillard et al. 2010). While alternative pathways could lead to apoptosis when nSMase2 function was blocked, nSMase2 was required for the mitogenic response to TNF-α. An additional recent publication claims that loss of ceramide transfer protein (CERT) augments EGF receptor signaling in breast cancer via its regulation of cellular pool of sphingomyelin (SM). Reduction of CERT expression led to reduced SM levels, thereby enhancing tumorigenesis (Heering et al. 2012). Furthermore, nSMase2 and ceramide generation could be instrumental in the spreading of oncogenic proteins and miRNAs via exosomal generation, as described below.

8 nSMase2-Dependent Ceramide Generation and Exosome Secretion: A Vehicle for Propagation of Oncogenes and miRNA

The conventional view of communication between cells is through gradients of soluble ligands, identified by the cell-associated receptors. However, recent studies demonstrated a novel form of interaction between cancer cells and their environment through shedding of membrane exosomes, which can fuse to cells in the surrounding area and cause stromal remodeling (Al-Nedawi et al. 2008; Bhowmick et al. 2004; Skog et al. 2008). Exosomes contain microvesicles that serve as "units" of information by containing a mass of biologically active protein and RNA species, including oncogenic receptors and microRNAs (miRNAs), short noncoding RNAs that modulate gene expression (Kim et al. 2009b; Liu et al. 2009). For example, it was demonstrated that a truncated oncogenic form of EGFR, EGFRvIII, was transferred from human brain glioma cells into neighboring glioma cells deficient in EGFRvIII via secretory membrane microvesicles (exosomes) (Al-Nedawi et al. 2008). In the lungs, however, no examples of exosomal oncoprotein transfers have (yet) been described. Instead, the bulk of studies in the role of exosomes in oncogenesis have been focused in miRNAs which become deregulated in numerous cancers, including of lung origin.

Some miRNAs are notably deregulated in lung cancer. For example, low expression of let-7a miRNA and high expression of miR-155 miRNA were associated with poor clinical outcome (Yanaihara et al. 2006). At the same time, there are miRNAs overexpressed in lung cancer that have key roles in modulating cancerous cell growth and tumorigenicity such as the EGFR (Liu et al. 2010). A recent publication by Kosaka et al. (2010) reported that exosomes may be used as a means for secretion of miRNAs: miRNAs secreted from donor cells could be absorbed and remain functional in recipient cells, suggesting an innovative mechanism of intercellular communication by miRNAs release. Consistently, Pegtel et al. (2010) also demonstrated that miRNAs secreted by Epstein–Barr virus (EBV)-infected cells were transferred to and acted in uninfected recipient cells. Importantly, Kosaka et al. (2010) demonstrated that miRNA incorporation into exosomes and secretion are ceramide-/nSMase2 dependent, which underscores a novel critical role of nSMase2 in regulating exosome-dependent miRNA dispersal.

Moreover, the molecular mechanism by which a change in lipid composition drives vesicle budding was recently demonstrated by Trajkovic et al. (2008) to be regulated via nSMase2 and, thusly, ceramide. Using mass spectrometric analyses, Trajkovic et al. demonstrated that ceramide levels were increased in secreted proteolipid protein-containing exosomes purified from cell culture medium. Moreover, disrupting nSMase2 expression by RNA interference or the use of specific inhibitors reduced secretion of these exosomes, suggesting that aggregation of ceramide-enriched microdomains leads to intraluminal vesicles formation.

Additionally, miRNA-containing microvesicles could be recovered from the plasma of patients with tumors, including patients with lung cancer, thus potentially serving as prognostic biomarkers (Al-Nedawi et al. 2009; Chen et al. 2008; Mitchell et al. 2008). Therefore, ceramide-dependent exosomal secretion may drive the transfer of oncoproteins and onco-miRNAs between subset of cancer cell populations and, therefore, necessitate the evaluation of ceramide and exosomes in the distribution oncogenes and their transforming phenotype in several cancer types, including lung cancer.

Though several groups continue to study the mechanism(s) of nSMase2 regulation in lung epithelial cells and other tissues, ceramide generation and its potential function(s) in oncogenic signaling remain poorly understood. However, within the setting of lung injury and lung cancer, recent studies by Goldkorn and colleagues (Filosto et al. 2011b) documented the proliferative effects of cigarette smoke-induced oxidative stress via the epidermal growth factor (EGF) receptor, a membrane-bound protein important for growth signaling, and possible potentiation of its activation by ceramide generation in the plasma membrane, as described in the following sections.

9 Cigarette Smoke-Induced Oxidative Stress and Its Effects on Oncogenic EGF Receptor Activation in Lung Cancer

In chronic smokers, the plasma membrane of airway epithelial cells works as the first line of defense against gaseous oxidants. Though this scenario depicts lung injury fairly well as previously discussed with ceramide, simultaneous studies on the EGF receptor (EGFR) portray the bipolar effects of cigarette smoke: Proliferative signaling becomes activated during oxidative stress in airway epithelial cells (Goldkorn et al. 2005; Ravid et al. 2002).

The EGFR is a transmembrane receptor tyrosine kinase (RTK) of the ErbB family, which includes ErbB1–4 and is the prototypical "master" regulator of cell growth in normal cells and tumorigenesis in lung cancer. The classical model of EGFR activation has been established wherein (1) binding of its extracellular ligand EGF (2) induces receptor dimerization leading to (3) activation of its intracellular tyrosine kinase (TK) domain and subsequent (4) auto-phosphorylation of specific tyrosines on the C-terminal "tail" of the receptor, ultimately (5) initiating a cascade of downstream proliferative signaling. Simultaneously, the activated EGFR is internalized in a clathrin-dependent manner and downregulated via lysosomal degradation. However, deregulation of EGFR function can happen for many reasons, including EGFR somatic mutations or its overexpression, leading to EGFR-driven lung tumorigenesis. Therefore, a collective effort by the field over the last two decades has led to better understanding the mechanism of EGFR kinase activation and its targeting by tyrosine kinase inhibitors.

Tyrosine kinase inhibitors (TKIs) are small molecules that inhibit TK activity by competitively binding to the ATP binding site of the EGFR. In the early 2000s, several TKIs—erlotinib (Tarceva™) and gefitinib (Iressa™)—were approved by the FDA (Food and Drug Administration) for the treatment of metastatic lung cancer, while several other TKIs are in preclinical trials (e.g., TKI AG1478) and currently in the drug development pipeline (Ellis et al. 2006; Gridelli et al. 2010; Watanabe et al. 2011; Wu et al. 2011). Unfortunately, despite earlier promising clinical success, these FDA-approved TKIs failed to benefit the majority of ailing lung cancer patients. Interestingly, however, encouraging responses to the TKIs were observed in a subset of non-small cell lung cancer (NSCLC) cases, typically presenting as adenocarcinomas in female never-smokers (Yatabe 2010). These responders were found to possess mutations in the EGFR tyrosine kinase domain that constitutively activate the receptor in a ligand-independent manner, which were thought to confer a selective growth advantage that could be efficiently neutralized by the TKIs (Lynch et al. 2004; Minna et al. 2004; Paez et al. 2004; Shigematsu and Gazdar 2006; Weinstein 2002). Most common, or "classic," among these somatic mutations are the exon 19 deletion mutation (Δ746–750) and the exon 21 single-point substitution mutation (L858R) (Mountzios et al. 2008; Parkin et al. 2005).

However, it has been recently observed that the same EGFR mutations are not limited to adenocarcinoma from female never-smokers. Indeed, a large number (40%) of these EGFR mutations are actually found in adenocarcinoma tumor specimens from men and people who are/were smokers (D'Angelo et al. 2011). Yet, these lung cancer patient groups of former and current smokers (harboring EGFR mutations, which are expectedly to be TKI sensitive) do not benefit from the TKI treatment. Moreover, anecdotal clinical observations suggest that lung cancer patients with smoking history tend to require up to a tenfold higher baseline dose of TKI to start during treatment. This indicates that smoking could affect the tumor's response to TKI, but the mechanisms involved are still mostly elusive.

In parallel with their ceramide studies, Goldkorn et al. investigated the effects of CS-derived reactive oxidants on the structure/function of EGFR in HAE cells in both normal and transformed (NSCLC) cells (Filosto et al. 2011b, 2012b; Goldkorn

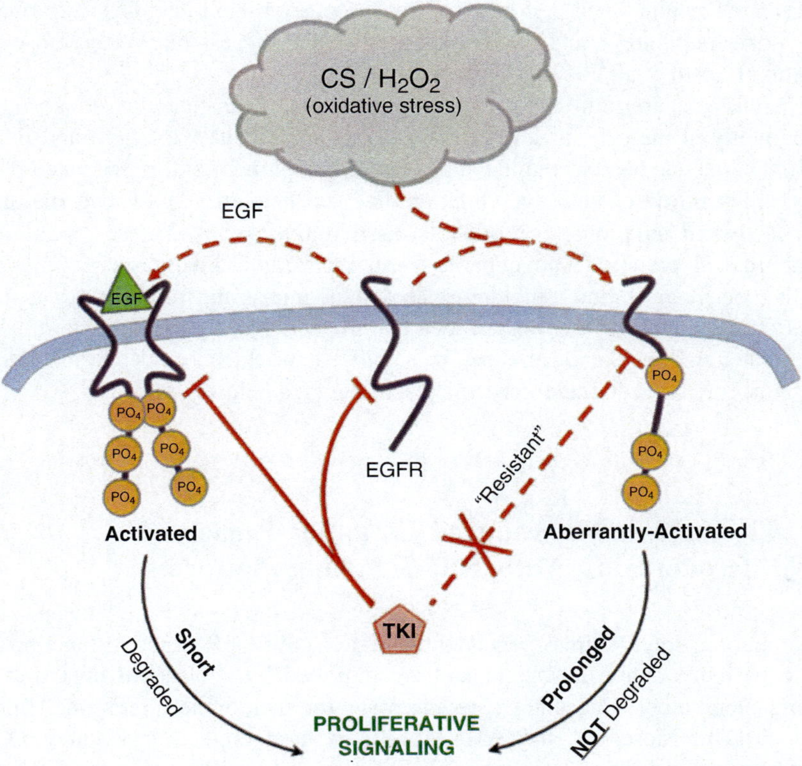

Fig. 3 Cigarette smoke (CS)-induced oxidative stress generates an aberrantly phosphorylated EGFR with a noncanonical activated conformation. This CS-activated EGFR is not readily degraded and has prolonged proliferation signaling and is "resistant" to tyrosine kinase inhibitors (TKIs) used clinically as therapy for EGFR-driven lung cancers

et al. 2005; Goldkorn and Filosto 2010; Khan et al. 2006, 2008; Ravid et al. 2002). Since then, it has been demonstrated that both CS and oxidative stress (H_2O_2) aberrantly activate the EGFR which can be suppressed in the presence of the antioxidant glutathione or catalase (Khan et al. 2008; Ravid et al. 2002). Specifically, the pattern of tyrosine-phosphorylation sites differs from that induced by the EGF ligand (Khan et al. 2008), resulting in an activated/signaling EGFR without the physiological negative feedback of its lysosomal degradation (Khan et al. 2006, 2008; Ravid et al. 2002). Goldkorn's group also showed that this abnormal EGFR activation impairs receptor dimerization and is not targeted/inhibited by TKIs (erlotinib, gefitinib, and AG1478) (Filosto et al. 2011b, 2012b) (Fig. 3).

Thus, it was recently demonstrated that a novel active conformation of the EGFR confers tumorigenesis and resistance to current TKIs following CS and oxidative stress exposure in HAE cells (Filosto et al. 2011b, 2012b). These findings have far-reaching implications suggesting that this oxidative stress-activated EGFR cannot be inhibited by clinical tyrosine kinase inhibitors developed for therapeutic intervention in non-small cell lung cancer, the predominant lung cancer variant in the smoker population, and may explain the sparse success rate of these drugs in patients (Filosto et al. 2012b).

Therefore, it appears that a dichotomous response to cigarette smoke occurs concurrently at the cellular level in airway epithelial cells (Goldkorn and Filosto 2010). On the one hand, ceramide, a pro-apoptotic sphingolipid, is generated by the induction of nSMase2 activity, while, on the other hand, the EGFR becomes aberrantly activated and promotes proliferative signaling. To explain these observations, Goldkorn et al. proposed a novel perhaps equally insidious function for ceramide, as the link between chronic smoking-related lung injury and tumorigenesis in the airway. That role was based on the idea that the novel EGFR conformation, which is generated under CS exposure, can be stabilized and thus become chronic only in ceramide-enriched cellular compartments, as detailed below.

10 Cigarette Smoke-Induced Ceramide-Enriched Microdomains: Mitogenic Signaling Platforms

Importantly, oxidative stress-dependent alterations of EGFR structure/function turn out to be temperature dependent and are inhibited by cholesterol uptake at the plasma membrane, suggesting a requirement for membrane alterations (Filosto et al. 2011b). Moreover, the TKI AG1478 is ineffective in quenching EGFR phosphorylation during H_2O_2-induced ox-stress in living cells while capable of inhibiting it in a crude membrane fraction of broken cells where the membrane structure was destroyed (Filosto et al. 2011b). This supports the idea that membrane fluidity/structure may be involved in either inducing or stabilizing the ox-stress-induced active conformation of EGFR.

In recent years, several studies demonstrated that sphingolipids in cell membranes are not homogenous but are sorted into specific domains mediated by interactions between sphingolipids and cholesterol (Zhang et al. 2009). In particular, the ceramide moiety of sphingomyelin binds to cholesterol via hydrophobic van der Waals interactions (Kolesnick et al. 2000). The culmination of these strong interactions and the high local concentration of sphingolipids and cholesterol lead to the formation of sphingolipids and cholesterol-enriched membrane domains known as lipid rafts. Cholesterol and some cholesterol precursors not only interact with sphingolipids in these rafts but also stabilize the structure of rafts by filling empty spaces between the bulky sphingolipids (Kolesnick et al. 2000; Stancevic and Kolesnick 2010; Xu et al. 2001; Zhang et al. 2009).

However, when ceramide becomes free from sphingomyelin, it can displace cholesterol from the rafts (Megha and London 2004). Thus, the generation of ceramide within membrane rafts may alter both their composition and structure. Furthermore, ceramide-enriched microdomains have the tendency to fuse and form larger macrodomains, also known as ceramide-enriched membrane platforms, with a diameter from a few hundred nanometers up to several micrometers (Cremesti et al.

2001). These ceramide-enriched membrane platforms have altered biophysical properties compared to the phospholipid membrane and cholesterol-enriched lipid rafts and may serve important functions in protein sorting in cells and redistribution of membrane receptors, such as the EGFR. Indeed, ceramide-enriched membrane platforms were shown to trap and to cluster receptor molecules (Cremesti et al. 2001; Grassme et al. 2001; Gulbins and Kolesnick 2003).

Given the critical role of lipid rafts in serving as signaling platforms for many membrane receptors, including the EGFR (Heering et al. 2012), alteration in membrane structure and fluidity following ceramide generation are likely to have an important effect on lung cancer cell growth. Intriguingly, cholesterol levels do not change during oxidative stress in HAE cells (Filosto et al. 2011b). Thus, Goldkorn et al. lately proposed that membrane ceramide generated as a result of exposure to ox-stress may displace cholesterol from the rafts, leading to the relocalization of EGFR to relatively rigid ceramide-enriched lipid platforms, which may stabilize aberrant EGFR signaling. Complementing this hypothesis, it was recently demonstrated that the CS-induced oxidative stress stabilizes aberrant interactions between the activated EGFR and Src, a non-receptor tyrosine kinase that is highly overexpressed in several types of cancers (Cai et al. 2011; Ceppi et al. 2012; Chung et al. 2009; Kim et al. 2009a). Furthermore, it was reported that phosphorylated caveolin-1 (Cav-1), a known substrate of Src, is strongly bound to EGFR under ox-stress (Khan et al. 2006, 2008). Notably, others have reported that Src stably interacts with ErbB2 (Kim et al. 2005), but not with WT EGFR, because of the structural difference in their respective kinase domains, and this Src binding was shown to confer elevated transformation ability (Marcotte et al. 2009). For these reasons, it was also proposed that such membrane alterations may also be involved in the aberrant perinuclear trafficking of EGFR and Src that is observed in HAE cells during exposure to oxidative stress (Filosto et al. 2011b; Khan et al. 2008).

Supporting this hypothesis, it was recently shown that during oxidative stress, activated EGFR and activated Src co-localize within ceramide-enriched membrane regions (Filosto et al. 2011b) (Fig. 4). At early time points of ox-stress exposure (15 min), active EGFR and elevated ceramide co-localize primarily in the plasma membrane of the cells. Interestingly, later on (30 min), this co-localization is observed mainly in a perinuclear region of the cells. Consistent with this latter observation, it was demonstrated that the ox-stress-activated EGFR, unlike the EGF-stimulated receptor, is not internalized via clathrin-coated pits. Instead, it is not degraded, remains active, and is trafficked via caveolae to the perinucleus because of strong association with phosphorylated Cav-1 (Khan et al. 2006, 2008). It has been recently shown that ceramide generation increases the recruitment of Cav-1 into caveolae (Kuebler et al. 2010; Yang et al. 2010). Taken together, ceramide generation may have a role in both stabilizing the aberrantly active EGFR and controlling the trafficking of EGFR via caveolae to the cell perinucleus that has been shown to promote tumorigenesis in lung epithelial cells under oxidative stress (Khan et al. 2008).

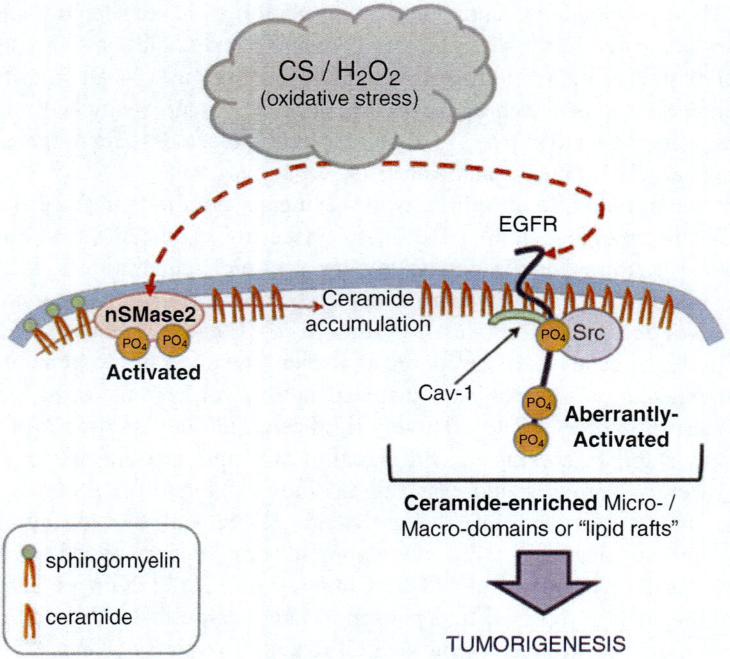

Fig. 4 Cigarette smoke (CS)-induced oxidative stress activates neutral sphingomyelinase 2 (nSMase2), leading to ceramide accumulation and subsequent formation of ceramide-enriched microdomains in the plasma membrane. Within these ceramide-enriched regions, the aberrantly activated EGFR is stabilized and promotes tumorigenesis in lung epithelial cells

11 Conclusions

Sphingolipids and their roles in lung cancer offer critically needed avenues for discovery of both prognostic and diagnostic biomarkers, thereby remaining a field requiring discovery. As described earlier, the best characterized sphingolipids in lung cancer research are sphingosine-1-phoshate and ceramide which traditionally have opposing roles in determining cell fate. However, there appears to be a very fine line in the distinct roles of specific sphingolipids as evidenced by recent findings that ceramide-1-phosphate has both proliferative and apoptotic effects depending on its concentration and context. Therefore, in this chapter, we summarized several years of investigation into ceramide, a canonically pro-apoptotic sphingolipid that is starting to gain momentum in cancer research for its potential roles in proliferation and cell-to-cell communication via exosomes. However, to provide context for this novel role of ceramide, we focused on cigarette smoke and its effects on airway epithelium as a model for ceramide's role in lung cancer.

Lung cancer and lung injury are common sequelae of chronic tobacco smoke exposure as it is well established by several epidemiological studies of smoker cohorts that lung cancer incidence is significantly increased in emphysema patients. Under exposure to cigarette smoke (CS) and oxidative stress, lung injury (apoptosis) and repair (proliferation) proceed simultaneously. Therefore, both "facets," apoptosis and proliferation, of the coin that underlie CS-mediated lung injury and lung cancer should be addressed together.

The studies presented herein clearly establish that these dichotomous signaling pathways are indeed activated under CS exposure. On the one hand, ceramide is generated by the activation of nSMase2 during CS-induced oxidative stress, eventually leading to apoptosis which is the dominant pathway in lung injury and COPD. On the other hand, the same stimulus aberrantly activates the EGFR, leading to prolonged proliferative signaling as observed in non-small cell lung cancer. Interestingly, recent studies provide evidence that these two CS-activated pathways may actually converge and interact. Specifically, Goldkorn et al. found that EGFR and Src are preferentially co-localized in ceramide-enriched regions of the plasma membrane under CS exposure, suggesting that nSMase2/ceramide may directly affect the aberrant EGFR/Src activation and mutual interaction, leading to enhanced tumorigenic signaling. Moreover, new findings suggest that CS exposure may induce resistance to EGFR-targeted therapy, the tyrosine kinase inhibitors (TKIs), solely through posttranslational molecular changes. These molecular alterations consist of an aberrant EGFR phosphorylation pattern accompanied by an aberrant conformation and the binding to Src during CS exposure. Furthermore, at the molecular level, these observed structural anomalies of the CS-activated EGFR are potentiated by excess ceramide and ceramide-enriched microdomains generated by the CS-activated nSMase2 in the plasma membrane of lung epithelial cells.

In summary, regard for sphingolipids in human health and disease has grown over the years. Despite the ongoing evolution of the field in cancer biology as presented here, there is an urgent and great need for investigation into the role of sphingolipids in lung cancer due to the plethora of promising observations reported. Therefore, we, the authors, hope that you, the readers, are convinced of the importance of sphingolipids in lung cancer.

References

Alberg AJ, Samet JM (2003) Epidemiology of lung cancer. Chest 123:21S–49S

Al-Nedawi K, Meehan B, Micallef J, Lhotak V, May L, Guha A et al (2008) Intercellular transfer of the oncogenic receptor EGFRvIII by microvesicles derived from tumour cells. Nat Cell Biol 10:619–624

Al-Nedawi K, Meehan B, Rak J (2009) Microvesicles: messengers and mediators of tumor progression. Cell Cycle 8:2014–2018

Bhowmick NA, Neilson EG, Moses HL (2004) Stromal fibroblasts in cancer initiation and progression. Nature 432:332–337

Caballero A, Torres-Duque CA, Jaramillo C, Bolivar F, Sanabria F, Osorio P et al (2008) Prevalence of COPD in five Colombian cities situated at low, medium, and high altitude (PREPOCOL study). Chest 133:343–349

Cai H, Babic I, Wei X, Huang J, Witte ON (2011) Invasive prostate carcinoma driven by c-Src and androgen receptor synergy. Cancer Res 71:862–872

Ceppi P, Rapa I, Lo Iacono M, Righi L, Giorcelli J, Pautasso M et al (2012) Expression and pharmacological inhibition of thymidylate synthase and Src kinase in nonsmall cell lung cancer. Int J Cancer 130:1777–1786

Chan C, Goldkorn T (2000) Ceramide path in human lung cell death. Am J Respir Cell Mol Biol 22:460–468

Chen X, Ba Y, Ma L, Cai X, Yin Y, Wang K et al (2008) Characterization of microRNAs in serum: a novel class of biomarkers for diagnosis of cancer and other diseases. Cell Res 18:997–1006

Chung BM, Dimri M, George M, Reddi AL, Chen G, Band V et al (2009) The role of cooperativity with Src in oncogenic transformation mediated by non-small cell lung cancer-associated EGF receptor mutants. Oncogene 28:1821–1832

Church DF, Pryor WA (1985) Free-radical chemistry of cigarette smoke and its toxicological implications. Environ Health Perspect 64:111–126

Clarke CJ, Guthrie JM, Hannun YA (2008) Regulation of neutral sphingomyelinase-2 (nSMase2) by tumor necrosis factor-alpha involves protein kinase C-delta in lung epithelial cells. Mol Pharmacol 74:1022–1032

Cremesti A, Paris F, Grassme H, Holler N, Tschopp J, Fuks Z et al (2001) Ceramide enables fas to cap and kill. J Biol Chem 276:23954–23961

D'Angelo SP, Pietanza MC, Johnson ML, Riely GJ, Miller VA, Sima CS et al (2011) Incidence of EGFR exon 19 deletions and L858R in tumor specimens from men and cigarette smokers with lung adenocarcinomas. J Clin Oncol 29:2066–2070

de Molina AR, de la Cueva A, Machado-Pinilla R, Rodriguez-Fanjul V, Del Pulgar TG, Cebrian A et al (2012) Acid ceramidase as a chemotherapeutic target to overcome resistance to the antitumoral effect of choline kinase alpha inhibition. Curr Cancer Drug Targets 12(6):617–24

de Torres JP, Bastarrika G, Wisnivesky JP, Alcaide AB, Campo A, Seijo LM et al (2007) Assessing the relationship between lung cancer risk and emphysema detected on low-dose CT of the chest. Chest 132:1932–1938

Devillard R, Galvani S, Thiers JC, Guenet JL, Hannun Y, Bielawski J et al (2010) Stress-induced sphingolipid signaling: role of type-2 neutral sphingomyelinase in murine cell apoptosis and proliferation. PLoS One 5:e9826

Ellis AG, Doherty MM, Walker F, Weinstock J, Nerrie M, Vitali A et al (2006) Preclinical analysis of the analinoquinazoline AG1478, a specific small molecule inhibitor of EGF receptor tyrosine kinase. Biochem Pharmacol 71:1422–1434

Esme H, Cemek M, Sezer M, Saglam H, Demir A, Melek H et al (2008) High levels of oxidative stress in patients with advanced lung cancer. Respirology 13:112–116

Filosto S, Fry W, Knowlton AA, Goldkorn T (2010) Neutral sphingomyelinase 2 (nSMase2) is a phosphoprotein regulated by calcineurin (PP2B). J Biol Chem 285:10213–10222

Filosto S, Castillo S, Danielson A, Franzi L, Khan E, Kenyon N et al (2011a) Neutral sphingomyelinase 2: a novel target in cigarette smoke-induced apoptosis and lung injury. Am J Respir Cell Mol Biol 44:350–360

Filosto S, Khan E, Tognon E, Becker C, Ashfaq M, Ravid T et al (2011b) EGF receptor exposed to oxidative stress acquires abnormal phosphorylation and aberrant activated conformation that impairs canonical dimerization. PLoS One 6:e23240

Filosto S, Ashfaq M, Chung S, Fry W, Goldkorn T (2012a) Neutral sphingomyelinase 2 activity and protein stability are modulated by phosphorylation of five conserved serines. J Biol Chem 287:514–522

Filosto S, Becker CR, Goldkorn T (2012b) Cigarette smoke induces aberrant EGF receptor activation that mediates lung cancer development and resistance to tyrosine kinase inhibitors. Mol Cancer Ther 11:795–804

Foley TD, Petro LA, Stredny CM, Coppa TM (2007) Oxidative inhibition of protein phosphatase 2A activity: role of catalytic subunit disulfides. Neurochem Res 32:1957–1964
Goldkorn T, Filosto S (2010) Lung injury and cancer: mechanistic insights into ceramide and EGFR signaling under cigarette smoke. Am J Respir Cell Mol Biol 43:259–268
Goldkorn T, Balaban N, Shannon M, Chea V, Matsukuma K, Gilchrist D et al (1998) H2O2 acts on cellular membranes to generate ceramide signaling and initiate apoptosis in tracheobronchial epithelial cells. J Cell Sci 111:3209–3220
Goldkorn T, Ravid T, Khan EM (2005) Life and death decisions: ceramide generation and EGF receptor trafficking are modulated by oxidative stress. Antioxid Redox Signal 7:119–128
Gomez-Munoz A, Kong JY, Salh B, Steinbrecher UP (2004) Ceramide-1-phosphate blocks apoptosis through inhibition of acid sphingomyelinase in macrophages. J Lipid Res 45:99–105
Grassme H, Jekle A, Riehle A, Schwarz H, Berger J, Sandhoff K et al (2001) CD95 signaling via ceramide-rich membrane rafts. J Biol Chem 276:20589–20596
Gridelli C, Maione P, Bareschino MA, Schettino C, Sacco PC, Ambrosio R et al (2010) Erlotinib in the treatment of non-small cell lung cancer: current status and future developments. Anticancer Res 30:1301–1310
Gulbins E, Kolesnick R (2003) Raft ceramide in molecular medicine. Oncogene 22:7070–7077
Hannun YA, Obeid LM (2011) Many ceramides. J Biol Chem 286:27855–27862
Heering J, Weis N, Holeiter M, Neugart F, Staebler A, Fehm TN et al (2012) Loss of the ceramide transfer protein augments EGF receptor signaling in breast cancer. Cancer Res 72(11):2855–66
Inokuchi J, Jimbo M, Momosaki K, Shimeno H, Nagamatsu A, Radin NS (1990) Inhibition of experimental metastasis of murine Lewis lung carcinoma by an inhibitor of glucosylceramide synthase and its possible mechanism of action. Cancer Res 50:6731–6737
Khan E, Heidinger J, Levy M, Lisanti M, Ravid T, Goldkorn T (2006) EGF receptor exposed to oxidative stress undergoes Src- and caveolin-1-dependent perinuclear trafficking. J Biol Chem 281:14486–14493
Khan EM, Lanir R, Danielson AR, Goldkorn T (2008) EGF receptor exposed to cigarette smoke is aberrantly activated and undergoes perinuclear trafficking. FASEB J 22:910–917
Kim H, Chan R, Dankort DL, Zuo D, Najoukas M, Park M et al (2005) The c-Src tyrosine kinase associates with the catalytic domain of ErbB-2: implications for ErbB-2 mediated signaling and transformation. Oncogene 24:7599–7607
Kim LC, Song L, Haura EB (2009a) Src kinases as therapeutic targets for cancer. Nat Rev Clin Oncol 6:587–595
Kim VN, Han J, Siomi MC (2009b) Biogenesis of small RNAs in animals. Nat Rev Mol Cell Biol 10:126–139
Kolesnick RN, Goni FM, Alonso A (2000) Compartmentalization of ceramide signaling: physical foundations and biological effects. J Cell Physiol 184:285–300
Kosaka N, Iguchi H, Yoshioka Y, Takeshita F, Matsuki Y, Ochiya T (2010) Secretory mechanisms and intercellular transfer of microRNAs in living cells. J Biol Chem 285:17442–17452
Kuebler WM, Yang Y, Samapati R, Uhlig S (2010) Vascular barrier regulation by PAF, ceramide, caveolae, and NO - an intricate signaling network with discrepant effects in the pulmonary and systemic vasculature. Cell Physiol Biochem 26:29–40
Lavrentiadou SN, Chan C, Ravid T, Tsaba A, van der Vliet A, Rasooly R et al (2001) Ceramide-mediated apoptosis in lung epithelial cells is regulated by GSH. Am J Respir Cell Mol Biol 25:676–684
Levy M, Castillo SS, Goldkorn T (2006) nSMase2 activation and trafficking are modulated by oxidative stress to induce apoptosis. Biochem Biophys Res Commun 344:900–905
Levy M, Khan E, Careaga M, Goldkorn T (2009) Neutral sphingomyelinase 2 is activated by cigarette smoke to augment ceramide-induced apoptosis in lung cell death. Am J Physiol Lung Cell Mol Physiol 297:125–133
Liu X, Sempere LF, Galimberti F, Freemantle SJ, Black C, Dragnev KH et al (2009) Uncovering growth-suppressive MicroRNAs in lung cancer. Clin Cancer Res 15:1177–1183

Liu X, Sempere LF, Ouyang H, Memoli VA, Andrew AS, Luo Y et al (2010) MicroRNA-31 functions as an oncogenic microRNA in mouse and human lung cancer cells by repressing specific tumor suppressors. J Clin Invest 120:1298–1309

Lynch TJ, Bell DW, Sordella R, Gurubhagavatula S, Okimoto RA, Brannigan BW et al (2004) Activating mutations in the epidermal growth factor receptor underlying responsiveness of non-small-cell lung cancer to gefitinib. N Engl J Med 350:2129–2139

Maceyka M, Harikumar KB, Milstien S, Spiegel S (2012) Sphingosine-1-phosphate signaling and its role in disease. Trends Cell Biol 22:50–60

Marcotte R, Zhou L, Kim H, Roskelly CD, Muller WJ (2009) c-Src associates with ErbB2 through an interaction between catalytic domains and confers enhanced transforming potential. Mol Cell Biol 29:5858–5871

Megha, London E (2004) Ceramide selectively displaces cholesterol from ordered lipid domains (rafts): implications for lipid raft structure and function. J Biol Chem 279:9997–10004

Merrill AH Jr (2011) Sphingolipid and glycosphingolipid metabolic pathways in the era of sphingolipidomics. Chem Rev 111:6387–6422

Minna JD, Gazdar AF, Sprang SR, Herz J (2004) Cancer. A bull's eye for targeted lung cancer therapy. Science 304:1458–1461

Mitchell PS, Parkin RK, Kroh EM, Fritz BR, Wyman SK, Pogosova-Agadjanyan EL et al (2008) Circulating microRNAs as stable blood-based markers for cancer detection. Proc Natl Acad Sci USA 105:10513–10518

Mitra P, Maceyka M, Payne SG, Lamour N, Milstien S, Chalfant CE et al (2007) Ceramide kinase regulates growth and survival of A549 human lung adenocarcinoma cells. FEBS Lett 581:735–740

Morales A, Lee H, Goni FM, Kolesnick R, Fernandez-Checa JC (2007) Sphingolipids and cell death. Apoptosis 12:923–939

Mountzios G, Fouret P, Soria JC (2008) Mechanisms of disease: signal transduction in lung carcinogenesis – a comparison of smokers and never-smokers. Nat Clin Pract Oncol 5:610–618

Mukhopadhyay A, Saddoughi SA, Song P, Sultan I, Ponnusamy S, Senkal CE et al (2009) Direct interaction between the inhibitor 2 and ceramide via sphingolipid-protein binding is involved in the regulation of protein phosphatase 2A activity and signaling. FASEB J 23:751–763

Mullen TD, Spassieva S, Jenkins RW, Kitatani K, Bielawski J, Hannun YA et al (2011) Selective knockdown of ceramide synthases reveals complex interregulation of sphingolipid metabolism. J Lipid Res 52:68–77

Nikolova-Karakashian M, Karakashian A, Rutkute K (2008) Role of neutral sphingomyelinases in aging and inflammation. Subcell Biochem 49:469–486

Nowak D, Antczak A, Krol M, Pietras T, Shariati B, Bialasiewicz P et al (1996) Increased content of H_2O_2 in the expired breath of cigarette smokers. Eur Respir J 9:652–657

Ogretmen B, Schady D, Usta J, Wood R, Kraveka JM, Luberto C et al (2001) Role of ceramide in mediating the inhibition of telomerase activity in A549 human lung adenocarcinoma cells. J Biol Chem 276:24901–24910

Oskouian B, Saba JD (2010) Cancer treatment strategies targeting sphingolipid metabolism. Adv Exp Med Biol 688:185–205

Paez JG, Janne PA, Lee JC, Tracy S, Greulich H, Gabriel S et al (2004) EGFR mutations in lung cancer: correlation with clinical response to gefitinib therapy. Science 304:1497–1500

Parkin DM, Bray F, Ferlay J, Pisani P (2005) Global cancer statistics, 2002. CA Cancer J Clin 55:74–108

Pegtel DM, Cosmopoulos K, Thorley-Lawson DA, van Eijndhoven MA, Hopmans ES, Lindenberg JL et al (2010) Functional delivery of viral miRNAs via exosomes. Proc Natl Acad Sci USA 107:6328–6333

Ponnusamy S, Meyers-Needham M, Senkal CE, Saddoughi SA, Sentelle D, Selvam SP et al (2010) Sphingolipids and cancer: ceramide and sphingosine-1-phosphate in the regulation of cell death and drug resistance. Future Oncol 6:1603–1624

Ravid T, Sweeney C, Gee P, Carraway KRK, Goldkorn T (2002) EGF receptor activation under oxidative stress fails to promote c-Cbl mediated down regulation. J Biol Chem 12:12
Ravid T, Tsaba A, Gee P, Rasooly R, Medina EA, Goldkorn T (2003) Ceramide accumulation precedes caspase-3 activation during apoptosis of A549 human lung adeno. cells. Am J Physiol Lung Cell Mol Physiol 284:L1082–1092
Rutkute K, Asmis RH, Nikolova-Karakashian MN (2007a) Regulation of neutral sphingomyelinase-2 by GSH: a new insight to the role of oxidative stress in aging-associated inflammation. J Lipid Res 48:2443–2452
Rutkute K, Karakashian AA, Giltiay NV, Dobierzewska A, Nikolova-Karakashian MN (2007b) Aging in rat causes hepatic hyperresposiveness to IL-1beta which is mediated by nSMase2. Hepatology 46:1166–1176
Schnitzer SE, Weigert A, Zhou J, Brune B (2009) Hypoxia enhances sphingosine kinase 2 activity and provokes sphingosine-1-phosphate-mediated chemoresistance in A549 lung cancer cells. Mol Cancer Res 7:393–401
Shigematsu H, Gazdar AF (2006) Somatic mutations of epidermal growth factor receptor signaling pathway in lung cancers. Int J Cancer 118:257–262
Skog J, Wurdinger T, van Rijn S, Meijer DH, Gainche L, Sena-Esteves M et al (2008) Glioblastoma microvesicles transport RNA and proteins that promote tumour growth and provide diagnostic biomarkers. Nat Cell Biol 10:1470–1476
Stancevic B, Kolesnick R (2010) Ceramide-rich platforms in transmembrane signaling. FEBS Lett 584:1728–1740
Sznajder JI, Fraiman A, Hall JB, Sanders W, Schmidt G, Crawford G et al (1989) Increased H_2O_2 in the expired breath of patients with acute hypoxemic respiratory failure. Chest 96:606–612
Trajkovic K, Hsu C, Chiantia S, Rajendran L, Wenzel D, Wieland F et al (2008) Ceramide triggers budding of exosome vesicles into multivesicular endosomes. Science 319:1244–1247
Uhlig S, Gulbins E (2008) Sphingolipids in the lungs. Am J Respir Crit Care Med 178:1100–1114
Watanabe S, Tanaka J, Ota T, Kondo R, Tanaka H, Kagamu H et al (2011) Clinical responses to EGFR-tyrosine kinase inhibitor retreatment in non-small cell lung cancer patients who benefited from prior effective gefitinib therapy: a retrospective analysis. BMC Cancer 11:1
Weinstein IB (2002) Cancer. Addiction to oncogenes–the Achilles heal of cancer. Science 297:63–64
Williams MD, Chance B (1983) Spontaneous chemiluminescence of human breath. Spectrum, lifetime, temporal distribution, and correlation with peroxide. J Biol Chem 258:3628–3631
Wu BX, Clarke CJ, Hannun YA (2010) Mammalian neutral sphingomyelinases: regulation and roles in cell signaling responses. Neuromolecular Med 12:320–330
Wu JY, Wu SG, Yang CH, Chang YL, Chang YC, Hsu YC et al (2011) Comparison of gefitinib and erlotinib in advanced NSCLC and the effect of EGFR mutations. Lung Cancer 72:205–212
Xu X, Bittman R, Duportail G, Heissler D, Vilcheze C, London E (2001) Effect of the structure of natural sterols and sphingolipids on the formation of ordered sphingolipid/sterol domains (rafts). Comparison of cholesterol to plant, fungal, and disease-associated sterols and comparison of sphingomyelin, cerebrosides, and ceramide. J Biol Chem 276:33540–33546
Yamada T, Bando H, Takeuchi S, Kita K, Li Q, Wang W et al (2011) Genetically engineered humanized anti-ganglioside GM2 antibody against multiple organ metastasis produced by GM2-expressing small-cell lung cancer cells. Cancer Sci 102:2157–2163
Yanaihara N, Caplen N, Bowman E, Seike M, Kumamoto K, Yi M et al (2006) Unique microRNA molecular profiles in lung cancer diagnosis and prognosis. Cancer Cell 9:189–198
Yang Y, Yin J, Baumgartner W, Samapati R, Solymosi EA, Reppien E et al (2010) Platelet-activating factor reduces endothelial nitric oxide production: role of acid sphingomyelinase. Eur Respir J 36:417–427
Yatabe Y (2010) EGFR mutations and the terminal respiratory unit. Cancer Metastasis Rev 29:23–36
Zhang Y, Li X, Becker KA, Gulbins E (2009) Ceramide-enriched membrane domains–structure and function. Biochim Biophys Acta 1788:178–183

Sphingolipids' Role in Radiotherapy for Prostate Cancer

Carla Hajj and Adriana Haimovitz-Friedman

Contents

Abstract There are several well-established mechanisms involved in radiation-induced cell death in mammalian cell systems. The p53-mediated apoptotic pathway is the most widely recognized mechanism (Lowe et al. Nature 362:847–849, 1993), although apoptosis has long been considered a less relevant mechanism of radiation-induced cell death (Steel, Acta Oncol 40:968–975, 2001; Brown and Wouters, Cancer Res 59:1391–1399, 1999; Olive and Durand, Int J Radiat Biol 71:695–707, 1997). We and others have recently focused instead on the emerging links between radiation, apoptosis, and ceramide and showed that ceramide is a

The authors have no conflict of interest or financial ties to disclose.

C. Hajj • A. Haimovitz-Friedman (✉)
Department of Radiation Oncology, Memorial Sloan-Kettering Cancer Center, 1275 York Avenue, New York, NY 10065, USA
e-mail: hajjc@mskcc.org; a-haimovitz-friedman@ski.mskcc.org

E. Gulbins and I. Petrache (eds.), *Sphingolipids in Disease*,
Handbook of Experimental Pharmacology 216,
DOI 10.1007/978-3-7091-1511-4_6, © Springer-Verlag Wien 2013

sphingolipid-derived second messenger capable of initiating apoptotic cascades in response to various stress stimuli, including radiation.

Ceramide, the backbone of all sphingolipids, is synthesized by a family of ceramide synthases (CerS), each using acyl-CoAs of defined chain length for *N*-acylation of the sphingoid long-chain base. Six mammalian CerS homologs have been cloned that demonstrated high selectivity towards acyl-CoAs (Lahiri et al. FEBS Lett 581:5289–5294, 2007), and more recently, it was shown that their activity can be modulated by dimer formation (Mesicek et al. Cell Signal 22:1300–1307, 2010; Laviad et al. J Biol Chem 283:5677–5684, 2008).

This de novo ceramide synthesis has been observed in irradiated cells through a pathway normally suppressed by ataxia telangiectasia-mutated (ATM) protein, a key component of the cellular response to DNA double-strand breaks (Liao et al. J Biol Chem 274:17908–17917, 1999). ATM is not the sole factor known to affect apoptotic potential by modulating CerS activity. Recent work has also implicated protein kinase Cα (PKCα) as a potential CerS activator (Truman et al. Cancer Biol Ther 8:54–63, 2009).

In this review, we summarize involvement of CerS in sphingolipid-mediated apoptosis in irradiated human prostate cancer cells and discuss future directions in this field.

Keywords Prostate cancer • Radioresistance • Ceramide synthase • Ceramide • PKCα

1 Radioresistance in Prostate Cancer: From Bedside to Bench

1.1 Local Relapse After Radiation Therapy

Prostate cancer is the most commonly diagnosed non-skin cancer in men in the United States. It is estimated that 217,730 men were diagnosed with prostate cancer during 2010. About 32,050 died of the disease the same year (Jemal et al. 2010). Radiotherapy (RT) is a widely used modality for men with prostate cancer, but although radiation is capable of permanently eradicating localized prostate tumors, nearly 30 % of patients treated with potentially curative doses relapse at the sites of the irradiated tumors (Scardino and Wheeler 1988; Crook et al. 1995; Zelefsky et al. 1998). More recently, continued progress in RT delivery techniques has improved outcome through dose escalation. Nonetheless, although higher radiation dose levels were consistently associated with improved biochemical control outcomes and reduction in distant metastases, the biochemical relapse rate was still greater than 60 % at 5 years in high-risk patients (Zelefsky et al. 2008). These data indicate that prostate tumors vary in sensitivity to ionizing radiation. Furthermore, clinical data show that in patients who relapse locally after RT, initial treatment eliminates the great majority of the tumor cells, whereas a small fraction of tumor clonogens survive the lethal effects of radiation and eventually repopulate the irradiated site.

This observation indicates that there are variations in clonal sensitivity to the lethal effects of radiation even within a given tumor. Thus far, there have been neither a criteria for predicting the presence or prevalence of radiation-resistant tumor clones nor an effective approach to modulate the radiation response of human prostate tumor cells. Improved understanding of pathways of radiation-induced cell death and signaling systems that regulate these pathways may yield opportunities for pharmacological modulation of radiation resistance in prostate cancer.

1.2 Human Prostate Cancer Cell Lines' Response to Radiation

Cell lines derived from human prostate cancer are regarded as relatively resistant to both radiation-induced clonogenic death and apoptotic death (Royai et al. 1996). The best-characterized human prostate cell lines include the PC-3, DU-145, CWR22-Rv1, and LNCaP cells that were established from metastatic human tumor lesions (Stone et al. 1978; Kaighn et al. 1979; Horoszewicz et al. 1983). In general, these cell lines are among the most radioresistant human tumor cells, as assessed by the clonogenic assay (Wollin et al. 1989; Leith et al. 1993; Leith 1994; DeWeese et al. 1998). However, the dose-survival data do indicate distinct differences between these cell lines, as expressed by the Dq, D_0, SF-2, and the linear quadratic α and β exponents. There are also differences in the apoptotic response to radiation. Several studies reported lack of apoptosis in PC-3 cells up to 72 h after exposure to doses of 10–30 Gy (Algan et al. 1996; Li and Franklin 1998) and an incidence of 10–15 % apoptosis in DU-145 cells at 72 h after 10–12 Gy (Algan et al. 1996; Bowen et al. 1998). However, one study reported 40 % apoptosis in PC-3 cells at 72 h after 20 Gy (Kyprianou et al. 1997). This study is also the only study that has thus far reported apoptosis in LNCaP cells, occurring at a rate of 35 % at 72 h after exposure to 20 Gy (Kyprianou et al. 1997). Altogether, these observations indicate clone-specific sensitivities of human prostate tumor cells to radiation. The pleiotropic nature of death pathways induced by radiation suggests that radiation resistance is likely to be regulated by a variety of mechanisms, each of which is associated with a specific death pathway. Whether radiation resistance of human prostate tumor clones is associated with a single mechanism or a spectrum of mechanisms is unknown. These data also suggest that an approach to reduce radiation resistance clinically might require the use of combinations of chemical and biological modifiers to cover a spectrum of resistance mechanisms that may operate concomitantly in prostate cancer.

1.3 Radiation-Induced DNA Damage

It is well accepted that radiation-induced cell death stems from lethal DNA damage. When cells are irradiated, X-rays induce numerous DNA breaks. All break types do not have the same biologic consequence as far as cell killing is concerned. Many DNA single-strand breaks are readily repaired using the opposite DNA strand as a template. Breaks in both strands, if well separated, are also readily repaired because they are handled independently. Breaks in both strands that are opposite or separated by only a few base pairs may lead to double-strand breaks (DSBs). There is increasing amount of evidence that DSBs rather than single-strand breaks lead to important biologic endpoints, including "mitotic catastrophe" and loss of replicative potential (Hall and Giaccia 2006a). Such lesions are produced in the DNA by direct interaction with X-rays or with reactive oxygen intermediates generated within the cell by the radiation. DNA DSBs, the most lethal form of ionizing radiation-induced damage, are repaired by nonhomologous end-joining (NHEJ) repair in the G1 phase of the cell cycle and homologous recombination (HR) repair pathway in the S/G2 phase of the cell cycle. Whereas most radiation-induced DNA DSBs are rapidly repaired by constitutively expressed DNA repair mechanisms, residual unrepaired or misrepaired breaks lead to genetic instability and to increased frequency of mutations and chromosomal aberrations. Lethal mutations or dysfunctional chromosomal aberrations eventually lead to either progeny cell death, usually after several mitotic cycles (also termed reproductive or postmitotic), or to p53-mediated apoptosis (Hall and Giaccia 2006b).

1.4 Radiation-Induced Apoptosis

Radiation-induced apoptosis has long been considered a less relevant mechanism in cell loss from normal tissues and tumors based on published data comparing apoptosis response and cell survival responses in tumor cells that have generally failed to find a causal relationship. Moreover, modulating apoptotic potential usually had little impact on cellular radiosensitivity (Steel 2001; Brown and Wouters 1999; Olive and Durand 1997; Kyprianou et al. 1997; Aldridge et al. 1995; Lock and Ross 1990). Apoptosis or programmed cell death is an intermitotic (interphase) inducible death pathway of sequential biochemical events that are constitutively expressed in an inactive form in most, if not all, mammalian cells. Also defined as a mechanism of cellular suicide, apoptosis occurs in response to a variety of physiological or environmental stresses impacting distinct cellular targets to initiate cell type-specific apoptotic signaling pathways. The various upstream signaling cascades converge downstream to activate a common final caspase-dependent effector mechanism eventually leading to activation of a calcium–magnesium-dependent endonuclease that cleaves the nuclear chromatin at selective inter-nucleosomal linker sites, thus dismantling the dying cell. Chromosomal fragmentation, cytoplasmic blebbing, and apoptotic bodies are consequently seen during apoptosis, which ultimately results in the condensation of the nucleus and shrinking of the cell. Apoptosis is characteristically different from cell necrosis in morphology and biochemistry, and its end result is cell death without inflammation of the surrounding tissue. After a few decades and after quite a few debates, the case against apoptosis is no longer successfully defended. Today, we acknowledge apoptosis, or programmed cell death, not only as the process leading to disorders of normal tissues (Orrenius 1995; Fadeel et al. 1999; Reed 2002; Mullauer et al. 2001) but also as a form of death in response to both chemotherapy and radiation therapy for cancer, in particular for hormone-dependent cancers such as prostate cancer (Wu 1996; Olson and Kornbluth 2001; Ameisen 2002; Meyn et al. 2009). As a matter of fact, facilitation of apoptosis in vivo has been shown to effectively increase the number of apoptotic cells in tumors (Dubray et al. 1998; Jansen et al. 2000), and the detected early apoptotic response correlates well with subsequent outcome (Symmans et al. 2000; Meyn et al. 1995; Ellis et al. 1997). Proapoptotic antibodies, anti-CD-95 antibodies, are considered to be used as new therapeutic agents for tumor treatment. Moreover, the emerging links between radiation, apoptosis, and ceramide suggest that ceramide-mediated apoptosis following ionizing radiation might explain part of normal tissue and tumor cell death.

2 Activation of the Ceramide Synthase Pathway in Response to Radiation

We have previously shown that DNA DSBs induce ceramide generation via the de novo ceramide synthesis pathway involving activation of CerS enzyme in bovine aortic endothelial cells (BAEC) (Liao et al. 1999). Actually, a variety of ceramide species syntheses are catalyzed by six CerS enzyme isotypes (Table 1). CerS are integral membrane proteins of the endoplasmic reticulum. Once activated, these enzymes catalyze the condensation of sphinganine and fatty acyl-CoA to form dihydroceramide, which is rapidly oxidized to ceramide. Originally, we showed that daunorubicin stimulates CerS and generates ceramide resulting in apoptosis in P388 murine leukemia cells and U937 human monoblastic leukemia cells (Bose et al. 1995). An obligatory role for CerS was thus defined, since its natural specific inhibitor, Fumonisin B1 (FB1), blocked daunorubicin-induced ceramide elevation and apoptosis. This study was the first to demonstrate that CerS activity is regulated in eukaryotes and constituted definitive evidence for a requirement for ceramide elevation in the induction of apoptosis.

In collaboration with Anthony Futerman's group [Weizmann Institute of Science, Israel (Lahiri et al. 2007)], we demonstrated that despite the high selectivity towards acyl-CoAs, mammalian CerS have a very similar K_m value towards sphinganine, strengthening the notion that the main biochemical difference between CerS is in their specificity towards acyl-CoAs. In addition, in these studies, conditions for assaying CerS activity were optimized, and a K_m value of all six mammalian CerS towards sphinganine in the low μM range (2–5 μM) was demonstrated, which is

Table 1 Distinct ceramide species confer discrete and even opposing signaling endpoints including apoptosis and cell survival (*VLCFA*: *Very Long Chain Fatty Acid*)

CerS/LASS	Fatty acyl-CoA substrate chain length	Tissue specific expression patterns
CerS1	C18:0-fatty acyl-CoA	Nervous system
CerS2	C24:0 and C24:1 (VLCFA)	Broad tissue distribution
CerS3	C24:0 and C24:1 (VLCFA)	Testis and keratinocyte
CerS4	C18:0; C20:0	Broad tissue distribution
CerS5	C16:0	Broad tissue distribution
CerS6	C16:0	Broad tissue distribution

significantly lower than that suggested using a variety of other assays and tissue sources.

Subsequently, we reported that ionizing radiation also induces de novo synthesis of ceramide in HeLa cells resulting in apoptosis by specifically activating CerS5 and CerS6 (Mesicek et al. 2010). Overexpression of CerS2 resulted in partial protection from RT-induced apoptosis, whereas overexpression of CerS5 generated the apoptogenic ceramide species C_{16}-ceramide and increased apoptosis in these cells. Knockdown studies determined that CerS2 is responsible for all observable RT-induced C(24:0) CerS activity, and while CerS5 and CerS6 each confer approximately 50 % of the C(16:0) CerS baseline synthetic activity, both are required for RT-induced activity. Additionally, co-immunoprecipitation studies suggest that CerS2, 5, and 6 might exist as heterocomplexes in HeLa cells, providing further insight into the regulation of CerS enzymes. Moreover, CerS were shown to have additional subcellular localizations, such as perinuclear membranes and the mitochondria-associated membrane (MAM). The interplay between long-chain C_{16}-ceramide and very long-chain $C_{24:1}$- and C_{24}-ceramides has come into recent spotlight regarding their roles in maintaining cellular homeostasis. These data add to the growing body of evidence demonstrating interplay among the CerS isozymes in a stress stimulus-, cell type-, and subcellular compartment-specific manner.

CerS can be phosphorylated (Sridevi et al. 2009) and glycosylated (Mizutani et al. 2006), and it was recently demonstrated that CerS can form both homo- and heterodimers (Laviad et al. 2008) and that the activity of one member of a heterodimer depends upon, and can be modulated by, the activity of the other member. These studies suggest a rapid and reversible mechanism of regulating CerS activity by dimer formation.

CerS engagement in mammalian apoptosis in response to ionizing radiation, drugs, and cytokines (Truman et al. 2009; Garcia-Bermejo et al. 2002; Canman and Kastan 1995; Kastan et al. 1992) is a cell type-specific phenomena. The biological significance of CerS as a transducer of apoptosis has also been confirmed in vivo, as CerS transactivates disease pathogenesis in gastric ulcer, ischemia/reperfusion, and emphysema rodent models of human disease (Lowe et al. 1993; Spiegel et al. 1996; Haimovitz-Friedman et al. 1997; Lin et al. 2000; Mathias et al. 1998; Pena et al. 1997). In these reports, epithelial apoptosis occurred 18–24 h after radiation exposure or about 6–10 h after CerS activation and was inhibited by FB1.

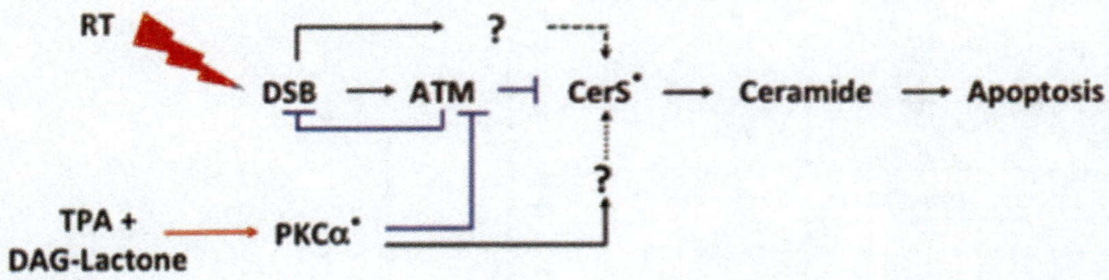

Fig. 1 Integration of radiation- and PKCα-induced and apoptosis (*Asterisk* denotes activated). TPA and DAG-Lactone simultaneously down-regulated ATM levels and enhance CerS activity in human prostate cancer cells *via* activation of PKCα. Radiation acts synergistically on CerS activation by inducing DNA DSBs, once the ATM inhibition is removed. This results in higher accumulation of ceramide levels in these cells and radiation-induced apoptosis both in vitro and in vivo [modified from Truman et al. (2009)]

3 Modulation of Ceramide Synthase Activity in Prostate Cancer Cells

There are several well-established mechanisms involved in radiation-induced apoptosis in mammalian cell systems. The p53-mediated pathway is the most widely recognized mechanism (Lowe et al. 1993). LNCaP and CWR22-Rv1 human prostate cell lines express wild-type p53 and exhibit extreme resistance to radiation-induced apoptosis. Yet, we found that these cells display high sensitivity to radiation-induced apoptosis both in vitro and in vivo when pretreated with TPA (12-*O*-tetradecanoylphorbol-13-acetate) or other PKC activators (Truman et al. 2009). Studies in our laboratory focus on ceramide-mediated apoptosis, a response to radiation that is distinct from the classical p53-mediated response. Our work indicates that the apoptotic pathway involved in LNCaP and CWR22-Rv1 radiosensitization is mediated by activation of CerS (Fig. 1), and not by other known pathways of ceramide-mediated apoptosis involving the activation of either acid sphingomyelinase (ASMase) or neutral sphingomyelinase (NSMase) (Garzotto et al. 1998, 1999).

3.1 PKCα Activation of Ceramide Synthase in Prostate Cancer Cells

TPA treatment of LNCaP cells activates PKCα and PKCδ to induce CerS activation (Garzotto et al. 1998) with rapid but progressive ceramide generation. This was followed by a delayed form of apoptosis that reached maximal levels at 48 h. Investigations into the mechanism of TPA-induced ceramide generation revealed that ASMase and NSMase activities were not enhanced. In contrast, TPA induced an increase in CerS activity that persisted for at least 16 h. Treatment with FB1 abrogated both TPA-induced ceramide production and

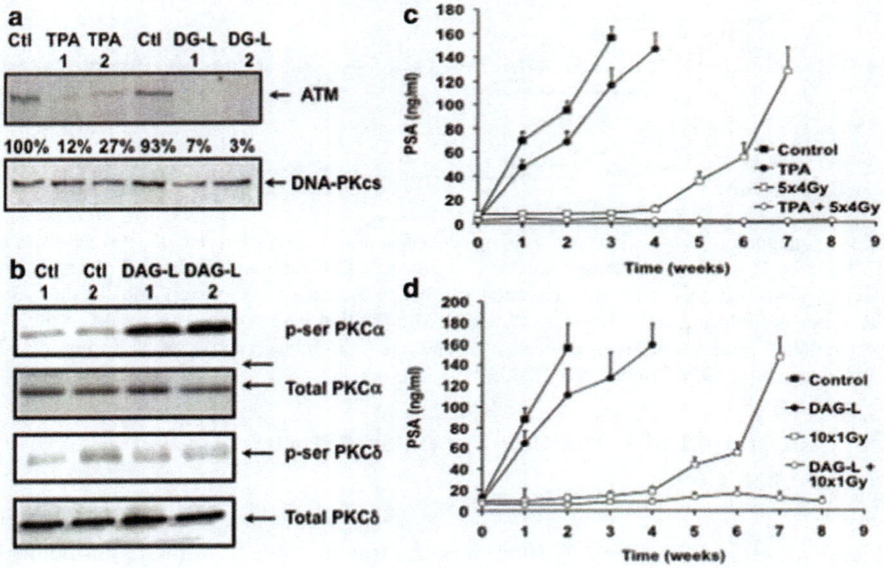

Fig. 2 Effect of PKCα stimulation on orthotopic prostate cancer model. (**a**) DAG-lactone was injected twice i.p. into orthotopically-implanted nude mice at 24 h intervals at a concentration of 12 mg/kg per day. Prostate tumors were removed 48 h after the first injection, homogenized and prepared for Western blot as described in the "Materials and Methods." The density of the bands was measured and is shown in comparison to the control. The results shown are representative of three separate experiments. (**b**) Western blots for prostate tumor samples were probed with either phospho-serine specific Abs to PKCα or PKCδ, and compared with total levels of PKCα or PKCδ as a control. The results are representative of two separate experiments. (**c**) PSA levels in TPA-treated mice. Mice were injected i.p. with 40 μg/kg TPA or DMSO solvent control for 5 days. Radiation-treated mice were locally irradiated with 4 Gy 16 h after each TPA injection for a total of 20 Gy. Control non-irradiated mice $n = 20$, radiated-treated control mice $n = 12$, TPA non-irradiated mice $n = 17$, and TPA-treated irradiated mice $n = 10$. (**d**) PSA levels in DAG-lactone treated mice. Mice were injected i.p. with 12 mg/kg per day of DAG-lactone or with DMSO solvent control for 10 days. Radiation-treated mice were locally irradiated at 1 Gy 16 h after each injection to a total of 10 Gy. Control non-irradiated mice $n = 14$, radiation-treated control mice $n = 14$, DAG-lactone treated mice $n = 13$, radiation-treated DAG-lactone mice $n = 12$. The values represent the mean ± standard errors (Truman et al. 2009)

apoptosis. Thus, CerS activation appears to be required for TPA-induced apoptosis in LNCaP cells (Garzotto et al. 1998).

Whereas LNCaP cells failed to respond to radiation with ceramide generation and apoptosis, pretreatment with TPA or diacylglycerol (DAG)-lactone, a potent and specific PKCα activator in LNCaP cells, converted this pattern, enabling radiation to signal CerS activation and apoptosis (Fig. 1) (Truman et al. 2005). The catalytic domain of activated PKCα is required for this event, since LNCaP cells transfected with a kinase-dead dominant-negative mutant of PKCα showed no increase in ceramide generation or in apoptosis levels, indicating that PKCα plays a role in CerS activation. Furthermore, treatment of nude mice with intravenous TPA or DAG-lactone elicited specific PKCα activation leading to downregulation of ataxia telangiectasia-mutated (ATM) protein in prostate tumors in vivo and consequently enhanced radiation-induced tumor response in orthotopically implanted LNCaP cells (Fig. 2). This represents the first description of a signaling-based therapy designed to overcome one form of radiation resistance expressed preferentially in human prostate cancer cells (Truman et al. 2009).

3.2 *ATM Negatively Regulates Ceramide Synthase Activity*

By use of (Liao et al. 1999) metabolic incorporation of ^{125}I-labeled 5-iodo-2′deoxyuridine ([^{125}I] dURd), which produces DNA DSBs and external beam irradiation, de novo ceramide synthesis was initiated by posttranslational activation of CerS in BAEC. In the same study, it was shown that ATM negatively regulates CerS activity (Liao et al. 1999). Subsequently, it was shown that TPA decreases ATM protein levels in LNCaP and CWR22Rv1 cells, an event crucial for the induction of apoptosis and mediating radiosensitization of these cells. The alteration in ATM protein levels appeared to be due to inhibition of ATM transcription via decreased binding of transcription factor Sp1 to the ATM promoter. The additional TPA-mediated effector in downregulating ATM transcription remains unknown, nor is there information on the precise mechanism by which TPA affects Sp1 binding. The catalytic domain of activated PKCα is required for this event, since LNCaP cells transfected with a kinase-dead dominant-negative mutant of PKCα showed no decrease in ATM protein. The enabling of CerS-mediated apoptosis in both androgen-sensitive (LNCaP, CWR22-Rv1) and androgen-insensitive (PC-3, DU-145) prostate cancer cells via reduction of ATM protein levels indicates that the interaction between ATM kinase and CerS may represent a generic mode of regulation of radiation-induced death in these cells (Fig. 1). Whereas ATM kinase normally represses CerS activity, a reduction in ATM protein by TPA or antisense ATM oligonucleotide AS-ATM-ODN treatment attenuates CerS repression, enabling ceramide synthesis and a proapoptotic state. However, the experiments with AS-ATM-ODN indicated that ATM reduction by itself is not sufficient to induce CerS activation and that a second signal, such as that provided by radiation, is required. Consistent with this observation, TPA mimicked radiation in inducing apoptosis in AS-ATM-ODN-treated LNCaP cells.

ATM inactivation by AS-ATM-ODN conferred extreme radiation hypersensitivity, since the level of apoptosis attained in AS-ATM-ODN-treated LNCaP exposed to 2 Gy was already at 72 % of the maximum effect observed with 20 Gy (32 ± 0.73 %), both significantly higher than the 3.4 ± 0.2 % observed after 20 Gy exposure of ODN-untreated cells (Truman et al. 2005). The mechanism by which ATM kinase regulates CerS remains unknown. Whereas ATM inhibition has been shown to affect progression through the G2/M checkpoint in some cell types (Shiloh 2001), our studies indicate that deregulation G2/M is not the mechanism of prostate cancer cell radiosensitization. The significant apoptotic response observed using clinically relevant doses of 1 Gy and 2 Gy in AS-ATM-ODN-treated cells suggests ATM as a potential molecular target for clinical application. Further development of AS reagents, siRNA, or small molecules aimed at ATM inactivation would appear warranted in the treatment of prostate cancer.

3.3 PKCα Downregulation of ATM in Prostate Cancer In Vivo

PKCα activation was shown to radiosensitize prostate cancer in vivo (Fig. 2). While fractionated radiation alone generated only a delayed tumor growth response in xenografts of human prostate tumors growing in nude mice, pretreatment with DAG-lactone followed by radiation resulted in permanent local tumor control. The detailed mechanistic pathway involved in this response remains unknown. We are currently trying to determine whether the radiation component of the apoptotic response in LNCaP cells is synergistic with the PKCα component or whether it promotes an autonomous hypersensitized response, from CerS, perhaps via direct activation of radiation-sensitive CerS homologs, whose activation is also enabled by PKCα-mediated ATM downregulation. Therefore, activation of PKCα regulates two main events necessary for the apoptotic response in these human prostate cells (1) ATM downregulation and (2) CerS activation. Since ATM is overexpressed in these cells, radiation is incapable of activating CerS unless a significant amount of ATM protein is downregulated (Fig. 1). In contrast to radiation-induced CerS activation, PKCα-mediated CerS activation does not depend on generation of DNA DSBs in these cells (Galvin and Haimovitz-Friedman, unpublished observations).

3.4 TNF-α and Fas Enhance Radiation-Induced Apoptosis in Prostate Cells

TNF-α was also shown to sensitize LNCaP prostate cancer cells to γ-radiation-induced apoptosis when added 24 h prior to radiation (Kimura et al. 1999). Simultaneous exposure of LNCaP cells to TNF-α and 8 Gy resulted in increased ceramide generation that correlated with a threefold increase in apoptotic cells within 72 h compared to TNF-α treatment alone. LNCaP cells could also be sensitized, although to a lesser degree, by the agonistic FAS antibody, CH-11 (Kimura and Gelmann 2000). In this study, TNF-α increased production of ceramide in LNCaP cells 48 h after exposure. Moreover, nontoxic levels of exogenous C_2-ceramide sensitized LNCaP cells to radiation similarly to TNF-α, suggesting that increased intracellular ceramide could explain the mechanism by which LNCaP cells were sensitized to radiation and even to chemotherapy. Further studies proved that TNF-α induced enhanced activation of the intrinsic apoptotic pathway and enhanced cell death, with or without γ-irradiation, yet CerS activity was not reported. Both TNF-α and γ-irradiation elevated levels of endogenous ceramide and activated the intrinsic cell death pathway in a synergistic fashion (Kimura et al. 1999, 2003).

Moreover, it was shown in LNCaP cells that androgen inhibits apoptosis induced by both TNF-α and by CD95 activation with or without concomitant irradiation. This was thought to be mediated by androgen blockade of caspase activation in both intrinsic and extrinsic cell death pathways, but whether SMase is involved in LNCaP cell apoptosis remains to be proven (Kimura et al. 2003). In addition, TNF-α and radiation induced a significant increase in sphingosine levels and markedly reduced sphingosine-1-phosphate (S1P).The increase in sphingosine levels either by exogenous sphingosine or by treatment with the sphingosine kinase (SphK) inhibitor induced apoptosis and also radiosensitized LNCaP cells in this study.

Increasing amount of data now suggest that the relative levels of proapoptotic sphingolipid metabolites, such as certain ceramide species and sphingosine, and the levels of the antiapoptotic sphingolipid metabolites, such as S1P, might play a role in determining the radiosensitivity of prostate cancer cells (Nava et al. 2000). In fact, SphK1 and S1P receptors are highly expressed in chemotherapy-resistant prostate cancer PC3 cells and are upregulated by anticancer drug camptothecin (Akao et al. 2006). Pchejetsky et al. (2008) showed that selective pharmacologic inhibition of SphK1 triggers apoptosis in LNCaP and PC3 cells, an effect reversed by SphK1 enforced expression. Nevertheless, further investigations are necessary to clarify the role of SphK1 and S1P in apoptosis in prostate cells.

3.5 *Modulation of Prostate Cancer Cell Response via Acid Ceramidase Activity*

Acid ceramidase (AC) converts ceramide into sphingosine and was found to be overexpressed in 60 % of primary prostate cancer tissues (Liu et al. 2009; Norris et al. 2006; Seelan et al. 2000). Conversely, upregulation of AC in prostate cancer cells conferred resistance to both chemo- and radiotherapy (Saad et al. 2007; Mahdy et al. 2009). Along the same line, autophagy was increased in prostate cancer cells overexpressing AC, thereby enhancing resistance to C_6-ceramide. This resistance was overcome via modulation of radiation effect by using AC inhibitors (Liu et al. 2009). Interestingly, in another hormone-regulated cancer, of the breast, C_6_ceramide and targeted inhibition of AC were also shown to induce synergistic decreases in the cancer cell growth (Flowers et al. 2012).

Resveratrol (3,5,4′-*trans*-trihydroxystilbene), a natural product from grapes and present in red wine, was shown to be synergistic with radiation in androgen-independent and otherwise radioresistant DU145 human prostate cancer cell line by promoting de novo ceramide biosynthesis in these cells, but no mechanism was suggested (Scarlatti et al. 2007).

4 Summary and Future Directions

In conclusion, both DNA DSB-induced damage and non-DNA DSB-induced damage contribute to cell killing of human prostate cancer cells after exposure to ionizing irradiation. The relative contribution from each mode of cell death may differ with dose and from one cell type to another, relative to their inherent and inducible capacities to overcome each of these types of lethal radiation damage and according to their microenvironment. Utilization of the SMase pathway for induction of apoptosis in response to cell death receptor signals and ionizing radiation has now been demonstrated in a large number of mammalian cells. An alternative mechanism to SMase-mediated generation of ceramide in response to stress is a pathway that involves de novo synthesis of ceramide, via activation of CerS. CerS activation was shown to be of particular importance in a variety of human prostate cancer cell lines (LNCaP, CWR22-Rv1, PC3, and DU145) that under normal conditions exhibit extreme resistance to radiation-induced apoptosis. Yet we found that these cells display high sensitivity to radiation-induced apoptosis both in vitro and in vivo when pretreated with specific PKCα activators. We have also demonstrated that PKCα activation suppresses ATM, derepressing CerS activity, thus enabling generation of apoptogenic ceramide.

Our in vivo data using the prostate orthotopic model suggest that development of radiation response modulators for human prostate cancer affecting the ATM/CerS pathway is essential to overcome radiation resistance in this cancer. Recently, a selective tissue and subcellular distribution of the six mammalian CerS isoforms, combined with distinct fatty acyl chain length substrate preferences, has been described (Table 1) (Mesicek et al. 2010). This implicates differential functions of specific ceramide species in cellular signaling in a stress stimulus-, cell type-, and subcellular compartment-specific manner. Understanding the contribution of the different CerSs homologs and their mode of activation within the different prostate cancer cell types may provide new therapeutic molecular targets to manipulate the radiation response of these cells. These studies may pave the way for future studies looking at DNA-damage-inducing chemotherapeutic drugs and radiation to be used in combination with specific PKCα activators and molecular targets within the sphingolipids metabolic pathways to combat metastatic prostate cancer. Using DAG-lactone as a radiosensitizer via activation of PKCα is currently being considered as a therapeutic approach for prostate cancer at our center.

References

Akao Y, Banno Y, Nakagawa Y, Hasegawa N, Kim TJ, Murate T, Igarashi Y, Nozawa Y (2006) High expression of sphingosine kinase 1 and S1P receptors in chemotherapy-resistant prostate cancer PC3 cells and their camptothecin-induced up-regulation. Biochem Biophys Res Commun 342:1284–1290

Aldridge DR, Arends MJ, Radford IR (1995) Increasing the susceptibility of the rat 208F fibroblast cell line to radiation-induced apoptosis does not alter its clonogenic survival dose–response. Br J Cancer 71:571–577

Algan O, Stobbe CC, Helt AM, Hanks GE, Chapman JD (1996) Radiation inactivation of human prostate cancer cells: the role of apoptosis. Radiat Res 146:267–275

Ameisen JC (2002) On the origin, evolution, and nature of programmed cell death: a timeline of four billion years. Cell Death Differ 9:367–393

Bose R, Verheij M, Haimovitz-Friedman A, Scotto K, Fuks Z, Kolesnick R (1995) Ceramide synthase mediates daunorubicin-induced apoptosis: an alternative mechanism for generating death signals. Cell 82:405–414

Bowen C, Spiegel S, Gelmann EP (1998) Radiation-induced apoptosis mediated by retinoblastoma protein. Cancer Res 58:3275–3281

Brown JM, Wouters BG (1999) Apoptosis, p53, and tumor cell sensitivity to anticancer agents. Cancer Res 59:1391–1399

Canman CE, Kastan MB (1995) Induction of apoptosis by tumor suppressor genes and oncogenes [Review]. Semin Cancer Biol 6:17–25

Crook JM, Perry GA, Robertson S, Esche BA (1995) Routine prostate biopsies following radiotherapy for prostate cancer: results for 226 patients. Urology 45:624–631

DeWeese TL, Shipman JM, Dillehay LE, Nelson WG (1998) Sensitivity of human prostatic carcinoma cell lines to low dose rate radiation exposure. J Urol 159:591–598

Dubray B, Breton C, Delic J, Klijanienko J, Maciorowski Z, Vielh P, Fourquet A, Dumont J, Magdelenat H, Cosset JM (1998) *In vitro* radiation-induced apoptosis and early response to low-dose radiotherapy in non-Hodgkin's lymphomas. Radiother Oncol 46:185–191

Ellis PA, Smith IE, McCarthy K, Detre S, Salter J, Dowsett M (1997) Preoperative chemotherapy induces apoptosis in early breast cancer. Lancet 349:849

Fadeel B, Orrenius S, Zhivotovsky B (1999) Apoptosis in human disease: a new skin for the old ceremony. Biochem Biophys Res Commun 266:699–717

Flowers M, Fabriás G, Delgado A, Casas J, Abad JL, Cabot MC (2012) C6-Ceramide and targeted inhibition of acid ceramidase induce synergistic decreases in breast cancer cell growth. Breast Cancer Res Treat 133:447–458

Garcia-Bermejo ML, Leekow FC, Fujii T, Wang Q, Blumberg PM, Ohba M, Kuroki J, Han K-C, Lee J, Marquez VE et al (2002) Diacylglycerol (DAG)-lactones, a new class of protein kinase C (PKC) agonists, induce apoptosis in LNCaP prostate cancer cells by selective activation of PKCα. J Biol Chem 277:645–655

Garzotto M, White-Jones M, Jiang Y, Ehleiter D, Liao WC, Haimovitz-Friedman A, Fuks Z, Kolesnick R (1998) 12-O-tetradecanoylphorbol-13-acetate-induced apoptosis in LNCaP cells is mediated through ceramide synthase. Cancer Res 58:2260–2264

Garzotto M, Haimovitz-Friedman A, Liao WC, White-Jones M, Huryk R, Heston DWW, Cardon-Cardo C, Kolesnick R, Fuks Z (1999) Reversal of radiation resistance in LNCaP cells by targeting apoptosis through ceramide synthase. Cancer Res 59:5194–5201

Haimovitz-Friedman A, Kolesnick RN, Fuks Z (1997) Ceramide signaling in apoptosis. Br Med Bull 53:539–553

Hall EJ, Giaccia AJ (2006a) DNA strand breaks and chromosomal aberrations. In: McAllister L, Bierig L, Barrett K (eds) Radiobiology for the radiologist. Lippincott, Williams & Wilkins, Philadelphia, PA, pp 16–29

Hall EJ, Giaccia AJ (2006b) Repair of radiation damage and the dose-rate effect. In: McAllister L, Bierig L, Barrett K (eds) Radiobiology for the radiologist. Lippincott, Williams & Wilkins, Philadelphia, PA, pp 85–105

Horoszewicz JS, Leong SS, Kawinski E, Karr JP, Rosenthal H, Chu TM, Mirand EA, Murphy GP (1983) LNCaP model of human prostatic carcinoma. Cancer Res 43:1809–1818

Jansen B, Wacheck V, Heere-Ress E, Schlagbauer-Wadl H, Hoeller C, Lucas T, Hoermann M, Hollenstein U, Wolff K, Pehamberger H (2000) Chemosensitization of malignant melanoma by BCL2 antisense therapy. Lancet 356:1728–1733

Jemal A, Siegel R, Xu J, Ward E (2010) Cancer statistics. CA Cancer J Clin 60:277–300
Kaighn ME, Narayan KS, Ohnuki Y, Lechner JF, Jones LW (1979) Establishment and characterization of a human prostatic carcinoma cell line (PC-3). Invest Urol 17:16–23
Kastan MB, Zhan Q, el-Deiry WS, Carrier F, Jacks T, Walsh WV, Plunkett BS, Vogelstein B, Fornace AJ Jr (1992) A mammalian cell cycle checkpoint pathway utilizing p53 and GADD45 is defective in ataxia-telangiectasia. Cell 71:587–597
Kimura K, Gelmann EP (2000) Tumor necrosis factor-alpha and Fas activate complementary Fas-associated death domain-dependent pathways that enhance apoptosis induced by gamma-irradiation. J Biol Chem 275:8610–8617
Kimura K, Bowen C, Spiegel S, Gelmann EP (1999) Tumor necrosis factor-alpha sensitizes prostate cancer cells to gamma-irradiation-induced apoptosis. Cancer Res 59:1606–1614
Kimura K, Markowski M, Edsall LC, Spiegel S, Gelmann EP (2003) Role of ceramide in mediating apoptosis of irradiated LNCaP prostate cancer cells. Cell Death Differ 10:240–248
Kyprianou N, King ED, Bradbury D, Rhee JG (1997) Bcl-2 over-expression delays radiation-induced apoptosis without affecting the clonogenic survival of human prostate cancer cells. Int J Cancer 70:341–348
Lahiri S, Lee H, Mesicek J, Fuks Z, Haimovitz-Friedman A, Kolesnick RN, Futerman AH (2007) Kinetic characterization of mammalian ceramide synthases: determination of K(m) values toward sphinganine. FEBS Lett 581:5289–5294
Laviad EL, Albee L, Pankova-Kholmyansky I, Epstein S, Park H, Merrill AH Jr, Futerman AH (2008) Characterization of ceramide synthase 2: tissue distribution, substrate specificity, and inhibition by sphingosine 1-phosphate. J Biol Chem 283:5677–5684
Leith JT (1994) *In vitro* radiation sensitivity of the LNCaP prostatic tumor cell line. Prostate 24:119–124
Leith JT, Quaranto L, Padfield G, Michelson S, Hercbergs A (1993) Radiobiological studies of PC-3 and DU-145 human prostate cancer cells: X-ray sensitivity *in vitro* and hypoxic fractions of xenografted tumors *in vivo*. Int J Radiat Oncol Biol Phys 25:283–287
Li WX, Franklin WA (1998) Radiation- and heat-induced apoptosis in PC-3 prostate cancer cells. Radiat Res 150:190–194
Liao W-C, Haimovitz-Friedman A, Persaud R, McLoughlin M, Ehleiter D, Zhang N, Gatei M, Lavin M, Kolesnick R, Fuks Z (1999) Ataxia Telangiectasia-mutated gene product inhibits DNA damage-induced apoptosis *via* ceramide synthase. J Biol Chem 274:17908–17917
Lin T, Genestier L, Pinkoski MJ, Castro A, Nicholas S, Mogil R, Paris F, Fuks Z, Schuchman EH, Kolesnick RN et al (2000) Role of acidic sphingomyelinase in Fas/CD95-mediated cell death. J Biol Chem 275:8657–8663
Liu X, Cheng JC, Turner LS, Elojeimy S, Beckham TH, Bielawska A, Keane TE, Hannun YA, Norris JS (2009) Acid ceramidase upregulation in prostate cancer: role in tumor development and implications for therapy. Expert Opin Ther Targets 13:1449–1458
Lock RB, Ross WE (1990) Inhibition of p34cdc2 kinase activity by etoposide or irradiation as a mechanism of G2 arrest in Chinese hamster ovary cells. Cancer Res 50:3761–3766
Lowe SW, Schmitt EM, Smith SW, Osborne BA, Jacks T (1993) p53 is required for radiation-induced apoptosis in mouse thymocytes [see comments]. Nature 362:847–849
Mahdy AE, Cheng JC, Li J, Elojeimy S, Meacham WD, Turner LS, Bai A, Gault CR, McPherson AS, Garcia N et al (2009) Acid ceramidase upregulation in prostate cancer cells confers resistance to radiation: AC inhibition, a potential radiosensitizer. Mol Ther 17:430–438
Mathias S, Pena LA, Kolesnick RN (1998) Signal transduction of stress *via* ceramide. Biochem J 335:465–480
Mesicek J, Lee H, Feldman T, Jiang X, Skobeleva A, Berdyshev EV, Haimovitz-Friedman A, Fuks Z, Kolesnick R (2010) Ceramide synthases 2, 5, and 6 confer distinct roles in radiation-induced apoptosis in HeLa cells. Cell Signal 22:1300–1307
Meyn RE, Stephens LC, Hunter NR, Milas L (1995) Apoptosis in murine tumors treated with chemotherapy agents. Anticancer Drugs 6:443–450

Meyn RE, Milas L, Ang KK (2009) The role of apoptosis in radiation oncology. Int J Radiat Biol 85:107–115

Mizutani Y, Kihara A, Igarashi Y (2006) LASS3 (longevity assurance homologue 3) is a mainly testis-specific (dihydro)ceramide synthase with relatively broad substrate specificity. Biochem J 398:531–538

Mullauer L, Gruber P, Sebinger D, Buch J, Wohlfart S, Chott A (2001) Mutations in apoptosis genes: a pathogenetic factor for human disease. Mutat Res 488:211–231

Nava VE, Cuvillier O, Edsall LC, Kimura K, Milstien S, Gelmann EP, Spiegel S (2000) Sphingosine enhances apoptosis of radiation-resistant prostate cancer cells. Cancer Res 60:4468–4474

Norris JS, Bielawska A, Day T, El-Zawahri A, ElOjeimy S, Hannun Y, Holman D, Hyer M, Landon C, Lowe S et al (2006) Combined therapeutic use of AdGFPFasL and small molecule inhibitors of ceramide metabolism in prostate and head and neck cancers: a status report. Cancer Gene Ther 13:1045–1051

Olive PL, Durand RE (1997) Apoptosis: an indicator of radiosensitivity in vitro? Int J Radiat Biol 71:695–707

Olson M, Kornbluth S (2001) Mitochondria in apoptosis and human disease. Curr Mol Med 1:91–122

Orrenius S (1995) Apoptosis: molecular mechanisms and implications for human disease. J Intern Med 237:529–536

Pchejetski D, Doumerc N, Golzio M, Naymark M, Teissié J, Kohama T, Waxman J, Malavaud B, Cuvillier O (2008) Chemosensitizing effects of sphingosine kinase-1 inhibition in prostate cancer cell and animal models. Mol Cancer Ther 7:1836–1845

Pena LA, Fuks Z, Kolesnick R (1997) Stress-induced apoptosis and the sphingomyelin pathway. Biochem Pharmacol 53:615–621

Reed JC (2002) Apoptosis-based therapies. Nat Rev Drug Discov 1:111–121

Royai R, Lange PH, Vessella R (1996) Preclinical models of prostate cancer. Semin Oncol 23:35–40

Saad AF, Meacham WD, Bai A, Anelli V, Elojeimy S, Mahdy AE, Turner LS, Cheng J, Bielawska A, Bielawski J et al (2007) The functional effects of acid ceramidase overexpression in prostate cancer progression and resistance to chemotherapy. Cancer Biol Ther 6:1455–1460

Scardino PT, Wheeler TM (1988) Local control of prostate cancer with radiotherapy: frequency and prognostic significance of positive results of postirradiation prostate biopsy. NCI Monogr 7:95–103

Scarlatti F, Sala G, Ricci C, Maioli C, Milani F, Minella M, Botturi M, Ghidoni R (2007) Resveratrol sensitization of DU145 prostate cancer cells to ionizing radiation is associated to ceramide increase. Cancer Lett 253:124–130

Schenck M, Carpinteiro A, Grassmé H, Lang F, Gulbins E (2007a) Ceramide: physiological and pathophysiological aspects. Arch Biochem Biophys 462:171–175

Seelan RS, Qian C, Yokomizo A, Bostwick DG, Smith DI, Liu W (2000) Human acid ceramidase is overexpressed but not mutated in prostate cancer. Genes Chromosomes Cancer 29:137–146

Shiloh Y (2001) ATM (ataxia telangiectasia mutated): expanding roles in the DNA damage response and cellular homeostasis. Biochem Soc Trans 29:661–666

Spiegel S, Foster D, Kolesnick R (1996) Signal transduction through lipid second messengers. Curr Opin Cell Biol 8:159–167

Sridevi P, Alexander H, Laviad EL, Pewzner-Jung Y, Hannink M, Futerman AH, Alexander S (2009) Ceramide synthase 1 is regulated by proteasomal mediated turnover. Biochim Biophys Acta 1793:1218–1227

Steel GG (2001) The case against apoptosis. Acta Oncol 40:968–975

Stone KR, Mickey DD, Wunderli H, Mickey GH, Paulson DF (1978) Isolation of a human prostate carcinoma cell line (DU 145). Int J Cancer 21:274–281

Symmans WF, Volm MD, Shapiro RL, Perkins AB, Kim AY, Demaria S, Yee HT, McMullen H, Oratz R, Klein P et al (2000) Paclitaxel-induced apoptosis and mitotic arrest assessed by serial fine-needle aspiration: implications for early prediction of breast cancer response to neoadjuvant treatment. Clin Cancer Res 6:4610–4617

Truman JP, Gueven N, Lavin M, Leibel S, Kolesnick R, Fuks Z, Haimovitz-Friedman A (2005) Down-regulation of ATM protein sensitizes human prostate cancer cells to radiation-induced apoptosis. J Biol Chem 280:23262–23272

Truman JP, Rotenberg SA, Kang JH, Lerman G, Fuks Z, Kolesnick R, Marquez VE, Haimovitz-Friedman A (2009) PKCalpha activation downregulates ATM and radio-sensitizes androgen-sensitive human prostate cancer cells *in vitro* and in vivo. Cancer Biol Ther 8:54–63

Wollin M, FitzGerald TJ, Santucci MA, Menon M, Longcope C, Reale F, Carlson J, Sakakeeny MA, Greenberger JS (1989) Radiosensitivity of human prostate cancer and malignant melanoma cell lines. Radiother Oncol 15:285–293

Wu J (1996) Apoptosis and angiogenesis: two promising tumor markers in breast cancer. Anticancer Res 16:2233–2239

Zelefsky MJ, Leibel SA, Gaudin PB, Kutcher GJ, Fleshner NE, Venkatramen ES, Reuter VE, Fair WR, Ling CC, Fuks Z (1998) Dose escalation with three-dimensional conformal radiation therapy affects the outcome in prostate cancer. Int J Radiat Oncol Biol Phys 41:491–500

Zelefsky MJ, Reuter VE, Fuks Z, Scardino P, Shippy A (2008) Influence of local tumor control on distant metastases and cancer related mortality after external beam radiotherapy for prostate cancer. J Urol 179:1368–1373 (discussion 1373)

Part II
Sphingolipids in Cardio-Reno-vascular Diseases

Sphingolipid Metabolism and Atherosclerosis

Xian-Cheng Jiang and Jing Liu

Contents

Abstract Atherosclerosis is the major cause of mortality in the developed countries. Although presently known risk factors have some predictive value for the disease, a major part of the variability in this process remains unexplained. It is extremely important to find new approaches for better understanding of the disease and for treating it. Exploration of the sphingolipid metabolism is one of these approaches. Sphingolipids are a large class of lipids with structural and signaling functions. Recent researches indicated that these lipids play important roles in the development of atherosclerosis. In this chapter, we summarized the major findings in the field.

Keywords Sphingolipid • Sphingomyelin • Ceramide • Sphingosine-1-phosphate • Atherosclerosis

X.-C. Jiang (✉) • J. Liu
Department of Cell Biology, SUNY Downstate Medical Center, 450 Clarkson Ave. Box 5, Brooklyn, NY 11203, USA

Molecular and Cellular Cardiology Program, VA New York Harbor Healthcare System, Brooklyn, NY 11209, USA
e-mail: XJiang@downstate.edu

E. Gulbins and I. Petrache (eds.), *Sphingolipids in Disease*,
Handbook of Experimental Pharmacology 216,
DOI 10.1007/978-3-7091-1511-4_7,

1 Sphingomyelin Synthesis in Atherosclerosis

Sphingomyelin synthesis is a major pathway of ceramide conversion. This is accomplished by sphingomyelin synthase (SMS) enzymes. SMS gene family consists of three members, SMS1-, SMS2-, and SMS-related protein (SMSr). SMS1 is found in the trans-Golgi complex, while SMS2 is predominantly found in the plasma membranes (Huitema et al. 2004; Yamaoka et al. 2004). SMSr, the third member of the gene family, has no SMS activity but catalyzes the synthesis of ceramide-phosphoethanolamine in the ER lumen (Huitema et al. 2004; Vacaru et al. 2009). SMS1 and SMS2 activity are coexpressed in a variety of tissues and cells with different ratios. SMS1 is the major SMS in macrophages (Li et al. 2012), while SMS2 is the major one in the liver (Liu et al. 2009a). Others and we have shown that SMS1 and SMS2 expression positively correlates with sphingomyelin levels in cells and lipid rafts (Miyaji et al. 2005; Van der Luit et al. 2007; Li et al. 2007).

To study the importance of sphingomyelin production, we generated knockout mice deficient in SMS1 and SMS2. Individual ablation of these enzymes decreases plasma sphingomyelin by about 50 % (Table 1), indicating these two enzymes contribute equally to plasma sphingomyelin levels. Further, SMS1 deficiency had no effect, while SMS2 deficiency increased plasma ceramide levels (Table 1). More importantly, SMS1, but not SMS2, knockout mice had fourfold to sevenfold increase in glucosylceramide and GM3 in the plasma. Tissue analysis revealed that SMS1 and SMS2 deficiencies significantly reduced cellular sphingomyelin levels. Similar to plasma, glucosylceramide and ganglioside GM3 levels were increased in the liver and macrophages of SMS1 deficient mice, but not in that of SMS2 deficient mice.

Based on these observations, at least two different mechanisms involved in the control of ceramide and sphingomyelin levels in the cells, related with atherosclerosis processes, and in the circulation were postulated. First involves vesicular trafficking that is well described for the transport of proteins from the Golgi to the plasma membrane. Second, at least in lipoprotein synthesizing tissues such as liver, intestine, and heart, sphingomyelins and ceramides could become part of newly synthesized triglyceride-rich apolipoprotein B-containing lipoproteins in the Golgi. There is significant evidence in the literature for this mechanism. It has been reported that the major carriers of ceramide and sphingomyelin in the plasma are very-low-density lipoproteins (VLDL) (Hammad et al. 2010). Studies in hamsters suggest that de novo sphingolipids are secreted via VLDL/LDL pathway by the liver (Memon et al. 1998, 1999). Moreover, isolated rat hepatocytes have been shown to secrete ceramides and sphingomyelins as part of apoB-containing particles (Merrill et al. 1995). Activation of serine palmitoyl transferase, the rate-limiting enzyme in the de novo sphingolipid synthesis by palmitic acid but not other fatty acids, elevates VLDL and LDL ceramides and sphingomyelins (Merrill et al. 1995). Despite evidence that hepatocytes secrete ceramides and sphingomyelin with VLDL, it remains to be determined how ceramides and sphingomyelin are

Table 1 Levels of various sphingolipids in wild-type and SMS knockout mice

	SM	PC	Cer	DHCer	S1P	Sa1P	Sph	GluCer	GM3
Plasma									
	nmol/ml	nmole/ml	ng/ml	ng/ml	ng/ml	ng/ml	ng/ml	ng/ml	ng/ml
Wild type	105 ± 5	1,443 ± 109	805 ± 33	263 ± 27	307 ± 15	122 ± 21	10 ± 1	3,714 ± 358	342 ± 22
Sms1 KO	52 ± 3*	1,322 ± 99	888 ± 80	300 ± 16	314 ± 29	123 ± 18	8 ± 1	25,705 ± 2,317*	1,996 ± 219*
Wild type	95 ± 3	1,301 ± 78	796 ± 55	307 ± 36	351 ± 30	103 ± 16	12 ± 2	3,209 ± 277	329 ± 34
Sms2 KO	47 ± 3*	1,232 ± 102	1,032 ± 50*	331 ± 29	323 ± 19	111 ± 21	10 ± 3	3,638 ± 421	367 ± 51
Liver									
	nmol/mg protein	nmole/mg protein	ng/mg protein	ng/mg protein			ng/mg protein	ng/mg protein	ng/mg protein
Wild type	11 ± 2	102 ± 38	247 ± 29	33 ± 4			4 ± 1	104 ± 10	22 ± 2
Sms1 KO	7 ± 1*	108 ± 21	184 ± 21*	29 ± 6			3 ± 1	396 ± 31*	120 ± 13*
Wild type	13 ± 3	122 ± 41	267 ± 31	39 ± 5			5 ± 2	122 ± 15	16 ± 3
Sms2 KO	8 ± 1*	130 ± 15	284 ± 42	33 ± 8			4 ± 1	109 ± 19	19 ± 5
Macrophages									
Wild type	63 ± 7	117 ± 22	985 ± 98	116 ± 33			16 ± 3	85 ± 17	151 ± 50
Sms1 KO	19 ± 5*	131 ± 19	1,027 ± 128	143 ± 36			35 ± 19*	1,019 ± 26*	1,255 ± 77*
Wild type	61 ± 6	157 ± 33	890 ± 109	86 ± 16			25 ± 3	111 ± 20	103 ± 12
Sms2 KO	51 ± 2*	139 ± 23	962 ± 91	96 ± 39			30 ± 5	130 ± 18	99 ± 7

Different sphingolipids were quantified by LC/MS/MS from tissues obtained from wild-type and SMS1 and SMS2 knockout mice. Statistical significance was evaluated using Student's *t*-test

Values are mean ± SD, $n = 6$. $^{*}P < 0.01$

SM sphingomyelin, *PC* phosphatidylcholine, *Cer* ceramide, *DHCer* dihydroceramide, *Glucer* glucosylceramide, *Sph* sphingosine, *S1P* sphingosine-1-phosphate, *Sa1P* sphinganine-1-phosphate, *GM3* ganglioside GM3

deposited in nascent lipoproteins. There are two candidate proteins that could play a role in this process. First, microsomal triglyceride transfer protein (Hussain et al. 2003), which transfers several lipids, can transfer sphingolipids and deposit them in nascent lipoproteins. Second, phospholipid transfer protein (Jiang et al. 2001) can transfer these lipids onto nascent lipoproteins. Alternatively, there might be other yet unidentified proteins that could specifically transfer sphingolipids to lipoproteins.

2 Sphingomyelin and Atherosclerosis

2.1 Plasma Sphingomyelin and Atherosclerosis

Sphingomyelin, which is the second most abundant phospholipid in mammalian plasma, appears in all major lipoproteins. Up to 18 % of total plasma phospholipid exists as sphingomyelin (Nilsson and Duan 2006) with the ratio of phosphatidylcholine/sphingomyelin varying widely among lipoprotein subclasses (Subbaiah et al. 1989). Atherogenic lipoproteins such as VLDL and LDL are sphingomyelin enriched (Nilsson and Duan 2006; Rodriguez et al. 1976). The sphingomyelin content of atherosclerotic lesions is higher than that of normal arterial tissue (Zilversmit et al. 1961).

Williams and Tabas have suggested that subendothelial retention and aggregation of atherogenic lipoproteins play a very important role in atherogenesis (Williams and Tabas 1995, 1998). Sphingomyelin-rich LDL retained in atherosclerotic lesions is acted on by an arterial-wall sphingomyelinase that appears to promote aggregation and retention, initiating the early phase of atherosclerosis development (Schissel et al. 1996). Plasma sphingomyelin levels in ApoE KO mice are fourfold higher than those in control mice (Jeong et al. 1998), and this may partially explain the increased atherosclerosis found in these animals (Plump et al. 1992). Indeed, chemical inhibition of sphingolipid biosynthesis significantly decreases plasma sphingomyelin levels, thus lessening atherosclerotic lesions in ApoE KO mice (Hojjati et al. 2005; Park et al. 2004). Moreover, human plasma sphingomyelin levels are also an independent risk factor for coronary heart disease (Jiang et al. 2000; Schlitt et al. 2006), and these levels are prognostic in patients with acute coronary syndrome (Schlitt et al. 2006). All these data suggest that plasma sphingomyelin plays a critical role in the development of atherosclerosis. However, studies in mice revealed that inhibition of sphingolipid de novo synthesis, leading to reduction of atherosclerosis, causes all tested sphingolipids to be significantly decreased (Hojjati et al. 2005; Park et al. 2004). Consequently, one could not rule out the effect of sphingolipids other than sphingomyelin on mouse atherogenicity.

2.2 *Membrane Sphingomyelin and Atherosclerosis*

A great deal of knowledge about membrane sphingomyelin and atherosclerosis is obtained from studies of macrophages. Macrophages are the most prominent cell types in atherosclerotic lesions and are associated with two hallmarks of the disease: foam cell formation due to excessive accumulation of cholesterol by macrophages (Ross 1993) and inflammation (Nilsson and Duan 2006; Libby 2002). Macrophage membrane sphingomyelin levels are closely related to macrophage cholesterol efflux and inflammatory responses.

2.2.1 Lipid Rafts

The interaction of sphingomyelin, cholesterol, and glycosphingolipid drives the formation of plasma membrane rafts (Simons and Ikonen 1997). Lipid rafts have been shown to be involved in cell signaling, lipid and protein sorting, and membrane trafficking (Simons and Ikonen 1997; Futerman and Hannun 2004; Holthuis et al. 2003). It is well known that both class A and B scavenger receptors (SR) are located in lipid rafts (Kim et al. 2004; Graf et al. 1999; Rhainds et al. 2004). Lipoprotein metabolism-related proteins, including LDL receptor-related protein (von Arnim et al. 2005), ATP-binding cassette transporter ABCA1 (Landry et al. 2006; Jessup et al. 2006), and ABCG1 (Jessup et al. 2006), are also associated with membrane rafts. Toll-like receptors (TLRs) and inflammatory response mediators are also located in the lipid rafts of cells, including macrophages (Lee et al. 2006; Wang et al. 2006a; Nakahira et al. 2006; Szabo et al. 2007). Lipid rafts may well have an impact on the development of atherosclerosis. The statins (cholesterol biosynthesis inhibitors) not only reduce cholesterol in the circulation but also significantly change the rafts' structure (Zhuang et al. 2005).

2.2.2 Macrophage Lipid Rafts and Cholesterol Efflux

In macrophages, ABCA1 exports cholesterol and phospholipid to lipid-free apolipoproteins, while ABCG1 and SR-BI export cholesterol to phospholipid-containing acceptors (Jessup et al. 2006). ABCA1-dependent cholesterol efflux requires aid from membrane lipid rafts (Gaus et al. 2004; Mendez et al. 2001). ABCG1 is mainly found intracellularly in the basal state, with little cell surface presentation. But on stimulation, for example, by liver X receptor (LXR) agonist treatment, ABCG1 redistributes to the plasma membranes and increases cholesterol mass efflux to HDL (Wang et al. 2006b). ABCA1 and ABCG1 cooperate in cholesterol efflux (Jessup et al. 2006). SR-BI facilitates cholesterol efflux from macrophages (Huang et al. 2003). Since ABCA1, ABCG1, and SR-BI are located in the plasma membranes, it is therefore conceivable that fundamental changes in sphingomyelin and glycosphingolipid levels of macrophage plasma membranes influence the functions of these proteins and alter cholesterol efflux. Indeed, reductions in cholesterol efflux from sphingomyelinase-deficient macrophages and induction of cholesterol efflux by ABCA1 from sphingomyelin-deficient Chinese hamster ovary (CHO) cells have been reported (Leventhal et al. 2001; Nagao et al. 2007). We found that SMS2 deficiency in macrophages diminishes plasma membrane sphingomyelin levels and causes significant induction of cholesterol efflux in vitro and in vivo (Liu et al. 2009b). Further serine palmitoyltransferase subunit 2 haploinsufficient, ($SPTLC2^{+/-}$) macrophages have significantly lower SM levels in plasma membrane and lipid rafts. This reduction enhanced cholesterol efflux, in vitro and in vivo, mediated by ABCA1 and ABCG1 (Chakraborty et al. 2013, J. Clin. Invest. In Press).

2.2.3 Macrophage Lipid Rafts and Inflammation

Macrophage toll-like receptors (TLRs) are critically involved in the inflammatory responses (Beutler et al. 2006; Takeda and Akira 2005). Each TLR has a unique extracellular domain that allows specific ligand recognition. The intracellular toll/ IL-1 receptor (TIR) domains share considerable homology, but there are enough differences to cause diverse adapter molecules to be used by some TLRs (Beutler et al. 2006; Takeda and Akira 2005; Miggin and O'Neill 2006; Miyake 2006). On ligand-induced stimulation, the TIR domain of TLRs associates with the TIR domain of their respective adaptor molecules to initiate intracellular signaling. MyD88 is a common TLR adaptor used by all TLRs, except for TLR3 (Miyake 2006). On stimulation with a specific ligand, the membrane-associated TLRs (such as TLR2 and TLR4 and other components of the TLR complex) are recruited into the lipid rafts (Triantafilou et al. 2004a, b). These rafts appear to provide a platform for the interaction of TLRs with their ligands in macrophages (Lee et al. 2006; Wang et al. 2006a; Nakahira et al. 2006; Szabo et al. 2007), initiating NFκB and MAP kinase activation and proinflammatory cytokine production, thus resulting in inflammatory responses.

Luberto et al. (Luberto et al. 2000) found that D609, a nonspecific SMS inhibitor, blocks TNFα- and phorbol ester-mediated NFκB activation that was concomitant with decreased levels of sphingomyelin and diacylglycerol. This did not affect the generation of ceramide, suggesting sphingomyelin and diacylglycerol derived from sphingomyelin synthesis are involved in NFκB activation. This effect may have been mediated by SMS1, since SMS1 KO mice had significantly decreased plasma, liver, and macrophage SM (59 %, 45 %, and 54 %, respectively), but had only a marginal effect on ceramide levels. Moreover, SMS1 deficiency significantly attenuated TLR4-mediated NFκB and MAP kinase activation after LPS treatment (Li et al. 2012). However, SMS2 deficiency also significantly attenuates NFκB activation (Hailemariam et al. 2008) and even SPTLC2 haploinsufficiency ($SPTLC2^{+/-}$) decreased inflammatory responses triggered by TLR4 and its downstream NFκB and MAP kinase pathways (Chakraborty and Jiang, unpublished observations).

In addition to TLRs, SMS1 has also been implicated in the regulation of lipid raft sphingomyelin level and raft functions such as FAS receptor clustering (Miyaji et al. 2005), endocytosis, and apoptosis (Van der Luit et al. 2007).

2.2.4 Macrophage Cholesterol Efflux and Inflammation

It is known that ABC transporter-mediated cholesterol efflux plays an important role in macrophage inflammation. Yvan-Charvet et al. (2008) reported that macrophage ABCA1 and ABCG1 deficiencies increase free cholesterol accumulation and increase signaling via TLRs. Zhu et al. reported that macrophage ABCA1 reduces MyD88-dependent TLR trafficking to lipid rafts by reduction of lipid raft cholesterol (Zhu et al. 2010). In addition, ABCA1 expression decreases cellular plasma membrane rigidity by reducing formation of tightly packed lipid rafts (Landry et al. 2006). SMS2 deficiency (Liu et al. 2009b; Hailemariam et al. 2008) or SPTLC2 haploinsufficiency-mediated (Chakraborty et al. 2013, J. Clin. Invest. In Press) sphingomyelin reduction in macrophage plasma membrane not only impairs inflammatory responses triggered by TLR4 pathway but also enhances reverse cholesterol transport mediated by ABC transporters. Further implicating the importance of SMS on inflammatory cells, low-density lipoprotein receptor-deficient (LDLR$^{-/-}$) mice transplanted with SMS2 KO- (Liu et al. 2009b) or SMS1 KO- (Li et al. 2012) or SPTLC2$^{+/-}$ (Chakraborty et al. 2013, J. Clin. Invest. In Press) bone marrow exhibit significantly less atherosclerotic lesions than that of controls.

3 Sphingosine-1-Phosphate and Atherosclerosis

Plasma sphingosine-1-phosphate (S1P) is mainly associated with HDL (Kimura et al. 2001; Sachinidis et al. 1999) and reduces atherosclerosis through inhibition of adhesion molecule expression and stimulation of endothelial nitric oxide synthase (eNOS) in endothelial cells and expression of monocyte chemoattractant peptide (MCP-1) in smooth muscle cells (Kimura et al. 2006; Tolle et al. 2008; Zhang et al. 2005). Recent study reported that those anti-atherogenic HDL actions on ECs are partly mediated by S1P (Christoffersen et al. 2011).

$S1P_1$–$S1P_3$ are widely expressed, especially rich in cardiovascular and immune system (Hla and Maciag 1990; Lee et al. 1999; Okazaki et al. 1993; Takuwa et al. 2008). $S1P_1$ and $S1P_3$ couple dominantly to Gi to lead to Rac activation and stimulation of chemotaxis of endothelial cells through rho and PI3K/Akt/Rac pathways (Okamoto et al. 1998, 2000; Sanchez et al. 2007; Sugimoto et al. 2003), whereas $S1P_2$ couples mainly to G12/13 to result in rho activation and phosphatase and tension homolog (PTEN)-dependent inhibition of migration (Sanchez et al. 2005). Recent studies from two groups both demonstrate that the area of the atherosclerotic plaque is dramatically reduced in $S1P_2$ KO/ApoE KO mice with reduced macrophage density, increased SMC density and eNOS phosphoration, and downregulation of

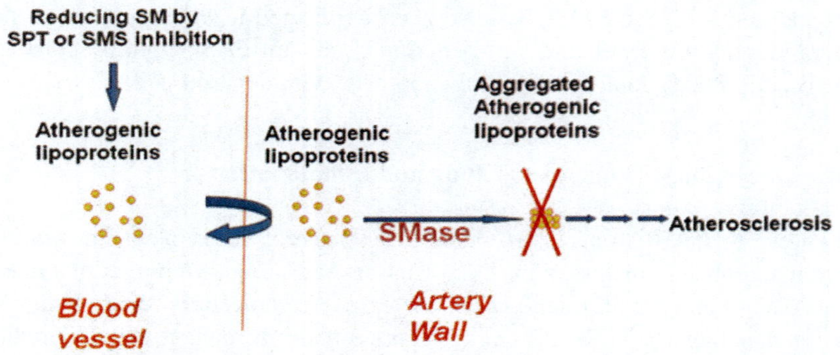

Fig. 1 The potential anti-atherogenic mechanism for plasma sphingomyelin reduction

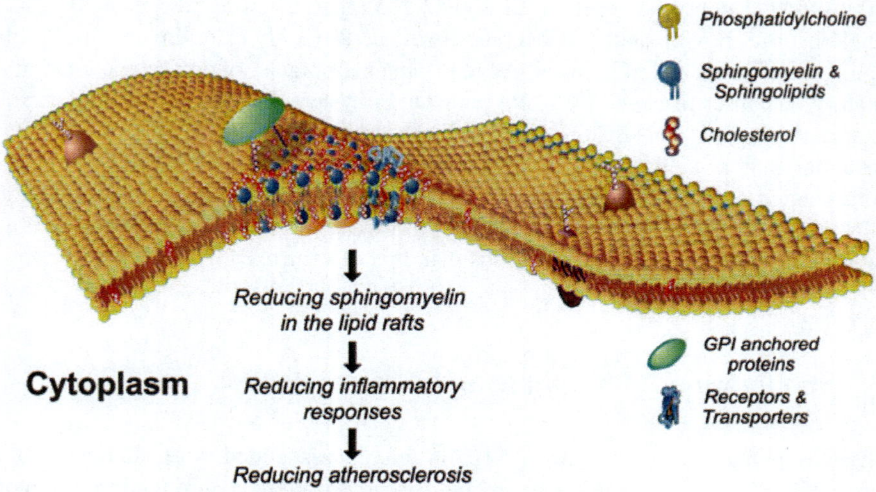

Fig. 2 The potential anti-atherogenic mechanism for membrane sphingomyelin reduction

proinflammatory cytokines TNF-a, IL-6, IFN-r, and MCP-1 compared with ApoE KO mice (Skoura et al. 2011; Wang et al. 2010). Macrophages from ApoE KO/S1P_2 KO mice have a significantly less accumulation of modified LDL through decreased uptake of oxidized LDLs and increased cholesterol efflux, further resulting in reducing foam cell formation and atherosclerotic lesions (Skoura et al. 2011; Wang et al. 2010). Thus, S1P_2 plays a distinct role in promoting atherosclerosis. Although S1P_3 deficiency does not affect lesion size, it shows dramatic reduction of the monocyte/macrophage content and increased the smooth muscle cell content in lesions of S1P_3 KO/ApoE KO mice (Rhainds et al. 2004). Proinflammatory cytokine TNF-α and MCP-1 both decrease in peritoneal macrophages in S1P_3 KO/ApoE KO mice (Keul et al. 2011). It suggests that S1P_3 plays a causal role in atherosclerosis.

FTY720, a synthetic analogue of S1P, functions as a potent agonist of four of the five G protein-coupled S1P receptors except $S1P_2$. Two groups report that FTY720 dramatically reduced atherosclerotic lesion volume in both ApoE KO mice (Keul et al. 2007) and LDL receptor KO mice (Nofer et al. 2007). The possible mechanism is that the drug stimulates $S1P_3$-mediated NO production, thus inhibiting the release of the monocyte chemokine MCP-1 by SMCs, resulting in suppressing recruitment of monocyte/macrophage into atherosclerotic lesions (Keul et al. 2007). Another group indicates that FTY720 treatment inhibits atherosclerosis through reducing splenocyte proliferation, interferon-γ levels, and proinflammatory cytokines in plasma (Nofer et al. 2007). The ratio between inflammatory M1- and anti-inflammatory M2-macrophages was decreased in LDL receptor KO mice treated with FTY720 (Nofer et al. 2007). In conclusion, though the role of S1P is complicated in the development of atherosclerosis, $S1P_2$ and $S1P_3$ both are proatherogenic factors.

4 Conclusions

Direct experimental evidence in mouse models has demonstrated that inhibition of sphingolipid de novo synthesis, sphingomyelin synthesis, and $S1P_2$- or $S1P_3$-mediated signaling pathways significantly decreased atherogenesis. There are three potential anti-atherogenic mechanisms for such inhibitions so far:

1. Decreasing sphingomyelin and ceramide contents in atherogenic lipoproteins could decrease the aggregation of these particles in aorta (Fig. 1).
2. Decreasing sphingomyelin levels on cell, especially macrophage, membrane lipid rafts could attenuate inflammatory responses and promote cholesterol efflux, thus preventing atherogenesis (Fig. 2).
3. Decreasing S1P-dependent stimulation of $S1P_2$ and $S1P_3$ could decrease atherosclerotic lesion formation and macrophage residency.

Much progress has been made from the description of sphingolipids in the atherosclerotic lesion to recognition of a causal relationship between distinct sphingolipids and atherogenesis. However, sphingolipid studies in atherosclerosis field are still in the infant stage, compared with cholesterol studies. Fundamental questions, including the delineation of molecular pathways, remain to be addressed.

Atherosclerosis is the major cause of mortality in the developed countries. Therapy aimed at lowering LDL cholesterol reduces only a small fraction (roughly 30 %) of the burden of atherosclerotic disease (Maher et al. 1997; Kjekshus and Pedersen 1995). It is extremely important to find new approaches for better understanding of the disease. Exploration of the sphingolipid metabolism is one of these approaches.

References

Beutler B, Jiang Z, Georgel P, Crozat K, Croker B, Rutschmann S, Du X, Hoebe K (2006) Genetic analysis of host resistance: Toll-like receptor signaling and immunity at large. Annu Rev Immunol 24:353–389

Christoffersen C, Obinata H, Kumaraswamy SB, Galvani S, Ahnstrom J, Sevvana M, Egerer-Sieber C, Muller YA, Hla T, Nielsen LB, Dahlback B (2011) Endothelium-protective sphingosine-1-phosphate provided by HDL-associated apolipoprotein M. Proc Natl Acad Sci USA 108:9613–9618

Futerman AH, Hannun YA (2004) The complex life of simple sphingolipids. EMBO Rep 5:777–782

Gaus K, Kritharides L, Schmitz G, Boettcher A, Drobnik W, Langmann T, Quinn CM, Death A, Dean RT, Jessup W (2004) Apolipoprotein A-1 interaction with plasma membrane lipid rafts controls cholesterol export from macrophages. FASEB J 18:574–576

Graf GA, Connell PM, van der Westhuyzen DR, Smart EJ (1999) The class B, type I scavenger receptor promotes the selective uptake of high density lipoprotein cholesterol ethers into caveolae. J Biol Chem 274:12043–12048

Hailemariam TK, Huan C, Liu J, Li Z, Roman C, Kalbfeisch M, Bui HH, Peake DA, Kuo MS, Cao G, Wadgaonkar R, Jiang XC (2008) Sphingomyelin synthase 2 deficiency attenuates NFkappaB activation. Arterioscler Thromb Vasc Biol 28:1519–1526

Hammad SM, Pierce JS, Soodavar F, Smith KJ, Al Gadban MM, Rembiesa B, Klein RL, Hannun YA, Bielawski J, Bielawska A (2010) Blood sphingolipidomics in healthy humans: impact of sample collection methodology. J Lipid Res 51:3074–3087

Hla T, Maciag T (1990) An abundant transcript induced in differentiating human endothelial cells encodes a polypeptide with structural similarities to G-protein-coupled receptors. J Biol Chem 265:9308–9313

Hojjati MR, Li Z, Zhou H, Tang S, Huan C, Ooi E, Lu S, Jiang XC (2005) Effect of myriocin on plasma sphingolipid metabolism and atherosclerosis in apoE-deficient mice. J Biol Chem 280:10284–10289

Holthuis JC, van Meer G, Huitema K (2003) Lipid microdomains, lipid translocation and the organization of intracellular membrane transport (Review). Mol Membr Biol 20:231–241

Huang ZH, Gu D, Lange Y, Mazzone T (2003) Expression of scavenger receptor BI facilitates sterol movement between the plasma membrane and the endoplasmic reticulum in macrophages. Biochemistry 42:3949–3955

Huitema K, van den Dikkenberg J, Brouwers JF, Holthuis JC (2004) Identification of a family of animal sphingomyelin synthases. EMBO J 23:33–44

Hussain MM, Shi J, Dreizen P (2003) Microsomal triglyceride transfer protein and its role in apoB-lipoprotein assembly. J Lipid Res 44:22–32

Jeong T, Schissel SL, Tabas I, Pownall HJ, Tall AR, Jiang X (1998) Increased sphingomyelin content of plasma lipoproteins in apolipoprotein E knockout mice reflects combined production and catabolic defects and enhances reactivity with mammalian sphingomyelinase. J Clin Invest 101:905–912

Jessup W, Gelissen IC, Gaus K, Kritharides L (2006) Roles of ATP binding cassette transporters A1 and G1, scavenger receptor BI and membrane lipid domains in cholesterol export from macrophages. Curr Opin Lipidol 17:247–257

Jiang XC, Paultre F, Pearson TA, Reed RG, Francis CK, Lin M, Berglund L, Tall AR (2000) Plasma sphingomyelin level as a risk factor for coronary artery disease. Arterioscler Thromb Vasc Biol 20:2614–2618

Jiang XC, Qin S, Qiao C, Kawano K, Lin M, Skold A, Xiao X, Tall AR (2001) Apolipoprotein B secretion and atherosclerosis are decreased in mice with phospholipid-transfer protein deficiency. Nat Med 7:847–852

Keul P, Tolle M, Lucke S, von Wnuck Lipinski K, Heusch G, Schuchardt M, van der Giet M, Levkau B (2007) The sphingosine-1-phosphate analogue FTY720 reduces atherosclerosis in apolipoprotein E-deficient mice. Arterioscler Thromb Vasc Biol 27:607–613

Keul P, Lucke S, von Wnuck Lipinski K, Bode C, Graler M, Heusch G, Levkau B (2011) Sphingosine-1-phosphate receptor 3 promotes recruitment of monocyte/macrophages in inflammation and atherosclerosis. Circ Res 108:314–323

Kim S, Watarai M, Suzuki H, Makino S, Kodama T, Shirahata T (2004) Lipid raft microdomains mediate class A scavenger receptor-dependent infection of Brucella abortus. Microb Pathog 37:11–19

Kimura T, Sato K, Kuwabara A, Tomura H, Ishiwara M, Kobayashi I, Ui M, Okajima F (2001) Sphingosine 1-phosphate may be a major component of plasma lipoproteins responsible for the cytoprotective actions in human umbilical vein endothelial cells. J Biol Chem 276:31780–31785

Kimura T, Tomura H, Mogi C, Kuwabara A, Damirin A, Ishizuka T, Sekiguchi A, Ishiwara M, Im DS, Sato K, Murakami M, Okajima F (2006) Role of scavenger receptor class B type I and sphingosine 1-phosphate receptors in high density lipoprotein-induced inhibition of adhesion molecule expression in endothelial cells. J Biol Chem 281:37457–37467

Kjekshus J, Pedersen TR (1995) Reducing the risk of coronary events: evidence from the Scandinavian Simvastatin Survival Study (4S). Am J Cardiol 76:64C–68C

Landry YD, Denis M, Nandi S, Bell S, Vaughan AM, Zha X (2006) ATP-binding cassette transporter A1 expression disrupts raft membrane microdomains through its ATPase-related functions. J Biol Chem 281:36091–36101

Lee MJ, Thangada S, Claffey KP, Ancellin N, Liu CH, Kluk M, Volpi M, Sha'afi RI, Hla T (1999) Vascular endothelial cell adherens junction assembly and morphogenesis induced by sphingosine-1-phosphate. Cell 99:301–312

Lee HK, Dunzendorfer S, Soldau K, Tobias PS (2006) Double-stranded RNA-mediated TLR3 activation is enhanced by CD14. Immunity 24:153–163

Leventhal AR, Chen W, Tall AR, Tabas I (2001) Acid sphingomyelinase-deficient macrophages have defective cholesterol trafficking and efflux. J Biol Chem 276:44976–44983

Li Z, Hailemariam TK, Zhou H, Li Y, Duckworth DC, Peake DA, Zhang Y, Kuo MS, Cao G, Jiang XC (2007) Inhibition of sphingomyelin synthase (SMS) affects intracellular sphingomyelin accumulation and plasma membrane lipid organization. Biochim Biophys Acta 1771:1186–1194

Li Z, Fan Y, Liu J, Li Y, Quan C, Bui HH, Kuo MS, Park TS, Cao G, Jiang XC (2012) Impact of sphingomyelin synthase 1 deficiency on sphingolipid metabolism and atherosclerosis in mice. Arterioscler Thromb Vasc Biol 32(7):1577–84

Libby P (2002) Inflammation in atherosclerosis. Nature 420:868–874

Liu J, Zhang H, Li Z, Hailemariam TK, Chakraborty M, Jiang K, Qiu D, Bui HH, Peake DA, Kuo MS, Wadgaonkar R, Cao G, Jiang XC (2009a) Sphingomyelin synthase 2 is one of the determinants for plasma and liver sphingomyelin levels in mice. Arterioscler Thromb Vasc Biol 29:850–856

Liu J, Huan C, Chakraborty M, Zhang H, Lu D, Kuo MS, Cao G, Jiang XC (2009b) Macrophage sphingomyelin synthase 2 deficiency decreases atherosclerosis in mice. Circ Res 105:295–303

Luberto C, Yoo DS, Suidan HS, Bartoli GM, Hannun YA (2000) Differential effects of sphingomyelin hydrolysis and resynthesis on the activation of NF-kappa B in normal and SV40-transformed human fibroblasts. J Biol Chem 275:14760–14766

Maher V, Sinfuego J, Chao P, Parekh J (1997) Primary prevention of coronary heart disease. What has WOSCOPS told us and what questions remain? West Of Scotland Coronary Prevention Study. Drugs 54:1–8

Memon RA, Holleran WM, Moser AH, Seki T, Uchida Y, Fuller J, Shigenaga JK, Grunfeld C, Feingold KR (1998) Endotoxin and cytokines increase hepatic sphingolipid biosynthesis and produce lipoproteins enriched in ceramides and sphingomyelin. Arterioscler Thromb Vasc Biol 18:1257–1265

Memon RA, Holleran WM, Uchida Y, Moser AH, Ichikawa S, Hirabayashi Y, Grunfeld C, Feingold KR (1999) Regulation of glycosphingolipid metabolism in liver during the acute phase response. J Biol Chem 274:19707–19713

Mendez AJ, Lin G, Wade DP, Lawn RM, Oram JF (2001) Membrane lipid domains distinct from cholesterol/sphingomyelin-rich rafts are involved in the ABCA1-mediated lipid secretory pathway. J Biol Chem 276:3158–3166

Merrill AH Jr, Lingrell S, Wang E, Nikolova-Karakashian M, Vales TR, Vance DE (1995) Sphingolipid biosynthesis de novo by rat hepatocytes in culture. Ceramide and sphingomyelin are associated with, but not required for, very low density lipoprotein secretion. J Biol Chem 270:13834–13841

Miggin SM, O'Neill LA (2006) New insights into the regulation of TLR signaling. J Leukoc Biol 80:220–226

Miyaji M, Jin ZX, Yamaoka S, Amakawa R, Fukuhara S, Sato SB, Kobayashi T, Domae N, Mimori T, Bloom ET, Okazaki T, Umehara H (2005) Role of membrane sphingomyelin and ceramide in platform formation for Fas-mediated apoptosis. J Exp Med 202:249–259

Miyake K (2006) Roles for accessory molecules in microbial recognition by Toll-like receptors. J Endotoxin Res 12:195–204

Nagao K, Takahashi K, Hanada K, Kioka N, Matsuo M, Ueda K (2007) Enhanced apoA-I-dependent cholesterol efflux by ABCA1 from sphingomyelin-deficient Chinese hamster ovary cells. J Biol Chem 282:14868–14874

Nakahira K, Kim HP, Geng XH, Nakao A, Wang X, Murase N, Drain PF, Sasidhar M, Nabel EG, Takahashi T, Lukacs NW, Ryter SW, Morita K, Choi AM (2006) Carbon monoxide differentially inhibits TLR signaling pathways by regulating ROS-induced trafficking of TLRs to lipid rafts. J Exp Med 203:2377–2389

Nilsson A, Duan RD (2006) Absorption and lipoprotein transport of sphingomyelin. J Lipid Res 47:154–171

Nofer JR, Bot M, Brodde M, Taylor PJ, Salm P, Brinkmann V, van Berkel T, Assmann G, Biessen EA (2007) FTY720, a synthetic sphingosine 1 phosphate analogue, inhibits development of atherosclerosis in low-density lipoprotein receptor-deficient mice. Circulation 115:501–508

Okamoto H, Takuwa N, Gonda K, Okazaki H, Chang K, Yatomi Y, Shigematsu H, Takuwa Y (1998) EDG1 is a functional sphingosine-1-phosphate receptor that is linked via a Gi/o to multiple signaling pathways, including phospholipase C activation, Ca2+ mobilization. Ras-mitogen-activated protein kinase activation, and adenylate cyclase inhibition. J Biol Chem 273:27104–27110

Okamoto H, Takuwa N, Yokomizo T, Sugimoto N, Sakurada S, Shigematsu H, Takuwa Y (2000) Inhibitory regulation of Rac activation, membrane ruffling, and cell migration by the G protein-coupled sphingosine-1-phosphate receptor EDG5 but not EDG1 or EDG3. Mol Cell Biol 20:9247–9261

Okazaki H, Ishizaka N, Sakurai T, Kurokawa K, Goto K, Kumada M, Takuwa Y (1993) Molecular cloning of a novel putative G protein-coupled receptor expressed in the cardiovascular system. Biochem Biophys Res Commun 190:1104–1109

Park TS, Panek RL, Mueller SB, Hanselman JC, Rosebury WS, Robertson AW, Kindt EK, Homan R, Karathanasis SK, Rekhter MD (2004) Inhibition of sphingomyelin synthesis reduces atherogenesis in apolipoprotein E-knockout mice. Circulation 110:3465–3471

Plump AS, Smith JD, Hayek T, Aalto-Setala K, Walsh A, Verstuyft JG, Rubin EM, Breslow JL (1992) Severe hypercholesterolemia and atherosclerosis in apolipoprotein E-deficient mice created by homologous recombination in ES cells. Cell 71:343–353

Rhainds D, Bourgeois P, Bourret G, Huard K, Falstrault L, Brissette L (2004) Localization and regulation of SR-BI in membrane rafts of HepG2 cells. J Cell Sci 117:3095–3105

Rodriguez JL, Ghiselli GC, Torreggiani D, Sirtori CR (1976) Very low density lipoproteins in normal and cholesterol-fed rabbits: lipid and protein composition and metabolism. Part 1. Chemical composition of very low density lipoproteins in rabbits. Atherosclerosis 23:73–83

Ross R (1993) The pathogenesis of atherosclerosis: a perspective for the 1990s. Nature 362:801–809

Sachinidis A, Kettenhofen R, Seewald S, Gouni-Berthold I, Schmitz U, Seul C, Ko Y, Vetter H (1999) Evidence that lipoproteins are carriers of bioactive factors. Arterioscler Thromb Vasc Biol 19:2412–2421

Sanchez T, Thangada S, Wu MT, Kontos CD, Wu D, Wu H, Hla T (2005) PTEN as an effector in the signaling of antimigratory G protein-coupled receptor. Proc Natl Acad Sci USA 102:4312–4317

Sanchez T, Skoura A, Wu MT, Casserly B, Harrington EO, Hla T (2007) Induction of vascular permeability by the sphingosine-1-phosphate receptor-2 (S1P2R) and its downstream effectors ROCK and PTEN. Arterioscler Thromb Vasc Biol 27:1312–1318

Schissel SL, Tweedie-Hardman J, Rapp JH, Graham G, Williams KJ, Tabas I (1996) Rabbit aorta and human atherosclerotic lesions hydrolyze the sphingomyelin of retained low-density lipoprotein. Proposed role for arterial-wall sphingomyelinase in subendothelial retention and aggregation of atherogenic lipoproteins. J Clin Invest 98:1455–1464

Schlitt A, Blankenberg S, Yan D, von Gizycki H, Buerke M, Werdan K, Bickel C, Lackner KJ, Meyer J, Rupprecht HJ, Jiang XC (2006) Further evaluation of plasma sphingomyelin levels as a risk factor for coronary artery disease. Nutr Metab 3:5

Simons K, Ikonen E (1997) Functional rafts in cell membranes. Nature 387:569–572

Skoura A, Michaud J, Im DS, Thangada S, Xiong Y, Smith JD, Hla T (2011) Sphingosine-1-phosphate receptor-2 function in myeloid cells regulates vascular inflammation and atherosclerosis. Arterioscler Thromb Vasc Biol 31:81–85

Subbaiah PV, Davidson MH, Ritter MC, Buchanan W, Bagdade JD (1989) Effects of dietary supplementation with marine lipid concentrate on the plasma lipoprotein composition of hypercholesterolemic patients. Atherosclerosis 79:157–166

Sugimoto N, Takuwa N, Okamoto H, Sakurada S, Takuwa Y (2003) Inhibitory and stimulatory regulation of Rac and cell motility by the G12/13-Rho and Gi pathways integrated downstream of a single G protein-coupled sphingosine-1-phosphate receptor isoform. Mol Cell Biol 23:1534–1545

Szabo G, Dolganiuc A, Dai Q, Pruett SB (2007) TLR4, ethanol, and lipid rafts: a new mechanism of ethanol action with implications for other receptor-mediated effects. J Immunol 178:1243–1249

Takeda K, Akira S (2005) Toll-like receptors in innate immunity. Int Immunol 17:1–14

Takuwa Y, Okamoto Y, Yoshioka K, Takuwa N (2008) Sphingosine-1-phosphate signaling and biological activities in the cardiovascular system. Biochim Biophys Acta 1781:483–488

Tolle M, Pawlak A, Schuchardt M, Kawamura A, Tietge UJ, Lorkowski S, Keul P, Assmann G, Chun J, Levkau B, van der Giet M, Nofer JR (2008) HDL-associated lysosphingolipids inhibit NAD(P)H oxidase-dependent monocyte chemoattractant protein-1 production. Arterioscler Thromb Vasc Biol 28:1542–1548

Triantafilou M, Brandenburg K, Kusumoto S, Fukase K, Mackie A, Seydel U, Triantafilou K (2004a) Combinational clustering of receptors following stimulation by bacterial products determines lipopolysaccharide responses. Biochem J 381:527–536

Triantafilou M, Morath S, Mackie A, Hartung T, Triantafilou K (2004b) Lateral diffusion of Toll-like receptors reveals that they are transiently confined within lipid rafts on the plasma membrane. J Cell Sci 117:4007–4014

Vacaru AM, Tafesse FG, Ternes P, Kondylis V, Hermansson M, Brouwers JF, Somerharju P, Rabouille C, Holthuis JC (2009) Sphingomyelin synthase-related protein SMSr controls ceramide homeostasis in the ER. J Cell Biol 185:1013–1027

Van der Luit AH, Budde M, Zerp S, Caan W, Klarenbeek JB, Verheij M, Van Blitterswijk WJ (2007) Resistance to alkyl-lysophospholipid-induced apoptosis due to downregulated sphingomyelin synthase 1 expression with consequent sphingomyelin- and cholesterol-deficiency in lipid rafts. Biochem J 401:541–549

von Arnim CA, Kinoshita A, Peltan ID, Tangredi MM, Herl L, Lee BM, Spoelgen R, Hshieh TT, Ranganathan S, Battey FD, Liu CX, Bacskai BJ, Sever S, Irizarry MC, Strickland DK, Hyman

BT (2005) The low density lipoprotein receptor-related protein (LRP) is a novel beta-secretase (BACE1) substrate. J Biol Chem 280:17777–17785

Wang R, Town T, Gokarn V, Flavell RA, Chandawarkar RY (2006a) HSP70 enhances macrophage phagocytosis by interaction with lipid raft-associated TLR-7 and upregulating p38 MAPK and PI3K pathways. J Surg Res 136:58–69

Wang N, Ranalletta M, Matsuura F, Peng F, Tall AR (2006b) LXR-induced redistribution of ABCG1 to plasma membrane in macrophages enhances cholesterol mass efflux to HDL. Arterioscler Thromb Vasc Biol 26:1310–1316

Wang F, Okamoto Y, Inoki I, Yoshioka K, Du W, Qi X, Takuwa N, Gonda K, Yamamoto Y, Ohkawa R, Nishiuchi T, Sugimoto N, Yatomi Y, Mitsumori K, Asano M, Kinoshita M, Takuwa Y (2010) Sphingosine-1-phosphate receptor-2 deficiency leads to inhibition of macrophage proinflammatory activities and atherosclerosis in apoE-deficient mice. J Clin Invest 120:3979–3995

Williams KJ, Tabas I (1995) The response-to-retention hypothesis of early atherogenesis. Arterioscler Thromb Vasc Biol 15:551–561

Williams KJ, Tabas I (1998) The response-to-retention hypothesis of atherogenesis reinforced. Curr Opin Lipidol 9:471–474

Yamaoka S, Miyaji M, Kitano T, Umehara H, Okazaki T (2004) Expression cloning of a human cDNA restoring sphingomyelin synthesis and cell growth in sphingomyelin synthase-defective lymphoid cells. J Biol Chem 279:18688–18693

Yvan-Charvet L, Welch C, Pagler TA, Ranalletta M, Lamkanfi M, Han S, Ishibashi M, Li R, Wang N, Tall AR (2008) Increased inflammatory gene expression in ABC transporter-deficient macrophages: free cholesterol accumulation, increased signaling via toll-like receptors, and neutrophil infiltration of atherosclerotic lesions. Circulation 118:1837–1847

Zhang B, Tomura H, Kuwabara A, Kimura T, Miura S, Noda K, Okajima F, Saku K (2005) Correlation of high density lipoprotein (HDL)-associated sphingosine 1-phosphate with serum levels of HDL-cholesterol and apolipoproteins. Atherosclerosis 178:199–205

Zhu X, Owen JS, Wilson MD, Li H, Griffiths GL, Thomas MJ, Hiltbold EM, Fessler MB, Parks JS (2010) Macrophage ABCA1 reduces MyD88-dependent Toll-like receptor trafficking to lipid rafts by reduction of lipid raft cholesterol. J Lipid Res 51:3196–3206

Zhuang L, Kim J, Adam RM, Solomon KR, Freeman MR (2005) Cholesterol targeting alters lipid raft composition and cell survival in prostate cancer cells and xenografts. J Clin Invest 115:959–968

Zilversmit DB, Mc CE, Jordan PH, Henly WS, Ackerman RF (1961) The synthesis of phospholipids in human atheromatous lesions. Circulation 23:370–375

Cardiovascular Effects of Sphingosine-1-Phosphate (S1P)

Bodo Levkau

Contents

Abstract Sphingosine-1-phosphate (S1P) regulates important functions in cardiac and vascular homeostasis. It has been implied to play causal roles in the pathogenesis of many cardiovascular disorders such as coronary artery disease, atherosclerosis, myocardial infarction, and heart failure. The majority of S1P in plasma is associated with high-density lipoproteins (HDL), and their S1P content has been shown to be responsible, at least in part, for several of the beneficial effects of HDL on cardiovascular risk. The attractiveness of S1P-based drugs for potential cardiovascular applications is increasing in the wake of the clinical approval of FTY720, but answers to important questions on the effects of S1P in cardiovascular biology and medicine must still be found. This chapter focuses on the current understanding of the role of S1P and its receptors in cardiovascular physiology, pathology, and disease.

B. Levkau (✉)
Institute of Pathophysiology, University Hospital Essen, University of Duisburg-Essen, Hufelandstrasse 55, 45122 Essen, Germany
e-mail: bodo.levkau@uni-due.de

E. Gulbins and I. Petrache (eds.), *Sphingolipids in Disease*,
Handbook of Experimental Pharmacology 216,
DOI 10.1007/978-3-7091-1511-4_8, © Springer-Verlag Wien 2013

Keywords High-density lipoproteins (HDL) • Apolipoproteins • Myocardial infarction • Preconditioning • Arterial tone • Atherosclerosis • Endothelial permeability • Coronary artery disease (CAD) • Macrophages • Heart rate • FTY720

1 Plasma S1P and HDL as Its Bioactive Carrier

The plasma concentrations of S1P in healthy individuals range between 200 and 1,000 nM and are considerably higher than those of any other extracellular fluid (Okajima 2002; Zhang et al. 2005; Sattler et al. 2010). The amphipathic nature of S1P precludes it from being present in a free, unbound form in plasma. There, it is bound to several lipoproteins, of which high-density lipoproteins (HDL) are the most important ones (~80–90 % of the plasma S1P), followed by low-density lipoproteins and very-low-density lipoproteins (together less than 10 %), with the remainder being bound to albumin (~10–20 %) (Murata et al. 2000; Sattler and Levkau 2009). HDL-bound S1P is biologically active: several biological effects of HDL can be attributed in part or entirely to the S1P contained within (Sattler and Levkau 2009). This has been revealed by the requirement of S1P receptor signaling for a number of HDL effects (Sattler and Levkau 2009). These are, among others, nitric oxide (NO) production through eNOS, NO-dependent vasodilatation, angiogenesis, Akt and Erk signaling, and certain aspects of the anti-oxidative, anti-apoptotic, and anti-inflammatory actions of HDL (Sattler and Levkau 2009). These observations are of utmost importance since plasma HDL cholesterol (HDL-C) levels are the best negative predictor of human cardiovascular disease (Barter et al. 2007) and raise the important question whether the S1P-content of HDL may be useful as a biomarker of human cardiovascular disease or may even constitute a therapeutic target. The first has proven to be the case: S1P contained in the HDL fraction from patients with coronary artery disease is lower than in healthy individuals and the S1P outside of HDL higher, respectively (Sattler et al. 2010). The second notion has yet to be addressed.

S1P binds with an extremely high affinity to HDL (Kontush et al. 2007). Accordingly, plasma S1P levels correlate positively with HDL-C as well as apolipoproteins AI and II as the two major apolipoproteins within HDL (Zhang et al. 2005). The binding partners of S1P within the HDL particle and their biochemical requirements for S1P binding have only begun to be characterized. Apolipoprotein M (ApoM) has been found to bind S1P in vitro (Sevvana et al. 2009) and in vivo (Christoffersen et al. 2011), and mice deficient for ApoM have ~90 % lower S1P concentrations contained in their HDL fraction (Christoffersen et al. 2011). In vitro, ApoM-deficient HDL was less potent in activating Akt and Erk signaling, promoting endothelial cell migration, and inducing endothelial adherens junction formation, respectively (Christoffersen et al. 2011). However, plasma S1P levels in ApoM-deficient mice were only ~50 % lower suggesting that other apolipoproteins and/or lipoproteins must also play a role. Another hint comes from a study that found no correlation between plasma S1P and ApoM levels in humans (Karuna et al. 2011), while such correlations very well exist for ApoAI and II (Zhang et al. 2005).

Only 1–2 % of the total plasma S1P is biologically active. However, as total plasma S1P concentrations (200–1,000 nM) are 20–100-fold higher than the K_d value of its receptors (Murata et al. 2000; Mandala et al. 2002), this is still sufficient for the maintenance of the "tonic," continuous signaling by S1P required for vascular integrity (Sanna et al. 2006; Camerer et al. 2009). As the largest part of S1P in plasma is biologically inactive and most of plasma S1P is associated with HDL, some part of the HDL-bound S1P must be inaccessible for signaling. How large it is and which part of the total S1P contained in HDL is actually serving signaling purposes is unknown. However, the capacity of HDL to take up additional S1P is 10-fold higher than already present within the particle as calculated per milligram of protein. The biophysical topography and signaling proficiency of S1P inside HDL are still a matter of investigation.

Erythrocytes are quantitatively the most important source of plasma S1P. They lack S1P-lyase and are extremely potent in synthesizing and releasing S1P from sphingosine (Pappu et al. 2007). Platelets, mast cells, and leukocytes, especially upon activation, can also produce and secrete S1P. In addition, vascular and lymphatic endothelial cells can both synthesize and release S1P (Pappu et al. 2007; Venkataraman et al. 2008). Inside the cell, S1P moves freely between membranes but requires specific transport mechanisms for its translocation to the outer leaflet of the cytoplasmic membrane (Hannun and Obeid 2008; van Meer et al. 2008). It is still unknown how S1P is transferred from there to the HDL particle. In any case, a direct contact between HDL and the cell membrane has been postulated as necessary. Both a passive transfer from erythrocytes (Bode et al. 2010) and an active transport via a yet unidentified erythrocyte ABC transporter without requirement for ATP hydrolysis (Kobayashi et al. 2009) have been proposed. Plasma S1P levels are unaltered in mice deficient for ABCA1, ABCA7, ABCC1 (Lee et al. 2007), or the sphingolipid transporter spinster homologue 2 (Spns2) (Kawahara et al. 2009; Osborne et al. 2008; Fukuhara et al. 2012), suggesting that they are dispensable for the maintenance of plasma S1P levels. In summary, plasma is the extracellular compartment with the highest S1P concentrations, but S1P bioavailability and signaling propensity are determined by its most important carrier in plasma, high-density lipoproteins.

2 Cardiac Development

There are five cognate G-protein-coupled receptors for S1P, $S1P_{1-5}$, with a K_d of 8–20 nM (Rivera and Chun 2008). The details on receptor binding and activation are complex as individual S1P receptors couple to one or more G-proteins with considerable overlap: of the main cardiovascular S1P receptors, $S1P_1$ is coupled to

$G_{i/o}$ (preferentially $G_{i\alpha 1}$ and $G_{i\alpha 3}$); $S1P_2$ to $G_{i/o}$, $G_{12/13}$, and G_q; and $S1P_3$ to $G_{i/o}$, G_q, or $G_{12/13}$ (Hannun and Obeid 2008; Rivera and Chun 2008). The coupling of one receptor to different G-proteins and the simultaneous activation of several S1P receptors underlie the variety of downstream signaling events (Spiegel and Milstien 2003). Vertical S1P signaling induces the activation of phospholipase C and Ca^{2+} mobilization via G_q, activation of Erks and PI3K, and inhibition of adenylate cyclase via G_i, as well as the activation of Rho/actin cytoskeleton assembly via $G_{12/13}$ (Chun et al. 2010). In addition, S1P receptors have been shown to transactivate tyrosine kinase receptors such as VEGFR, PDGFR, and EGFR (Olivera and Spiegel 1993; Hobson et al. 2001; Waters et al. 2003; Tanimoto et al. 2002) as well as G-protein-coupled receptors such as the CXCR4 (Walter et al. 2007; Ryser et al. 2008; Kimura et al. 2004). A crossactivation of the TGF-β receptor type II by Smads activation has also been described (Xin et al. 2004). S1P receptor signaling occurs at the full spectrum of the S1P binding curve: low concentrations transactivate the PDGFR, those at the steep slope turn off lymphocyte recirculation, while high concentrations (50–100-fold higher than the K_d) lead to receptor desensitization and degradation (Rosen et al. 2008).

The S1P receptors that are predominantly expressed in the adult rodent heart are $S1P_1$, $S1P_2$, and $S1P_3$. $S1P_1$ is approximately 4–6-fold more abundant than the other two (Alewijnse et al. 2004). Adult human hearts express similar amounts of $S1P_1$ and $S1P_3$ and much less $S1P_2$. $S1P_1$ is strongly expressed in ventricular, septal, and atrial cardiomyocytes and in endothelial cells of cardiac vessels (Mazurais et al. 2002). In the developing mouse heart, $S1P_{1-3}$ are expressed from E8.5 to E12.5 together with $S1P_4$ (Wendler and Rivkees 2006). Signaling through S1P affects heart development in zebra fish: the *miles apart* (mil) mutation, an orthologue to the mammalian $S1P_2$, results in impaired fusion of the bilateral heart tubes leading to cardia bifida (Kupperman et al. 2000). Interestingly, its expression is required in extracardiac tissue and not in the precardiac cells that normally migrate to the midline and fuse to build the definitive heart tube (Kupperman et al. 2000). The sphingolipid transporter Spns2 identified in zebra fish was shown to cause the *two of hearts* mutation (Kawahara et al. 2009; Osborne et al. 2008). This indicates that S1P signaling is involved in early cardiac morphogenesis. S1P signaling is also important during early heart development. It influences the migration, differentiation, and survival of embryonic cardiac cells (Wendler and Rivkees 2006). The maintenance of a specific concentration range of S1P is apparently necessary for cardiac cushion development as a decrease as well as an increase in S1P hamper the process, with low S1P leading to cell death and high S1P inhibiting cell migration and mesenchymal cell formation (Wendler and Rivkees 2006). In mice, $S1P_2$ and $S1P_3$ knockouts display no overt cardiac phenotype (Ishii et al. 2002; Liu et al. 2000; MacLennan et al. 2001). Knockout mice for $S1P_1$ die during embryogenesis between embryonic day E12.5 and E14.5 due to the vascular hemorrhage thus precluding the investigation of $S1P_1$ in cardiac development.

The enzymes responsible for S1P production and degradation are also present and active in the heart. Sphingosine kinases 1 and 2 (Sphk1 and Sphk2) are both expressed and enzymatically active in the heart (Argraves et al. 2004). Their cardiac expression has been found as early as E8.5 (Wendler and Rivkees 2006), with Sphk2 but not Sphk1 activity decreasing with age (Vessey et al. 2009). S1P phosphatase 1 (SPP1) is also expressed in the adult heart (Johnson et al. 2003; Mandala et al. 2000) and throughout the early stages of cardiac development (Wendler and Rivkees 2006). The gene for S1P lyase (Sgpl1) is transcriptionally active in the rodent heart (Zhou and Saba 1998), and the enzyme is active there (Bandhuvula et al. 2011), respectively. S1P lyase is expressed much more abundantly in cardiomyocytes than in cardiac fibroblasts (Bandhuvula et al. 2011). In the vasculature, Sphk1 is present in arteries, but its expression differs dependent on the vascular bed: e.g., the mRNA is expressed 40–80-fold higher in cerebral arteries than in the aorta or mesenteric arteries (Salomone et al. 2010). The mRNA encoding SPP1 has been shown to be expressed in Hamster gracilis muscle resistance arteries (Peter et al. 2008).

3 Vascular Development

Vascular smooth muscle cells (VSMC) express mainly $S1P_2$, less $S1P_3$, and even less $S1P_1$ (Alewijnse and Peters 2008). However, VSMC from different vascular beds and different differentiation states may differ substantially in their S1P receptor abundance (Kluk and Hla 2001). Endothelial cells express predominantly $S1P_1$, less $S1P_3$, and only little $S1P_2$ under normal circumstances (Alewijnse and Peters 2008; Ozaki et al. 2003; Saba and Hla 2004), but again, in endothelial pathologies as in hypoxic retinopathy, $S1P_2$ functions become crucial (Skoura and Hla 2009). Endothelial cells of different origins (arteries, capillaries, veins, lymphatics) and different arteries (aortic, cerebral, coronary, renal, and mesenteric) also differ in their relative S1P receptor expression (Levkau 2008; Igarashi and Michel 2009).

$S1P_1$ knockout mice die during embryogenesis between E12.5 and E14.5 due to vascular hemorrhage because of the inability of their vascular smooth muscle cells (VSMC) to surround and support the developing vasculature (Liu et al. 2000). The recruitment of VSMC requires $S1P_1$ in endothelial cells but not VSMC suggesting that the paracrine effects play the decisive role, which is confirmed by the appearance of same phenotype in endothelial-specific $S1P_1$ knockout mice (Allende and Proia 2002). Identical defects are also displayed by knockout mice for both Sphk1 and Sphk2 that completely lack S1P (Mizugishi et al. 2005). These studies demonstrate the important role of S1P in vascular morphogenesis. The mechanistic basis for the defects could be deduced from the effects S1P has been shown to have in vascular cells: it stimulates endothelial proliferation, migration, and angiogenesis; protects against apoptosis; and guards vascular integrity (Rosen et al. 2009; Hla 2004). It is also a potent chemoattractant for endothelial cells (Hla et al. 2001) and promotes directed migration, vascular differentiation, and capillary network formation (Liu et al. 2000; Lee et al. 1999). Small GTPases play an important role as mediators of these effects: Rac1 activation by $S1P_1$ induces focal contact assembly, membrane ruffling, and cortical actin formation via G_i, while RhoA activation by $S1P_3$ promotes stress fiber assembly via G_q (Skoura and Hla 2009).

$S1P_2$ and $S1P_3$ play more discrete roles in vascular development that are not evident in the single knockouts but rather reveal themselves in the vascular defects present in $S1P_1$, $S1P_2$, and $S1P_3$ triple knockouts, where the embryos die earlier (between E10.5 and E11.5) and exhibit a more severe phenotype (Kono et al. 2004). The underlying mechanisms are incompletely understood. In vitro, $S1P_2$ appears to have both anti-migratory and anti-angiogenic effects in endothelial cells: pharmacological $S1P_2$ blockade enhances the pro-migratory response of S1P elicited by $S1P_1$, while overexpression of $S1P_2$ inhibits S1P-induced migration by suppressing Rac1 activity (Ryu et al. 2002). $S1P_2$ is a hypoxia-regulated gene and its deletion makes mice resistant to hypoxia-triggered neoangiogenesis suggesting $S1P_2$ to be important in ischemia-driven retinopathy (Skoura et al. 2007). Vice versa, $S1P_2$ activates hypoxia-inducible factor 1 (HIF-1) in a hypoxia-independent manner and HIF-1-regulated genes such as VEGF (Michaud et al. 2009). Tumor-associated angiogenesis is another S1P-responsive process: an antibody that neutralizes S1P has been shown to reduce tumor angiogenesis and block tumor progression in several xenograft and allograft tumor models (Visentin et al. 2006), while its humanized version inhibited retinal and choroidal neovascularization in oxygen-induced ischemic retinopathy (Xie et al. 2009).

4 Cardiac Function and Heart Rate

Blood pressure and cardiac function are not altered in $S1P_2$-deficient mice (Lorenz et al. 2007), and blood pressure is normal in $S1P_3$-deficient mice (Levkau et al. 2004). The embryonically lethal phenotype of the global $S1P_1$ knockout has precluded the assessment of cardiac function in vivo, and no data are yet available on a cardiomyocyte-specific deletion of $S1P_1$. In cardiomyocytes in vitro, S1P induces protein synthesis and cellular hypertrophy (Robert et al. 2001). Via $S1P_1$, it exerts negative inotropic effects in response to adrenergic stimulation through G_i and the inhibition of cAMP formation (Landeen et al. 2008; Means et al. 2007). Furthermore, S1P activates the MAPK and STAT3 in cardiomyocytes (Frias et al. 2009), which have both been implicated in the physiological hypertrophic response (Kunisada et al. 1998). Cardiomyocytes from $S1P_3$ but not $S1P_2$ receptor knockout mice cannot activate PLC in response to S1P (Means and Brown 2009). As G_q signaling and pressure overload-induced hypertrophy are closely associated, it has been suggested that the $S1P_3/G_q$/PLC pathway may contribute to cardiac hypertrophy, but this has not been corroborated experimentally (Means and Brown 2009). S1P levels have not been found changed in the peripheral circulation of patients with symptomatic heart failure (Knapp et al. 2012), and it is currently unknown whether pharmacological alteration of S1P receptor signaling may be effective in the setting of heart failure (Mann 2012).

S1P exerts several effects on ion currents in cardiomyocytes in vitro. It stimulates the inward rectifier potassium current ($I_{K.ACh}$) resulting in the reduction of spontaneous pacing rate (Guo et al. 1999) and inhibits the isoproterenol-induced increase in L-type calcium channels ($I_{Ca,L}$) currents, respectively, as well as the hyperpolarization-activated inward current (I_f) thereby attenuating the positive chronotropic effects of β-adrenergic stimuli in sinoatrial node cells and ventricular myocytes (Landeen et al. 2008; Guo et al. 1999). In vivo, S1P receptors are involved in the control of heart rate. FTY720 phosphate induces a transient bradycardia in mice and men (Cohen 2011; Sanna et al. 2004; Budde et al. 2002). Experimental studies suggest that both $S1P_1$ and $S1P_3$ agonism prior to receptor internalization are involved in the process (Sanna et al. 2004; Forrest et al. 2004; Koyrakh et al. 2005). However, species differences including S1P receptor subtype differences have also been observed, which may differentially affect heart rate in mouse (Sanna et al. 2004) and humans (Cohen 2011). These differences may in part explain the absence of related cardiac abnormalities in humans exposed for up to 7 years to FTY720.

5 Arterial Tone

S1P plays an important role in the regulation of arterial tone (Levkau 2008; Igarashi and Michel 2009). It acts on S1P receptors on endothelial and VSMC in concert with other vasoactive substances and affects the contractile mechanisms of VSMC as a second messenger. Exogenous application of S1P to isolated vessels promotes vasoconstriction in mesenteric, cerebral, and coronary arteries but not in femoral and carotid arteries or aorta and does so via direct effects on VSMC (Igarashi and Michel 2009; Salomone et al. 2008). An explanation may be the different expression of S1P receptors (e.g., aorta versus cerebral arteries) (Coussin et al. 2002) as well as of Sphk1 (40–80-fold higher in cerebral mesenteric arteries or aorta) (Salomone et al. 2010) although this has not been examined systematically.

S1P has been shown to induce vasoconstriction in the basilar artery of wild-type and $S1P_2$-deficient but not $S1P_3$-deficient mice suggesting $S1P_3$ to be the responsible receptor (Salomone et al. 2008). In cerebral arteries, a $S1P_3$ antagonist (Murakami et al. 2010) and FTY720 phosphate (Salomone et al. 2008) induced vasoconstriction in a $S1P_3$-dependent manner, while in coronary arteries $S1P_3$ activation decreased myocardial perfusion (Levkau et al. 2004). $S1P_2$ knockout mice exhibited a decrease in resting vascular tone as well as in the contractile response to α-adrenergic stimulation (Lorenz et al. 2007). They had an elevated regional blood flow together with decreased mesenteric and renal resistance without any effects on systemic blood pressure, suggesting that $S1P_2$ regulates homeostatic arterial tone in the mesenterium and kidney (Lorenz et al. 2007). In line with this, S1P has been shown to be produced and secreted by VSMC, to engage $S1P_2$ in an autocrine/paracrine manner, and to be important for the myogenic response of resistance arteries (the vasoconstrictive reaction in response to increased transmural pressure as an adaptory mechanism to maintain constant blood flow to tissues) (Peter et al. 2008; Bolz et al. 2003). This myogenic response involves a pool of extracellular S1P generated by Sphk and degraded by the SPP1 after import into the cell by the cystic fibrosis transmembrane conductance regulator (CFTR) (Peter et al. 2008). The increase in myogenic tone of cerebral arteries that went along with a decreased cerebral perfusion in a model of heart failure has been attributed to a decreased CFTR-mediated S1P uptake and degradation (Meissner et al. 2012). While the vasoconstrictive effects of S1P are mediated by its actions on VSMC, its effects on endothelial cells cause the opposite. S1P is an important stimulator of NO production in endothelial cells through activation of the eNOS (Igarashi and Michel 2008, 2009; Dantas et al. 2003). This has physiological consequences as S1P can vasodilate precontracted aortae of mice and rats ex vivo (Salomone et al. 2008; Nofer et al. 2004) and reverse the endothelin-induced elevation of mean arterial blood pressure after intra-aortic injection in vivo (Nofer et al. 2004). Thus S1P effects on endothelial cells counteract those on VSMC especially when arterial tone is already elevated, making S1P an important factor in the fine-tuning of vascular tone (Levkau 2008).

Besides receptor-dependent effects of S1P on vascular tone, also receptor-independent ones have been described, in which it acts as an intracellular second messenger (Salomone et al. 2008). Sphk1-produced S1P is generated by various G-protein-coupled receptor agonists such as 5-hydroxytryptamine and activates Ca^{2+}-sensitizing mechanisms such as the RhoA/Rho kinase pathway leading to increased myosin light chain phosphorylation and contraction (Salomone et al. 2008). Accordingly, Sphk inhibition and Sphk1 knockout mice have blunted vasoconstriction responses to several GPCR agonists and even KCl. This is S1P receptor-independent as the contraction to the prostanoid U46619 and endothelin 1 in basilar arteries from both $S1P_2$ and $S1P_3$ knockout mice was normal (Salomone et al. 2008), and Sphk inhibition blocked vasoconstriction in cerebral arteries from $S1P_3$ knockout mice (Salomone et al. 2010). These studies suggest that intracellular S1P actions may be important mediators of the vasoconstriction response induced by many G-protein-coupled receptors.

6 Myocardial Infarction and Ischemic Preconditioning

S1P plays an important protective role during myocardial ischemia/reperfusion injury and takes part in the processes of pre- and postconditioning (Karliner 2006). In isolated mouse hearts subjected to ischemia/reperfusion, administration of S1P improved hemodynamics, reduced creatinine kinase release, and diminished infarct size (Vessey et al. 2006). In a mouse model of acute myocardial infarction, S1P administration also reduced infarct size (Theilmeier et al. 2006). A decline of myocardial Sphk activity that was accompanied by a decrease in S1P levels has been described to occur during ischemia, and Sphk1-deficient hearts had larger infarct sizes when subjected to ischemia/reperfusion injury (Karliner 2009). To the opposite, heterozygous S1P lyase knockout hearts exhibited smaller infarct sizes and better functional recovery after ischemia/reperfusion (Bandhuvula et al. 2011). Pharmacological inhibition of the S1P lyase using tetrahydroxybutylimidazole also reduced infarct size and enhanced hemodynamic recovery (Bandhuvula et al. 2011).

In vitro, S1P rescued isolated cardiomyocytes from hypoxia through $S1P_1$ and G_i-dependent activation of Akt as shown with the $S1P_1$ receptor agonist SEW2871 (Zhang et al. 2007; Hofmann et al. 2009). A different study has implied $S1P_2$ together with $S1P_3$ in Akt and Erk activation in cardiomyocytes, whereas $S1P_1$ exclusively decreased cAMP accumulation (Means et al. 2008). Coronary ischemia/reperfusion injury in vivo was altered neither in $S1P_2$- nor $S1P_3$-deficient mice (Theilmeier et al. 2006; Means et al. 2008) but greatly enhanced in the double knockouts. In these, Akt activation by ischemia/reperfusion as well as exogenous S1P was abolished (Means et al. 2007). Exogenous application of S1P prior to ischemia/reperfusion decreased infarct size (Theilmeier et al. 2006) and its cardioprotective effect was dependent on $S1P_3$ and nitric oxide generation (NO) (Theilmeier et al. 2006). Several endothelial S1P effects potentially relevant in the setting of ischemia/reperfusion are also known to be intrinsic to NO: endothelial barrier sealing, inhibition of neutrophil adhesion, and prevention of microvascular leakage (Schulz et al. 2004; Bolli 2001). Thus increased NO production in the microcirculation following endothelial $S1P_3$ signaling may play a crucial role in the rather mild, inflammation-dominated injury setting of short ischemia/reperfusion. However, with longer ischemia periods, a reperfusion injury dominated by cardiomyocyte death becomes the leading cause, and the direct anti-apoptotic effects of S1P on cardiomyocytes (Theilmeier et al. 2006; Karliner et al. 2001; Jin et al. 2004) may become increasingly important. In a model of permanent coronary occlusion, Sphk activity was reduced and $S1P_1$ expression decreased, respectively, early after infarction in the remote, uninfarcted myocardium (Yeh et al. 2009). $S1P_1$ activation with oral SEW2871 during the first 2 weeks after infarction resulted in improved myocardial function (Yeh et al. 2009), suggesting that impaired S1P signaling plays a role in the development of ischemic cardiomyopathy. However, caution should be exercised with agonists such as SEW2871 as it induced irreversible tachyarrhythmias during reperfusion (Tsukada et al. 2007) and increased ventricular tachyarrhythmias and mortality when administered before reperfusion (Hofmann et al. 2010). Peculiarly, it protected against reperfusion arrhythmias when administered 24 h prior to ischemia (Hofmann et al. 2010). When FTY720 was administered in isolated rat hearts during reperfusion, it attenuated the rise in left ventricular end-diastolic pressure and improved the recovery of left ventricular developed pressure (LVDP) but did not ameliorate infarct size; SEW2871 was not able to improve recovery and even increased LVDP in this setting (Hofmann et al. 2009). In vivo, FTY720 did not alter infarct size—neither when given immediately at reperfusion nor 24 h before (Hofmann et al. 2010). The underlying mechanisms are unknown.

Pathological remodeling of the ventricle after myocardial infarction is the cause for cardiac dilation and ischemic cardiomyopathy. It is governed by the transformation of cardiac fibroblasts into myofibroblasts and their deposition of extracellular matrix. Sphk activity is 10-fold higher in cardiac fibroblasts than in adult mouse cardiomyocytes and promoted hypoxia-induced proliferation while dampening the pro-inflammatory response (Kacimi et al. 2007). S1P was shown to induce myofibroblast transformation and increase collagen expression in a $S1P_2$-dependent manner (Gellings Lowe et al. 2009). Also, the pro-fibrotic cytokine TGF-β activated Sphk1 (Gellings Lowe et al. 2009). Administration of a S1P-neutralizing antibody in a mouse coronary ligation model attenuated macrophage and mast cell infiltration into the infracted zone and reduced perivascular fibrosis within the non-infarcted myocardium, suggesting that S1P promotes pathological fibrosis (Lowe et al. 2006). The effect of S1P blockade on remodeling and cardiac function after myocardial infarction in vivo has yet to be addressed.

Ischemic preconditioning is a process where brief periods of ischemia render the heart more resistant to subsequent prolonged periods of ischemia and reperfusion (Vessey et al. 2006). S1P has been shown to participate in the process: ischemic preconditioning was associated with an increase in cardiac S1P levels and prevented the decline of Sphk enzyme activity during ischemia (Jin et al. 2004), while pharmacological inhibition of Sphk by dimethylsphingosine abolished the benefit of ischemic preconditioning (Lecour et al. 2002), and no preconditioning occurred in Sphk1-deficient hearts (Jin et al. 2005). Vice versa, adenoviral overexpression of Sphk1 protected against myocardial injury and prevented hemodynamic deterioration during ischemia/reperfusion injury in rat hearts (Duan et al. 2007). The complex mechanisms involved in ischemic preconditioning serve all to inhibit the mitochondrial permeability transition pore (mPTP), which is the critical determinant of lethal reperfusion injury in the heart (Hausenloy and Yellon 2007). S1P may be participating in several aspects of preconditioning. The cytokine TNFα is involved in the endogenous protection exhibited by ischemic preconditioning (Skyschally et al. 2007) and activates both sphingomyelinase and Sphk (Dettbarn et al. 1994; Oral et al. 1997; Thielmann et al. 2002; Hannun and Bell 1989; Coroneos et al. 1995). Thus Sphk may be a mediator of the beneficial TNFα effect, especially as inhibition of ceramidase by *N*-oleoylethanolamine (NOE) abolished both ischemic and TNFα-mediated preconditioning (Lecour et al. 2002; Skyschally et al. 2007). The anesthetic isoflurane, a well-known preconditioning agent in many organs including the heart (Ludwig et al. 2004), protects the kidney against ischemia/reperfusion injury via Sphk activation and S1P production (Kim et al. 2007). εPKC is required for ischemic preconditioning (Gray et al. 2004). In εPKC-deficient mice, the beneficial effects of Sphk-activating agents such as GM-1 were absent and the Sphk1 activation through preconditioning lost, respectively, (Karliner 2006; Jin et al. 2002, 2004), suggesting that Sphk activation is upstream of εPKC. As application of S1P was protective in εPKC-deficient hearts (Jin et al. 2002) and compensated for the adverse effects of Sphk inhibition or deficiency on infarct size and preconditioning (Karliner 2006; Jin et al. 2004), exogenous actions of S1P are apparently sufficient for protection. The most obvious candidates to mediate them are S1P receptors. However, their identity and involvement is currently unknown.

Sphk2 has also been demonstrated to protect against ischemia/reperfusion injury and contribute to ischemic preconditioning as shown by the larger myocardial injury and absence of preconditioning, respectively, in Sphk2-deficient hearts, albeit the mechanism paying a role there is apparently different from that by Sphk1 (Vessey et al. 2011). The lack of preconditioning in the absence of Sphk2 may simply be due to its necessity (besides that of Sphk1) to achieve enough S1P for preconditioning. However, another explanation has been put forward as well. Sphk2 mediates S1P synthesis in mitochondria, and a mitochondrial S1P pool appears to be important for proper assembly of the respiratory chain (Strub et al. 2011). Sphk2-deficient mice have decreased mitochondrial cytochrome oxidase activity and reduced respiration, which leads to increased reactive oxygen species generation that may thus facilitate the opening of the mitochondrial permeability transition pore to increase reperfusion injury.

Finally, S1P and Sphk1 are important not only in ischemic pre- but also postconditioning (Jin et al. 2008). Postconditioning is a phenomenon where brief periods of ischemia/reperfusion are administered at the beginning of the reperfusion period after prolonged ischemia and act in a cardioprotective manner (Zhao et al. 2003).

7 Arterial Remodeling and Response to Injury

Arterial remodeling takes place in the arterial wall both during the development of native atherosclerosis as well as the restenotic response of the artery after balloon angioplasty or stenting. The dedifferentiation of vascular smooth muscle cells is a fundamental phenomenon that underlies the morphological changes of the arterial wall. During this process, VSMC with a contractile quiescent phenotype typical for cells from the media transdifferentiate into the VSMC of synthetic and migratory phenotype characteristic for lesions. The causal stimulus for this process is vascular injury. As a consequence, medial VSMC downregulate expression of contractile genes, migrate to the newly forming VSMC rich neointima, and proliferate. In this process, the transcriptional repression of VSMC-specific differentiation genes plays a crucial role such as SMα-actin, smooth muscle myosin heavy chain (SMMHC), and SM22α, which all have a CArG box in their promoters to which serum response factor (SRF) and myocardin or myocardin-related transcription factors (MRTFs) bind (Wamhoff et al. 2008). S1P has been shown to increase the expression of these genes through a RhoA/MRTF-dependent pathway, with S1P receptors playing an important role (Wamhoff et al. 2008): $S1P_2$ promoted and $S1P_1$/$S1P_3$ inhibited S1P-induced activation of SMα-actin and SMMHC via L-type voltage-gated Ca^{2+} channels and RhoA/Rho kinase-dependent SRF enrichment of CArG box promoter regions (Wamhoff et al. 2008; Grabski et al. 2009). Finally, VSMC from different vascular beds and different phenotypes (medial versus neointimal VSMC) (Kluk and Hla 2001) have different S1P receptor expression levels: pup-intimal VSMCs express higher levels of $S1P_1$ than adult-medial VSMCs (Kluk and Hla 2001) and respond differently to S1P.

S1P has been shown to inhibit VSMC migration toward potent chemotactic stimuli such as PDGF-BB in mouse, rat, and human VSMC in vitro via inhibition of Rac1 activation by $S1P_2$ (Ryu et al. 2002; Bornfeldt et al. 1995; Okamoto et al. 2000). $S1P_2$-deficient VSMCs migrate faster to S1P and PDGF-BB than wild-type cells in vitro (Shimizu et al. 2007; Wang et al. 2010). Most studies suggest that S1P itself is not chemotactic for VSMC (Ryu et al. 2002; Bornfeldt et al. 1995; Okamoto et al. 2000), although one study has found this to be the case (Wang et al. 2010). Thus $S1P_2$ may serve as an important brake on migration that becomes visible only with potent chemotactic stimuli. Studies on the proliferation effects of S1P on VSMC are difficult to interpret: no effect has been described for S1P both in human and mouse VSMC (Bornfeldt et al. 1995; Keul et al. 2010) but a stimulatory one for rat VSMC, where it was abolished by a $S1P_1/S1P_3$ inhibitor (Wamhoff et al. 2008). In vitro, $S1P_2$-deficient VSMC did not show altered proliferation in one study (Shimizu et al. 2007) but an enhanced one in another (Wang et al. 2010). $S1P_3$-deficient VSMCs exhibit no alterations in proliferation or migration (Keul et al. 2010).

S1P receptor signaling might be affecting the arterial remodeling process by any or all of these means. Administration of a $S1P_1/S1P_3$ antagonist inhibited neointima formation after carotid balloon injury in rats (Wamhoff et al. 2008), while neointima formation was greatly enhanced in $S1P_2^{-/-}$ and $S1P_3^{-/-}$ mice after carotid ligation (Shimizu et al. 2007; Keul et al. 2010). In the case of $S1P_2$, S1P signaling may be preventing both VSMC migration from the media into the intima and their proliferative response (Grabski et al. 2009). The reasons for increased neointima formation in $S1P_3^{-/-}$ mice after carotid ligation are yet unknown (Keul et al. 2010). A plausible reason for the observed differences among mouse stains and rats is the dependence of the extent of arterial remodeling dependent on the genetic background: FVB mice with a tendency to a larger neointima formation have lower $S1P_2$ and higher $S1P_1$ expression in the arterial wall in comparison to C57/Bl6 mice which show a minimal response (Shimizu et al. 2007). Another reason may be the regulation of S1P receptor expression after vascular injury: in the rat carotid injury model, there is an early induction of $S1P_1$ and $S1P_3$ expression (Wamhoff et al. 2008), while no major differences have been observed in the carotid ligation model (Grabski et al. 2009). Nevertheless, S1P is an important player in arterial remodeling and the response to injury although the mechanisms still need further exploration.

8 Atherosclerosis

The best evidence that S1P plays a causal role in atherosclerosis stems from studies using FTY720 in different mouse models of atherosclerosis and S1P receptor knockouts on an atherosclerosis-susceptible background, respectively. Two of the three studies on FTY720 have shown it attenuate the development of atherosclerosis in apolipoprotein E-deficient (ApoE) mice (Keul et al. 2007) and LDL-receptor-deficient (LDL-R) mice (Nofer et al. 2007), respectively. The third study observed no effect, but the higher FTY720 dose used there leads to a hypercholesterolemia of unknown origin (Klingenberg et al. 2007) that may have antagonized any protective effects. The causal mechanism of how FTY720 attenuates atherosclerosis is unknown. Each of the many potentially atheroprotective effects of FTY720 could serve the purpose: it stimulates NO production (Tolle et al. 2005); inhibits the generation of reactive oxygen species; reduces the production and release of inflammatory chemokines such as MCP-1, TNFα, IL-6, IL-12, and IFNγ (Tolle et al. 2008; Nofer et al. 2007; Keul et al. 2007); and is a potent immunosuppressant that biases T cell immune responses toward an atheroprotective T_H2 (Muller et al. 2005; Hansson and Libby 2006).

Several studies have looked at the role of $S1P_2$ and $S1P_3$ in atherosclerosis by using their knockouts crossbred to ApoE-deficient mice (Skoura et al. 2011; Keul et al. 2011). Deficiency for $S1P_2$ resulted in a reduction in atherosclerosis, a reduced macrophage, and increased VSMC content, respectively, of lesions (Skoura et al. 2011). This could be attributed to effects of $S1P_2$ on macrophages as atherosclerosis was attenuated to the same extent in bone marrow chimera. $S1P_2$ on macrophages acts as a chemorepellent receptor in vitro: $S1P_2$-deficient mice recruit more macrophages in a peritonitis model, and ability of S1P to inhibit macrophage migration to C5a and CXCL12 is abrogated in vitro (Michaud et al. 2010). This may prevent monocyte/macrophage emigration from lesions. In contrast, $S1P_3$ promoted macrophage recruitment during peritonitis and chemotactic for S1P (Keul et al. 2011). Although $S1P_3$ deficiency did not affect the size of atherosclerotic lesions, it greatly reduced their macrophage content, with both hematopoietic and non-hematopoietic cells being involved (Keul et al. 2011). In contrast, $S1P_1$ does not appear to play a role in macrophage recruitment as shown in $S1P_1$ hematopoietic chimeras and myeloid-specific deletion of $S1P_1$ (Michaud et al. 2010; Keul et al. 2011). Effects of S1P on macrophage polarization have been described but remain controversial. One study showed reduced serum IL-1β and IL-18 but unchanged TNFα in $S1P_2$-deficient mice challenged with LPS (Skoura et al. 2011). Another has suggested that $S1P_1$ inhibits and $S1P_2$ promotes, respectively, LPS/IFN-γ-stimulated cytokine production (Hughes et al. 2008), but others have not been able to confirm it (Keul et al. 2011). Possible explanations could be the different macrophage sources as well as differences in macrophage activation and differentiation.

The permeability of the endothelial lining for lipids, lipoproteins, and inflammatory cells is a major factor in atherosclerosis. Erythrocyte-derived plasma S1P and endothelial $S1P_1$ are required for the maintenance of physiological endothelial barrier. Mice that lack S1P in plasma display vascular leakage that can be alleviated by administration of erythrocytes and $S1P_1$ receptor agonists (Camerer et al. 2009), while Sphk1-deficient mice exhibit a similar although less severe leakage despite unaltered plasma S1P levels (Li et al. 2008). In contrast, $S1P_1$ antagonists cause vascular leakage and edema (Sanna et al. 2006; Foss et al. 2007; Rosen et al. 2008). Endothelial cells constitutively produce and secrete S1P in response to laminar shear stress (Venkataraman et al. 2008) and are induced to do so by LPS or PAR-1 during inflammatory lung injury to sustain and restore, respectively, barrier function via $S1P_1$ (McVerry and Garcia 2005; Tauseef et al. 2008). The barrier-protective functions of activated protein C (APC) can be partially attributed to its activation of Sphk1 and $S1P_1$ (Feistritzer and Riewald 2005; Finigan et al. 2005), and

PAF-induced generalized vascular leakage and death can be ameliorated by administration of $S1P_1$ agonists (Camerer et al. 2009). Mechanistically, S1P promotes adherens junction assembly (Singleton et al, 2005; Mehta et al. 2005) via $S1P_1$ (Lee et al. 1999; Lee et al. 2006) while weakening them via $S1P_2$, Rho-ROCK, and PTEN (Sanchez et al. 2007).

Monocyte adhesion to activated endothelium and their differentiation into macrophages and foam cells is the next key step in the pathogenesis of atherosclerosis. The data on the effects of S1P on these processes are controversial. S1P suppressed the inflammatory cell adherence to TNFα-activated aortic endothelium (Bolick et al. 2005) and abrogate along with $S1P_1$ agonists to monocytic cell adhesion to aortic endothelium via NO-mediated suppression of NF-κB (Whetzel et al. 2006). In contrast, exogenous S1P stimulated VCAM-1 and E-selectin via NF-κB, and endogenous S1P generated by Sphk1 after TNFα stimulation promoted the expression of adhesion molecules (Kimura et al. 2006; Rizza et al. 1999; Xia et al. 1998; Shimamura et al. 2004). Sphk1 has been implicated in the induction of COX-2 by TNFα and the production of the inflammatory prostaglandin PGE_2 (Pettus et al. 2003), and its overexpression leads to higher VCAM-1 expression both under basal and TNFα-stimulated conditions (Limaye et al. 2009). The knock down of S1P phosphatase and S1P lyase also augmented prostaglandin production (Pettus et al. 2003). The amounts of S1P used in the different studies may in part explain such differences as high S1P concentrations are known to inhibit the same processes that low concentrations activate (Rivera and Chun 2008): S1P induced VCAM-1 and E-selectin expression at 5–20 μmol/L (Rizza et al. 1999; Xia et al. 1998) and lead to higher monocyte adhesion (Whetzel et al. 2006), but nanomolar concentrations inhibited VCAM-1 in the same study (Whetzel et al. 2006). In addition, endogenously produced S1P may exert different effects than exogenously applied one, and the effects individual S1P receptors have in the process are not characterized yet.

In summary, the overall role of S1P in the pathogenesis of atherosclerosis is still unknown as it has both atherogenic and atheroprotective effects. Studies that modulate endogenous S1P levels in the context of atherosclerosis are still missing but most likely to answer this question.

9 S1P in Human Cardiovascular Disease

The S1P levels in plasma are altered in patients with coronary artery disease (CAD). There is a positive association between serum S1P and the severity of CAD as determined by the scoring of coronary stenosis (Deutschman et al. 2003). When comparing plasma S1P and S1P contained in HDL among healthy individuals, patients with stable CAD, and in patients with acute myocardial infarction, HDL-C-normalized plasma S1P was found to be higher in CAD than in controls and even higher in AMI (Sattler et al. 2010; Yeboah et al. 2010). In the first hours after AMI, S1P in plasma and HDL increased to levels even higher than controls. In contrast, S1P contained in HDL from stable CAD patients was lower than in controls, suggesting that their HDL may have an uptake defect for S1P. The non-HDL-bound plasma S1P pool was able to discriminate patients with MI and stable CAD from controls (Sattler et al. 2010), fulfilling the definition of a biomarker.

In cardiovascular medicine, it is well established that several beneficial properties of HDL are impaired in patients with CAD such as their anti-inflammatory, anti-oxidative, and vasodilatory potential thus defining the term "dysfunctional HDL" (Kontush and Chapman 2006). However, the molecular and biochemical determinants of HDL dysfunctionality have remained elusive. In recent years, the S1P content of HDL has been shown responsible for several of the beneficial HDL functions such as nitric oxide-mediated vasodilation, anti-oxidative, anti-apoptotic, and anti-inflammatory effects and direct cardioprotection against reperfusion injury (Sattler and Levkau 2009). Accordingly, the lower amount of HDL-bound S1P in CAD patients may comprise an important part of their HDL dysfunction, a notion that expects experimental proof. If this was the case, any intervention aimed at increasing the S1P concentration in HDL may be of clinical benefit.

In summary, there is now ample evidence that S1P and its receptors play important roles in cardiovascular homeostasis and disease. Experimental evidence from in vitro and in vivo studies suggests that pharmacological interference with S1P signaling may offer novel approaches to cardiovascular diseases. However, a number of potentially important but incompletely understood functions of S1P in cardiovascular diseases still remain: $S1P_1$ gene polymorphisms have been associated with CAD and stroke, but their functional role is unknown (Luke et al. 2009; Shiffman et al. 2008); plasma S1P and HDL-S1P are altered in CAD and myocardial infarction, but their biological consequences remain unclear (Sattler et al. 2010); S1P effects on cardiac function, arrhythmogenesis, and heart failure are only beginning to be understood (Mann 2012), and the role of S1P in atherosclerosis remains still enigmatic. Nevertheless, the potential attractiveness of S1P-based drugs for cardiovascular applications remains high especially as FTY720 has been approved by the FDA and the EMEA for the treatment of relapsing multiple sclerosis. Thus if such drugs were to prove effective in the treatment of any cardiovascular diseases or provide additional benefit to established therapies, their implementation would be greatly expedited.

References

Alewijnse AE, Peters SL (2008) Sphingolipid signalling in the cardiovascular system: good, bad or both? Eur J Pharmacol 585:292–302

Alewijnse AE, Peters SL, Michel MC (2004) Cardiovascular effects of sphingosine-1-phosphate and other sphingomyelin metabolites. Br J Pharmacol 143:666–684

Allende ML, Proia RL (2002) Sphingosine-1-phosphate receptors and the development of the vascular system. Biochim Biophys Acta 1582:222–227

Argraves KM, Wilkerson BA, Argraves WS, Fleming PA, Obeid LM, Drake CJ (2004) Sphingosine-1-phosphate signaling promotes critical migratory events in vasculogenesis. J Biol Chem 279:50580–50590

Bandhuvula P, Honbo N, Wang GY, Jin ZQ, Fyrst H et al (2011) S1P lyase: a novel therapeutic target for ischemia-reperfusion injury of the heart. Am J Physiol Heart Circ Physiol 300: H1753–H1761

Barter P, Gotto AM, LaRosa JC, Maroni J, Szarek M et al (2007) HDL cholesterol, very low levels of LDL cholesterol, and cardiovascular events. N Engl J Med 357:1301–1310

Bode C, Sensken SC, Peest U, Beutel G, Thol F et al (2010) Erythrocytes serve as a reservoir for cellular and extracellular sphingosine 1-phosphate. J Cell Biochem 109:1232–1243

Bolick DT, Srinivasan S, Kim KW, Hatley ME, Clemens JJ et al (2005) Sphingosine-1-phosphate prevents tumor necrosis factor-{alpha}-mediated monocyte adhesion to aortic endothelium in mice. Arterioscler Thromb Vasc Biol 25:976–981

Bolli R (2001) Cardioprotective function of inducible nitric oxide synthase and role of nitric oxide in myocardial ischemia and preconditioning: an overview of a decade of research. J Mol Cell Cardiol 33:1897–1918

Bolz SS, Vogel L, Sollinger D, Derwand R, Boer C et al (2003) Sphingosine kinase modulates microvascular tone and myogenic responses through activation of RhoA/Rho kinase. Circulation 108:342–347

Bornfeldt KE, Graves LM, Raines EW, Igarashi Y, Wayman G et al (1995) Sphingosine-1-phosphate inhibits PDGF-induced chemotaxis of human arterial smooth muscle cells: spatial and temporal modulation of PDGF chemotactic signal transduction. J Cell Biol 130:193–206

Budde K, Schmouder RL, Brunkhorst R, Nashan B, Lucker PW et al (2002) First human trial of FTY720, a novel immunomodulator, in stable renal transplant patients. J Am Soc Nephrol 13:1073–1083

Camerer E, Regard JB, Cornelissen I, Srinivasan Y, Duong DN et al (2009) Sphingosine-1-phosphate in the plasma compartment regulates basal and inflammation-induced vascular leak in mice. J Clin Invest 119:1871–1879

Christoffersen C, Obinata H, Kumaraswamy SB, Galvani S, Ahnstrom J et al (2011) Endothelium-protective sphingosine-1-phosphate provided by HDL-associated apolipoprotein M. Proc Natl Acad Sci USA 108:9613–9618

Chun J, Hla T, Lynch KR, Spiegel S, Moolenaar WH (2010) International Union of Basic and Clinical Pharmacology. LXXVIII. Lysophospholipid receptor nomenclature. Pharmacol Rev 62:579–587

Cohen JACJ (2011) Mechanisms of fingolimod's efficacy and adverse effects in multiple sclerosis. Ann Neurol 69:759–777

Coroneos E, Martinez M, McKenna S, Kester M (1995) Differential regulation of sphingomyelinase and ceramidase activities by growth factors and cytokines. Implications for cellular proliferation and differentiation. J Biol Chem 270:23305–23309

Coussin F, Scott RH, Wise A, Nixon GF (2002) Comparison of sphingosine 1-phosphate-induced intracellular signaling pathways in vascular smooth muscles: differential role in vasoconstriction. Circ Res 91:151–157

Dantas AP, Igarashi J, Michel T (2003) Sphingosine 1-phosphate and control of vascular tone. Am J Physiol Heart Circ Physiol 284:H2045–H2052

Dettbarn CA, Betto R, Salviati G, Palade P, Jenkins GM, Sabbadini RA (1994) Modulation of cardiac sarcoplasmic reticulum ryanodine receptor by sphingosine. J Mol Cell Cardiol 26:229–242

Deutschman DH, Carstens JS, Klepper RL, Smith WS, Page MT et al (2003) Predicting obstructive coronary artery disease with serum sphingosine-1-phosphate. Am Heart J 146:62–68

Duan HF, Wang H, Yi J, Liu HJ, Zhang QW et al (2007) Adenoviral gene transfer of sphingosine kinase 1 protects heart against ischemia/reperfusion-induced injury and attenuates its postischemic failure. Hum Gene Ther 18:1119–1128

Feistritzer C, Riewald M (2005) Endothelial barrier protection by activated protein C through PAR1-dependent sphingosine 1-phosphate receptor-1 crossactivation. Blood 105:3178–3184
Finigan JH, Dudek SM, Singleton PA, Chiang ET, Jacobson JR et al (2005) Activated protein C mediates novel lung endothelial barrier enhancement: role of sphingosine 1-phosphate receptor transactivation. J Biol Chem 280:17286–17293
Forrest MSS, Hajdu R et al (2004) Immune cell regulation and cardiovascular effects of sphingosine 1-phosphate receptor agonists in rodents are mediated via distinct receptor subtypes. J Pharmacol Exp Ther 309:758–768
Foss FW Jr, Snyder AH, Davis MD, Rouse M, Okusa MD et al (2007) Synthesis and biological evaluation of gamma-aminophosphonates as potent, subtype-selective sphingosine 1-phosphate receptor agonists and antagonists. Bioorg Med Chem 15:663–677
Frias MA, James RW, Gerber-Wicht C, Lang U (2009) Native and reconstituted HDL activate Stat3 in ventricular cardiomyocytes via ERK1/2: role of sphingosine-1-phosphate. Cardiovasc Res 82:313–323
Fukuhara S, Simmons S, Kawamura S, Inoue A, Orba Y et al (2012) The sphingosine-1-phosphate transporter Spns2 expressed on endothelial cells regulates lymphocyte trafficking in mice. J Clin Invest 122:1416–1426
Gellings Lowe N, Swaney JS, Moreno KM, Sabbadini RA (2009) Sphingosine-1-phosphate and sphingosine kinase are critical for transforming growth factor-beta-stimulated collagen production by cardiac fibroblasts. Cardiovasc Res 82:303–312
Grabski AD, Shimizu T, Deou J, Mahoney WM Jr, Reidy MA, Daum G (2009) Sphingosine-1-phosphate receptor-2 regulates expression of smooth muscle alpha-actin after arterial injury. Arterioscler Thromb Vasc Biol 29:1644–1650
Gray MO, Zhou HZ, Schafhalter-Zoppoth I, Zhu P, Mochly-Rosen D, Messing RO (2004) Preservation of base-line hemodynamic function and loss of inducible cardioprotection in adult mice lacking protein kinase C epsilon. J Biol Chem 279:3596–3604
Guo J, MacDonell KL, Giles WR (1999) Effects of sphingosine 1-phosphate on pacemaker activity in rabbit sino-atrial node cells. Pflugers Arch 438:642–648
Hannun YA, Bell RM (1989) Functions of sphingolipids and sphingolipid breakdown products in cellular regulation. Science 243:500–507
Hannun YA, Obeid LM (2008) Principles of bioactive lipid signalling: lessons from sphingolipids. Nat Rev Mol Cell Biol 9:139–150
Hansson GK, Libby P (2006) The immune response in atherosclerosis: a double-edged sword. Nat Rev Immunol 6:508–519
Hausenloy DJ, Yellon DM (2007) Reperfusion injury salvage kinase signalling: taking a RISK for cardioprotection. Heart Fail Rev 12:217–234
Hla T (2004) Physiological and pathological actions of sphingosine 1-phosphate. Semin Cell Dev Biol 15:513–520
Hla T, Lee MJ, Ancellin N, Paik JH, Kluk MJ (2001) Lysophospholipids–receptor revelations. Science 294:1875–1878
Hobson JP, Rosenfeldt HM, Barak LS, Olivera A, Poulton S et al (2001) Role of the sphingosine-1-phosphate receptor EDG-1 in PDGF-induced cell motility. Science 291:1800–1803
Hofmann U, Burkard N, Vogt C, Thoma A, Frantz S et al (2009) Protective effects of sphingosine-1-phosphate receptor agonist treatment after myocardial ischaemia-reperfusion. Cardiovasc Res 83:285–293
Hofmann U, Hu K, Walter F, Burkard N, Ertl G et al (2010) Pharmacological pre- and post-conditioning with the sphingosine-1-phosphate receptor modulator FTY720 after myocardial ischaemia-reperfusion. Br J Pharmacol 160:1243–1251
Hughes JE, Srinivasan S, Lynch KR, Proia RL, Ferdek P, Hedrick CC (2008) Sphingosine-1-phosphate induces an antiinflammatory phenotype in macrophages. Circ Res 102:950–958
Igarashi J, Michel T (2008) S1P and eNOS regulation. Biochim Biophys Acta 1781:489–495
Igarashi J, Michel T (2009) Sphingosine-1-phosphate and modulation of vascular tone. Cardiovasc Res 82:212–220

Ishii I, Ye X, Friedman B, Kawamura S, Contos JJ et al (2002) Marked perinatal lethality and cellular signaling deficits in mice null for the two sphingosine 1-phosphate (S1P) receptors, S1P(2)/LP(B2)/EDG-5 and S1P(3)/LP(B3)/EDG-3. J Biol Chem 277:25152–25159

Jin ZQ, Zhou HZ, Zhu P, Honbo N, Mochly-Rosen D et al (2002) Cardioprotection mediated by sphingosine-1-phosphate and ganglioside GM-1 in wild-type and PKC epsilon knockout mouse hearts. Am J Physiol Heart Circ Physiol 282:H1970–H1977

Jin ZQ, Goetzl EJ, Karliner JS (2004) Sphingosine kinase activation mediates ischemic preconditioning in murine heart. Circulation 110:1980–1989

Jin ZQ, Huang Y, Vessey DA, Karliner JS (2005) A sphingosine kinase 1 mutation sensitizes the myocardium to ischemia/reperfusion injury and abrogates ischemic preconditioning. Circ Res 97:1204

Jin ZQ, Karliner JS, Vessey DA (2008) Ischaemic postconditioning protects isolated mouse hearts against ischaemia/reperfusion injury via sphingosine kinase isoform-1 activation. Cardiovasc Res 79:134–140

Johnson KR, Johnson KY, Becker KP, Bielawski J, Mao C, Obeid LM (2003) Role of human sphingosine-1-phosphate phosphatase 1 in the regulation of intra- and extracellular sphingosine-1-phosphate levels and cell viability. J Biol Chem 278:34541–34547

Kacimi R, Vessey DA, Honbo N, Karliner JS (2007) Adult cardiac fibroblasts null for sphingosine kinase-1 exhibit growth dysregulation and an enhanced proinflammatory response. J Mol Cell Cardiol 43:85–91

Karliner JS (2006) Toward solving the riddle: the enigma becomes less mysterious. Circ Res 99:465–467

Karliner JS (2009) Sphingosine kinase and sphingosine 1-phosphate in cardioprotection. J Cardiovasc Pharmacol 53:189–197

Karliner JS, Honbo N, Summers K, Gray MO, Goetzl EJ (2001) The lysophospholipids sphingosine-1-phosphate and lysophosphatidic acid enhance survival during hypoxia in neonatal rat cardiac myocytes. J Mol Cell Cardiol 33:1713–1717

Karuna R, Park R, Othman A, Holleboom AG, Motazacker MM et al (2011) Plasma levels of sphingosine-1-phosphate and apolipoprotein M in patients with monogenic disorders of HDL metabolism. Atherosclerosis 219:855–863

Kawahara A, Nishi T, Hisano Y, Fukui H, Yamaguchi A, Mochizuki N (2009) The sphingolipid transporter spns2 functions in migration of zebrafish myocardial precursors. Science 323:524–527

Keul P, Sattler K, Levkau B (2007) HDL and its sphingosine-1-phosphate content in cardioprotection. Heart Fail Rev 12:301–306

Keul P, Tolle M, Lucke S, von Wnuck LK, Heusch G et al (2007) The sphingosine-1-phosphate analogue FTY720 reduces atherosclerosis in apolipoprotein E-deficient mice. Arterioscler Thromb Vasc Biol 27:607–613

Keul P, Lucke S, von Wnuck LK, Bode C, Graler M et al (2010) Sphingosine-1-phosphate receptor macrophages in inflammation/macrophages in inflammation and atherosclerosis. Circ Res 108:314–323

Keul P, Lucke S, von Wnuck LK, Bode C, Graler M et al (2011) Sphingosine-1-phosphate receptor 3 promotes recruitment of monocyte/macrophages in inflammation and atherosclerosis. Circ Res 108:314–323

Kim M, Kim N, D'Agati VD, Emala CW Sr, Lee HT (2007) Isoflurane mediates protection from renal ischemia-reperfusion injury via sphingosine kinase and sphingosine-1-phosphate-dependent pathways. Am J Physiol Renal Physiol 293:F1827–F1835

Kimura T, Boehmler AM, Seitz G, Kuci S, Wiesner T et al (2004) The sphingosine 1-phosphate receptor agonist FTY720 supports CXCR4-dependent migration and bone marrow homing of human CD34+ progenitor cells. Blood 103:4478–4486

Kimura T, Tomura H, Mogi C, Kuwabara A, Damirin A et al (2006) Role of scavenger receptor class B type I and sphingosine 1-phosphate receptors in high density lipoprotein-induced inhibition of adhesion molecule expression in endothelial cells. J Biol Chem 281:37457–37467

Kluk MJ, Hla T (2001) Role of the sphingosine 1-phosphate receptor EDG-1 in vascular smooth muscle cell proliferation and migration. Circ Res 89:496–502

Klingenberg R, Nofer JR, Rudling M, Bea F, Blessing E et al (2007) Sphingosine-1-phosphate analogue FTY720 causes lymphocyte redistribution and hypercholesterolemia in ApoE-deficient mice. Arterioscler Thromb Vasc Biol 27:2392–2399

Knapp M, Baranowski M, Lisowska A, Musial W (2012) Decreased free sphingoid base concentration in the plasma of patients with chronic systolic heart failure. Adv Med Sci 57 (1):100–105

Kobayashi N, Yamaguchi A, Nishi T (2009) Characterization of the ATP-dependent sphingosine 1-phosphate transporter in rat erythrocytes. J Biol Chem 284:21192–21200

Kono M, Mi Y, Liu Y, Sasaki T, Allende ML et al (2004) The sphingosine-1-phosphate receptors S1P1, S1P2, and S1P3 function coordinately during embryonic angiogenesis. J Biol Chem 279:29367–29373

Kontush A, Chapman MJ (2006) Functionally defective high-density lipoprotein: a new therapeutic target at the crossroads of dyslipidemia, inflammation, and atherosclerosis. Pharmacol Rev 58:342–374

Kontush A, Therond P, Zerrad A, Couturier M, Negre-Salvayre A et al (2007) Preferential sphingosine-1-phosphate enrichment and sphingomyelin depletion are key features of small dense HDL3 particles: relevance to antiapoptotic and antioxidative activities. Arterioscler Thromb Vasc Biol 27:1843–1849

Koyrakh LRM, Brinkmann V, Wickman K (2005) The heart rate decrease caused by acute FTY720 administration is mediated by the G protein-gated potassium channel I. Am J Transplant 5:529–536

Kunisada K, Tone E, Fujio Y, Matsui H, Yamauchi-Takihara K, Kishimoto T (1998) Activation of gp130 transduces hypertrophic signals via STAT3 in cardiac myocytes. Circulation 98:346–352

Kupperman E, An S, Osborne N, Waldron S, Stainier DY (2000) A sphingosine-1-phosphate receptor regulates cell migration during vertebrate heart development. Nature 406:192–195

Landeen LK, Dederko DA, Kondo CS, Hu BS, Aroonsakool N et al (2008) Mechanisms of the negative inotropic effects of sphingosine-1-phosphate on adult mouse ventricular myocytes. Am J Physiol Heart Circ Physiol 294:H736–H749

Lecour S, Smith RM, Woodward B, Opie LH, Rochette L, Sack MN (2002) Identification of a novel role for sphingolipid signaling in TNF alpha and ischemic preconditioning mediated cardioprotection. J Mol Cell Cardiol 34:509–518

Lee MJ, Thangada S, Claffey KP, Ancellin N, Liu CH et al (1999) Vascular endothelial cell adherens junction assembly and morphogenesis induced by sphingosine-1-phosphate. Cell 99:301–312

Lee JF, Zeng Q, Ozaki H, Wang L, Hand AR et al (2006) Dual roles of tight junction-associated protein, zonula occludens-1, in sphingosine 1-phosphate-mediated endothelial chemotaxis and barrier integrity. J Biol Chem 281:29190–29200

Lee YM, Venkataraman K, Hwang SI, Han DK, Hla T (2007) A novel method to quantify sphingosine 1-phosphate by immobilized metal affinity chromatography (IMAC). Prostaglandins Other Lipid Mediat 84:154–162

Levkau B (2008) Sphingosine-1-phosphate in the regulation of vascular tone: a finely tuned integration system of S1P sources, receptors, and vascular responsiveness. Circ Res 103:231–233

Levkau B, Hermann S, Theilmeier G, van der Giet M, Chun J et al (2004) High-density lipoprotein stimulates myocardial perfusion in vivo. Circulation 110:3355–3359

Limaye V, Xia P, Hahn C, Smith M, Vadas MA et al (2009) Chronic increases in sphingosine kinase-1 activity induce a pro-inflammatory, pro-angiogenic phenotype in endothelial cells. Cell Mol Biol Lett 14:424–441

Liu Y, Wada R, Yamashita T, Mi Y, Deng CX et al (2000) Edg-1, the G protein-coupled receptor for sphingosine-1-phosphate, is essential for vascular maturation. J Clin Invest 106:951–961

Li X, Stankovic M, Bonder CS, Hahn CN, Parsons M et al (2008) Basal and angiopoietin-1-mediated endothelial permeability is regulated by sphingosine kinase-1. Blood 111:3489–3497

Lorenz JN, Arend LJ, Robitz R, Paul RJ, MacLennan AJ (2007) Vascular dysfunction in S1P2 sphingosine 1-phosphate receptor knockout mice. Am J Physiol Regul Integr Comp Physiol 292:R440–R446

Lowe NGSJ, Vekich JA, Patel HH, Roth DM, Sabbadini RA (2006) S1P-stimulated cardiac fibrosis is attenuated by a novel anti-S1P monoclonal antibody (Abstract 981). Circulation 114:II179

Ludwig L, Weihrauch D, Kersten JR, Pagel PS, Warltier DC (2004) Protein kinase C translocation and Src protein tyrosine kinase activation mediate isoflurane-induced preconditioning in vivo: potential downstream targets of mitochondrial adenosine triphosphate-sensitive potassium channels and reactive oxygen species. Anesthesiology 100:532–539

Luke MM, O'Meara ES, Rowland CM, Shiffman D, Bare LA et al (2009) Gene variants associated with ischemic stroke: the cardiovascular health study. Stroke 40:363–368

MacLennan AJ, Carney PR, Zhu WJ, Chaves AH, Garcia J et al (2001) An essential role for the H218/AGR16/Edg-5/LP(B2) sphingosine 1-phosphate receptor in neuronal excitability. Eur J Neurosci 14:203–209

Mandala SM, Thornton R, Galve-Roperh I, Poulton S, Peterson C et al (2000) Molecular cloning and characterization of a lipid phosphohydrolase that degrades sphingosine-1-phosphate and induces cell death. Proc Natl Acad Sci USA 97:7859–7864

Mandala S, Hajdu R, Bergstrom J, Quackenbush E, Xie J et al (2002) Alteration of lymphocyte trafficking by sphingosine-1-phosphate receptor agonists. Science 296:346–349

Mann DL (2012) Sphingosine 1-phosphate as a therapeutic target in heart failure: more questions than answers. Circulation 125(22):2692–2694

Mazurais D, Robert P, Gout B, Berrebi-Bertrand I, Laville MP, Calmels T (2002) Cell type-specific localization of human cardiac S1P receptors. J Histochem Cytochem 50:661–670

McVerry BJ, Garcia JG (2005) In vitro and in vivo modulation of vascular barrier integrity by sphingosine 1-phosphate: mechanistic insights. Cell Signal 17:131–139

Means CK, Brown JH (2009) Sphingosine-1-phosphate receptor signalling in the heart. Cardiovasc Res 82:193–200

Means CK, Xiao CY, Li Z, Zhang T, Omens JH et al (2007) Sphingosine 1-phosphate S1P2 and S1P3 receptor-mediated Akt activation protects against in vivo myocardial ischemia-reperfusion injury. Am J Physiol Heart Circ Physiol 292:H2944–H2951

Means CK, Miyamoto S, Chun J, Brown JH (2008) S1P1 receptor localization confers selectivity for Gi-mediated cAMP and contractile responses. J Biol Chem 283:11954–11963

Meissner A, Yang J, Kroetsch JT, Sauve M, Dax H et al (2012) TNFalpha-mediated down-regulation of CFTR drives pathological S1P signaling in a mouse model of heart failure. Circulation 125(22):2739–2750

Mehta D, Konstantoulaki M, Ahmmed GU, Malik AB (2005) Sphingosine 1-phosphate-induced mobilization of intracellular Ca2+ mediates rac activation and adherens junction assembly in endothelial cells. J Biol Chem 280:17320–17328

Michaud MD, Robitaille GA, Gratton JP, Richard DE (2009) Sphingosine-1-phosphate: a novel nonhypoxic activator of hypoxia-inducible factor-1 in vascular cells. Arterioscler Thromb Vasc Biol 29:902–908

Michaud J, Im DS, Hla T (2010) Inhibitory role of sphingosine 1-phosphate receptor 2 in macrophage recruitment during inflammation. J Immunol 184:1475–1483

Mizugishi K, Yamashita T, Olivera A, Miller GF, Spiegel S, Proia RL (2005) Essential role for sphingosine kinases in neural and vascular development. Mol Cell Biol 25:11113–11121

Muller H, Hofer S, Kaneider N, Neuwirt H, Mosheimer B et al (2005) The immunomodulator FTY720 interferes with effector functions of human monocyte-derived dendritic cells. Eur J Immunol 35:533–545

Murakami A, Takasugi H, Ohnuma S, Koide Y, Sakurai A et al (2010) Sphingosine 1-phosphate (S1P) regulates vascular contraction via S1P3 receptor: investigation based on a new S1P3 receptor antagonist. Mol Pharmacol 77:704–713

Murata N, Sato K, Kon J, Tomura H, Yanagita M et al (2000) Interaction of sphingosine 1-phosphate with plasma components, including lipoproteins, regulates the lipid receptor-mediated actions. Biochem J 352(Pt 3):809–815
Nofer JR, van der Giet M, Tolle M, Wolinska I, von Wnuck LK et al (2004) HDL induces NO-dependent vasorelaxation via the lysophospholipid receptor S1P3. J Clin Invest 113:569–581
Nofer JR, Bot M, Brodde M, Taylor PJ, Salm P et al (2007) FTY720, a synthetic sphingosine 1 phosphate analogue, inhibits development of atherosclerosis in low-density lipoprotein receptor-deficient mice. Circulation 115:501–508
Okajima F (2002) Plasma lipoproteins behave as carriers of extracellular sphingosine 1-phosphate: is this an atherogenic mediator or an anti-atherogenic mediator? Biochim Biophys Acta 1582:132–137
Okamoto H, Takuwa N, Yokomizo T, Sugimoto N, Sakurada S et al (2000) Inhibitory regulation of Rac activation, membrane ruffling, and cell migration by the G protein-coupled sphingosine-1-phosphate receptor EDG5 but not EDG1 or EDG3. Mol Cell Biol 20:9247–9261
Olivera A, Spiegel S (1993) Sphingosine-1-phosphate as second messenger in cell proliferation induced by PDGF and FCS mitogens. Nature 365:557–560
Oral H, Dorn GW 2nd, Mann DL (1997) Sphingosine mediates the immediate negative inotropic effects of tumor necrosis factor-alpha in the adult mammalian cardiac myocyte. J Biol Chem 272:4836–4842
Osborne N, Brand-Arzamendi K, Ober EA, Jin SW, Verkade H et al (2008) The spinster homolog, two of hearts, is required for sphingosine 1-phosphate signaling in zebrafish. Curr Biol 18:1882–1888
Ozaki H, Hla T, Lee MJ (2003) Sphingosine-1-phosphate signaling in endothelial activation. J Atheroscler Thromb 10:125–131
Pappu R, Schwab SR, Cornelissen I, Pereira JP, Regard JB et al (2007) Promotion of lymphocyte egress into blood and lymph by distinct sources of sphingosine-1-phosphate. Science 316:295–298
Peter BF, Lidington D, Harada A, Bolz HJ, Vogel L et al (2008) Role of sphingosine-1-phosphate phosphohydrolase 1 in the regulation of resistance artery tone. Circ Res 103:315–324
Pettus BJ, Bielawski J, Porcelli AM, Reames DL, Johnson KR et al (2003) The sphingosine kinase 1/sphingosine-1-phosphate pathway mediates COX-2 induction and PGE2 production in response to TNF-alpha. FASEB J 17:1411–1421
Rivera R, Chun J (2008) Biological effects of lysophospholipids. Rev Physiol Biochem Pharmacol 160:25–46
Rizza C, Leitinger N, Yue J, Fischer DJ, Wang DA et al (1999) Lysophosphatidic acid as a regulator of endothelial/leukocyte interaction. Lab Invest 79:1227–1235
Robert P, Tsui P, Laville MP, Livi GP, Sarau HM et al (2001) EDG1 receptor stimulation leads to cardiac hypertrophy in rat neonatal myocytes. J Mol Cell Cardiol 33:1589–1606
Rosen H, Gonzalez-Cabrera P, Marsolais D, Cahalan S, Don AS, Sanna MG (2008) Modulating tone: the overture of S1P receptor immunotherapeutics. Immunol Rev 223:221–235
Rosen H, Gonzalez-Cabrera PJ, Sanna MG, Brown S (2009) Sphingosine 1-phosphate receptor signaling. Annu Rev Biochem 78:743–768
Ryser MF, Ugarte F, Lehmann R, Bornhauser M, Brenner S (2008) S1P(1) overexpression stimulates S1P-dependent chemotaxis of human CD34+ hematopoietic progenitor cells but strongly inhibits SDF-1/CXCR4-dependent migration and in vivo homing. Mol Immunol 46:166–171
Ryu Y, Takuwa N, Sugimoto N, Sakurada S, Usui S et al (2002) Sphingosine-1-phosphate, a platelet-derived lysophospholipid mediator, negatively regulates cellular Rac activity and cell migration in vascular smooth muscle cells. Circ Res 90:325–332
Saba JD, Hla T (2004) Point-counterpoint of sphingosine 1-phosphate metabolism. Circ Res 94:724–734

Salomone S, Potts EM, Tyndall S, Ip PC, Chun J et al (2008) Analysis of sphingosine 1-phosphate receptors involved in constriction of isolated cerebral arteries with receptor null mice and pharmacological tools. Br J Pharmacol 153:140–147

Salomone S, Soydan G, Ip PC, Hopson KM, Waeber C (2010) Vessel-specific role of sphingosine kinase 1 in the vasoconstriction of isolated basilar arteries. Pharmacol Res 62:465–474

Sanchez T, Skoura A, Wu MT, Casserly B, Harrington EO, Hla T (2007) Induction of vascular permeability by the sphingosine-1-phosphate receptor-2 (S1P2R) and its downstream effectors ROCK and PTEN. Arterioscler Thromb Vasc Biol 27:1312–1318

Sanna MG, Liao J, Jo E, Alfonso C, Ahn MY et al (2004) Sphingosine 1-phosphate (S1P) receptor subtypes S1P1 and S1P3, respectively, regulate lymphocyte recirculation and heart rate. J Biol Chem 279:13839–13848

Sanna MG, Wang SK, Gonzalez-Cabrera PJ, Don A, Marsolais D et al (2006) Enhancement of capillary leakage and restoration of lymphocyte egress by a chiral S1P1 antagonist in vivo. Nat Chem Biol 2:434–441

Sattler K, Levkau B (2009) Sphingosine-1-phosphate as a mediator of high-density lipoprotein effects in cardiovascular protection. Cardiovasc Res 82:201–211

Sattler KJ, Elbasan S, Keul P, Elter-Schulz M, Bode C et al (2010) Sphingosine 1-phosphate levels in plasma and HDL are altered in coronary artery disease. Basic Res Cardiol 105:821–832

Schulz R, Kelm M, Heusch G (2004) Nitric oxide in myocardial ischemia/reperfusion injury. Cardiovasc Res 61:402–413

Sevvana M, Ahnstrom J, Egerer-Sieber C, Lange HA, Dahlback B, Muller YA (2009) Serendipitous fatty acid binding reveals the structural determinants for ligand recognition in apolipoprotein M. J Mol Biol 393:920–936

Shiffman D, O'Meara ES, Bare LA, Rowland CM, Louie JZ et al (2008) Association of gene variants with incident myocardial infarction in the Cardiovascular Health Study. Arterioscler Thromb Vasc Biol 28:173–179

Shimamura K, Takashiro Y, Akiyama N, Hirabayashi T, Murayama T (2004) Expression of adhesion molecules by sphingosine 1-phosphate and histamine in endothelial cells. Eur J Pharmacol 486:141–150

Shimizu T, Nakazawa T, Cho A, Dastvan F, Shilling D et al (2007) Sphingosine 1-phosphate receptor 2 negatively regulates neointimal formation in mouse arteries. Circ Res 101:995–1000

Singleton PA, Dudek SM, Chiang ET, Garcia JG (2005) Regulation of sphingosine 1-phosphate-induced endothelial cytoskeletal rearrangement and barrier enhancement by S1P1 receptor, PI3 kinase, Tiam1/Rac1, and alpha-actinin. FASEB J 19:1646–1656

Skoura A, Hla T (2009) Regulation of vascular physiology and pathology by the S1P2 receptor subtype. Cardiovasc Res 82:221–228

Skoura A, Michaud J, Im DS, Thangada S, Xiong Y et al (2011) Sphingosine-1-phosphate receptor-2 function in myeloid cells regulates vascular inflammation and atherosclerosis. Arterioscler Thromb Vasc Biol 31:81–85

Skoura A, Sanchez T, Claffey K, Mandala SM, Proia RL, Hla T (2007) Essential role of sphingosine 1-phosphate receptor 2 in pathological angiogenesis of the mouse retina. J Clin Invest 117:2506–2516

Skyschally A, Gres P, Hoffmann S, Haude M, Erbel R et al (2007) Bidirectional role of tumor necrosis factor-alpha in coronary microembolization: progressive contractile dysfunction versus delayed protection against infarction. Circ Res 100:140–146

Spiegel S, Milstien S (2003) Sphingosine-1-phosphate: an enigmatic signalling lipid. Nat Rev Mol Cell Biol 4:397–407

Strub GM, Paillard M, Liang J, Gomez L, Allegood JC et al (2011) Sphingosine-1-phosphate produced by sphingosine kinase 2 in mitochondria interacts with prohibitin 2 to regulate complex IV assembly and respiration. FASEB J 25:600–612

Tanimoto T, Jin ZG, Berk BC (2002) Transactivation of vascular endothelial growth factor (VEGF) receptor Flk-1/KDR is involved in sphingosine 1-phosphate-stimulated phosphorylation of Akt and endothelial nitric-oxide synthase (eNOS). J Biol Chem 277:42997–43001

Tauseef M, Kini V, Knezevic N, Brannan M, Ramchandaran R et al (2008) Activation of sphingosine kinase-1 reverses the increase in lung vascular permeability through sphingosine-1-phosphate receptor signaling in endothelial cells. Circ Res 103:1164–1172

Theilmeier G, Schmidt C, Herrmann J, Keul P, Schafers M et al (2006) High-density lipoproteins and their constituent, sphingosine-1-phosphate, directly protect the heart against ischemia/reperfusion injury in vivo via the S1P3 lysophospholipid receptor. Circulation 114:1403–1409

Thielmann M, Dorge H, Martin C, Belosjorow S, Schwanke U et al (2002) Myocardial dysfunction with coronary microembolization: signal transduction through a sequence of nitric oxide, tumor necrosis factor-alpha, and sphingosine. Circ Res 90:807–813

Tolle M, Levkau B, Keul P, Brinkmann V, Giebing G et al (2005) Immunomodulator FTY720 Induces eNOS-dependent arterial vasodilatation via the lysophospholipid receptor S1P3. Circ Res 96:913–920

Tolle M, Pawlak A, Schuchardt M, Kawamura A, Tietge UJ et al (2008) HDL-associated lysosphingolipids inhibit NAD(P)H oxidase-dependent monocyte chemoattractant protein-1 production. Arterioscler Thromb Vasc Biol 28:1542–1548

Tsukada YT, Sanna MG, Rosen H, Gottlieb RA (2007) S1P1-selective agonist SEW2871 exacerbates reperfusion arrhythmias. J Cardiovasc Pharmacol 50:660–669

van Meer G, Voelker DR, Feigenson GW (2008) Membrane lipids: where they are and how they behave. Nat Rev Mol Cell Biol 9:112–124

Venkataraman K, Lee YM, Michaud J, Thangada S, Ai Y et al (2008) Vascular endothelium as a contributor of plasma sphingosine 1-phosphate. Circ Res 102:669–676

Vessey DA, Kelley M, Li L, Huang Y, Zhou HZ et al (2006) Role of sphingosine kinase activity in protection of heart against ischemia reperfusion injury. Med Sci Monit 12:BR318–BR324

Vessey DA, Kelley M, Li L, Huang Y (2009) Sphingosine protects aging hearts from ischemia/reperfusion injury: superiority to sphingosine 1-phosphate and ischemic pre- and post-conditioning. Oxid Med Cell Longev 2:146–151

Vessey DA, Jin Z, Kelley M, Honbo N, Zhang J, Karliner JS (2011) A sphingosine kinase form 2 knockout sensitizes mouse myocardium to ischemia/reoxygenation injury and diminishes responsiveness to ischemic preconditioning. Oxid Med Cell Longev 2011:1–8, Article ID 961059

Visentin B, Vekich JA, Sibbald BJ, Cavalli AL, Moreno KM et al (2006) Validation of an anti-sphingosine-1-phosphate antibody as a potential therapeutic in reducing growth, invasion, and angiogenesis in multiple tumor lineages. Cancer Cell 9:225–238

Walter DH, Rochwalsky U, Reinhold J, Seeger F, Aicher A et al (2007) Sphingosine-1-phosphate stimulates the functional capacity of progenitor cells by activation of the CXCR4-dependent signaling pathway via the S1P3 receptor. Arterioscler Thromb Vasc Biol 27:275–282

Wamhoff BR, Lynch KR, Macdonald TL, Owens GK (2008) Sphingosine-1-phosphate receptor subtypes differentially regulate smooth muscle cell phenotype. Arterioscler Thromb Vasc Biol 28:1454–1461

Wang F, Okamoto Y, Inoki I, Yoshioka K, Du W et al (2010) Sphingosine-1-phosphate receptor-2 deficiency leads to inhibition of macrophage proinflammatory activities and atherosclerosis in apoE-deficient mice. J Clin Invest 120:3979–3995

Waters C, Sambi B, Kong KC, Thompson D, Pitson SM et al (2003) Sphingosine 1-phosphate and platelet-derived growth factor (PDGF) act via PDGF beta receptor-sphingosine 1-phosphate receptor complexes in airway smooth muscle cells. J Biol Chem 278:6282–6290

Wendler CC, Rivkees SA (2006) Sphingosine-1-phosphate inhibits cell migration and endothelial to mesenchymal cell transformation during cardiac development. Dev Biol 291:264–277

Whetzel AM, Bolick DT, Srinivasan S, Macdonald TL, Morris MA et al (2006) Sphingosine-1 phosphate prevents monocyte/endothelial interactions in type 1 diabetic NOD mice through activation of the S1P1 receptor. Circ Res 99:731–739

Xia P, Gamble JR, Rye KA, Wang L, Hii CS et al (1998) Tumor necrosis factor-alpha induces adhesion molecule expression through the sphingosine kinase pathway. Proc Natl Acad Sci USA 95:14196–14201

Xie B, Shen J, Dong A, Rashid A, Stoller G, Campochiaro PA (2009) Blockade of sphingosine-1-phosphate reduces macrophage influx and retinal and choroidal neovascularization. J Cell Physiol 218:192–198
Xin C, Ren S, Kleuser B, Shabahang S, Eberhardt W et al (2004) Sphingosine 1-phosphate cross-activates the Smad signaling cascade and mimics transforming growth factor-beta-induced cell responses. J Biol Chem 279:35255–35262
Yeboah J, McNamara C, Jiang XC, Tabas I, Herrington DM et al (2010) Association of plasma sphingomyelin levels and incident coronary heart disease events in an adult population: multi-ethnic study of atherosclerosis. Arterioscler Thromb Vasc Biol 30:628–633
Yeh CC, Li H, Malhotra D, Huang MC, Zhu BQ et al (2009) Sphingolipid signaling and treatment during remodeling of the uninfarcted ventricular wall after myocardial infarction. Am J Physiol Heart Circ Physiol 296:H1193–H1199
Zhang B, Tomura H, Kuwabara A, Kimura T, Miura S et al (2005) Correlation of high density lipoprotein (HDL)-associated sphingosine 1-phosphate with serum levels of HDL-cholesterol and apolipoproteins. Atherosclerosis 178:199–205
Zhang J, Honbo N, Goetzl EJ, Chatterjee K, Karliner JS, Gray MO (2007) Signals from type 1 sphingosine 1-phosphate receptors enhance adult mouse cardiac myocyte survival during hypoxia. Am J Physiol Heart Circ Physiol 293:H3150–H3158
Zhao ZQ, Corvera JS, Halkos ME, Kerendi F, Wang NP et al (2003) Inhibition of myocardial injury by ischemic postconditioning during reperfusion: comparison with ischemic preconditioning. Am J Physiol Heart Circ Physiol 285:H579–H588
Zhou J, Saba JD (1998) Identification of the first mammalian sphingosine phosphate lyase gene and its functional expression in yeast. Biochem Biophys Res Commun 242:502–507

Cross Talk Between Ceramide and Redox Signaling: Implications for Endothelial Dysfunction and Renal Disease

Pin-Lan Li and Yang Zhang

Contents

Abstract Recent studies have demonstrated that cross talk between ceramide and redox signaling modulates various cell activities and functions and contributes to the development of cardiovascular diseases and renal dysfunctions. Ceramide triggers the generation of reactive oxygen species (ROS) and increases oxidative stress in many mammalian cells and animal models. On the other hand, inhibition of ROS-generating enzymes or treatment of antioxidants impairs sphingomyelinase activation and ceramide production. As a mechanism, ceramide-enriched signaling platforms, special cell membrane rafts (MR) (formerly lipid rafts), provide an

P.-L. Li (✉) • Y. Zhang
Department of Pharmacology and Toxicology, Medical College of Virginia Campus, Virginia Commonwealth University, Richmond, VA 23298, USA
e-mail: pli@vcu.edu

E. Gulbins and I. Petrache (eds.), *Sphingolipids in Disease*,
Handbook of Experimental Pharmacology 216,
DOI 10.1007/978-3-7091-1511-4_9, © Springer-Verlag Wien 2013

important microenvironment to mediate the cross talk of ceramide and redox signaling to exert a corresponding regulatory role on cell and organ functions. In this regard, activation of acid sphingomyelinase and generation of ceramide mediate the formation of ceramide-enriched membrane platforms, where trans-membrane signals are transmitted or amplified through recruitment, clustering, assembling, or integration of various signaling molecules. A typical such signaling platform is MR redox signaling platform that is centered on ceramide production and aggregation leading to recruitment and assembling of NADPH oxidase to form an active complex in the cell plasma membrane. This redox signaling platform not only conducts redox signaling or regulation but also facilitates a feedforward amplification of both ceramide and redox signaling. In addition to this membrane MR redox signaling platform, the cross talk between ceramide and redox signaling may occur in other cell compartments. This book chapter focuses on the molecular mechanisms, spatial–temporal regulations, and implications of this cross talk between ceramide and redox signaling, which may provide novel insights into the understanding of both ceramide and redox signaling pathways.

Keywords Sphingolipid • Membrane rafts • Free radicals • Signalosomes • Redoxosome • Caveolae • Acid sphingomyelinase • Lysosome • Methyl-β-cyclodextrin

1 Introduction

Redox signaling is a fundamental signaling mechanism in cell biology which importantly participates in a variety of cellular activities including cell proliferation (Burdon 1996; Cai 2006; Nicco et al. 2005), differentiation (Del Prete et al. 2008; Hansberg et al. 1993; Kusmartsev and Gabrilovich 2003; Sasaki et al. 2009; Sauer et al. 2001), and apoptosis (Hildeman 2004; Liu et al. 2009; Mates and Sanchez-Jimenez 2000; Perrone et al. 2008; Wolf 2005). Abnormal redox signaling is frequently involved in various pathophysiological processes such as senescence (Colavitti and Finkel 2005), inflammation (Azad et al. 2008; Muller-Peddinghaus 1989; Yamamoto et al. 2009), hypoxia (Bell and Chandel 2007; Guzy and Schumacker 2006; Kietzmann and Gorlach 2005; MacFarlane et al. 2008), and ischemia/reperfusion (Goswami et al. 2007; Szocs 2004; Toledo-Pereyra et al. 2004), which contribute to the progression of almost all diseases, from cardiovascular ones such as shock (Flowers and Zimmerman 1998; Gendzwill 2007a, b), hypertension (Delles et al. 2008; Hirooka 2008; Ong et al. 2008; Paravicini and Touyz 2008; Puddu et al. 2008; Zeng et al. 2009), and atherosclerosis (Kojda and Harrison 1999; Patel et al. 2000), to metabolic ones such as diabetes mellitus (Bagi et al. 2009; Ksiazek and Wisniewska 2001), to neurodegenerative ones such as Alzheimer's disease (Casadesus et al. 2004; Perry et al. 1998), infectious diseases (Jamaluddin et al. 2009; Mashimo et al. 2006; Ochsendorf 1998; Sun et al. 2008), and cancer (Azad et al. 2009; Oyagbemi et al. 2009; Weinberg and Chandel 2009).

Currently, it is of high interest to explore how redox signaling is regulated under both physiological and pathological conditions.

Despite extensive studies, the precise mechanisms for rapid activation of redox enzymes by different stimuli are still poorly understood. Redox enzymes such as nicotinamide adenine dinucleotide phosphate (NADPH) oxidase, unlike G-protein-coupled enzymes, are not coupled with any specific receptors. Therefore, a previous undefined mechanism must exist to bridge receptor activation and redox signaling. Recent studies in non-phagocytes revealed that membrane raft (MR, formerly lipid raft) signaling platforms might be responsible for activation of death receptors, in particular CD95 and tumor necrosis factor receptor 1 (TNFR1). These death receptors were found to be localized in MRs, and these receptors in MRs can interact to stabilize the MRs and allow raft aggregation (i.e., MRs clustering), by which many raftophilic molecules are invariably recruited into the complex with the MRs, producing massive signaling and effector responses. Furthermore, it has been well documented that ROS or oxidative stress is a downstream mechanism of receptor clustering in MR platforms. This MR-associated ROS generation, downstream of CD95 and TNFR1, may be of high importance in the early alterations of cell functions during activation of death receptors and in induction of apoptosis (Dumitru et al. 2007; Morgan et al. 2007). Indeed, recently collected evidence support the view that MR signaling platforms are an important mechanism for initiating and transmitting redox signaling in mammalian cells (Li and Gulbins 2007; Oakley et al. 2009; Yang and Rizzo 2007; Zhang and Li 2010). Given the central role of ceramide and related signaling in the formation or regulation of MR redox platforms through clustering of ceramide-enriched microdomains, the cross talk between ceramide and redox signaling is emerging as an important cellular signaling mechanism that mediates the regulation of cellular activities.

Ceramide is generated by several enzymatic pathways in mammalian cells. Two major pathways are the sphingomyelinase (SMase) pathway that generates ceramide from sphingomyelin (SM) by the activities of SMase and the de novo synthesis pathway that synthesizes ceramide from serine and palmitoyl-CoA by the activity of ceramide synthase. The biophysical properties of ceramide molecules predict a tight interaction of ceramide molecules with each other, resulting in the formation of stable and tightly packed ceramide-enriched membrane microdomains that spontaneously fuse to form large ceramide-enriched membrane macrodomains or platforms. Among SMases, acid SMase (ASMase) has been considered as the major enzyme responsible for the formation of ceramide-enriched membrane platforms. Recently, we and others have reported that various death factors bind to their receptors in or around individual MRs and stimulate ASMase to produce ceramide from SM, leading to the formation of ceramide-enriched membrane platforms. In such platforms, NADPH oxidase subunits such as gp91^{phox} and p47^{phox} and other redox molecules are aggregated, clustered, and/or recruited, leading to signal transduction by increase or scavenging of $O_2^{\bullet -}$ or ROS. Such redox signaling associated with ceramide-enriched membrane platforms has been found to contribute to the regulation of a variety of cellular activities and organ functions and lead to pathological changes such as endothelial dysfunction, cell apoptosis, and phagosomal action in neutrophils or macrophages (Jin et al. 2008b; Li et al. 2007;

Zhang et al. 2006). Given the focus of this chapter on the cross talk between ceramide and redox signaling, we will discuss the role of ceramide in the regulation of MR or nonraft redox signaling and vice versa as well as summarize some evidence related to physiological and pathological relevances of this cross talk. Given that there are a lot of discussions about the basic knowledge of ceramide signaling in other chapters, here, we first provide some background information regarding current knowledge of redox signaling and regulation.

2 Redox Signaling and Oxidative Stress

2.1 Reactive Oxygen Species

Reactive oxygen species (ROS) is a collective term that often includes not only the oxygen radicals such as superoxide ($O_2^{\bullet -}$), hydroxyl (OH^-), peroxyl (RO_2), alkoxyl ($RO^\bullet$), and hydroperoxyl ($HO_2^\bullet$) but also non-radicals such as hydrogen peroxide (H_2O_2), hypochlorous acid (HOCl), ozone (O_3), singlet oxygen (ΔgO_2), and peroxynitrite ($ONOO^-$). Since these oxygen derivatives, whether they are radicals or non-radicals, are very reactive, they can oxidize or reduce other molecules in living cells or tissues. Therefore, in general, redox signaling is often referred to as the signaling induced by ROS (Halliwell and Cross 1994; Stadtman 2004; Tang et al. 2002). The most important ROS is $O_2^{\bullet -}$, which is unstable and short-lived because it has an unpaired electron, and it is highly reactive with a variety of cellular molecules, including proteins and DNA. $O_2^{\bullet -}$ is reduced to H_2O_2 by superoxide dismutase (SOD), and both $O_2^{\bullet -}$ and H_2O_2 can diffuse from their sites of generation to other cellular locations. H_2O_2 is further reduced to generate the highly reactive $^\bullet OH$ through the Haber–Weiss or Fenton reaction under pathological conditions. In contrast to $O_2^{\bullet -}$ and H_2O_2, $^\bullet OH$ is highly reactive and, therefore, causes primarily local damage. In addition, $O_2^{\bullet -}$ can also interact with NO to form another reactive oxygen free radical, $ONOO^{\bullet -}$. Under physiological conditions, O_2^- preferably produces H_2O_2 via the dismutation reaction. However, when excess $O_2^{\bullet -}$ is produced, a substantial amount of $O_2^{\bullet -}$ reacts with NO to produce $ONOO^{\bullet -}$. Taken together, these ROS constitute a redox regulatory network that controls cellular activity and function.

2.2 Redox Signaling and Injury

It has been reported that ROS can be produced as a basic signaling messenger to maintain cell or organ functions or increasingly generated or released in response to various stimuli. Meanwhile, these active molecules are constantly scavenged by the endogenous antioxidant systems, mainly composed of the enzyme-mediated pathways as SOD, catalase, glutathione peroxidase, glutathione-*S*-transferase, thioredoxin/thioredoxin reductase, and other peroxidases. In addition, direct reactions between ROS and different molecules may also result in antioxidant actions such as the interactions between ROS and NO, –SH, vitamin E, β-carotene, ceruloplasmin, ferritin, transferrin, hemoglobin, and ascorbates. Being tightly regulated under normal conditions, intracellular and extracellular ROS are maintained at very low levels (less than 1 % of produced ROS). If the generation of ROS exceeds its removal by scavengers, the intracellular and extracellular levels of ROS will increase, leading to oxidative stress and a progression of various pathophysiological processes and respective diseases. If the level of ROS increases to even higher levels, its damaging effects, to DNAs, proteins, lipids, and glycols, become inevitable. These damaging effects of ROS are often tightly correlated together and share a common redox system responsible for the generation and scavenging of ROS molecules.

2.3 *ROS-Generating Systems*

Among four common ROS including $O_2^{\bullet-}$, H_2O_2, HO^-, and $ONOO^-$, $O_2^{\bullet-}$ has been considered as the progenitor of other common ROS. The production of $O_2^{\bullet-}$ and related regulation in biological systems have been intensively studied. In mammalian cells, many pathways are involved in the production of $O_2^{\bullet-}$, including NADPH oxidase, xanthine oxidase, mitochondrial respiration chain, cytochrome P450, lipoxygenase, cyclooxygenase, peroxisomes, and NO synthase uncoupling. Some nonenzymatic derivatives of $O_2^{\bullet-}$ may be formed via photolysis, heme protein + Fe, and auto-oxidation reactions. Among these pathways, NADPH oxidase has been reported to be a major source of $O_2^{\bullet-}$ for the redox regulation in some cells such as vascular endothelial and smooth muscle cells (Griendling et al. 2000). It is estimated that this nonmitochondrial NADPH oxidase-derived $O_2^{\bullet-}$ constitutes more than 95 % of the production of $O_2^{\bullet-}$ in these cells, especially when they are stimulated (Mohazzab et al. 1994; Rajagopalan et al. 1996).

3 Interactions of Ceramide and Redox Signaling Pathways

There is accumulating evidence that ceramide induces the activation of ROS-generating enzymes, including NADPH oxidase, xanthine oxidase, NO synthase, and the mitochondrial respiratory chain (Corda et al. 2001; Lecour et al. 2006). In particular, ceramide has been shown to activate NADPH oxidase and to increase the production of $O_2^{\bullet-}$ in a variety of mammalian cells, including human aortic smooth muscle cells, endothelial cells (ECs), and macrophages (Bhunia et al. 1997; Zhang et al. 2007, 2008). Because many stimuli activate NADPH oxidase by translocation and aggregation of its subunits, it has been proposed that ceramide may mediate the fusion of small raft domains to ceramide-enriched membrane platforms, thereby clustering subunits of NADPH oxidase, assembling them into an active enzyme complex and producing of $O_2^{\bullet-}$. In addition, ceramide has also been shown to interact with the mitochondrial electron transport chain leading to the generation of ROS (Corda et al. 2001). Given the crucial role of NADPH oxidase in the normal regulation of cell functions and as one of the most important redox signaling pathways (Dworakowski et al. 2006; Finkel and Holbrook 2000), we will focus on the cross talk between ceramide signaling pathway and NADPH oxidase-derived redox regulation.

3.1 Ceramide-Enriched Microdomains in MR Redox Signaling

Ceramide belongs to a highly hydrophobic lipid family, which consists of fatty acids with carbon chains in variable lengths (2–28 carbons) and sphingosine (Mathias and Kolesnick 1993). SMase is released to the cell surface or extracellular space in an autocrine or paracrine manner that hydrolyzes cell-surface SM, inducing cell–cell communications and exerting remote action through blood circulation. More importantly, SMase may act on SM incorporated in the MR area of cell membranes and thereby produce ceramide locally (Cremesti et al. 2002). Ceramide molecules spontaneously bind to each other to form microdomains that fuse to large ceramide-enriched membrane platforms (Brown and London 1998; Grassme et al. 2001). In such ceramide-enriched platforms, redox molecules such as NADPH oxidase subunits or cofactors can be aggregated to assemble into active NADPH oxidase complex, producing $O_2^{\bullet-}$ to conduct signaling (Jin et al. 2008b; Zhang et al. 2006).

This ceramide-enriched redox signaling platform has been found to be formed in response to different death receptor ligands such as CD95 ligand, TNF-α, or endostatin (Jin et al. 2008a, b; Zhang et al. 2006). Furthermore, ultraviolet irradiation also induces the formation of ceramide-enriched platforms that mediate ROS production (Chatterjee and Wu 2001). We recently demonstrated that a rapid movement and consequent fusion of lysosomes to supply ASMase into the MR area of cell membranes occur in response to various stimuli (Jin et al. 2007, 2008a). This lysosome fusion is critical for the formation of ceramide-enriched platforms and therefore determines MR redox signaling in different cells, in particular in ECs (Jin et al. 2008a, b).

Further studies have revealed that sortilin, a glycoprotein responsible for transferring ASMase from the Golgi apparatus to lysosomes, is also important in initiating the movement of lysosomes and promoting their fusion to the cell membrane in ECs (Bao et al. 2010a, b). Sortilin is a 95-kDa glycoprotein, which has been reported to play an important role in targeting or transferring proteins to lysosomes (Ni and Morales 2006). Its Vps10p domain in the luminal region may be the binding site for the saposin-like motif of ASMase, while its cytoplasmic tail containing an acidic cluster-dileucine motif binds the monomeric adaptor protein GGA and is structurally similar to the cytoplasmic domain of M6P. The coupled sortilin-1 and ASMase work together to promote the movement of lysosomes toward the cell membrane, which, in turn, leads to MRs clustering and NADPH oxidase activation in ECs. This ASMase-dependent clustering of receptors was also

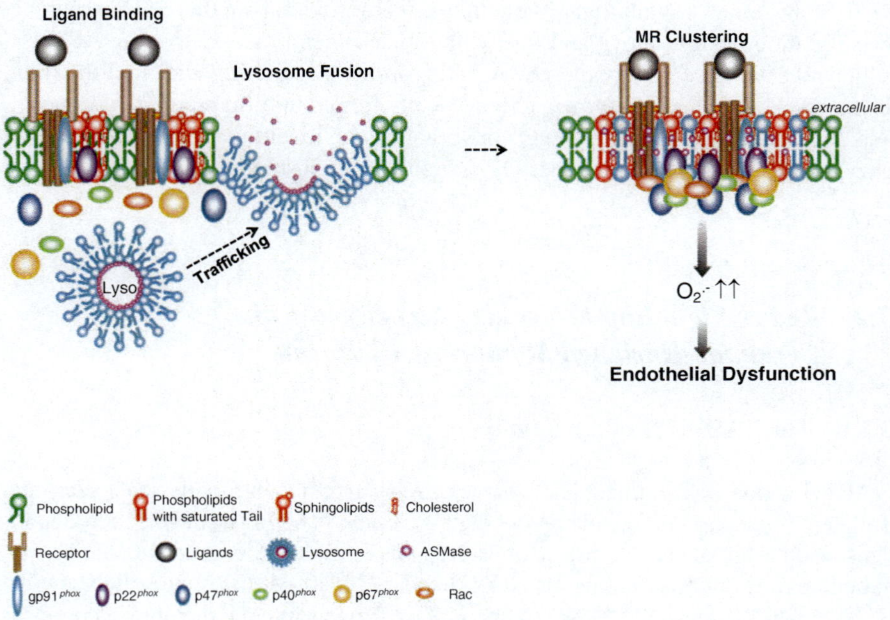

Fig. 1 Lysosome biogenesis and fusion to cell membrane to form ceramide-enriched redox signaling platforms. ASMase is synthesized from the ER and transported through Golgi apparatus to lysosomes. These lysosomes can be mobilized to traffic and fuse into cell membrane, where ASMase is activated and ceramide produced, resulting in MRs clustering and formation of ceramide-enriched platforms (adapted from Xia M, et al. *Cardiovasc Res* (2011) 89 (2): 401–409)

observed for other receptors such as CD20, CD40, TNFR, and epidermal growth factor receptors (EGFR) (Rodighiero et al. 2004). In addition, the SNARE (soluble *N*-ethylmaleimide-sensitive factor attachment protein receptor)-centered exocytic machinery was found to be involved in MR clustering to form redox signaling platforms. SNAREs comprise a superfamily of small, mostly membrane-anchored proteins, which mediate membrane fusion between organelles or from organelles to cell plasma membranes (Gerst 1999). In particular, SNARE-mediated membrane fusion plays an essential role in the secretory pathway of various eukaryotic cells, which is named as the SNARE or SNARE-centered exocytic machinery (Blank et al. 2002). It seems that SNARE as a membrane fusion facilitator is also present in MR redox signaling platforms although its major function is to help lysosome fusion (Zhang and Li 2010).

A comprehensive working model for the mediation of MR clustering and the formation of MR signaling platforms in arterial ECs is presented in Fig. 1; this model emphasizes the derivation of membrane ASMase as being from lysosomes, which target ASMase when it is synthesized from ER and transported through Golgi apparatus. Many mature lysosomes with ASMase are proximal to the cell membrane. When a receptor such as death receptor is activated by a ligand binding to it or by other stimulations, these lysosomes proximal to the cell membrane become mobilized to move and fuse with the cell membrane, activating ASMase and synthesizing ceramide, thereby resulting in MRs clustering and the formation of ceramide-enriched platforms. These MR platforms, in turn, recruit, translocate, and aggregate NOX and its subunits or cofactors and assemble them into an active enzyme complex, which produces $O_2^{\bullet-}$, promoting transmembrane signaling.

3.2 Redox Signaling Molecules Associated with Ceramide-Enriched Membrane Platforms

3.2.1 The NADPH Oxidase Family

NADPH oxidase, identified and characterized first in neutrophils, catalyzes the 1-electron reduction of oxygen producing $O_2^{\bullet-}$ using NADPH as the electron donor. This neutrophil oxidase consists of at least five subunits: two membrane-bound subunits gp91^{phox} (also known as NOX2) and p22^{phox} and three cytosolic subunits p47^{phox}, p40^{phox}, and p67^{phox}. NOX2 and p22^{phox} form an integral membrane complex termed cytochrome b_{558}, and the other four subunits, p47^{phox}, p67^{phox}, p40^{phox}, and the small GTPase Rac, localize in the cytosol in resting cells (Babior et al. 2002). In the classic model of phagocytic-type NADPH oxidase, activation involves translocation of the four cytosolic proteins to the cell membrane and interactions with the membrane spanning subunits p22^{phox} and NOX2, resulting in the transfer of the NADPH electron to oxygen molecules and the generation of $O_2^{\bullet-}$ (Babior et al. 2002; Dang et al. 2002).

NOX protein family consists of six other homologues of NOX2 (gp91^{phox}) catalytic subunits, namely, NOX1, NOX3, NOX4, NOX5, DUOX1 (dual oxidases1), and DUOX2, which determine ROS production in non-phagocytes (Cheng et al. 2001). NOX2 is also expressed in non-phagocytes, including neurons, cardiac cells, skeletal muscle cells, liver cells, ECs, B lymphocytes, epithelial cells, and hematopoietic cells (Piccoli et al. 2005). The structure and function of non-phagocytic NOX are very similar to NOX2. They can also catalyze a single-electron reduction of molecular oxygen, generating $O_2^{\bullet-}$ and other ROS. It is interesting to note that almost all NOXs were demonstrated to have some structural or functional link to or relationship with MRs (Ushio-Fukai et al. 2001; Zhang et al. 2006; Zuo et al. 2004, 2005). Given that NOX activation requires many cofactors to work together, MRs provide a wonderful platform for NOX and the other NADPH oxidase subunits and cofactors to assemble and then work as an active enzymatic complex. However, the driving force or actual physical platform for NADPH oxidase assembly as an active enzyme complex is still unknown. As noted above, the ceramide-enriched membrane macrodomains or platforms may represent an important mechanism mediating this assembly or activation process of NADPH oxidase.

3.2.2 Superoxide Dismutase

Recently, proteomic analysis demonstrated that membrane SOD (SOD1) is present in MR fractions (Zhai et al. 2009), a fact consistent with previous reports that SOD1 is detectable in MRs (Siafakas et al. 2006). Reported SOD1 levels, for example, in MR fractions were much higher than that in other areas of the plasma membrane. These results support the view that aggregation of MRs may play an important role for SOD1 actions. It is assumed that localization and subsequent aggregation of SOD1 in MRs could affect cellular functions as well as the interplay between different cell types, as MRs are rich in receptors and the signaling molecules necessary for cell–cell communications (Zhai et al. 2009). Indeed, a more recent study has reported that H_2O_2, generated extracellularly by extracellular SOD, anchored to ECs surface via the heparin-binding domain (HBD), enhances VEGF-induced VEGF receptor 2 (VEGFR2) autophosphorylation in caveolin-enriched MRs, but not in non-caveolar MRs. The HBD of endothelial SOD is required for its localization in plasma membrane MRs, suggesting that localization of endothelial SOD in caveolae/MRs via HBD can serve as an important mechanism by which SOD-derived extracellular H_2O_2 efficiently promotes VEGFR2 signaling in ECs and postnatal angiogenesis (Oshikawa et al. 2010).

3.2.3 Catalase

In neutrophils, recent proteomic analysis (Feuk-Lagerstedt et al. 2007) has also found catalase in MR fractions that play critical roles in redox signaling by cleavage of H_2O_2. Although some studies have demonstrated that MR-associated catalase may be related to peroxisome biogenesis, the function of this catalase association with MRs remains still largely unknown. It is possible that MRs in hepatic peroxisomal membrane cells are able to help catalase sorting and distribution to different compartments of these cells, assigning them an important role in hepatocyte proliferation and lipid metabolism. Given that hepatic caveolin-1 plays an important role in liver regeneration and lipid metabolism, caveolae with catalase may be critically involved in this liver regeneration and lipid metabolism. However, recent studies found that the absence of caveolin-1 did not affect the peroxisomal location of catalase in mouse liver. It seems caveolin-1 is not required for peroxisome biogenesis, whereas other types of peroxisomal MRs are required (Woudenberg et al. 2010). Obviously, more research and thinking need to be invested into the formation and function of MR-associated catalase complexes.

3.2.4 Thioredoxin

Although it is not yet extensively studied, thioredoxin (TRX) has also been reported as a MR-associated protein. In some reports, MRs have been shown to mediate the effects of TRX. There is convincing evidence that MRs may mediate the actions of

TRX on leukocyte–EC interaction related to redox regulation during inflammation. TRX is a ubiquitous protein with a redox-active disulfide that functions in concert with NADPH and TRX reductase to control the redox state of cysteine residues of different oxidant-targeted proteins. Given the antioxidant role of TRX, the MR-mediated role of TRX in the interaction between leukocytes and ECs may importantly regulate inflammatory responses through counteracting oxidative stress and ROS. In addition, TRX can be internalized into the cells through MR-mediated endocytosis. In particular, a TRX mutant, TRX-C35S (with replacement of cysteine 35 by serine), was found to bind rapidly to the cell surface and be internalized into the cells through MRs in the plasma membrane. This indicates that the cysteine at the active site of TRX is important for the internalization and signal transduction of extracellular TRX through MRs (Hara et al. 2007; Kondo et al. 2007).

3.2.5 Transient Receptor Protein C3 and C4 (TRPC3 and TRPC4)-Redox Sensors

MRs have also been reported to promote molecules aggregation, gating, or activation producing their downstream impact on redox sensing or enhancement of effector responses to redox signaling. Among these molecules, a currently identified redox-sensitive protein-transient receptor protein (TRP) is particularly noteworthy. TRPs are a family of voltage-independent nonspecific cation-permeable channels. Evidence exists that TRPC3 and TRPC4 localize or relocalize in MRs and can form a TRPC3–TRPC4 complex with different properties from their respective homomeric channels, which are redox sensitive (Poteser et al. 2006). Perhaps these TRP channels are directly gated or influenced by the formation of MR platforms, and therefore, their redox-sensing function is altered. Indeed, the TRPC3 channel activity is increased by cholesterol loading of the cell membrane when TRPC3 is overexpressed. This increased channel activity may lead to enhanced redox sensitivity of the channels, exerting an important redox regulation or resulting in pathologic consequences in different cells (Poteser et al. 2006).

3.3 Redox Regulation of Ceramide-Enriched Membrane Platform Formation

There is increasing evidence that the formation of MR-derived signaling platforms can also be regulated by redox molecules. For example, the formation of ceramide-enriched membrane platforms in the membrane of coronary arterial ECs can be reduced by SOD but increased by $O_2^{\bullet -}$ donor or generating systems (Zhang et al. 2007). H_2O_2 was also found to activate pro-survival signaling pathways, including activation of PI3 kinase/Akt and extracellular signal-regulated kinases (ERK)1/2 by changes in MRs behaviors (Yang et al. 2006a). Exogenous administration of xanthine/xanthine oxidase, a $O_2^{\bullet-}$ generating system, has demonstrated a dramatic increase in MRs clustering and ceramide-enriched membrane platform formation in the membrane of ECs (Qiu et al. 2003; Zhang et al. 2007). Furthermore, ROS in T lymphocytes were also shown to enhance MR signaling, and blockade of ROS production by the SOD-mimic MnTBAP reduced the localization of several signaling molecules such as LAT, phospho-LAT, and PLC-gamma in MRs fractions. Treatment of T cells with the ROS synthesizer, *tert*-butyl hydrogen peroxide (TBHP), greatly enhanced MR formation and the distribution of phospho-LAT into MRs. Moreover, lipid peroxides were found to promote the formation of larger rafts or platforms on the membrane, and photooxidation, at the lipid double bonds, caused raft enlargement (Ayuyan and Cohen 2006). These observations corroborate and reinforce the conclusion that ROS are able to enhance MR clustering or formation of macrodomains and must contribute to the formation of MR platforms (Lu et al. 2007). In addition to the direct regulation of SMase/ceramide pathway, various ROS were found to influence MR signaling or function through their actions on many other MR constituents such as caveolin-1, cholesterol, and related raft proteins (Dumitru et al. 2007; Morgan et al. 2007).

3.3.1 ROS Interact with Caveolin-1

Biochemical and morphological experiments have shown that at least two subtypes of lipid microdomains are present in mammalian cells: caveolar and noncaveolar MRs. The size of noncaveolar MRs is 50–100 nm or even smaller, and each of them contains 10–30 protein molecules. Caveolae, cave-like plasma membrane subdomains, are considered as another subtype of lipid domains. Caveolin-1 is the major protein component of caveolae, and its polymerization forms a rigid scaffold that maintains the characteristic cave-like morphology. In addition to its structural function, caveolin-1 has several important regulatory activities through direct interaction with other functional proteins and signaling molecules. Caveolin-1 is subject to two types of posttranslational modification that might be critical for regulating its intracellular activity and localization, namely, phosphorylation and palmitoylation. Recent studies have indicated that both phosphorylation and palmitoylation of caveolin-1 can be regulated by ROS and ultimately affect caveolar functions. In vascular ECs, H_2O_2 causes increased tyrosine phosphorylation of caveolins (Brown and London 1998). In addition, Parat and colleagues showed that exogenous H_2O_2 did not alter the intracellular localization of caveolin-1 in ECs, but it inhibited the trafficking of newly synthesized caveolin-1 to MRs (Brown et al. 1998). They further demonstrated that H_2O_2 did not alter the rate of caveolin-1 depalmitoylation but rather decreased the "on-rate" of palmitoylation (Brown et al. 1998). Functional studies substantiated that caveolin-1 is a sensitive target of oxidative stress and that the oxidation of caveolar membrane cholesterol causes the translocation of caveolin-1 from the plasma membrane to the Golgi apparatus (Grassme et al. 2001). In a separate study, treatment of ECs with ROS caused a release of caveolin-1 from membranes and also a decrease in the number of caveolae detected by electron microscopy (van den Elzen et al. 2005). In summary, these results suggest that oxidative stress modulates caveolin-1 function and cellular levels, which may ultimately affect caveolar function and plasma membrane composition, namely, alteration in the ratio of caveolar vs. noncaveolar MRs or membrane signaling platforms.

3.3.2 ROS Interact with Cholesterol

It has been known that the formation of MRs is driven by tight packing between cholesterol and sphingomyelin and other sphingolipids. Oxysterols are derivatives of cholesterol that contain a second oxygen atom as a carbonyl, hydroxyl, or epoxide group (Morita et al. 2004). Cytotoxic oxysterols formed by nonspecific oxidative mechanisms can affect many cellular processes that contribute to the pathogenesis of disease. According to their biophysical properties, which can be distinct from those of cholesterol, oxysterols can promote or inhibit the formation of membrane microdomains or MRs (Byfield et al. 2006; Scheel-Toellner et al. 2004). For instance, the activities of receptor tyrosine kinases such as the epidermal growth factor (EGF) receptor and insulin receptors, which are found in MR/caveolae, can be modulated by changes in cellular cholesterol content. EGF stimulation-induced phosphatidylinositol 4,5-bisphosphate turnover is inhibited by depletion of cholesterol, but the effects of repletion with different oxysterols varied according to their structure. Turnover was not restored by 25-hydroxycholesterol, while 7-ketocholesterol and 5α, 6α-epoxycholesterol restored turnover (Morita et al. 2004). Likewise, desmosterol, another oxysterol, impaired raft-dependent signaling via the insulin receptor, while nonraft-dependent protein secretion was not affected (Morita et al. 2004). Therefore, these studies suggest that ROS may oxidize cholesterol to various oxysterols, which affect the stability of MRs and formation of MR-derived signaling platforms.

3.4 Redox Regulation of SMase Activity

The role of SMase in the formation of ceramide-enriched membrane platforms has been extensively studied. Recently, several studies have indicated that the generation of ROS may be involved in the activation of the enzyme in response to various stimuli (Charruyer et al. 2005; Dumitru and Gulbins 2006; Malaplate-Armand et al. 2006; Scheel-Toellner et al. 2004). Scheel-Toellner and colleagues demonstrated that ASMase activation, ceramide generation, and CD95 clustering play a crucial role in the spontaneous apoptosis of neutrophils; apoptosis was substantially delayed in Asm-deficient mice (Scheel-Toellner et al. 2004). Based on the observations that the intracellular redox balance changes in aging neutrophils, the authors investigated the possibility that ROS may be involved in ASMase activation. They demonstrated that pretreating neutrophils with the antioxidants

N-acetylcysteine (NAC) and desferrioxamine substantially inhibited the events downstream of ASMase, such as ceramide generation and CD95 clustering, indicating that ROS release is required for ASM activation (Scheel-Toellner et al. 2004). Similarly, pretreatment with the antioxidant, pyrrolidine dithiocarbamate (PDTC), abolished ASMase activation by ultraviolet (UV)-C light in U937 cells (Charruyer et al. 2005).

In other studies, neuronal stimulation with soluble oligomers of amyloid-beta peptide was found to result in the release of ROS and the subsequent activation of SMases (Malaplate-Armand et al. 2006). Treatment of these neurons with antioxidant molecules and a cPLA2-specific inhibitor or antisense oligonucleotide led to inhibition of SMase activation and subsequent apoptosis, suggesting that amyloid-beta oligomers induce neuronal death by activating both NSMase and ASMase via a redox-sensitive cPLA2-arachidonic acid pathway (Malaplate-Armand et al. 2006). In addition, Dumitru and colleagues have also demonstrated the involvement of ROS in TRAIL-induced activation of ASMase and apoptosis (Dumitru and Gulbins 2006). Stimulation with TRAIL/DR5 led to the activation of ASMase and the subsequent formation of ceramide-enriched membrane platforms, DR5 clustering, and consequent apoptosis. Pretreatment with antioxidants NAC and Tiron substantially inhibited TRAIL-induced ASMase activation, ceramide/DR5 clustering, and apoptosis, demonstrating that ROS play a crucial role in TRAIL-associated signaling pathway (Dumitru and Gulbins 2006). Finally, studies investigating the cellular effects of Cu^{2+} showed that Cu^{2+} also promotes the ROS-dependent activation of ASMase and leads to the death of hepatocytes (Dumitru and Gulbins 2006). It was shown that the accumulation of Cu^{2+}, as occurred in Wilson disease, activates ASMase in hepatocytes and triggers the release of ceramide in these cells. This process results in Cu^{2+}-induced hepatocyte death, which can be prevented by a deficiency in ASMase (Lang et al. 2007).

One of the mechanisms by which ASMase activity is regulated by ROS has been described by Qiu and coworkers (Qiu et al. 2003). It has been proposed that C-terminal cysteine (Cys629) plays a crucial role in the enzymatic activity of recombinant human ASMase (rhASM). The loss of the free sulfhydryl group on this amino acid results in activation of the enzyme, and this loss of free sulfhydryl group may be due to copper-promoted dimerization of rhASM by C-terminal cysteine, thiol-specific chemical modification of this cysteine to form a mixed disulfide bond or a sulfur-carbon linkage, deletion of this cysteine by carboxypeptidase or recombinant DNA technology, and site-specific mutation to change the cysteine to a serine residue. Because zinc is required for ASMase activity, the effect of C-terminal cysteine modification on the activation of ASMase may be associated with zinc coordination. It is known that zinc coordinates with a water molecule to produce an optimal structure for catalysis of ASMase. This zinc coordination model is essentially identical to the "cysteine switch" activation mechanism described previously for the matrix metalloproteinase family (Van Wart and Birkedal-Hansen 1990).

Another redox regulatory mechanism of various SMases has been described in a number of studies (Bezombes et al. 2002; Hernandez et al. 2000; Mansat-de Mas et al. 1999), which is related to the effect of glutathione (GSH) (Liu and Hannun 1997; Martin et al. 2007). It has been demonstrated that the GSH/GSSG ratio is critical to such redox regulation of SMase activity. In this regard, the influence of glutathione, its analogs, and individual fragments on the activity of various SMase isoforms have been studied (Liu and Hannun 1997; Liu et al. 1998). In particular, the inhibitory effect of GSH on the neutral Mg^{2+}-dependent SMase was shown to be associated with the g-glutamyl-group of GSH. Since the effect of GSH is accompanied by decrease in diene conjugate and diene ketone levels, the ability of GSH to inhibit oxidative processes in the cell due to its antioxidative properties may be mainly responsible for its effect to inhibit SMase activity. In other words, ROS-mediated oxidation of NSMase or AMSase may enhance their activity (Tsyupko et al. 2001). In addition, the intracellular GSH concentration is thought to be involved in the regulation of SMase activity by increase in its expression (Yoshimura et al. 1999).

3.5 *Feedforward Amplifying Mechanism*

MRs signaling platforms usually contain different proteins including different signaling molecules and cross-linkers or enzymes (Simons and Ikonen 1997; Simons and Toomre 2000). The formation of MR platforms activates, facilitates, and/or amplifies signal transductions. As mentioned above, if MR clustering forms a ceramide-enriched membrane platform, the ceramide production or enrichment is mainly from SMase-catalyzed cleavage of SM cholines in individual MRs (Gulbins and Kolesnick 2003; Hoekstra et al. 2003). Redox regulation of SMase, MRs clustering, and ceramide-enriched platform formation suggest that MRs and ROS may constitute an amplification system of redox signals and ceramide signaling cell membranes, which insures the efficiency of signal transduction. The formation of such feedforward amplifying loop for MR redox and ceramide signaling may also be responsible for the tempospatial regulation of a complex signalosome that precisely and efficiently control cell function. If the activity of this regulatory loop is excessively enhanced, excessive production of both ROS and ceramide may result in the progress and development of different diseases or pathological processes.

It should be noted that ceramide-enriched membrane platforms might also be formed without the presence of classically defined MRs simply through a fusion of several ceramide molecules. These ceramide molecules can come from MRs or other membrane fractions. The clustering of receptor molecules within ceramide-enriched membrane platforms might well have several important functions such as the aggregation in close proximity of many receptor molecules (Gulbins and Grassme 2002), the facilitation of the transactivation of signaling molecules associating or interacting with a receptor, and the amplification of the specific signals generated by activated receptors. However, in some studies the formation of ceramide or ceramide platforms may not play roles in signaling, but contribute to the scrambling of the cell membrane as shown at the erythrocyte surface. It is shown that the eryptosis may be linked to apoptotic pathways via ceramide, which may be causally cross talked to local oxidative stress. This may represent another type of MR redox signaling in erythrocytes (Lang et al. 2006, 2010).

4 Functional Relevance of the Cross Talk

It has been known that the biological responses to cellular or tissue ROS levels are very different and vary from physiological to pathological reactions. Therefore, the cross talk between ceramide and redox signaling may be implicated in various cell and organ functions, depending upon the amount of ROS and ceramide. With respect to ROS, when small amounts of ROS are produced, they may mediate physiological redox signaling, but when large amounts of ROS are produced, which refer to increased oxidative stress, cell/tissue damage may occur, resulting in cellular apoptosis and necrosis and ultimately causing various systemic or organ-based diseases. In regard to ceramide, largely increased ceramide production through such feedwarding mechanism may also contribute to the development of different diseases. This section will focus on the functional relevance of ceramide–redox signaling to the regulation of endothelial function and renal glomerular and tubular functions.

4.1 Regulation of Endothelial Function

Ceramide was demonstrated to increase endothelial $O_2^{\bullet -}$ in the endothelium of isolated small coronary arteries, which was blocked by different NADPH oxidase inhibitors such as *N*-vanillylnonanamide, apocynin, and diphenyleneiodonium. By analysis of the enzyme activity, ceramide was found to significantly stimulate the activity of NADPH oxidase in ECs, which was prevented by NADPH oxidase inhibitors, but not by inhibitors of NOS, xanthine oxidase, and mitochondrial electron transport chain enzymes. In addition, inhibition of NADPH oxidase by different NADPH oxidase inhibitors largely prevented ceramide-induced and $O_2^{\bullet -}$-mediated impairment of endothelium-dependent relaxation to agonists in small bovine coronary arteries (Zhang et al. 2001). These studies very clearly indicate that NADPH oxidases mediate dysfunction of ECs induced by ceramide. In additional studies, ceramide-induced activation of NADPH oxidase was associated with a rapid translocation of $p47^{phox}$ to the cytoplasmic membrane. As discussed above, $p47^{phox}$ translocation is a crucial step leading to activation of NADPH oxidase in phagocytes, these data suggest that $p47^{phox}$ translocation may initiate ceramide-induced activation of NADPH oxidase in coronary ECs. However, the signaling mechanisms that initiate $p47^{phox}$ translocation are unclear. It has been suggested that TNF-α activates PKC-ζ, which in turn phosphorylates $p47^{phox}$, thereby inducing the translocation of this subunit to the membrane where it associates with gp91*phox* to form the active enzyme complex (Frey et al. 2002). It is possible that ceramide employs this kinase to regulate NADPH oxidase in ECs. On the other hand, it might be also possible that ceramide-enriched membrane platforms recruit the subunits of NADPH oxidase to assemble and activate the oxidase at the cell membrane after treatment with TNF-α.

Although it is very attractive to speculate that MRs and ceramide-enriched membrane platforms are involved in the homeostasis of ECs and the response of these cells to cytokines, little is known about the role of these domains for the regulation of vascular endothelial functions. Recently, work in our laboratory has tested whether MR clustering and trafficking on the cell membrane of ECs are associated with ceramide production and action (Jin et al. 2007, 2008a; Zhang et al. 2006, 2007). It was found that ASMase and ceramide are of importance in CD95 ligand-induced formation of MR clusters on the EC membrane. We also demonstrated that ceramide-mediated clustering of MRs is involved in the regulation of $O_2^{\bullet -}$ production in coronary ECs via NADPH oxidase. This effect was associated with the recruitment and aggregation of the NADPH oxidase subunits gp91*phox* and p47*phox* in MRs. It was shown that silencing the ASMase gene by siRNA reduced CD95 ligand-induced gp91phox aggregation in MR clusters and p47*phox* translocation and completely inhibited CD95 ligand-induced $O_2^{\bullet -}$ production in these cells. In isolated small bovine coronary arteries, transfection of ASMase siRNA markedly attenuated CD95 ligand-induced inhibition of endothelium-dependent vasorelaxation (a response to bradykinin) by 60 % (Zhang et al. 2007). The results suggest that ASMase, the release of ceramide, and MR-derived ceramide-enriched membrane platforms are involved in the activation of NADPH oxidase in response to cytokines in coronary ECs, consequently leading to endothelial dysfunction (Jin et al. 2007; Zhang et al. 2006, 2007).

Recently, the MR redox signaling platform associated with NADPH oxidase has been demonstrated to be responsible for endothelial dysfunction induced by various stimuli such as death receptor activation, homocysteine, cytokines, or adipokines (Jin et al. 2008b; Xia et al. 2011; Zhang et al. 2006). As a commonly used functional study, endothelium-dependent vasodilation (EDVD) response in isolated perfused arteries was intensively tested. It was found that various stimulations which led to the formation of MR redox signaling platforms such as CD95 ligand, endostatin, homocysteine, and visfatin all led to impairment of EDVD. This impairment was homeostatically recovered by NADPH oxidase inhibition using apocynin, M-β-CD, filipin, or ASMase siRNA, suggesting that MR redox signaling platforms with NADPH oxidase participate in the impairment of endothelial function (Jin et al. 2007, 2008b; Zhang et al. 2006, 2007). This MR redox enhancement in endothelial injury and dysfunction may be intimately involved in the pathophysiology of diverse cardiovascular diseases such as atherosclerosis, hypertension, shock, and ischemia/reperfusion injuries.

In addition, the formation or enhancement of MR redox signaling platforms may contribute to macrophage reprogramming, foam cell formation, and cell deformability. Induction of lipid oxidation through ROS was found to amplify foam cell formation through oxidized low-density lipoprotein (Ox-LDL) uptake and a subsequent clustering of ceramide-enriched lipid domains (Morita et al. 2004). In addition, Ox-LDL was found to affect cell-surface turnover of ceramide-backbone sphingolipids and apoE-mediated uptake, by low-density lipoprotein receptor-related protein (MRP) family members, leading, in turn, to cell-surface expansion of ceramide-enriched domains and activation of apoE-/MRP1-/CD1-mediated antigen presentation (van den Elzen et al. 2005). On the other hand, high-density lipoprotein (HDL)-mediated lipid efflux can disrupt MRs and prevent foam cell formation. It has been suggested that MR redox signaling or regulation plays an important role in the formation of foam cells and thus in the progression of atherosclerosis (Schmitz and Grandl 2007).

In addition to the role in alterations of macrophage behavior, MR redox signaling may also play an important role in cell deformability, thereby initiating or promoting atherogenesis (Levitan and Gooch 2007). Studies have demonstrated that disruption of MRs by oxidants such as Ox-LDL altered the cytoskeletal structure, including the extent of polymerization, stabilization, cross-linking, and membrane association (Byfield et al. 2006). These molecular alterations may increase the force generated by the cytoskeleton, resulting in a stiffening of the cytoskeleton and hence stiffening of the cell and plasma membrane. Increased force in the cytoskeleton and its downstream increased stiffness may also elevate membrane tension and thereby influence the activity of various mechanosensitive ion channels. Direct evidence suggests that Ox-LDL can disrupt MRs, resulting in a series of pathological changes in the biomechanical properties of vascular ECs and ultimately induce endothelial dysfunction and atherogenesis (Blair et al. 1999; Byfield et al. 2006).

As summarized in Fig. 2, ASMase-mediated ceramide signaling and ROS-based redox signaling interact through a MR signaling platforms. As a molecular cross talk, this interaction of ceramide and ROS may play an important role in the regulation of endothelial function. When various death factors or other stimuli act on ECs, ASMase located in situ or translocated from lysosomes or lysosome-like vesicles is activated to produce ceramide from SM, resulting in the formation of a number of ceramide-enriched membrane signaling platforms. In these platforms, ASMase, NADPH oxidase subunits such as gp91^{phox} and p47^{phox}, and other proteins are aggregated and activated, producing $O_2^{\bullet -}$. $O_2^{\bullet -}$ reacts with NO to decrease NO bioavailability and to produce peroxynitrite ($ONOO^-$). Increased $ONOO^-$ uncouples NOS to produce more $O_2^{\bullet -}$ but less NO. $O_2^{\bullet -}$ or ROS may feedforward enhance MRs clustering by enhancement of ASMase and alteration of MR clustering process, forming positive amplifications. All these together constitute a redox signaling network resulting in endothelial dysfunction and impairment of endothelium-dependent vasodilation, which may be the basis for difference cardiovascular diseases such as atherosclerosis, coronary artery disease, hypertension, and peripheral arterial disease.

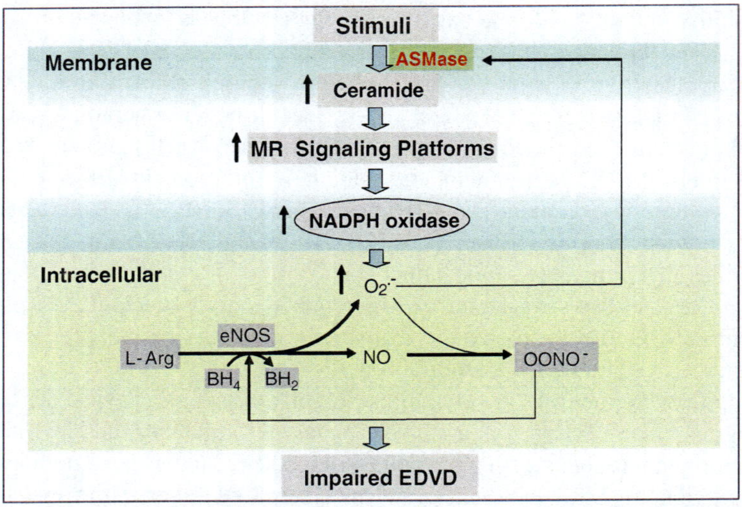

Fig. 2 Ceramide-enriched redox signaling platforms associated with NADPH oxidase in endothelial dysfunction. Upon ASMase stimulation, ceramide is released to promote MR clustering and form ceramide-enriched MR platforms, with aggregation and assembling of NADPH oxidase subunits and other proteins such as Rac GTPase. Then, NADPH oxidase is activated to produce $O_2^{\bullet-}$, which reacts with NO to produce $ONOO^-$ resulting in endothelial dysfunction in coronary arteries. Further, NADPH oxidase-derived $O_2^{\bullet-}$ regulates ASMase activation and ceramide production in a feedforward manner

4.2 Regulation of Renal Function

Recent studies have indicated that ceramide may be implicated in the regulation of kidney function and seems to be involved in renal glomerular and tubular pathology (Kaushal et al. 1998; Ueda et al. 2000; Yi et al. 2004; Yin et al. 1997). More recently, our group demonstrated that ceramide importantly contributes to the development of chronic glomerular injury associated with hyperhomocysteinemia, and thereby ceramide may serve as an important mechanism of end-stage renal disease (Yi et al. 2004, 2007, 2009b). Several studies that employed TLC and HPLC analysis reported the detection of ceramide in the kidney, leading to the hypothesis that ceramide might be involved already in the regulation of normal renal function (Kaushal et al. 1998; Ueda et al. 2000; Yi et al. 2004; Yin et al. 1997). To determine whether ceramide also participates in the development of chronic renal failure, we employed a model of hyperhomocysteinemia-induced renal injury. These studies revealed that hyperhomocysteinemia significantly increased ceramide levels in the renal cortex from rats. Likewise, treatment of cultured mesangial cells with L-homocysteine resulted in a concentration-dependent increase in ceramide. Evidence for a de novo synthesis of ceramide by L-homocysteine was provided in studies that employed fumonisin B1 and myriocin, inhibitors of the de novo synthesis pathway of ceramide. These inhibitors prevented

L-homocysteine-induced ceramide formation in mesangial cells as well as in vivo in the kidney and attenuated glomerular injury and proteinuria (Yi et al. 2004, 2009b). These data provide direct evidence that the ceramide pathway is critically involved in L-homocysteine-induced glomerular injury and glomerular sclerosis.

Further mechanistic studies have demonstrated that L-homocysteine stimulated ceramide production in different glomerular cells such as glomerular capillary ECs, podocytes, and mesangial cells and that ceramides appear to be an important regulator of the function of glomerular filtration membrane, which is consistent with previous results that ceramide may be involved in the regulation of normal renal function (Kaushal et al. 1998; Ueda et al. 2000). It was also found that blockade of ceramide production in hyperhomocysteinimic rats substantially inhibited the enhancement of NADPH oxidase activity and production of $O_2^{\bullet -}$ in the kidney (Yi et al. 2004). Although translocation of p47^{phox}, seen in ECs, was not shown to occur in L-homocysteine- or ceramide-induced activation of NADPH oxidase in rat mesangial cells (Yi et al. 2004), in podocytes and glomerular capillary ECs, homocysteine was shown to induce the formation of MR redox signaling platforms associated with NADPH oxidase (Yi et al. 2009a; Zhang et al. 2010). Perhaps, the transformation of small MRs to ceramide-enriched membrane platforms results in a clustering of NADPH oxidase molecules, producing redox signaling or injury in these glomerular cells, resulting in local oxidative stress and ultimate glomerular injury. This oxidative stress mediated by NADPH oxidase has been indicated to play an important role in progressive glomerular injuries or glomerulosclerosis associated with hyperhomocysteinemia and other diseases such as diabetes and hypertension (Eid et al. 2009; Fujimoto et al. 2008; Yi et al. 2004). It is now known that the formation of MR redox platforms and ROS production is a major mechanism responsible for hyperhomocysteinemia-induced enhancement of glomerular permeability, thereby producing glomerular injuries and consequent sclerosis, which is associated with the regulation of microtubule stability. It seems that the early injurious effects of hyperhomocysteinemia and other pathogenic factors acting on NADPH oxidase are associated with the formation of redox signaling platforms via MR clustering and consequent increases in glomerular permeability due to disruption of microtubule networks in the glomerular filtration membrane (Yi et al. 2007; Zhang et al. 2010).

In addition to their role in the regulation of renal glomerular function, MRs-associated NADPH oxidase may maintain an inactive state of this enzyme in human renal proximal tubule (RPT) cells. Disruption of such inactive MRs may result in their activation (Han et al. 2008). Different cells use MRs to conduct redox signaling in different ways. As Li et al. have reported, NADPH oxidase-dependent ROS production is differentially regulated in MRs and non-MR compartments of RPT epithelial cells (Yi et al. 2009b). This differential regulation or MR-associated inactive NADPH oxidase is mainly attributed to the action of the neurotransmitter dopamine. Dopamine is an essential neurotransmitter involved, mainly via its peripheral receptors, in the control of blood pressure, sodium balance, and various renal and adrenal functions (Jose et al. 2002). As G-protein-coupled receptors, dopamine receptors are associated with both caveolar and noncaveolar MRs (Allen et al. 2007; Lingwood et al. 2009; Simons and Ikonen 1997; Yu et al. 2004). It has been shown that D_1-like receptors can exert an inhibitory action on ROS production in VSM and RPT cells (White and Sidhu 1998; Yang et al. 2006b; Yasunari et al. 2000). However, the molecular mechanisms involved still remain unknown. By sucrose density gradient ultracentrifugation and analysis of NADPH oxidase isoforms and subunits in MRs, it was found that the majority of membrane proteins was in non-MR fractions; only a small portion of proteins were in MR fractions. The D_1-like receptor agonist, fenoldopam, decreases NOX2 and Rac1 proteins in MRs, albeit to a greater extent in hypertensive than normotensive rats. Fenoldopam decreased the amount of NOX2 that co-immunoprecipitated with $p67^{phox}$ in cells from normotensive rats. These observations suggest that fenoldopam causes a redistribution of NOX2, NOX4, and Rac1 from MRs and to non-MR fractions. Further studies have shown that disruption of MRs results in the reactivation of NADPH oxidase that was destroyed by antioxidants and the silencing of NOX2 or NOX4. Perhaps this explains why in human RPT cells, MRs maintain NADPH oxidase in an inactive state (Han et al. 2008).

5 Concluding Remarks

In summary, there is no doubt that redox signaling through NADPH oxidase and other ROS producing or scavenging systems is correlated with the unique membrane structures known as MRs, where MRs serve as platforms to aggregate the membrane spanning or cytosolic components of the enzymes subunits or cofactors. In particular, MRs clustering may be a major mechanism for the assembly of NADPH oxidase subunits and cofactors into an active enzyme complex. Such MR redox platforms produce $O_2^{\bullet -}$ and thereby conduct redox signaling with compartmentalization and amplifications in response to different receptor bindings or other stimuli. It is well known that the formation of this MR redox signaling platform associated with NADPH oxidase is associated with activation of ASMase and production of ceramide. On the other hand, ROS production may enhance ASMase activity or alter MR components to promote MRs clustering. It is clear that ROS-based redox signaling and ceramide producing system and consequent signaling interact under control condition or upon different stimuli, constituting a temperospatial cross talk between two signaling pathways. Such cross talk may be significantly implicated in the regulation of organ functions. As an example, the regulation of endothelial function and renal glomerular or tubular functions is closely associated with ceramide–redox cross talk, and their interplay, if excessively enhanced, may result in endothelial dysfunction and renal glomerular or tubular dysfunction, leading to various cardiovascular and renal diseases. In perspective, it is imperative to develop new in vivo research strategies that are able to address the contribution of ceramide–redox interaction to organ functions and related regulatory mechanisms. More studies may also be needed to translate experimental results related to MR redox signaling platforms and ceramide–redox cross talk to clinical use.

Acknowledgment The works cited in the authors' laboratory were supported by the National Institutes of Health grants (HL075316 and HL057244).

References

Allen JA, Halverson-Tamboli RA, Rasenick MM (2007) Lipid raft microdomains and neurotransmitter signalling. Nat Rev 8:128–140

Ayuyan AG, Cohen FS (2006) Lipid peroxides promote large rafts: effects of excitation of probes in fluorescence microscopy and electrochemical reactions during vesicle formation. Biophys J 91:2172–2183

Azad N, Rojanasakul Y, Vallyathan V (2008) Inflammation and lung cancer: roles of reactive oxygen/nitrogen species. J Toxicol Environ Health 11:1–15

Azad MB, Chen Y, Gibson SB (2009) Regulation of autophagy by reactive oxygen species (ROS): implications for cancer progression and treatment. Antioxid Redox Signal 11:777–790

Babior BM, Lambeth JD, Nauseef W (2002) The neutrophil NADPH oxidase. Arch Biochem Biophys 397:342–344

Bagi Z, Feher A, Beleznai T (2009) Preserved coronary arteriolar dilatation in patients with type 2 diabetes mellitus: implications for reactive oxygen species. Pharmacol Rep 61:99–104

Bao JX, Jin S, Zhang F, Wang ZC, Li N, Li PL (2010a) Activation of membrane NADPH oxidase associated with lysosome-targeted acid sphingomyelinase in coronary endothelial cells. Antioxid Redox Signal 12:703–712

Bao JX, Xia M, Poklis JL, Han WQ, Brimson C, Li PL (2010b) Triggering role of acid sphingomyelinase in endothelial lysosome-membrane fusion and dysfunction in coronary arteries. Am J Physiol Heart Circ Physiol 298:H992–H1002

Bell EL, Chandel NS (2007) Mitochondrial oxygen sensing: regulation of hypoxia-inducible factor by mitochondrial generated reactive oxygen species. Essays Biochem 43:17–27

Bezombes C, de Thonel A, Apostolou A, Louat T, Jaffrezou JP, Laurent G et al (2002) Overexpression of protein kinase Czeta confers protection against antileukemic drugs by inhibiting the redox-dependent sphingomyelinase activation. Mol Pharmacol 62:1446–1455

Bhunia AK, Han H, Snowden A, Chatterjee S (1997) Redox-regulated signaling by lactosylceramide in the proliferation of human aortic smooth muscle cells. J Biol Chem 272:15642–15649

Blair A, Shaul PW, Yuhanna IS, Conrad PA, Smart EJ (1999) Oxidized low density lipoprotein displaces endothelial nitric-oxide synthase (eNOS) from plasmalemmal caveolae and impairs eNOS activation. J Biol Chem 274:32512–32519

Blank U, Cyprien B, Martin-Verdeaux S, Paumet F, Pombo I, Rivera J et al (2002) SNAREs and associated regulators in the control of exocytosis in the RBL-2H3 mast cell line. Mol Immunol 38:1341–1345

Brown DA, London E (1998) Structure and origin of ordered lipid domains in biological membranes. J Membr Biol 164:103–114

Brown DL, Doubilet PM, Miller FH, Frates MC, Laing FC, DiSalvo DN et al (1998) Benign and malignant ovarian masses: selection of the most discriminating gray-scale and Doppler sonographic features. Radiology 208:103–110

Burdon RH (1996) Control of cell proliferation by reactive oxygen species. Biochem Soc Trans 24:1028–1032

Byfield FJ, Tikku S, Rothblat GH, Gooch KJ, Levitan I (2006) OxLDL increases endothelial stiffness, force generation, and network formation. J Lipid Res 47:715–723

Cai H (2006) A new mechanism for flow-mediated vasoprotection? Focus on "lung endothelial cell proliferation with decreased shear stress is mediated by reactive oxygen species". Am J Physiol 290:C35–C36
Casadesus G, Smith MA, Zhu X, Aliev G, Cash AD, Honda K et al (2004) Alzheimer disease: evidence for a central pathogenic role of iron-mediated reactive oxygen species. J Alzheimers Dis 6:165–169
Charruyer A, Grazide S, Bezombes C, Muller S, Laurent G, Jaffrezou JP (2005) UV-C light induces raft-associated acid sphingomyelinase and JNK activation and translocation independently on a nuclear signal. J Biol Chem 280:19196–19204
Chatterjee M, Wu S (2001) Cell line dependent involvement of ceramide in ultraviolet light-induced apoptosis. Mol Cell Biochem 219:21–27
Cheng G, Cao Z, Xu X, van Meir EG, Lambeth JD (2001) Homologs of gp91phox: cloning and tissue expression of Nox3, Nox4, and Nox5. Gene 269:131–140
Colavitti R, Finkel T (2005) Reactive oxygen species as mediators of cellular senescence. IUBMB Life 57:277–281
Corda S, Laplace C, Vicaut E, Duranteau J (2001) Rapid reactive oxygen species production by mitochondria in endothelial cells exposed to tumor necrosis factor-alpha is mediated by ceramide. Am J Respir Cell Mol Biol 24:762–768
Cremesti AE, Goni FM, Kolesnick R (2002) Role of sphingomyelinase and ceramide in modulating rafts: do biophysical properties determine biologic outcome? FEBS Lett 531:47–53
Dang PM, Cross AR, Quinn MT, Babior BM (2002) Assembly of the neutrophil respiratory burst oxidase: a direct interaction between p67PHOX and cytochrome b558 II. Proc Natl Acad Sci USA 99:4262–4265
Del Prete A, Zaccagnino P, Di Paola M, Saltarella M, Oliveros Celis C, Nico B et al (2008) Role of mitochondria and reactive oxygen species in dendritic cell differentiation and functions. Free Radic Biol Med 44:1443–1451
Delles C, Miller WH, Dominiczak AF (2008) Targeting reactive oxygen species in hypertension. Antioxid Redox Signal 10:1061–1077
Dumitru CA, Gulbins E (2006) TRAIL activates acid sphingomyelinase via a redox mechanism and releases ceramide to trigger apoptosis. Oncogene 25:5612–5625
Dumitru CA, Zhang Y, Li X, Gulbins E (2007) Ceramide: a novel player in reactive oxygen species-induced signaling? Antioxid Redox Signal 9:1535–1540
Dworakowski R, Anilkumar N, Zhang M, Shah AM (2006) Redox signalling involving NADPH oxidase-derived reactive oxygen species. Biochem Soc Trans 34:960–964
Eid AA, Gorin Y, Fagg BM, Maalouf R, Barnes JL, Block K et al (2009) Mechanisms of podocyte injury in diabetes: role of cytochrome P450 and NADPH oxidases. Diabetes 58:1201–1211
Feuk-Lagerstedt E, Movitz C, Pellme S, Dahlgren C, Karlsson A (2007) Lipid raft proteome of the human neutrophil azurophil granule. Proteomics 7:194–205
Finkel T, Holbrook NJ (2000) Oxidants, oxidative stress and the biology of ageing. Nature 408:239–247
Flowers F, Zimmerman JJ (1998) Reactive oxygen species in the cellular pathophysiology of shock. New Horizons 6:169–180
Frey RS, Rahman A, Kefer JC, Minshall RD, Malik AB (2002) PKCzeta regulates TNF-alpha-induced activation of NADPH oxidase in endothelial cells. Circ Res 90:1012–1019
Fujimoto S, Satoh M, Horike H, Hatta H, Haruna Y, Kobayashi S et al (2008) Olmesartan ameliorates progressive glomerular injury in subtotal nephrectomized rats through suppression of superoxide production. Hypertens Res 31:305–313
Gendzwill A (2007a) Reactive oxygen species and vascular hyporeactivity in septic shock. Part II–Scavengers and vascular hyporeactivity in septic shock. Pol Merkur Lekarski 23:284–287
Gendzwill A (2007b) Reactive oxygen species and vascular hyporeactivity in septic shock. Part I–Reactive oxygen species and vascular hyporeactivity. Pol Merkur Lekarski 23:280–283

Gerst JE (1999) SNAREs and SNARE regulators in membrane fusion and exocytosis. Cell Mol Life Sci 55:707–734

Goswami SK, Maulik N, Das DK (2007) Ischemia-reperfusion and cardioprotection: a delicate balance between reactive oxygen species generation and redox homeostasis. Ann Med 39:275–289

Grassme H, Jekle A, Riehle A, Schwarz H, Berger J, Sandhoff K et al (2001) CD95 signaling via ceramide-rich membrane rafts. J Biol Chem 276:20589–20596

Griendling KK, Sorescu D, Ushio-Fukai M (2000) NAD(P)H oxidase: role in cardiovascular biology and disease. Circ Res 86:494–501

Gulbins E, Grassme H (2002) Ceramide and cell death receptor clustering. Biochim Biophys Acta 1585:139–145

Gulbins E, Kolesnick R (2003) Raft ceramide in molecular medicine. Oncogene 22:7070–7077

Guzy RD, Schumacker PT (2006) Oxygen sensing by mitochondria at complex III: the paradox of increased reactive oxygen species during hypoxia. Exp Physiol 91:807–819

Halliwell B, Cross CE (1994) Oxygen-derived species: their relation to human disease and environmental stress. Environ Health Perspect 102(Suppl 10):5–12

Han W, Li H, Villar VA, Pascua AM, Dajani MI, Wang X et al (2008) Lipid rafts keep NADPH oxidase in the inactive state in human renal proximal tubule cells. Hypertension 51:481–487

Hansberg W, de Groot H, Sies H (1993) Reactive oxygen species associated with cell differentiation in Neurospora crassa. Free Radic Biol Med 14:287–293

Hara T, Kondo N, Nakamura H, Okuyama H, Mitsui A, Hoshino Y et al (2007) Cell-surface thioredoxin-1: possible involvement in thiol-mediated leukocyte-endothelial cell interaction through lipid rafts. Antioxid Redox Signal 9:1427–1437

Hernandez OM, Discher DJ, Bishopric NH, Webster KA (2000) Rapid activation of neutral sphingomyelinase by hypoxia-reoxygenation of cardiac myocytes. Circ Res 86:198–204

Hildeman DA (2004) Regulation of T-cell apoptosis by reactive oxygen species. Free Radic Biol Med 36:1496–1504

Hirooka Y (2008) Role of reactive oxygen species in brainstem in neural mechanisms of hypertension. Auton Neurosci 142:20–24

Hoekstra D, Maier O, van der Wouden JM, Slimane TA, van ISC (2003) Membrane dynamics and cell polarity: the role of sphingolipids. J Lipid Res 44:869–877

Jamaluddin M, Tian B, Boldogh I, Garofalo RP, Brasier AR (2009) Respiratory syncytial virus infection induces a reactive oxygen species-MSK1-phospho-Ser-276 RelA pathway required for cytokine expression. J Virol 83:10605–10615

Jin S, Yi F, Li PL (2007) Contribution of lysosomal vesicles to the formation of lipid raft redox signaling platforms in endothelial cells. Antioxid Redox Signal 9:1417–1426

Jin S, Yi F, Zhang F, Poklis JL, Li PL (2008a) Lysosomal targeting and trafficking of acid sphingomyelinase to lipid raft platforms in coronary endothelial cells. Arterioscler Thromb Vasc Biol 28:2056–2062

Jin S, Zhang Y, Yi F, Li PL (2008b) Critical role of lipid raft redox signaling platforms in endostatin-induced coronary endothelial dysfunction. Arterioscler Thromb Vasc Biol 28:485–490

Jose PA, Eisner GM, Felder RA (2002) Role of dopamine receptors in the kidney in the regulation of blood pressure. Curr Opin Nephrol Hypertens 11:87–92

Kaushal GP, Singh AB, Shah SV (1998) Identification of gene family of caspases in rat kidney and altered expression in ischemia-reperfusion injury. Am J Physiol 274:F587–F595

Kietzmann T, Gorlach A (2005) Reactive oxygen species in the control of hypoxia-inducible factor-mediated gene expression. Semin Cell Dev Biol 16:474–486

Kojda G, Harrison D (1999) Interactions between NO and reactive oxygen species: pathophysiological importance in atherosclerosis, hypertension, diabetes and heart failure. Cardiovasc Res 43:562–571

Kondo N, Ishii Y, Kwon YW, Tanito M, Sakakura-Nishiyama J, Mochizuki M et al (2007) Lipid raft-mediated uptake of cysteine-modified thioredoxin-1: apoptosis enhancement by inhibiting the endogenous thioredoxin-1. Antioxid Redox Signal 9:1439–1448

Ksiazek K, Wisniewska J (2001) The role of glucose and reactive oxygen species in the development of vascular complications of diabetes mellitus. Przegl Lek 58:915–918

Kusmartsev S, Gabrilovich DI (2003) Inhibition of myeloid cell differentiation in cancer: the role of reactive oxygen species. J Leukoc Biol 74:186–196

Lang F, Lang KS, Lang PA, Huber SM, Wieder T (2006) Mechanisms and significance of eryptosis. Antioxid Redox Signal 8:1183–1192

Lang PA, Schenck M, Nicolay JP, Becker JU, Kempe DS, Lupescu A et al (2007) Liver cell death and anemia in Wilson disease involve acid sphingomyelinase and ceramide. Nat Med 13:164–170

Lang F, Gulbins E, Lang PA, Zappulla D, Foller M (2010) Ceramide in suicidal death of erythrocytes. Cell Physiol Biochem 26:21–28

Lecour S, Van der Merwe E, Opie LH, Sack MN (2006) Ceramide attenuates hypoxic cell death via reactive oxygen species signaling. J Cardiovasc Pharmacol 47:158–163

Levitan I, Gooch KJ (2007) Lipid rafts in membrane-cytoskeleton interactions and control of cellular biomechanics: actions of oxLDL. Antioxid Redox Signal 9:1519–1534

Li PL, Gulbins E (2007) Lipid rafts and redox signaling. Antioxid Redox Signal 9:1411–1415

Li PL, Zhang Y, Yi F (2007) Lipid raft redox signaling platforms in endothelial dysfunction. Antioxid Redox Signal 9:1457–1470

Lingwood D, Kaiser HJ, Levental I, Simons K (2009) Lipid rafts as functional heterogeneity in cell membranes. Biochem Soc Trans 37:955–960

Liu B, Hannun YA (1997) Inhibition of the neutral magnesium-dependent sphingomyelinase by glutathione. J Biol Chem 272:16281–16287

Liu B, Andrieu-Abadie N, Levade T, Zhang P, Obeid LM, Hannun YA (1998) Glutathione regulation of neutral sphingomyelinase in tumor necrosis factor-alpha-induced cell death. J Biol Chem 273:11313–11320

Liu CY, Lee CF, Wei YH (2009) Role of reactive oxygen species-elicited apoptosis in the pathophysiology of mitochondrial and neurodegenerative diseases associated with mitochondrial DNA mutations. J Formosan Med Assoc 108:599–611

Lu SP, Lin Feng MH, Huang HL, Huang YC, Tsou WI, Lai MZ (2007) Reactive oxygen species promote raft formation in T lymphocytes. Free Radic Biol Med 42:936–944

MacFarlane PM, Wilkerson JE, Lovett-Barr MR, Mitchell GS (2008) Reactive oxygen species and respiratory plasticity following intermittent hypoxia. Respir Physiol Neurobiol 164:263–271

Malaplate-Armand C, Florent-Bechard S, Youssef I, Koziel V, Sponne I, Kriem B et al (2006) Soluble oligomers of amyloid-beta peptide induce neuronal apoptosis by activating a cPLA2-dependent sphingomyelinase-ceramide pathway. Neurobiol Dis 23:178–189

Mansat-de Mas V, Bezombes C, Quillet-Mary A, Bettaieb A, D'Orgeix AD, Laurent G et al (1999) Implication of radical oxygen species in ceramide generation, c-Jun N-terminal kinase activation and apoptosis induced by daunorubicin. Mol Pharmacol 56:867–874

Martin SF, Sawai H, Villalba JM, Hannun YA (2007) Redox regulation of neutral sphingomyelinase-1 activity in HEK293 cells through a GSH-dependent mechanism. Arch Biochem Biophys 459:295–300

Mashimo M, Nishikawa M, Higuchi K, Hirose M, Wei Q, Haque A et al (2006) Production of reactive oxygen species in peripheral blood is increased in individuals with Helicobacter pylori infection and decreased after its eradication. Helicobacter 11:266–271

Mates JM, Sanchez-Jimenez FM (2000) Role of reactive oxygen species in apoptosis: implications for cancer therapy. Int J Biochem Cell Biol 32:157–170

Mathias S, Kolesnick R (1993) Ceramide: a novel second messenger. Adv Lipid Res 25:65–90

Mohazzab KM, Kaminski PM, Wolin MS (1994) NADH oxidoreductase is a major source of superoxide anion in bovine coronary artery endothelium. Am J Physiol 266:H2568–H2572

Morgan MJ, Kim YS, Liu Z (2007) Lipid rafts and oxidative stress-induced cell death. Antioxid Redox Signal 9:1471–1483

Morita SY, Kawabe M, Sakurai A, Okuhira K, Vertut-Doi A, Nakano M et al (2004) Ceramide in lipid particles enhances heparan sulfate proteoglycan and low density lipoprotein receptor-related protein-mediated uptake by macrophages. J Biol Chem 279:24355–24361

Muller-Peddinghaus R (1989) Reactive oxygen species and inflammation. Dtsch Tierarzti Wochenschr 96:210–212

Ni X, Morales CR (2006) The lysosomal trafficking of acid sphingomyelinase is mediated by sortilin and mannose 6-phosphate receptor. Traffic 7:889–902

Nicco C, Laurent A, Chereau C, Weill B, Batteux F (2005) Differential modulation of normal and tumor cell proliferation by reactive oxygen species. Biomed Pharmacother 59:169–174

Oakley FD, Abbott D, Li Q, Engelhardt JF (2009) Signaling components of redox active endosomes: the redoxosomes. Antioxid Redox Signal 11:1313–1333

Ochsendorf FR (1998) Infection and reactive oxygen species. Andrologia 30(Suppl 1):81–86

Ong SL, Zhang Y, Whitworth JA (2008) Reactive oxygen species and glucocorticoid-induced hypertension. Clin Exp Pharmacol Physiol 35:477–482

Oshikawa J, Urao N, Kim HW, Kaplan N, Razvi M, McKinney R et al (2010) Extracellular SOD-derived H2O2 promotes VEGF signaling in caveolae/lipid rafts and post-ischemic angiogenesis in mice. PLoS One 5:e10189

Oyagbemi AA, Azeez OI, Saba AB (2009) Interactions between reactive oxygen species and cancer: the roles of natural dietary antioxidants and their molecular mechanisms of action. Asian Pac J Cancer Prev 10:535–544

Paravicini TM, Touyz RM (2008) NADPH oxidases, reactive oxygen species, and hypertension: clinical implications and therapeutic possibilities. Diabetes Care 31(Suppl 2):S170–S180

Patel RP, Moellering D, Murphy-Ullrich J, Jo H, Beckman JS, Darley-Usmar VM (2000) Cell signaling by reactive nitrogen and oxygen species in atherosclerosis. Free Radic Biol Med 28:1780–1794

Perrone GG, Tan SX, Dawes IW (2008) Reactive oxygen species and yeast apoptosis. Biochim Biophys Acta 1783:1354–1368

Perry G, Castellani RJ, Hirai K, Smith MA (1998) Reactive oxygen species mediate cellular damage in Alzheimer disease. J Alzheimers Dis 1:45–55

Piccoli C, Ria R, Scrima R, Cela O, D'Aprile A, Boffoli D et al (2005) Characterization of mitochondrial and extra-mitochondrial oxygen consuming reactions in human hematopoietic stem cells. Novel evidence of the occurrence of NAD(P)H oxidase activity. J Biol Chem 280:26467–26476

Poteser M, Graziani A, Rosker C, Eder P, Derler I, Kahr H et al (2006) TRPC3 and TRPC4 associate to form a redox-sensitive cation channel. Evidence for expression of native TRPC3-TRPC4 heteromeric channels in endothelial cells. J Biol Chem 281:13588–13595

Puddu P, Puddu GM, Cravero E, Rosati M, Muscari A (2008) The molecular sources of reactive oxygen species in hypertension. Blood Press 17:70–77

Qiu H, Edmunds T, Baker-Malcolm J, Karey KP, Estes S, Schwarz C et al (2003) Activation of human acid sphingomyelinase through modification or deletion of C-terminal cysteine. J Biol Chem 278:32744–32752

Rajagopalan S, Kurz S, Munzel T, Tarpey M, Freeman BA, Griendling KK et al (1996) Angiotensin II-mediated hypertension in the rat increases vascular superoxide production via membrane NADH/NADPH oxidase activation. Contribution to alterations of vasomotor tone. J Clin Invest 97:1916–1923

Rodighiero S, De Simoni A, Formenti A (2004) The voltage-dependent nonselective cation current in human red blood cells studied by means of whole-cell and nystatin-perforated patch-clamp techniques. Biochim Biophys Acta 1660:164–170

Sasaki H, Yamamoto H, Tominaga K, Masuda K, Kawai T, Teshima-Kondo S et al (2009) NADPH oxidase-derived reactive oxygen species are essential for differentiation of a mouse macrophage cell line (RAW264.7) into osteoclasts. J Med Invest 56:33–41

Sauer H, Wartenberg M, Hescheler J (2001) Reactive oxygen species as intracellular messengers during cell growth and differentiation. Cell Physiol Biochem 11:173–186

Scheel-Toellner D, Wang K, Assi LK, Webb PR, Craddock RM, Salmon M et al (2004) Clustering of death receptors in lipid rafts initiates neutrophil spontaneous apoptosis. Biochem Soc Trans 32:679–681

Schmitz G, Grandl M (2007) Role of redox regulation and lipid rafts in macrophages during Ox-LDL-mediated foam cell formation. Antioxid Redox Signal 9:1499–1518

Siafakas AR, Wright LC, Sorrell TC, Djordjevic JT (2006) Lipid rafts in Cryptococcus neoformans concentrate the virulence determinants phospholipase B1 and Cu/Zn superoxide dismutase. Eukaryot Cell 5:488–498

Simons K, Ikonen E (1997) Functional rafts in cell membranes. Nature 387:569–572

Simons K, Toomre D (2000) Lipid rafts and signal transduction. Nat Rev Mol Cell Biol 1:31–39

Stadtman ER (2004) Role of oxidant species in aging. Curr Med Chem 11:1105–1112

Sun G, Xu X, Wang Y, Shen X, Chen Z, Yang J (2008) Mycoplasma pneumoniae infection induces reactive oxygen species and DNA damage in A549 human lung carcinoma cells. Infect Immun 76:4405–4413

Szocs K (2004) Endothelial dysfunction and reactive oxygen species production in ischemia/reperfusion and nitrate tolerance. Gen Physiol Biophys 23:265–295

Tang XL, Takano H, Rizvi A, Turrens JF, Qiu Y, Wu WJ et al (2002) Oxidant species trigger late preconditioning against myocardial stunning in conscious rabbits. Am J Physiol Heart Circ Physiol 282:H281–H291

Toledo-Pereyra LH, Lopez-Neblina F, Toledo AH (2004) Reactive oxygen species and molecular biology of ischemia/reperfusion. Ann Transplant 9:81–83

Tsyupko AN, Dudnik LB, Evstigneeva RP, Alessenko AV (2001) Effects of reduced and oxidized glutathione on sphingomyelinase activity and contents of sphingomyelin and lipid peroxidation products in murine liver. Biochemistry 66:1028–1034

Ueda N, Kaushal GP, Shah SV (2000) Apoptotic mechanisms in acute renal failure. Am J Med 108:403–415

Ushio-Fukai M, Hilenski L, Santanam N, Becker PL, Ma Y, Griendling KK et al (2001) Cholesterol depletion inhibits epidermal growth factor receptor transactivation by angiotensin II in vascular smooth muscle cells: role of cholesterol-rich microdomains and focal adhesions in angiotensin II signaling. J Biol Chem 276:48269–48275

van den Elzen P, Garg S, Leon L, Brigl M, Leadbetter EA, Gumperz JE et al (2005) Apolipoprotein-mediated pathways of lipid antigen presentation. Nature 437:906–910

Van Wart HE, Birkedal-Hansen H (1990) The cysteine switch: a principle of regulation of metalloproteinase activity with potential applicability to the entire matrix metalloproteinase gene family. Proc Natl Acad Sci USA 87:5578–5582

Weinberg F, Chandel NS (2009) Reactive oxygen species-dependent signaling regulates cancer. Cell Mol Life Sci 66:3663–3673

White BH, Sidhu A (1998) Increased oxidative stress in renal proximal tubules of the spontaneously hypertensive rat: a mechanism for defective dopamine D1A receptor/G-protein coupling. J Hypertens 16:1659–1665

Wolf G (2005) Role of reactive oxygen species in angiotensin II-mediated renal growth, differentiation, and apoptosis. Antioxid Redox Signal 7:1337–1345

Woudenberg J, Rembacz KP, van den Heuvel FA, Woudenberg-Vrenken TE, Buist-Homan M, Geuken M et al (2010) Caveolin-1 is enriched in the peroxisomal membrane of rat hepatocytes. Hepatology 51:1744–1753

Xia M, Zhang C, Boini KM, Thacker AM, Li PL (2011) Membrane raft-lysosome redox signalling platforms in coronary endothelial dysfunction induced by adipokine visfatin. Cardiovasc Res 89:401–409

Yamamoto S, Shimizu S, Mori Y (2009) Involvement of TRPM2 channel in amplification of reactive oxygen species-induced signaling and chronic inflammation. Nippon Yakurigaku Zasshi 134:122–126

Yang B, Rizzo V (2007) TNF-alpha potentiates protein-tyrosine nitration through activation of NADPH oxidase and eNOS localized in membrane rafts and caveolae of bovine aortic endothelial cells. Am J Physiol Heart Circ Physiol 292:H954–H962

Yang B, Oo TN, Rizzo V (2006a) Lipid rafts mediate H_2O_2 prosurvival effects in cultured endothelial cells. FASEB J 20:1501–1503

Yang Z, Asico LD, Yu P, Wang Z, Jones JE, Escano CS et al (2006b) D5 dopamine receptor regulation of reactive oxygen species production, NADPH oxidase, and blood pressure. Am J Physiol Regul Integr Comp Physiol 290:R96–R104

Yasunari K, Kohno M, Kano H, Minami M, Yoshikawa J (2000) Dopamine as a novel antioxidative agent for rat vascular smooth muscle cells through dopamine D(1)-like receptors. Circulation 101:2302–2308

Yi F, Zhang AY, Janscha JL, Li PL, Zou AP (2004) Homocysteine activates NADH/NADPH oxidase through ceramide-stimulated Rac GTPase activity in rat mesangial cells. Kidney Int 66:1977–1987

Yi F, Chen QZ, Jin S, Li PL (2007) Mechanism of homocysteine-induced Rac1/NADPH oxidase activation in mesangial cells: role of guanine nucleotide exchange factor Vav2. Cell Physiol Biochem 20:909–918

Yi F, Jin S, Zhang F, Xia M, Bao JX, Hu J et al (2009a) Formation of lipid raft redox signalling platforms in glomerular endothelial cells: an early event of homocysteine-induced glomerular injury. J Cell Mol Med 13:3303–3314

Yi F, Xia M, Li N, Zhang C, Tang L, Li PL (2009b) Contribution of guanine nucleotide exchange factor Vav2 to hyperhomocysteinemic glomerulosclerosis in rats. Hypertension 53:90–96

Yin T, Sandhu G, Wolfgang CD, Burrier A, Webb RL, Rigel DF et al (1997) Tissue-specific pattern of stress kinase activation in ischemic/reperfused heart and kidney. J Biol Chem 272:19943–19950

Yoshimura S, Banno Y, Nakashima S, Hayashi K, Yamakawa H, Sawada M et al (1999) Inhibition of neutral sphingomyelinase activation and ceramide formation by glutathione in hypoxic PC12 cell death. J Neurochem 73:675–683

Yu P, Yang Z, Jones JE, Wang Z, Owens SA, Mueller SC et al (2004) D1 dopamine receptor signaling involves caveolin-2 in HEK-293 cells. Kidney Int 66:2167–2180

Zeng C, Villar VA, Yu P, Zhou L, Jose PA (2009) Reactive oxygen species and dopamine receptor function in essential hypertension. Clin Exp Hypertens 31:156–178

Zhai J, Strom AL, Kilty R, Venkatakrishnan P, White J, Everson WV et al (2009) Proteomic characterization of lipid raft proteins in amyotrophic lateral sclerosis mouse spinal cord. FEBS J 276:3308–3323

Zhang C, Li PL (2010) Membrane raft redox signalosomes in endothelial cells. Free Radic Res 44(8):831–842

Zhang DX, Zou AP, Li PL (2001) Ceramide reduces endothelium-dependent vasodilation by increasing superoxide production in small bovine coronary arteries. Circ Res 88:824–831

Zhang AY, Yi F, Zhang G, Gulbins E, Li PL (2006) Lipid raft clustering and redox signaling platform formation in coronary arterial endothelial cells. Hypertension 47:74–80

Zhang AY, Yi F, Jin S, Xia M, Chen QZ, Gulbins E et al (2007) Acid sphingomyelinase and its redox amplification in formation of lipid raft redox signaling platforms in endothelial cells. Antioxid Redox Signal 9:817–828

Zhang Y, Li X, Carpinteiro A, Gulbins E (2008) Acid sphingomyelinase amplifies redox signaling in Pseudomonas aeruginosa-induced macrophage apoptosis. J Immunol 181:4247–4254

Zhang C, Hu JJ, Xia M, Boini KM, Brimson C, Li PL (2010) Redox signaling via lipid raft clustering in homocysteine-induced injury of podocytes. Biochim Biophys Acta 1803:482–491

Zuo L, Ushio-Fukai M, Hilenski LL, Alexander RW (2004) Microtubules regulate angiotensin II type 1 receptor and Rac1 localization in caveolae/lipid rafts: role in redox signaling. Arterioscler Thromb Vasc Biol 24:1223–1228

Zuo L, Ushio-Fukai M, Ikeda S, Hilenski L, Patrushev N, Alexander RW (2005) Caveolin-1 is essential for activation of Rac1 and NAD(P)H oxidase after angiotensin II type 1 receptor stimulation in vascular smooth muscle cells: role in redox signaling and vascular hypertrophy. Arterioscler Thromb Vasc Biol 25:1824–1830

Part III
Sphingolipids in Inflammation, Infection and Lung Diseases

Sphingolipids in Lung Endothelial Biology and Regulation of Vascular Integrity

Taimur Abbasi and Joe G.N. Garcia

Contents

Abstract Of the multiple and diverse homeostatic events that involve the lung vascular endothelium, participation in preserving vascular integrity and therefore organ function is paramount. We were the first to show that the lipid growth factor and angiogenic factor, sphingosine-1-phosphate, is a critical agonist involved in regulation of human lung vascular barrier function (Garcia et al. J Clin Invest, 2011). Utilizing both in vitro models and preclinical murine, rat, and canine models of acute and chronic inflammatory lung injury, we have shown that S1Ps, as well as multiple S1P analogues such as FTY720 and ftysiponate, serve as protective agents

T. Abbasi
Department of Medicine, Institute for Personalized Respiratory Medicine, The University of Illinois at Chicago, Chicago, IL 60612, USA

J.G.N. Garcia (✉)
Earl Bane Professor of Medicine, Pharmacology & Bioengineering, Chicago, IL 60612, USA
e-mail: jggarcia@uic.edu

E. Gulbins and I. Petrache (eds.), *Sphingolipids in Disease*,
Handbook of Experimental Pharmacology 216,
DOI 10.1007/978-3-7091-1511-4_10, © Springer-Verlag Wien 2013

limiting the disruption of the vascular EC monolayer in the pulmonary microcirculation and attenuate parenchymal accumulation of inflammatory cells and high protein containing extravasated fluid, thereby reducing interstitial and alveolar edema. The vasculo-protective mechanism of these therapeutic effects occurs via ligation of specific G-protein-coupled receptors and an intricate interplay of S1P with other factors (such as MAPKS, ROCKs, Rho, Rac1) with rearrangement of the endothelial cytoskeleton to form strong cortical actin rings in the cell periphery and enhanced cell-to-cell and cell-to-matrix tethering dynamics. This cascade leads to reinforcement of focal adhesions and paracellular junctional complexes via cadherin, paxillin, catenins, and zona occludens. S1P through its interaction with Rac and Rho influences the cytoskeletal rearrangement indicated in the later stages of angiogenesis as a stabilizing force, preventing excessive vascular permeability. These properties translate into a therapeutic potential for acute and chronic inflammatory lung injuries. S1P has potential for providing a paradigm shift in the approach to disruption of critical endothelial gatekeeper function, loss of lung vascular integrity, and increased vascular permeability, defining features of acute lung injury (ALI), and may prove to exhibit an intrinsically protective role in the pulmonary vasculature ameliorating agonist- or sepsis-induced pulmonary injury and vascular leakage.

Keywords Endothelial cells • S1P • Rac • Rho • Cytoskeleton

1 Introduction

The lung vascular endothelium is a critical participant in multiple homeostatic events that preserve the lung's physiologic function. These range from maintenance of vascular tone, blood coagulation, inflammation, angiogenesis, cell homeostasis, and the maintenance of lung fluid balance. For example, the lung endothelium is essential to adequate gas transfer and tissue oxygenation but serves as a primary cellular target in the profound physiologic derangement that accompanies acute and chronic inflammatory lung injuries. Via disruption of critical endothelial gatekeeper function, the loss of lung vascular integrity results in parenchymal accumulation of leukocytes and extravascular lung water and increased vascular permeability, defining features of acute lung injury (ALI) that contribute to the increased morbidity and mortality of this devastating syndrome. Consequently, there is substantial interest in the development and utilization of clinically effective agents that either interfere with or prevent lung vascular endothelial cell (EC) barrier dysfunction, restore EC barrier integrity, reduce alveolar flooding, and improve respiratory mechanics (Schuchardt et al. 2011; Snider et al. 2010; Belvitch and Dudek 2012; Bazzoni and Dejana 2004; Jacobson and Garcia 2007; Komarova et al. 2007).

The vascular endothelium is composed of a single layer of ECs and an underlying layer of extracellular matrix and serves as a semipermeable barrier, regulating exchange of gases, water, and solutes across it (Donati and Bruni 2006; Belvitch and Dudek 2012; Bazzoni and Dejana 2004). The transport of fluids and macromolecules across the endothelium is regulated by three main processes. Capillary beds contain fenestrae that are small diaphragm covered pores in the endothelium allowing the passive diffusion based transport of the particles based on the size and charge selective characteristics of the particle. In contrast to diffusion, transcytosis is an active process that involves the fusion of endocytic vesicles with the luminal endothelial membrane in response to ligation with cell surface glycoproteins. The third process, and the most important in pathological conditions, is the paracellular pathway, which involves the transit of particles through the gaps between ECs (McVerry and Garcia 2004). The pulmonary vascular and microvascular network is extremely extensive and is composed of non-fenestrated type of endothelium and is the major hub of fluid and solute exchange. Prolonged disruption of this homeostatic exchange is an indication of inflammatory pathological conditions such as acute lung injury and sepsis (Snider et al. 2010; Donati and Bruni 2006; Belvitch and Dudek 2012; Bazzoni and Dejana 2004; English et al. 2001).

The class of cell membrane components termed the sphingolipids (Schuchardt et al. 2011), aptly named in 1884 after the mysterious Sphinx of Greek mythology, is recognized as integral components of cellular function and modulators of EC function. This chapter focuses on the role of sphingolipids as critical integral participants in numerous vascular physiological and biological responses (Hla and Brinkmann 2011; Snider et al. 2010) with a particular focus on lung vascular barrier regulation (Lucke and Levkau 2010). The cumulative effects of these actions are the reduction of barrier permeability resulting in decreased fluid collection, diminished inflammatory cell migration, and vascular cellular disruption, ultimately leading to enhancement of barrier integrity. We outline the molecular mechanisms of both homeostatic and pathobiologic lung vascular barrier regulation and barrier restoration processes that occur in response to the bioactive sphingolipid growth factor, sphingosine-1-phosphate (S1P), and ligation of the family of S1P G-protein-coupled receptors and the translation of this information into novel barrier-modulatory therapeutic strategies (Lucke and Levkau 2010; Donati and Bruni 2006; Belvitch and Dudek 2012; Bazzoni and Dejana 2004; Bode et al. 2010; Boguslawski et al. 2002; Rosen et al. 2007).

2 Sphingosine-1-Phosphate Biosynthesis in Lung Endothelium

As noted elsewhere in this book, over 300 different sphingolipids have been identified, each having a unique head group and a common long chain sphingoid base backbone of ceramide. Sphingosine-1-phosphate (S1P) is a sphingolipid resulting from the phosphorylation of sphingosine, a product of sphingomyelinase catabolism of sphingomyelin, catalyzed by sphingosine kinase (SphK) (Fig. 1). The catabolism of ceramide also may lead to the production of S1P (Schuchardt et al. 2011; Lucke and Levkau 2010). Various enzymes are involved in this process of S1P genesis and may take one of two routes: de novo generation or via a salvage pathway. The endoplasmic reticulum is the main site for the de novo pathway and involves various intermediates such as serine, palmitoyl coenzyme A, and fatty acids, leading to the formation of

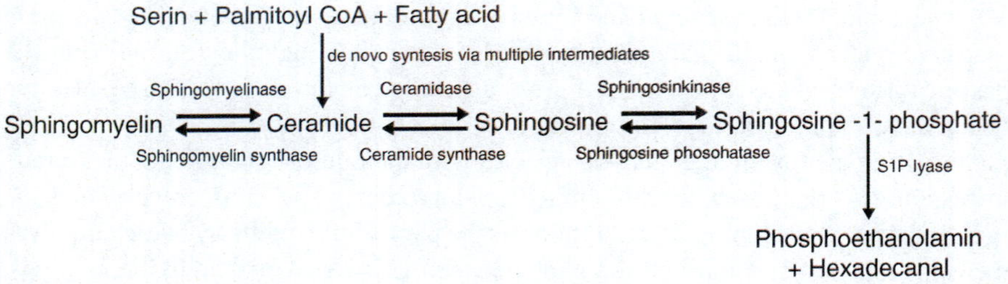

Fig. 1 Ceramide is either formed de novo from serine, palmitoyl coA, and fatty acid or via breakdown of membrane sphingomyelin. Ceramide is further converted to sphingosine, which can be phosphorylated to generate S1P. Degradation of S1P could be reversible by dephosphorylation or irreversible by S1P lyase. S1P, sphingosine-1-phosphate [Modified from Schuchardt et al. (2011)]

ceramide. The remaining steps ultimately leading to S1P production occur in the Golgi apparatus and involve the phosphorylation of sphingosine by SphK which exists in two isoforms (SphK1, SphK2) exhibiting a similar structure but distinct catalytic properties, cellular distribution, and expression traits (Schuchardt et al. 2011; Lucke and Levkau 2010; Waeber et al. 2004). The salvage pathway involves the recycling of sphingolipids and ceramide by catabolic actions of glucocerebrosidases and sphingomyelinases and the action of ceramidases that breakdown ceramides to sphingosine and finally the formation of S1P, which then exerts its influence through the S1P receptors (Schuchardt et al. 2011; Snider et al. 2010). All cells in the body have the ability to produce S1P during the course of sphingomyelin metabolism; however, erythrocytes and ECs are very prominent source of S1P present in plasma (Fig. 2). The transfer of the S1P from the intracellular environment of its genesis to the extracellular environment of its action is facilitated by ATP-binding cassette (ABC)-type transporters, although additional mechanisms must exist as experiments in mice lacking these transporters showed unchanged levels of S1P in the plasma. The concentration of S1P in the plasma ranges between 200 and 1,000 nM, primarily bound to HDL but also to albumin, LDL, and VLDL. Even though the total quantity of S1P in the plasma is greater than the concentration of its receptors, the biologically active available fraction of S1P is only about 1–2 %. This indicates that the major portion of the S1P in the plasma is present as a buffered form or in a condensed form, explaining the enormous capacity of the HDL to transport S1P in the plasma, buffer it, and to carry away excess S1P produced at sites of inflammation. This relationship is also involved in many HDL–S1P biological activities such as NO-induced vasodilatation, antioxidant, anti-apoptotic, angiogenic, and anti-inflammatory responses. A likely explanation is that the HDL transports S1P such that only a limited portion of its cargo is exposed to S1P receptors, potentially intricate process involving HDL engagement with its own receptors before S1P docking with S1P receptors. This mechanism is supported by the observation that experimental exogenous loading of HDL with S1P enhances inhibition of apoptosis in EC induced by oxidized LDLs (Schuchardt et al. 2011; Lucke and Levkau 2010; Hla and Brinkmann 2011; Donati and Bruni 2006; Zhao et al. 2007; Berdyshev et al. 2011; Venkataraman et al. 2006;

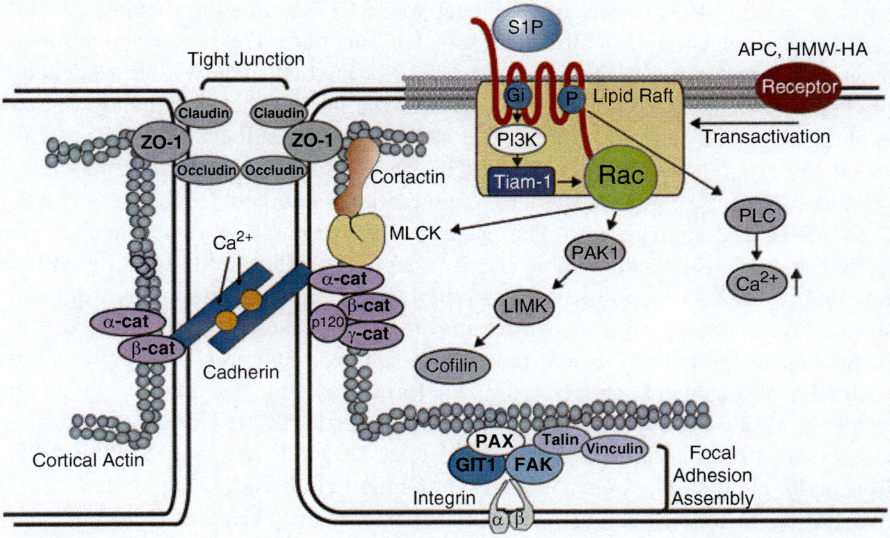

Fig. 2 Regulation of vascular permeability by S1P/ SIP1 signaling. Binding of S1P to the SIP_1 receptor stimulates the G_i-dependent recruitment of PI3 kinase, Tiam1, and Rac1 to lipid rafts (CEM), which serves to activate Rac1 in a G_i-PI3K-Tiam1-dependent manner. In addition, S1P induces an increase in intracellular Ca^{2+} concentration via a G_i-PLC pathway with additional activation of Rac1. After the activation of Rac1, S1P induces a series of profound events including adherens junction and tight junction assembly, cytoskeletal reorganization, and formation of focal adhesions that combine to enhance vascular barrier function. Furthermore, the transactivation of $S1P_1$ signaling by other barrier-enhancing agents is recently recognized as a common mechanism for promoting endothelial barrier function. *TJ* tight junction, *AJ* adherens junction, *S1P* sphingosine-1-phosphate, *SIP1* sphingosine-1-phosphate receptor 1, *PI3K* phosphoinositide 3-kinase, *Tiam 1* t-lymphoma invasion and metastasis gene 1, *Rac1* Rho family of GTPase Rac1, *PAK1* p21-activated protein kinase 1, *LIMK* LIM kinase, *PLC* phospholipase C, *ZO-1* zona occluden protein-1, *nmMLCK* non-muscle myosin light-chain kinase, *VE-Cad* vascular endothelial cadherin, *a-Cat* a-Catenin, *β-Cat* β-Catenin, *Vin* vinculin, *Pax* paxillin, *FAK* focal adhesion kinase, *GIT2* G-protein-coupled receptor kinase interactor-1, *ECM* extracellular matrix, *APC* activated protein C, *HMW-HA* high molecular weight hyaluronan [Modified from Wang and Dudek (2009)]

Yang et al. 1999; Yatomi et al. 1995; Waeber et al. 2004; Boguslawski et al. 2002; Saba and Hla 2004).

3 S1P Signal Transduction via Lung Endothelial S1P Receptors and Rho Family GTPases

The neglected status of S1P changed with the discovery and study of the five G-protein-coupled S1P cell surface receptors (S1PR1–S1PR5), formerly known as endothelial differentiation gene or *Edg* receptors, resulting in the appreciation of S1P as a multifunctional bioactive signaling lipid molecule. Although in mouse embryonic fibroblasts devoid of S1P receptors, S1P continued to mediate growth and survival, the majority of S1P-associated functions are performed through receptor-mediated signaling. The various S1PRs interact mainly with α subunits of G proteins: Gi, Gq, and G 12/13 to initiate signaling associated with S1P (Snider et al. 2010; Garcia et al. 2001; Waeber et al. 2004; Rosenfeldt et al. 2003; Lee et al. 1998; Liu et al. 2000; An et al. 2000; Ohmori et al. 2001). These receptors have prominent effects on the vasculature, promoting endothelial cell mitogenesis, chemotaxis, and angiogenesis. The primary S1P receptors expressed in ECs are S1PR1 and S1PR3, which exhibit distinct coupling to Rho family GTPases (Rho, Rac, and Cdc42), a group of regulatory molecules that channel signals from cell surface receptors to their downstream effectors including actin cytoskeletal components. Rac activity is associated with adherens junctional assembly, cytoskeletal rearrangement, and lamellipodia formation (Lucke and Levkau 2010; Snider et al. 2010; Garcia et al. 2001; Berdyshev et al. 2011; Gosens et al. 2004; Lee et al. 1996, 1998, 1999; Birukova et al. 2004; Rosen and Goetzl 2005; Rosenfeldt et al. 2003; Amano et al. 1996; Ohmori et al. 2001). We demonstrated that S1PR1 is the critical S1P receptor for barrier enhancement (McVerry and Garcia 2004; Sun et al. 2009) mediated by the small GTPase Rac1, inducing a signaling pathway that leads to cytoskeletal rearrangement with increased cortical peripheral actin resulting in increased EC junctional integrity and focal adhesion strength (Garcia et al. 2001; Adyshev et al. 2011; Wang and Dudek 2009; Lee et al. 1996; Singleton et al. 2005) (Fig. 3). Ligation of S1PR1 increases Rac GTPase activity resulting in increased endoplasmic reticulum-derived cytosolic calcium (Worthylake et al. 2001), processes critical to S1P-mediated vascular effects. S1P activates Rac GTPase in pertussis toxin-sensitive fashion to aid in the enhancement of vascular integrity via the construction of adherens junctions and cytoskeletal modifications and lamellipodia formation (Fig. 3). Microinjection of dominative-negative Rac into EC reduces S1P-induced VE-cadherin and β-catenin accumulation at the cell–cell junctions, translocation of cortactin, and the polymerization of cortical actin. Overexpression of activated Rac GTPase institutes changes in the cortical actin similar to those produced by S1P ligation of S1PR1 (Adyshev et al. 2011; Ebnet et al. 2000; Wojciak-Stothard et al. 2006; Corada et al. 1999; Carmeliet et al. 1999; Mitra et al. 2005; Owen et al. 2007; Chae et al. 2004; Waterman-Storer et al. 1999). In addition to lamellipodia, there is increased actin polymerization at the cell periphery (i.e., the cortical actin ring) that occurs with increased force driven by the actin-binding proteins, cortactin, and nmMLCK, which also translocate to this spatially defined region. Like lamellipodia formation, Rac GTPase-dependent increases in cortical actin follow, after exposure to multiple EC barrier-enhancing or maintaining functions, with cortactin directly interacting with nmMLCK, an association which is increased by p60 Src tyrosine phosphorylation of either cortactin or nmMLCK (Parizi et al. 2000). Rac activation is in conjunction with Akt-mediated phosphorylation events known to be involved in EC proliferation and migration (Singleton et al. 2007) and EC barrier enhancement (McVerry and Garcia 2004; Krump-Konvalinkova et al. 2005; Mehta et al. 2005; Arce et al. 2008; Birukova et al. 2004, 2007; Ryu et al. 2002).

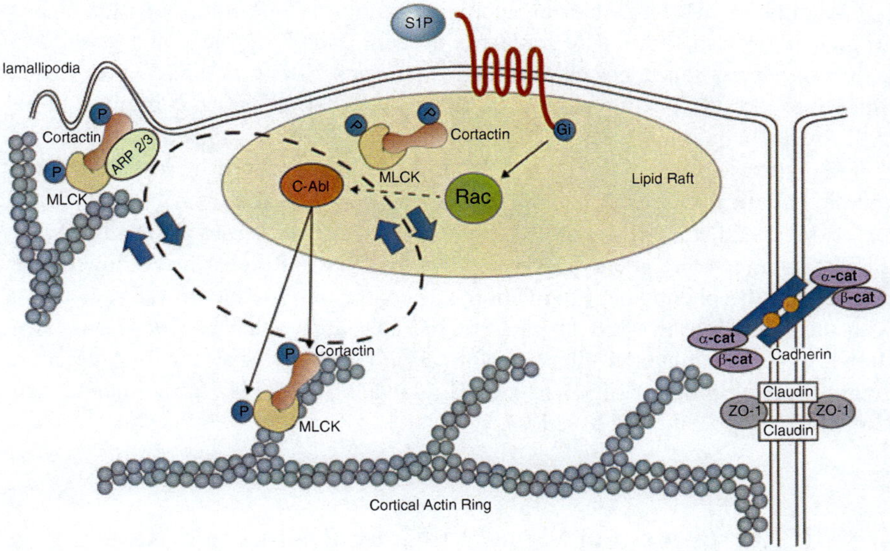

Fig. 3 S1P regulates enhanced EC barrier function. Ligation of the S1PR1 G_i protein-coupled receptor by S1P rapidly (within 1–5 min) activates Rac and recruits signaling molecules and cytoskeletal effectors such as c-Abl, cortactin, and nmMLCK to lipid rafts (or CEMs). Tyrosine phosphorylation of these molecules is observed both in lipid rafts and at the EC periphery in association with cortical actin and lamellipodia formation. This activated complex likely interacts with Arp 2/3 machinery to produce lamellipodia protrusion at the cell periphery, which serves to increase overlap between adjacent ECs. The initiation and precise sequence of events responsible for these protein movements are unclear, but within 5 min after S1P stimulation, these proteins are found simultaneously distributed in lipid rafts, cortical actin structures, and peripheral membrane ruffling/lamellipodia (indicated by the *bidirectional circle*). S1P also induces adherens junction (AJ) and tight junction (TJ) assembly that serve to further strengthen the endothelial barrier. Multiple other signaling and cytoskeletal effector molecules participate in this process as reviewed elsewhere (Wang and Dudek 2009). *MLCK* non-muscle myosin light-chain kinase, *VE-cad* vascular endothelial cadherin, *ZO-1* zona occluden protein-1 [Modified from Belvitch and Dudek (2012)]

The cortical cytoskeletal rearrangement produced by the S1P involves specific signaling sequences that bind to the p21-associated Ser/Thr kinase (PAK), an important downstream Rac target (Shikata et al. 2003b; Arce et al. 2008), as its binding to Rac results in the phosphorylation and activation of LIM kinase and the subsequent inactivation of the LIM kinase target, cofilin (Shikata et al. 2003b; Arce et al. 2008). Cofilin is an actin-binding protein with actin-severing capabilities leading to actin disassembly, events resulting in EC barrier enhancement. PAK and cofilin allow polymerization–depolymerization cycling to occur and thus facilitate rearrangement of actin from primarily transcytoplasmic to primarily cortical in a spatially distinct organization as a cortical actin cellular ring, findings which are integral to EC barrier function (Schuchardt et al. 2011). Transfection of EC with PAK-1 dominant negative construct significantly reduces S1P-mediated increases in the cortical restructuring, whereas adenoviral-mediated cofilin overexpression prominently attenuates the barrier stabilizing actions of S1P. Rac activation is also critical to the translocation of cortactin, an actin-binding protein that stimulates actin polymerization and stabilizes actin filaments. (Belvitch and Dudek 2012; Shikata et al. 2003b; Singleton et al. 2005; Mitra et al. 2005; Narumiya et al. 1997) And via binding to MLCK localizes to the site of cortical actin polymerization (Belvitch and Dudek 2012; Sun et al. 2009). S1P induced increases in trans-endothelial electrical resistance (TER) responses, a reflection of barrier integrity, are also reduced when Rac expression is experimentally reduced using siRNA. Of the remaining two members the Rho family of GTPases, Rho has been shown to be involved in the phenotypic modulation and contraction in smooth muscles along with the formation of stress fibers and focal adhesions, whereas Cdc42 has been shown to be a regulator in filopodia formation (Adyshev et al. 2011; Arce et al. 2008; Wojciak-Stothard et al. 2006; Birukova et al. 2004; Wysolmerski and Lagunoff 1990; Ohmori et al. 2001).

4 S1P Recruitment of Signaling Molecules to Lipid Rafts

Lipid rafts are a complex aggregation of various proteins and lipids with sphingolipids and cholesterol embedded within the plasma membrane. Integral components of the lipid rafts include the ganglioside GM1 and caveolin-1 that are directly involved in lipid raft-mediated cell signaling and in S1P-directed vascular barrier regulation as disruption of the liquid-ordered phase of lipid rafts results in inhibition of S1P-mediated barrier enhancement. The dependence of S1P on raft-mediated signaling to maintain and enhance vascular barrier integrity was confirmed through two-dimensional gel electrophoresis (2-DE) immunoblots of phosphotyrosine proteins that demonstrated an attenuation in levels of phosphotyrosine induced by S1P when the membrane raft formation is interrupted. Mass spectrometry identified over 200 proteins in membrane rafts with S1P inducing recruitment of >20 barrier-regulatory phosphotyrosine proteins established such as focal adhesion kinase (FAK), cortactin, p85α phosphatidylinositol 3-kinase (p85α PI3K), myosin light-chain kinase (nmMLCK), filamin A/C, and the non-receptor tyrosine kinase, c-Abl (McVerry et al. 2004). In addition, S1P-induced signaling in human lung EC caveolin-rich microdomains (CEMs) identified additional upstream effectors that contribute to the barrier regulatory properties including Rac, PAK, cofilin, S1P1, S1P3, PI3 kinase catalytic subunits p110α/β, Tiam1 (T-cell lymphoma invasion and metastasis-inducing protein 1), and α-actinin 1/4 to CEMs (Quadri et al. 2003; Vestweber 2008; Venkiteswaran et al. 2002; Parizi et al. 2000; Spiegel and Milstien 2003; Rosenfeldt et al. 2003). Experimental reduction or inhibition of each upstream effector components derails the PI3 kinase-Tiam-Rac1 pathway and impairs S1P-induced cytoskeletal rearrangement imperative for barrier enhancement. Disruption of the CEM using methyl-β-cyclodextrin (MβCD), a cholesterol exhausting agent, hampers the recruitment of S1PR and PI3 kinase p110α/β by S1P to CEMs and negatively impacts EC barrier enhancement. Similarly, reductions in expression of S1PR1, Tiam1, or PI3k all yield the same negative results on barrier dynamics. Finally, advanced studies employing quantitative proteomic analysis (iTRAQ) revealed additional proteins involved with S1P in barrier maintenance such as myristoylated alanine-rich protein kinase C substrate (MARCKS) and MARCKS-related protein (MRP) whose silencing also attenuates S1P-mediated EC barrier enhancement. Endoplasmic reticulum Ca^{2+} also plays a vital role in the S1P-mediated Rac activation. S1P produces an increase in the intracellular Ca^{2+} concentration via the Gβ-dependent pathway, which can be blocked by inhibition of Gβ, phospholipase C (PLC), or inositol triphosphate receptors, thus preventing the increase in Ca^{2+}, Rac activation, adherens junction assembly, and ultimately barrier enhancement (Mehta et al. 2005; Singleton et al. 2005; Brinkmann et al. 2002; Argraves et al. 2004; Rosen et al. 2007).

5 S1PR1 Signaling to the Lung Endothelial Cytoskeleton and Restoration of Vascular Integrity

As noted above, we demonstrated that S1PR1 is the critical S1P receptor for barrier enhancement (Berdyshev et al. 2011; Adyshev et al. 2011; Arce et al. 2008) mediated by Rac1 GTPase-mediated signaling pathway, leading to cytoskeletal rearrangement with increased cortical peripheral actin resulting in increased EC junctional integrity and focal adhesion strength (Berdyshev et al. 2011; Adyshev et al. 2011; Arce et al. 2008; Birukova et al. 2004) (Fig. 3). Consistent with the conceptual framework that barrier regulation is intimately linked to the cytoskeleton, changes in the actin cytoskeleton were essential for S1P-mediated barrier enhancement as cytochalasin B, an actin depolymerizing agent, and latrunculin B, which inhibits actin polymerization; each prevent the barrier-enhancing effects of S1P. While increases in MLC phosphorylation within stress fibers are critical to barrier disruption (Fig. 4), MLC phosphorylation is also a key element in S1P-mediated barrier enhancement and occurs in a peripheral distribution within the cortical actin ring (Schuchardt et al. 2011), providing strength to this spatially directed scaffolding force and enhancing cell–cell tethering as we described via atomic force microscopy (Belvitch and Dudek 2012). Immunofluorescence studies demonstrated that overexpressed GFP-nmMLCK distributes along cytoplasmic actin fibers, but rapidly translocates to the cortical regions of the cell after S1P treatment, rapidly catalyzing MLC phosphorylation. In addition, confocal microscopy studies show EC challenged with S1P demonstrates colocalization of nmMLCK with the key actin-binding and EC barrier-regulatory protein, cortactin (Belvitch and Dudek 2012; Garcia et al. 2001; Sammani et al. 2010) (Fig. 4). Cortactin is involved in stimulating actin polymerization (Owen et al. 2007) and cortical actin rearrangement (Belvitch and Dudek 2012), and tyrosine phosphorylation of cortactin is seen after stimuli which cause cytoskeletal rearrangement (Belvitch and Dudek 2012). The C-terminal SH3 region of phosphorylated cortactin directly interacts with nmMLCK at higher rates than non-phosphorylated cortactin,

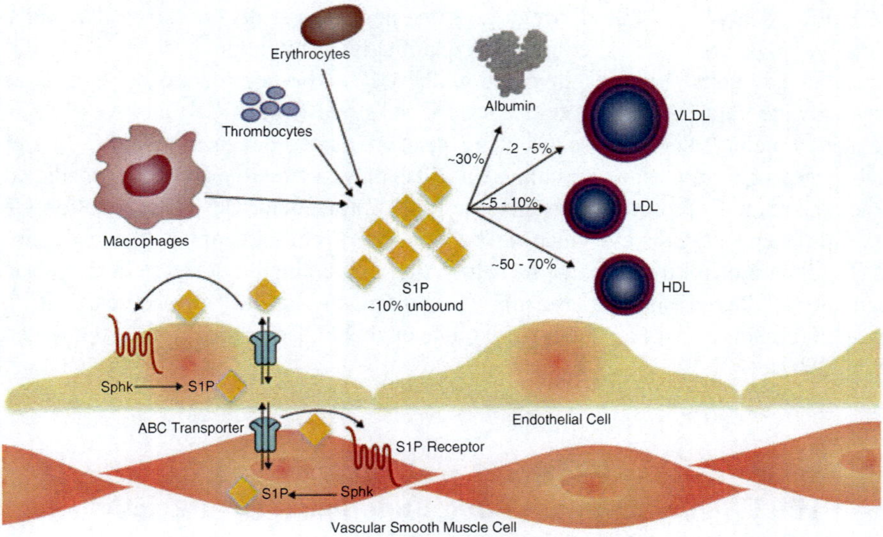

Fig. 4 Secretion of S1P by erythrocytes, platelets, macrophages, and endothelium. Once secreted, most of the S1P is uptaken by serum albumin or various serum lipoproteins. Intracellular-produced S1P in ECs or vascular smooth muscle cells could be transported across the membrane by ABC transporters. *HDL* high-density lipoprotein, *LDL* low-density lipoprotein, *Sphk* sphingosine kinase, *S1P* sphingosine-1-phosphate, *VLDL* very low-density lipoprotein [Modified from Belvitch and Dudek (2012)]

and the interaction of cortactin and nmMLCK decreases cortactin-stimulated actin polymerization (Belvitch and Dudek 2012; Sun et al. 2009; Garcia et al. 2001; Sammani et al. 2010; Shikata et al. 2003b; Singleton et al. 2005) and is essential to S1P barrier protection. A cortactin-blocking peptide (CBP), which competitively blocks the cortactin SH3 site and nmMLCK interaction, did not affect S1P-induced cortactin translocation or cortical actin ring formation, but significantly attenuated S1P-induced barrier enhancement. Immunofluorescence showed that S1P as well as other barrier-enhancing agents (such as HGF) produced rapid translocation of cortactin to the EC periphery, an effect not seen when EC are treated with the barrier-disrupting agent, thrombin (Narumiya et al. 1997). Thus, tyrosine phosphorylation of cortactin is not necessary for peripheral translocation of cortactin after S1P but is necessary for S1P-induced barrier enhancement. $p60^{src}$ is not involved in this pathway, but other tyrosine kinases, such as c-abl, are likely involved (Garcia et al. 2001; Sammani et al. 2010; Shikata et al. 2003b; Singleton et al. 2005). Thus, cortactin, nmMLCK, and the intracellular location of phosporylated MLC are all critically important in the barrier enhancement induced by S1P (Wang and Dudek 2009; Vouret-Craviari et al. 2002; Xu et al. 2007; Vestweber 2008; Venkiteswaran et al. 2002; Narumiya et al. 1997; Parizi et al. 2000; Wysolmerski and Lagunoff 1990; Rosenfeldt et al. 2003; Zondag et al. 1998). Antisense oligonucleotide techniques to create depleted cortactin states result in a significant blunting of EC barrier response to S1P. Similarly studies also showed elevated TER readings in EC following overexpression of wild-type cortactin after administration of S1P. EC with a reduced expression of cortactin, when exposed to S1P, showed AFM elasticity patterns that closely resembled the elasticity patterns of unstimulated EC. It has been observed that in order to achieve peak S1P-induced barrier enhancement phosphorylation of cortactin at three essential tyrosine residues (Tyr421, Tyr466 and Tyr 482) is required. This phosphorylation is mediated through Src. When parallel studies are done in mutant strains of cortactin that lack these tyrosine residues, significantly lower TER patterns are obtained. As with cortactin, S1P-stimulated rapid displacement of nmMLCK to areas of active membrane ruffling acting in concert with the relocated cortactin combined actions essential for the optimal action of S1P. Thus, cortactin and nmMLCK play a critical role in the enhancement of vascular barrier integrity in response to S1P (Donati and Bruni 2006; Sun et al. 2009; Zhang et al. 1997).

6 S1P Signaling to Lung Endothelial Adherens Junctions and Focal Contacts

Highly specialized proteins serve as points of contact between the individual ECs and between the ECMs with three main types of junctional points of contact between the ECs: the adherens junctions (AJ or zonula adherens), tight junctions (TJ or zonula occludens), and gap junctions (GJ). AJ and TJ have very few junctional proteins in common and institute intracellular adhesion through the formation of paracellular zipper-like structures along their transmembrane adhesion sites (Bazzoni and Dejana 2004; Furuse et al. 1994; Lee et al. 2006; Ebnet et al. 2000; Mehta et al. 2002, 2005). The major structural protein of AJ in the EC is the vascular endothelial cadherin (VE-cadherin) composed of five extracellular cadherin domains and employing a Ca^2-dependent mechanism to mediate homophilic interaction between adjacent ECs. The VE-cadherin cytoplasmic tail is similar to other cadherins and avidly binds to β-catenin or plakoglobin (ɣ-catenin). β-catenin or plakoglobin actively bind to α-catenin, an actin-binding protein, firmly anchoring the AJ to the actin cytoskeleton. α-catenin can also bind α-actinin and vinculin, further adding stability to the AJ complexes. An additional non-actin-binding player that also assists in AJ stability is p120, binding to a membrane proximal domain of VE-cadherin. VE-cadherin is essential in barrier function and normal vascular development as evident from VE-cadherin knockout mice, which do not survive embryonically due to immature vascular development. In mouse model, exposure to anti-VE-cadherin antibodies results in increased pulmonary vascular permeability; similarly, overexpression of VE-cadherin lacking mutant, or the chelation of extracellular Ca^{2+} by EDTA, also yields the same results along with EC barrier disruption. Severing interaction between β-catenin and VE-cadherin negatively alters the AJ junction formation and its adhesive strength with adjacent cells and the actin cytoskeleton. Studies in human EC show a substantial increase in the levels of VE-cadherin and β-catenin at the cell–cell contact region to enhance AJ assembly and interaction. Both protein complexes are critical to maintaining EC barrier function and signal transduction from within each cell to its surrounding matrix and neighboring cells (Sun et al. 2009; Bazzoni and Dejana 2004; Furuse et al. 1994; Lee et al. 1996; Ebnet et al. 2000; Mehta et al. 2002, 2005). Co-immunoprecipitation studies revealed the increased association of VE-cadherin with FAK and paxillin in S1P-challenged EC with enhancement of VE-cadherin interaction with α-catenin and β-catenin associated with the increased formation of FAK-β-catenin protein complexes. Depletion of β-catenin using specific siRNA resulted in complete loss of S1P effects on VE-cadherin association with FAK and paxillin rearrangement. These results demonstrate that S1P-induced endothelial barrier enhancement involves β-catenin-linked adherens junction/focal adhesion interaction (Hla and Brinkmann 2011; Sun et al. 2009; Bazzoni and Dejana 2004; Furuse et al. 1994; Lee et al. 1996; Quadri and Bhattacharya 2007; Shikata et al. 2003a, b; Venkiteswaran et al. 2002; Corada et al. 1999). Experiments involving human embryonic kidney cell lines show that overexpression of S1P increases the expression values of P and E-cadherin, but not α- and β-cadherin, to promote cell–cell aggregation through a Ca^2-based mechanism. Similarly S1P silencing leads to an underexpression of both VE-cadherin and platelet endothelial cell adhesion molecule-1 (PECAM-1). Despite all this evidence regarding the important role of VE-cadherin in assisting in AJ junctional stability, a direct link between VE-cadherin and S1P-mediated barrier regulation is still vague; furthermore, recent studies have shown that VE-cadherin may not be involved in the rapid and immediate effects of S1P on the barrier regulation as shown through TER studies but may play a role in the delayed onset-sustained effects of S1P on barrier enhancement. Thus, the S1P/S1P1 signaling pathways not only plays a crucial role in the translocation of cadherin molecules and AJ assembly but also oversees the expression of integral junctional molecules (Bazzoni and Dejana 2004; Furuse et al. 1994; Gosens et al. 2004; Ebnet et al. 2000; Mehta et al. 2005; Quadri and Bhattacharya 2007; Corada et al. 1999; Carmeliet et al. 1999; Birukova et al. 2007; Narumiya et al. 1997; Vestweber et al. 2009).

The influence of S1P in relation is not limited to just AJ but also includes tight junctions (TJ). The TJ are positioned on the outer leaflets of the lateral membranes between adjacent cells. TJ regulates the movement of solutes across intercellular spaces (barrier function) and the movement of membrane proteins between the apical and basolateral domains of the plasma membrane (fence function). The TJ consists of a complex of proteins including claudins, occludins, and junctional adhesion molecules (JAM). The TJ anchors into the actin cytoskeleton through the interaction and binding of the occludins, claudins, and JAM with the zona occludens proteins (ZO-1, ZO-2, or ZO-3). Following stimulation by S1P, ZO-1 is reassigned to the lamellipodia and to the cell–cell junctions via the S1P1/Gi/Akt/Rac pathway, whereas the barrier-enhancing actions of S1P are blunted by siRNA-induced downregulation of ZO-1 expression. So like the AJ, the TJ also plays an important role in barrier regulation in association with S1P (Donati and Bruni 2006; Bazzoni and Dejana 2004; Furuse et al. 1994; Gosens et al. 2004; Ebnet et al. 2000; Mehta et al. 2005; Mitra et al. 2005; Narumiya et al. 1997).

7 S1P Signaling to Lung Endothelial Focal Adhesions

Focal adhesions (FA) are a specific set of cellular sites that help anchor cells to the underlying ECM and play an essential role in the maintenance of the endothelial monolayer by adhering the cells to their underlying substrate and providing bidirectional signaling between the ECM and the EC cytoskeleton. Focal adhesions bridge the intracellular and extracellular space and are composed of extracellular matrix proteins, transmembrane proteins, and cytoplasmic focal adhesion plaques (Snider et al. 2010; Ebnet et al. 2000). Focal adhesions facilitate communication between the actin cytoskeleton and the extracellular space. FA is an amalgamation of integrin proteins, actin-binding structural proteins such as vinculin, talin, and α-actinin, adaptor proteins such as paxillin, and focal adhesion kinases (FAK). The integrins exist as multiple α- and β-glycoprotein chains that non-covalently link in parallel arrays to form different heterodimers. These different dimers then specify the ECM-binding target such as collagen, laminin, or fibronectin. FAK is a highly conserved cytoplasmic tyrosine kinase that is involved in the engagement of integrins and the assembly of FA through the breakdown of numerous downstream signals. S1P initiates a series of rapid signaling events in the EC membrane that are realized by downstream effectors into cytoskeletal changes and variations in barrier function. FAK also participates in these events at multiple levels and, with other effector molecules, is recruited to lipid rafts or CEMs and undergo phosphorylation and activation. These CEM signaling platforms are integral for S1P-mediated barrier regulation. S1P promotes FAK phosphorylation at specific sites (Y576), leading to FA disruption and relocation toward the periphery of the cell. S1P also mediates a temporary association between G-protein-coupled receptor kinase-interacting protein 1 (GITI1) and paxillin and induces relocation of the GIT2-paxillin complex toward the cortical ring of the cell (Mehta and Malik 2006; Peng et al. 2004; Ebnet et al. 2000; Quadri and Bhattacharya 2007; Narumiya et al. 1997; Dudek et al. 2007; Yuan et al. 2005; Birukova et al. 2004; Wysolmerski and Lagunoff 1990). In contrast to S1P, thrombin induces FAK phosphorylation at specific sites (Y397, Y576, Y925), resulting in translocation of FA proteins to stress fiber ends and EC barrier disruption. Inhibition of Src with Src-specific inhibitor PP2 abolishes S1P-induced FAK phosphorylation and migration of FA proteins that is potentially unique for specific signaling pathways (Mehta et al. 2005; Shikata et al. 2003b; Singleton et al. 2005; Mitra et al. 2005; Spiegel and Milstien 2003; Miura et al. 2000; Rosenfeldt et al. 2001; Mehta and Malik 2006).

8 Transactivation of S1PR1 in Lung Endothelial Signal Transduction

Activated protein C (APC) has the ability to induce cytoskeletal changes that resemble those produced by S1P. APC itself is an anti-inflammatory protein, and its recombinant human version is used in the clinical settings of severe sepsis. APC quickly increases the phosphorylation of endothelial MLC upon the attenuation of endothelial protein C receptor (EPCR) and induces robust actin-phospho-MLC, restructuring at the periphery of the cell and at the same time reducing the formation of central stress fibers. S1P1 is also phosphorylated by APC on the threonine residue 236 through an EPCR and PI3-kinase/AKT-dependent pathway to mediate Rac1-dependent cytoskeletal changes. On the other hand, ligation of S1P1 results in a blunting of the APC-mediated barrier enhancements, especially against thrombin. All this points toward the importance of APC and EPCR in the transactivation of S1P1-mediated signaling in barrier integrity enhancement (Tani et al. 2007; Xu et al. 2007; Zhang et al. 1997; Wang et al. 2009).

Our earlier studies were the first to link the angiogenic properties of S1P and S1P-receptor ligation to vascular barrier regulation and demonstrated that physiologic doses of S1P induce EC activation, marked cytoskeletal rearrangement, and stabilization of lung EC barrier function in vitro (Belvitch and Dudek 2012; Camp et al. 2009). This novel function for S1P was of particular relevance to clinical medicine as thrombocytopenia is well known to be associated with increased vascular leak (Peng et al. 2004; Shea et al. 2010), and while the mechanism of this effect was unknown, we demonstrated that activated platelets are an important source of S1P and directly enhance barrier function via S1P1 ligation (Arce et al. 2008). Platelets contain significant levels of sphingosine kinase but reduced levels of sphingosine lyase, thereby serving as enriched sources for the barrier-promoting S1P (Arce et al. 2008). Prior to the last decade, permeability-reducing strategies primarily consisted of cAMP augmentation, producing only modest barrier enhancement (Peng et al. 2004; Arce et al. 2008; Shea et al. 2010). More recently, a number of barrier-promoting agents have been identified that share common signal transduction mechanisms that are distinct from cAMP signals and target the endothelial actin cytoskeleton to facilitate barrier-restorative processes. We have conceptualized a paradigm whereby barrier recovery after edemagenic agonists involves development of a cortical actin ring to anchor cellular junctions and a carefully choreographed (but poorly understood) gap-closing process via formation of Rac GTPase-dependent lamellipodial protrusions into the paracellular space between activated ECs. Within these lamellipodia, signals are transduced to actin-binding proteins (nmMLCK and cortactin) and phosphorylated MLCs in spatial-specific cellular locations. Lamellipodia also require formation of focal adhesions (regulated by the cytoskeleton) critical to establishment of linkage of the actin cytoskeleton to target effectors that restore cell–cell adhesion and cell–matrix adhesion (Tani et al. 2007; Wojciak-Stothard et al. 2006; Xu et al. 2007; Birukova et al. 2004; Zhang et al. 1997; Wang et al. 2009; Hla 2004; Kihara and Igarashi 2008; Bode et al. 2010; Boguslawski et al. 2002; Fyrst and Saba 2010; Futerman and Riezman 2005; English et al. 2001).

Another molecule that plays a role in the transactivation of S1P1 signaling is the glycosaminoglycan (GAG) called hyaluronan (HA). HA is composed of a high molecular weight HA (HMW-HA) and a low molecular weight HA (LMW-HA). HMW-HA has the ability to induce S1P1 phosphorylation via AKT to enhance barrier integrity. However, LMW-HA promotes Src and ROCK 1/2-mediated phosphorylation of S1P3 to cause barrier disruption. Blocking of S1P1 or S1P3 mutes the actions of HMW-HA and LMW-HA, respectively, on the endothelium in terms of barrier stability. Keeping the negative effects of S1P3 on the endothelium in mind, a novel approach toward enhancing barrier protection would be the blocking of S1P3. One such protein understudy is methylnaltrexone (MNTX), which is mu-opioid receptor antagonist that inhibits the actions of S1P3, and provides barrier protection against thrombin and LPS (Donati and Bruni 2006; McVerry and Garcia 2004; Tani et al. 2007; Xu et al. 2007; Owen et al. 2007; Singleton et al. 2006, 2007; Zhang et al. 1997; Wang et al. 2009; Hla 2004; Kihara and Igarashi 2008; Waterman-Storer et al. 1999).

9 S1PR3 Signaling to the Lung Endothelial Cytoskeleton and Loss of Vascular Integrity

Interestingly, S1P at elevated concentrations (>5 μM) results in S1PR3-dependent RhoA-mediated signaling and increased barrier permeability (Adyshev et al. 2011; Gosens et al. 2004; Narumiya et al. 1997; Birukova et al. 2004). A "conversation" elicited by ligation of G-protein-coupled receptors takes place between transmembrane components (such as large and small GTPases) and cytoskeletal proteins in membrane domains such as caveolin-enriched lipid rafts and Rac GTPase-dependent lamellipodia. These pathways induce EC cytoskeletal rearrangement resulting in enhanced junctional linkages between ECs as well as increased linkage of the cytoskeleton with the underlying extracellular matrix. These events provide the conceptual underpinning for the molecular targeting of these permeability-reducing therapeutic strategies. Barrier-restorative agonists (detailed in reference) evoke a carefully choreographed resolution of inflammation-mediated paracellular gaps by promoting the formation of lamellipodia, which protrude into the paracellular space driven by actin polymerizations with focal contacts, which reseal the gaps between activated ECs (Fig. 3). These lamellipodia contain a variety of actin-binding proteins such as the Ca^{2+}/calmodulin-dependent non-muscle myosin light-chain kinase (nmMLCK), which triggers myosin ATPase activity, actin polymerization, and EC tension development. Increases in vascular permeability must ultimately be attributed to loss or disruption of endothelial intercellular junctions, in combination with a breakdown of the tethering forces characteristic of cell–cell or cell–matrix interactions, which result in paracellular vascular leakage (Zhang et al. 1997; Jaillard et al. 2005; Bode et al. 2010; Boguslawski et al. 2002; Fyrst and Saba 2010; Futerman and Riezman 2005).

Disruption of pulmonary barrier integrity and edema are cardinal characteristics of pulmonary pathologies. Airway administration of S1P induces the pulmonary vasculature leak phenomenon by disruption of epithelial tight junctions that is mediated through the S1P3 receptors. The disruption of tight junctions is accompanied by the development of gaps in the paracellular spaces and a disintegration of cytoplasmic plaques associated with tight junctions along with other integral membrane organizer proteins. This effect can be experimentally compounded by the addition of the proinflammatory cytokine TNF. The integral role of S1P3 receptors can be gauged in S1P3-null mice as they are resistant to S1P-induced vascular leakage, indicating a probable protective role of S1P3 antagonism. This role of S1P3 is in sharp contrast to that of S1P1 receptors that enhance the stability of vascular endothelial barrier. Interestingly, IV infusion of S1P inhibits pulmonary leakage in the pulmonary endothelium, when it is exposed to thrombin or LPS. This is again quite the opposite, as the effect of S1P when introduced via IT injection on pulmonary epithelium is to induce vascular leakage leading to acute pulmonary edema. Immunoreactivity studies in mice (WT and S1P3-null mice) have shown the presence of S1P3 receptors on all pulmonary epithelial surfaces and not on the pulmonary endothelium. When these mice were exposed to S1P, disruption was mainly seen in the epithelial tight junctions, indicating that S1P-induced, and S1P3-mediated, vascular disruption was epithelial in nature. In this complex interplay, Rho signaling has been assumed to play a very integral role in the molecular regulation of tight junctional integrity. S1P3 fosters actin cytoskeleton rearrangement by the activation of Gα12 and Gα13 via Rho signaling. Rho activity has been linked to the regulation of the cellular tight junctions as well as seen in the overexpression of constitutively active RhoA which results in a disarray of the structured tight junction morphology. This effect is mediated through ROCK1 (Rho-associated, coiled-coil containing protein kinase 1) which oversees the formation of stress fibers by maintaining the active state of myosin light chains by inactivating myosin light-chain phosphatase (MLCP), and it also stimulates LIM kinases to exert their effect on cofilin, by inhibiting it, ultimately resulting in a reorganization of actin cytoskeleton (Gon et al. 2005; Gosens et al. 2004; Birukova et al. 2004; Worthylake et al. 2001; Wang et al. 2009; Hla 2004; Fyrst and Saba 2010; Ohashi et al. 2000; Amano et al. 1996).

10 S1P and Lung Angiogenesis

The interface between EC barrier regulation and angiogenesis is an exciting area of vascular biology. New blood vessel formation, or angiogenesis, is a complex process involving EC activation, migration, maturation, and remodeling. These events may occur in a variety of contexts including during normal development and growth, in response to wound healing, or as part of the pathogenesis of a number of cancers and autoimmune diseases (Birukova et al. 2004; Boguslawski et al. 2002). Our initial studies determined that S1P is the most potent EC chemotactic agent present in serum and is ultimately involved in angiogenesis and vascular hemostasis through its ability to evoke various cell-specific responses (Singleton et al. 2007; Birukova et al. 2004; Bode et al. 2010; Boguslawski et al. 2002). In the setting of coagulation, S1P is abundantly released from platelets and, via its pleiotropic effects, potentially contributes to new blood vessel formation. This is evidenced by in vivo studies that establish S1P as remarkably effective in avian chorioallantoic membranes, in Matrigel-implanted plugs in mice (McVerry et al.

2004; Bazzoni and Dejana 2004; Corada et al. 1999; Argraves et al. 2004; Liu et al. 2000; English et al. 2001) and in the avascular mouse cornea. In contrast to VEGF-induced increases in EC permeability, we were the first to report that another angiogenic factor (S1P) can also produce EC barrier restoration and enhancement. S1P strongly enhances TER across human EC monolayers and significantly attenuates thrombin-induced barrier disruption while rapidly restoring barrier integrity in the isolated perfused murine lung (McVerry and Garcia 2004; Peng et al. 2004; Sammani et al. 2010; Mathew et al. 2011; McVerry et al. 2004). A single intravenous dose of S1P, given 1 h after intratracheal endotoxin administration, produced highly significant reductions in multiple indices of inflammatory lung injury, including vascular leak, as demonstrated in both murine (Peng et al. 2004; Sammani et al. 2010; Mathew et al. 2011; McVerry et al. 2004; English et al. 2001) and canine models of ALI (Quadri et al. 2003; Paik et al. 2004; Liu et al. 2001). Furthermore, S1P is a major serum component released by platelets and represents a key mechanism by which platelets nurture the microcirculation and preserve vascular integrity (McVerry and Garcia 2004; Peng et al. 2004; Sammani et al. 2010; Mathew et al. 2011; McVerry et al. 2004; Singleton et al. 2007; Shea et al. 2010; Sanna et al. 2006; English et al. 2001).

11 In Vivo Effects of S1P in Preclinical Models of Human Lung Disease

One of the main pathophysiological mechanisms involved in the genesis of various vascular disease conditions is endothelial dysfunction. Conditions ranging from atherosclerosis, hypertension, pulmonary hypertension, and cerebrovascular disease have a basis in endothelial barrier imbalance and dysregulation. Devastating inflammatory conditions such as acute lung injury (ALI) are characterized by significant and prolonged vascular permeability. Experimentally the intratracheal administration of lipopolysaccharide (LPS) has been used to mimic the clinical presentation of ALI in murine lung models. LPS induces all the responses associated with ALI including inflammatory lung injury, thickening of the alveolar wall, neutrophilic migration into the lung interstitium, and alveolar space. However, after appropriate spacing, the delivery of S1P prominently reduces the inflammatory landscape and attenuates neutrophilic migration into the LPS-exposed lung parenchyma. Similar effects have also been achieved using a S1P analogue FTY720 in reducing inflammatory changes associated with LPS administration. Experiments on both murine and canine models have yielded promising results into the therapeutic potential of S1P in countering the devastating effects of barrier dysfunction-related pathologies of the vasculature. However, given studies implicating gene expression in the modulation of ALI, the exact manner in which S1P imparts protection may involve the reduction of inflammatory and/or innate immunity gene expression (Donati and Bruni 2006; McVerry and Garcia 2004; Sammani et al.

2010; Mathew et al. 2011; McVerry et al. 2004; Komarova et al. 2007; Mansoor and Melendez 2008; Chae et al. 2004; Brinkmann et al. 2002).

In addition to the abundant in vitro data describing the EC barrier-enhancing effect of S1P, the potential utility of S1P in restoring lung water balance in patients with inflammatory injury was underscored in studies involving small and large animal models of ALI in which S1P provided dramatic attenuation of LPS-mediated lung inflammation and permeability. Mice treated with S1P had significantly less histologic evidence of inflammatory changes/lung injury, with decreased neutrophil alveolitis on BAL and decreased lung myeloperoxidase (MPO) activity. Interestingly, mice treated with S1P after intratracheal LPS also showed an attenuated renal inflammatory response compared to control, measured by tissue MPO activity and Evans blue dye extravasation as a measure of capillary leak. S1P also protected against intrabronchial LPS-induced ALI and concomitant VILI in a canine model, with decreased shunt fraction, decreased BAL protein, decreased extravascular lung water, and improved oxygenation (Kihara and Igarashi 2008). Use of a large animal canine model allowed investigation of regional lung changes in ALI and the effect of S1P on these changes. Computed tomography (CT) scans of animals subjected to LPS/VILI found that animals treated with S1P had dramatic improvement in alveolar air content (with decreased edema) in all lung regions (Kihara and Igarashi 2008). Additional in vivo studies found that S1P protects against ventilator-induced lung injury (VILI) in a murine model as assessed by Evans blue dye extravasation (Sammani et al. 2010; Mathew et al. 2011; McVerry et al. 2004; Mansoor and Melendez 2008; Kihara and Igarashi 2008; Chae et al. 2004; Brinkmann et al. 2002; Rogers et al. 1989).

As a result, we are excited about the potential utility of FTY720, an unphosphorylated S1P analogue and a derivative of the natural immunosuppressant myriocin that has been recently described to cause peripheral lymphopenia by inhibiting cellular egress from lymphoid tissues. We demonstrated FTY to induce delayed endothelial barrier enhancement through a Gi-coupled receptor and to protect against murine inflammatory lung injury. Thus, targeting S1PR1 activation, either directly or via S1PR1 transactivation by agonists such as activated protein C and high molecular weight hyaluronan (both robustly barrier protective), or antagonism of S1PR3 as with methylnaltrexone appears to be promising strategies for attenuating the vascular leak associated with ALI (Peng et al. 2004; Sammani et al. 2010; McVerry et al. 2004; Camp et al. 2009; Yuan et al. 2005; Chae et al. 2004; Paik et al. 2004; Wu 2005; Waeber et al. 2004; Shea et al. 2010). MLC phosphorylation is also a key element in S1P-mediated barrier enhancement and occurs in a peripheral distribution within the cortical actin ring (Belvitch and Dudek 2012; Camp et al. 2009), providing strength to this spatially directed scaffolding force and enhancing cell–cell tethering as we described via atomic force microscopy. Immunofluorescence studies have also demonstrated this via the overexpression of Green Fluorescent Protein–nmMLCK at sites of barrier remodelling. (Birukova et al. 2004; Boguslawski et al. 2002). These observations serve to highlight the importance of the cellular location of cytoskeletal proteins in small and large animal models of ALI in which S1P provided dramatic attenuation of LPS-mediated lung inflammation and permeability. Mice treated with

S1P had significantly less histologic evidence of inflammatory changes/lung injury, with decreased neutrophil alveolitis on bronchoalveolar lavage (BAL) and decreased lung myeloperoxidase (MPO) activity (Peng et al. 2004; Mathew et al. 2011; McVerry et al. 2004; Camp et al. 2009; Jacobson and Garcia 2007; Birukova et al. 2004, 2007; Dudek et al. 2007; Brinkmann et al. 2002). Similarly, mice treated with S1P after intratracheal LPS also showed an attenuated renal inflammatory response compared to control, measured by tissue MPO activity and Evan's blue dye extravasation as a measure of capillary leak. S1P also protected against intrabronchial LPS-induced ALI and concomitant VILI in a canine model, with decreased shunt fraction, decreased BAL protein, decreased extravascular lung water, and improved oxygenation˙ Use of a large animal canine model allowed investigation of regional lung changes in ALI and the effect of S1P on these changes. Computed tomography scans of animals subjected to LPS/VILI found that animals treated with S1P had dramatic improvement in alveolar air content (with decreased edema) in all lung regions. Additional in vivo studies found that S1P protects against VILI in a murine model as assessed by Evan's blue dye extravasation (Peng et al. 2004; Mathew et al. 2011; McVerry et al. 2004; Camp et al. 2009; Chae et al. 2004; Paik et al. 2004; Wu 2005; Waeber et al. 2004; Shea et al. 2010; Rosen and Goetzl 2005; Rosenfeldt et al. 2003).

We have also evaluated a potential role for S1P in ameliorating lung ischemia–reperfusion (IR) injury, a common sequelae of lung transplantation, which is characterized by alveolar damage, edema, and inflammation in donor lungs and is a significant cause of transplant failure. Utilizing a rat model of IR injury (pulmonary artery ligation and reperfusion), we determined that rats pretreated with S1P exhibited reduced lung vascular permeability and inflammation compared to controls. Lung myeloperoxidase activity, an index of parenchymal leukocyte infiltration, and levels of IL-6, IL-1b, and IL-2 were also attenuated in S1P-treated animals exposed to IR injury. Together, these findings suggest that S1P may serve as an effective permeability-reducing agent in diverse conditions that share an element of lung inflammatory burden (Peng et al. 2004; Sammani et al. 2010; Mathew et al. 2011; McVerry et al. 2004; Camp et al. 2009; Chae et al. 2004; Paik et al. 2004; Wu 2005; Waeber et al. 2004; Shea et al. 2010; Sanna et al. 2006).

Clinically significant radiation-induced lung injury (RILI) is a common toxicity in patients administered thoracic radiotherapy. Although the molecular etiology is poorly understood, we previously characterized a murine model of RILI in which alterations in lung barrier integrity surfaced as a potentially important pathobiologic event and genome-wide lung gene mRNA levels identified dysregulation of sphingolipid metabolic pathway genes. We hypothesized that sphingolipid signaling components serve as modulators and novel therapeutic targets of RILI. Sphingolipid involvement in murine RILI was confirmed by radiation-induced increases in lung expression of sphingosine kinase (SphK) isoforms 1 and 2 and increases in the ratio of ceramide to sphingosine-1-phosphate (S1P) and dihydro-S1P (DHS1P) levels in plasma, bronchoalveolar lavage fluid, and lung tissue. Mice with a targeted deletion of SphK1 (SphK1(−/−)) or with reduced expression of S1P receptors (S1PR1(+/−), S1PR2(−/−), and S1PR3(−/−)) exhibited marked RILI susceptibility. Finally, studies of 3 potent vascular barrier-protective S1P analogues, FTY720, (S)-FTY720-

phosphonate (fTyS), and SEW-2871, identified significant RILI attenuation and radiation-induced gene dysregulation by the phosphonate analogue, fTyS (0.1 and 1 mg/kg i.p., 2× per week) and to a lesser degree by SEW-2871 (1 mg/kg i.p., 2× per week), compared with those in controls. These results support the targeting of S1P signaling as a novel therapeutic strategy in RILI (Peng et al. 2004; Mathew et al. 2011; McVerry et al. 2004).

S1P-induced Rac activation, and cytoskeletal rearrangement produce increased linkage of actin to VE-cadherin and β-catenin, both important AJ components, as well as S1P-induced phosphorylation of focal adhesion-related proteins paxillin and focal adhesion kinase (FAK), with translocation of these proteins to the EC periphery, further implicating S1P-induced cell–cell adhesive changes as part of the mechanism of S1P-induced barrier enhancement (Quadri et al. 2003; Worthylake et al. 2001; Zhang et al. 1997; Spiegel and Milstien 2003; Kihara and Igarashi 2008). Despite profound attractiveness as a therapeutic agent targeting the endothelium in high permeability states, S1P has several attributes which limit its potential utility as a permeability-reducing strategy. With an affinity for ligation of the S1P3 receptor (S1PR3), intratracheal S1P has been implicated as a cause of pulmonary edema via endothelial/epithelial barrier disruption. S1P also causes bradycardia via ligation of cardiac S1P3 receptors. These findings generated increased interest in FTY720, a derivative of the natural immunosuppressant myriocin and a recently described immunosuppressive agent that causes peripheral lymphopenia by inhibiting cellular egress from lymphoid tissues. FTY720 is structurally similar (but not identical) to S1P and is phosphorylated by sphingosine kinase to FTY720-phosphate, which is an agonist at S1P receptors. This characteristic prompted investigation of the effect of FTY720 on EC barrier function. FTY720 did not have superior efficacy compared to mycophenolate mofetil in preventing renal transplant rejection, but it is in Phase III clinical trials as an immunosuppressant in multiple sclerosis patients. The clinical availability of FTY720 makes it attractive as a potential mediator of EC barrier function in patients with ALI. An in vivo study demonstrated that intraperitoneal FTY720 protected against intratracheal LPS in a murine model of ALI, as measured by Evans blue dye extravasation. Like S1P, FTY720 causes increased TER measurements in pulmonary EC, an effect which is abolished by pertussis toxin (implicating G_i-coupled receptor activation), and requires the generation of signaling components within membrane lipid rafts. Interestingly, however, the mechanism of FTY720-induced EC barrier enhancement diverges from the mechanism described for S1P in several ways including the delayed kinetics of the rise in TER compared to S1P. Decreased expression of the S1P1 receptor prevented S1P-induced increase in TER but did not affect FTY720-induced TER increases. Unlike S1P, FTY720 did not result in threonine phosphorylation of the S1P1 receptor, nor did inhibition of PI3 kinase prevent FTY720-induced EC barrier enhancement. Furthermore, FTY720 did not cause the increased intracellular calcium, the MLC phosphorylation, or the cytoskeletal rearrangement seen in response to S1P (Waeber et al. 2004). Downregulation of Rac or cortactin using siRNAs attenuated the barrier-enhancing effect of S1P, but not that of FTY720. Although FTY720 is an S1P receptor agonist, its mechanism of barrier enhancement is distinct from that of

S1P and does not require the S1P1 receptor (Camp et al. 2009; Dudek et al. 2007; Chae et al. 2004; Paik et al. 2004; Wu 2005; Waeber et al. 2004; Shea et al. 2010; Liu et al. 2001).

12 Summary

Sphingosine-1-phosphate is a biologically active lipid growth factor secreted by erythrocytes, activated platelets, and other cells including endothelium that is integral to numerous vascular biological mechanisms ranging from cell differentiation, proliferation, motility, angiogenesis, and barrier regulation. S1P is the natural ligand for five G-protein-coupled receptors and is an integral intracellular secondary messenger exhibiting an intrinsically protective role in the pulmonary vasculature ameliorating agonist- or sepsis-induced pulmonary injury and vascular leakage. These vasculo-protective mechanisms involve an intricate interplay of S1P with other factors (such as MAPKS, ROCKs, Rho, Rac1) with rearrangement of the endothelial cytoskeleton to form strong cortical actin rings in the cell periphery and enhanced cell-to-cell and cell-to-matrix tethering dynamics. This cascade leads to reinforcement of focal adhesions and paracellular junctional complexes via cadherin, paxillin, catenins, and zona occludens. S1P through its interaction with Rac and Rho influence the cytoskeletal rearrangement indicated in the later stages of angiogenesis as a stabilizing force, preventing excessive vascular permeability. These properties translate into a therapeutic potential for S1P, as shown by its analogue FTY720, as a protective agent limiting the disruption of the vascular EC monolayer in the pulmonary microcirculation that results in accumulation of protein and inflammatory cell-filled fluid in the interstitial and alveolar compartments leading to pulmonary edema and ultimately respiratory failure.

References

Adyshev DM, Moldobaeva NK, Elangovan VR, Garcia JG, Dudek SM (2011) Differential involvement of ezrin/radixin/moesin proteins in sphingosine 1-phosphate-induced human pulmonary endothelial cell barrier enhancement. Cell Signal 23(12):2086–2096

Amano M, Ito M, Kimura K, Fukata Y, Chihara K, Nakano T, Matsuura Y, Kaibuchi K (1996) Phosphorylation and activation of myosin by Rho-associated kinase (Rho-kinase). J Biol Chem 271(34):20246–20249

An S, Zheng Y, Bleu T (2000) Sphingosine 1-phosphate-induced cell proliferation, survival, and related signaling events mediated by G protein-coupled receptors Edg3 and Edg5. J Biol Chem 275(1):288–296

Arce FT, Whitlock JL, Birukova AA, Birukov KG, Arnsdorf MF, Lal R, Garcia JG, Dudek SM (2008) Regulation of the micromechanical properties of pulmonary endothelium by S1P and thrombin: role of cortactin. Biophys J 95(2):886–894

Argraves KM, Wilkerson BA, Argraves WS, Fleming PA, Obeid LM, Drake CJ (2004) Sphingosine-1-phosphate signaling promotes critical migratory events in vasculogenesis. J Biol Chem 279(48): 50580–50590

Bazzoni G, Dejana E (2004) Endothelial cell-to-cell junctions: molecular organization and role in vascular homeostasis. Review. Physiol Rev 84(3):869–901

Belvitch P, Dudek SM (2012) Role of FAK in S1P-regulated endothelial permeability. Microvasc Res 83(1):22–30

Berdyshev EV, Gorshkova I, Usatyuk P, Kalari S, Zhao Y, Pyne NJ, Pyne S, Sabbadini RA, Garcia JG, Natarajan V (2011) Intracellular S1P generation is essential for S1P-induced motility of human lung endothelial cells: role of sphingosine kinase 1 and S1P lyase. PLoS One 6(1):e16571

Birukova AA, Smurova K, Birukov KG, Usatyuk P, Liu F, Kaibuchi K, Ricks-Cord A, Natarajan V, Alieva I, Garcia JG, Verin AD (2004) Microtubule disassembly induces cytoskeletal remodeling and lung vascular barrier dysfunction: role of Rho-dependent mechanisms. J Cell Physiol 201(1):55–70

Birukova AA, Malyukova I, Poroyko V, Birukov KG (2007) Paxillin-beta-catenin interactions are involved in Rac/Cdc42-mediated endothelial barrier-protective response to oxidized phospholipids. Am J Physiol Lung Cell Mol Physiol 293(1):L199–L211

Bode C, Sensken SC, Peest U, Beutel G, Thol F, Levkau B, Li Z, Bittman R, Huang T, Tölle M, van der Giet M, Gräler MH (2010) Erythrocytes serve as a reservoir for cellular and extracellular sphingosine 1-phosphate. J Cell Biochem 109(6):1232–1243

Boguslawski G, Grogg JR, Welch Z, Ciechanowicz S, Sliva D, Kovala AT, McGlynn P, Brindley DN, Rhoades RA, English D (2002) Migration of vascular smooth muscle cells induced by sphingosine 1-phosphate and related lipids: potential role in the angiogenic response. Exp Cell Res 274(2):264–274

Brinkmann V, Davis MD, Heise CE, Albert R, Cottens S, Hof R, Bruns C, Prieschl E, Baumruker T, Hiestand P, Foster CA, Zollinger M, Lynch KR (2002) The immune modulator FTY720 targets sphingosine 1-phosphate receptors. J Biol Chem 277(24):21453–21457

Camp SM, Bittman R, Chiang ET, Moreno-Vinasco L, Mirzapoiazova T, Sammani S, Lu X, Sun C, Harbeck M, Roe M, Natarajan V, Garcia JG, Dudek SM (2009) Synthetic analogs of FTY720 [2-amino-2-(2-[4-octylphenyl]ethyl)-1,3-propanediol] differentially regulate pulmonary vascular permeability in vivo and in vitro. J Pharmacol Exp Ther 331(1):54–64

Carmeliet P, Lampugnani MG, Moons L, Breviario F, Compernolle V, Bono F, Balconi G, Spagnuolo R, Oosthuyse B, Dewerchin M, Zanetti A, Angellilo A, Mattot V, Nuyens D, Lutgens E, Clotman F, de Ruiter MC, Gittenberger-de Groot A, Poelmann R, Lupu F, Herbert JM, Collen D, Dejana E (1999) Targeted deficiency or cytosolic truncation of the VE-cadherin gene in mice impairs VEGF-mediated endothelial survival and angiogenesis. Cell 98(2): 147–157

Chae SS, Proia RL, Hla T (2004) Constitutive expression of the S1P1 receptor in adult tissues. Prostaglandins Other Lipid Mediat 73(1–2):141–150

Corada M, Mariotti M, Thurston G, Smith K, Kunkel R, Brockhaus M, Lampugnani MG, Martin-Padura I, Stoppacciaro A, Ruco L, McDonald DM, Ward PA, Dejana E (1999) Vascular endothelial-cadherin is an important determinant of microvascular integrity in vivo. Proc Natl Acad Sci USA 96(17):9815–9820

Donati C, Bruni P (2006) Sphingosine 1-phosphate regulates cytoskeleton dynamics: implications in its biological response. Biochim Biophys Acta 1758(12):2037–2048

Dudek SM, Camp SM, Chiang ET, Singleton PA, Usatyuk PV, Zhao Y, Natarajan V, Garcia JG (2007) Pulmonary endothelial cell barrier enhancement by FTY720 does not require the S1P1 receptor. Cell Signal 19(8):1754–1764

Ebnet K, Schulz CU, Meyer Zu Brickwedde MK, Pendl GG, Vestweber D (2000) Junctional adhesion molecule interacts with the PDZ domain-containing proteins AF-6 and ZO-1. J Biol Chem 275(36):27979–27988

English D, Garcia JG, Brindley DN (2001) Platelet-released phospholipids link haemostasis and angiogenesis. Review. Cardiovasc Res 49(3):588–599

Furuse M, Itoh M, Hirase T, Nagafuchi A, Yonemura S, Tsukita S, Tsukita S (1994) Direct association of occludin with ZO-1 and its possible involvement in the localization of occludin at tight junctions. J Cell Biol 127(6 Pt 1):1617–1626

Futerman AH, Riezman H (2005) The ins and outs of sphingolipid synthesis. Review. Trends Cell Biol 15(6):312–318

Fyrst H, Saba JD (2010) An update on sphingosine-1-phosphate and other sphingolipid mediators. Review. Nat Chem Biol 6(7):489–497, Erratum in Nat Chem Biol. 2010;6(9):689

Garcia JG, Liu F, Verin AD, Birukova A, Dechert MA, Gerthoffer WT, Bamberg JR, English D (2001) Sphingosine 1-phosphate promotes endothelial cell barrier integrity by Edg-dependent cytoskeletal rearrangement. J Clin Invest 108(5):689–701

Gon Y, Wood MR, Kiosses WB, Jo E, Sanna MG, Chun J, Rosen H (2005) S1P3 receptor-induced reorganization of epithelial tight junctions compromises lung barrier integrity and is potentiated by TNF. Proc Natl Acad Sci USA 102(26):9270–9275

Gosens R, Schaafsma D, Meurs H, Zaagsma J, Nelemans SA (2004) Role of Rho-kinase in maintaining airway smooth muscle contractile phenotype. Eur J Pharmacol 483(1):71–78

Hla T (2004) Physiological and pathological actions of sphingosine 1-phosphate. Review. Semin Cell Dev Biol 15(5):513–520

Hla T, Brinkmann V (2011) Sphingosine 1-phosphate (S1P): physiology and the effects of S1P receptor modulation. Neurology 76(8 Suppl 3):S3–S8

Jacobson JR, Garcia JG (2007) Novel therapies for microvascular permeability in sepsis. Review. Curr Drug Targets 8(4):509–514

Jaillard C, Harrison S, Stankoff B, Aigrot MS, Calver AR, Duddy G, Walsh FS, Pangalos MN, Arimura N, Kaibuchi K, Zalc B, Lubetzki C (2005) Edg8/S1P5: an oligodendroglial receptor with dual function on process retraction and cell survival. J Neurosci 25(6):1459–1469

Kihara A, Igarashi Y (2008) Production and release of sphingosine 1-phosphate and the phosphorylated form of the immunomodulator FTY720. Review. Biochim Biophys Acta 1781(9):496–502

Komarova YA, Mehta D, Malik AB (2007) Dual regulation of endothelial junctional permeability. Review. Sci STKE 412:re8

Krump-Konvalinkova V, Yasuda S, Rubic T, Makarova N, Mages J, Erl W, Vosseler C, Kirkpatrick CJ, Tigyi G, Siess W (2005) Stable knock-down of the sphingosine 1-phosphate receptor S1P1 influences multiple functions of human endothelial cells. Arterioscler Thromb Vasc Biol 25(3):546–52

Lee MJ, Evans M, Hla T (1996) The inducible G protein-coupled receptor edg-1 signals via the G(i)/mitogen-activated protein kinase pathway. J Biol Chem 271(19):11272–11279

Lee MJ, Van Brocklyn JR, Thangada S, Liu CH, Hand AR, Menzeleev R, Spiegel S, Hla T (1998) Sphingosine-1-phosphate as a ligand for the G protein-coupled receptor EDG-1. Science 279(5356):1552–1555

Lee MJ, Thangada S, Claffey KP, Ancellin N, Liu CH, Kluk M, Volpi M, Sha'afi RI, Hla T (1999) Vascular endothelial cell adherens junction assembly and morphogenesis induced by sphingosine-1-phosphate. Cell 99(3):301–312

Lee JF, Zeng Q, Ozaki H, Wang L, Hand AR, Hla T, Wang E, Lee MJ (2006) Dual roles of tight junction-associated protein, zonula occludens-1, in sphingosine 1-phosphate-mediated endothelial chemotaxis and barrier integrity. J Biol Chem 281(39):29190–29200

Liu Y, Wada R, Yamashita T, Mi Y, Deng CX, Hobson JP, Rosenfeldt HM, Nava VE, Chae SS, Lee MJ, Liu CH, Hla T, Spiegel S, Proia RL (2000) Edg-1, the G protein-coupled receptor for sphingosine-1-phosphate, is essential for vascular maturation. J Clin Invest 106(8):951–961

Liu F, Verin AD, Wang P, Day R, Wersto RP, Chrest FJ, English DK, Garcia JG (2001) Differential regulation of sphingosine-1-phosphate- and VEGF-induced endothelial cell chemotaxis. Involvement of G(ialpha2)-linked Rho kinase activity. Am J Respir Cell Mol Biol 24(6):711–719

Lucke S, Levkau B (2010) Endothelial functions of sphingosine-1-phosphate. Cell Physiol Biochem 26(1):87–96

Mansoor M, Melendez AJ (2008) Recent trials for FTY720 (fingolimod): a new generation of immunomodulators structurally similar to sphingosine. Rev Recent Clin Trials 3(1):62–69

Mathew B, Jacobson JR, Berdyshev E, Huang Y, Sun X, Zhao Y, Gerhold LM, Siegler J, Evenoski C, Wang T, Zhou T, Zaidi R, Moreno-Vinasco L, Bittman R, Chen CT, LaRiviere PJ, Sammani S, Lussier YA, Dudek SM, Natarajan V, Weichselbaum RR, Garcia JG (2011) Role of sphingolipids in murine radiation-induced lung injury: protection by sphingosine 1-phosphate analogs. FASEB J 25(10):3388–3400

McVerry BJ, Garcia JG (2004) Endothelial cell barrier regulation by sphingosine 1-phosphate. Review. J Cell Biochem 92(6):1075–1085

McVerry BJ, Peng X, Hassoun PM, Sammani S, Simon BA, Garcia JG (2004) Sphingosine 1-phosphate reduces vascular leak in murine and canine models of acute lung injury. Am J Respir Crit Care Med 170(9):987–993

Mehta D, Malik AB (2006) Signaling mechanisms regulating endothelial permeability. Physiol Rev 86(1):279–367, Review

Mehta D, Tiruppathi C, Sandoval R, Minshall RD, Holinstat M, Malik AB (2002) Modulatory role of focal adhesion kinase in regulating human pulmonary arterial endothelial barrier function. J Physiol 539(Pt 3):779–789

Mehta D, Konstantoulaki M, Ahmmed GU, Malik AB (2005) Sphingosine 1-phosphate-induced mobilization of intracellular Ca2+ mediates rac activation and adherens junction assembly in endothelial cells. J Biol Chem 280(17):17320–17328

Mitra SK, Hanson DA, Schlaepfer DD (2005) Focal adhesion kinase: in command and control of cell motility. Review. Nat Rev Mol Cell Biol 6(1):56–68

Miura Y, Yatomi Y, Rile G, Ohmori T, Satoh K, Ozaki Y (2000) Rho-mediated phosphorylation of focal adhesion kinase and myosin light chain in human endothelial cells stimulated with sphingosine 1-phosphate, a bioactive lysophospholipid released from activated platelets. J Biochem 127(5):909–914

Narumiya S, Ishizaki T, Watanabe N (1997) Rho effectors and reorganization of actin cytoskeleton. Review. FEBS Lett 410(1):68–72

Ohashi K, Nagata K, Maekawa M, Ishizaki T, Narumiya S, Mizuno K (2000) Rho-associated kinase ROCK activates LIM-kinase 1 by phosphorylation at threonine 508 within the activation loop. J Biol Chem 275(5):3577–3582

Ohmori T, Yatomi Y, Okamoto H, Miura Y, Rile G, Satoh K, Ozaki Y (2001) G(i)-mediated Cas tyrosine phosphorylation in vascular endothelial cells stimulated with sphingosine 1-phosphate: possible involvement in cell motility enhancement in cooperation with Rho-mediated pathways. J Biol Chem 276(7):5274–5280

Owen KA, Pixley FJ, Thomas KS, Vicente-Manzanares M, Ray BJ, Horwitz AF, Parsons JT, Beggs HE, Stanley ER, Bouton AH (2007) Regulation of lamellipodial persistence, adhesion turnover, and motility in macrophages by focal adhesion kinase. J Cell Biol 179(6):1275–1287

Paik JH, Skoura A, Chae SS, Cowan AE, Han DK, Proia RL, Hla T (2004) Sphingosine 1-phosphate receptor regulation of N-cadherin mediates vascular stabilization. Genes Dev 18(19):2392–2403

Parizi M, Howard EW, Tomasek JJ (2000) Regulation of LPA-promoted myofibroblast contraction: role of Rho, myosin light chain kinase, and myosin light chain phosphatase. Exp Cell Res 254(2):210–220

Peng X, Hassoun PM, Sammani S, McVerry BJ, Burne MJ, Rabb H, Pearse D, Tuder RM, Garcia JG (2004) Protective effects of sphingosine 1-phosphate in murine endotoxin-induced inflammatory lung injury. Am J Respir Crit Care Med 169(11):1245–1251

Quadri SK, Bhattacharya J (2007) Resealing of endothelial junctions by focal adhesion kinase. Am J Physiol Lung Cell Mol Physiol 292(1):L334–L342

Quadri SK, Bhattacharjee M, Parthasarathi K, Tanita T, Bhattacharya J (2003) Endothelial barrier strengthening by activation of focal adhesion kinase. J Biol Chem 278(15):13342–13349

Rogers DF, Boschetto P, Barnes PJ (1989) Plasma exudation. Correlation between Evans blue dye and radiolabeled albumin in guinea pig airways in vivo. J Pharmacol Methods 21(4):309–315

Rosen H, Goetzl EJ (2005) Sphingosine 1-phosphate and its receptors: an autocrine and paracrine network. Nat Rev Immunol 5(7):560–570

Rosen H, Sanna MG, Cahalan SM, Gonzalez-Cabrera PJ (2007) Tipping the gatekeeper: S1P regulation of endothelial barrier function. Trends Immunol 28(3):102–107

Rosenfeldt HM, Hobson JP, Maceyka M, Olivera A, Nava VE, Milstien S, Spiegel S (2001) EDG-1 links the PDGF receptor to Src and focal adhesion kinase activation leading to lamellipodia formation and cell migration. FASEB J 15(14):2649–2659

Rosenfeldt HM, Amrani Y, Watterson KR, Murthy KS, Panettieri RA Jr, Spiegel S (2003) Sphingosine 1-phosphate stimulates contraction of human airway smooth muscle cells. FASEB J 17(13):1789–1799

Ryu Y, Takuwa N, Sugimoto N, Sakurada S, Usui S, Okamoto H, Matsui O, Takuwa Y (2002) Sphingosine-1-phosphate, a platelet-derived lysophospholipid mediator, negatively regulates cellular Rac activity and cell migration in vascular smooth muscle cells. Circ Res 90(3): 325–332

Saba JD, Hla T (2004) Point-counterpoint of sphingosine 1-phosphate metabolism. Circ Res 94(6): 724–734

Sammani S, Moreno-Vinasco L, Mirzapoiazova T, Singleton PA, Chiang ET, Evenoski CL, Wang T, Mathew B, Husain A, Moitra J, Sun X, Nunez L, Jacobson JR, Dudek SM, Natarajan V, Garcia JG (2010) Differential effects of sphingosine 1-phosphate receptors on airway and vascular barrier function in the murine lung. Am J Respir Cell Mol Biol 43(4):394–402

Sanna MG, Wang SK, Gonzalez-Cabrera PJ, Don A, Marsolais D, Matheu MP, Wei SH, Parker I, Jo E, Cheng WC, Cahalan MD, Wong CH, Rosen H (2006) Enhancement of capillary leakage and restoration of lymphocyte egress by a chiral S1P1 antagonist in vivo. Nat Chem Biol 2(8): 434–441

Schuchardt M, Tölle M, Prüfer J, van der Giet M (2011) Pharmacological relevance and potential of sphingosine 1-phosphate in the vascular system. Br J Pharmacol 163(6):1140–1162. doi:10.1111/j.1476-5381.2011.01260.x

Shea BS, Brooks SF, Fontaine BA, Chun J, Luster AD, Tager AM (2010) Prolonged exposure to sphingosine 1-phosphate receptor-1 agonists exacerbates vascular leak, fibrosis, and mortality after lung injury. Am J Respir Cell Mol Biol 43(6):662–673

Shikata Y, Birukov KG, Birukova AA, Verin A, Garcia JG (2003a) Involvement of site-specific FAK phosphorylation in sphingosine-1 phosphate- and thrombin-induced focal adhesion remodeling: role of Src and GIT. FASEB J 17(15):2240–2249

Shikata Y, Birukov KG, Garcia JG (2003b) S1P induces FA remodeling in human pulmonary endothelial cells: role of Rac, GIT1, FAK, and paxillin. J Appl Physiol 94(3):1193–1203

Singleton PA, Dudek SM, Chiang ET, Garcia JG (2005) Regulation of sphingosine 1-phosphate-induced endothelial cytoskeletal rearrangement and barrier enhancement by S1P1 receptor, PI3 kinase, Tiam1/Rac1, and alpha-actinin. FASEB J 19(12):1646–1656

Singleton PA, Lingen MW, Fekete MJ, Garcia JG, Moss J (2006) Methylnaltrexone inhibits opiate and VEGF-induced angiogenesis: role of receptor transactivation. Microvasc Res 72(1–2):3–11

Singleton PA, Moreno-Vinasco L, Sammani S, Wanderling SL, Moss J, Garcia JG (2007) Attenuation of vascular permeability by methylnaltrexone: role of mOP-R and S1P3 transactivation. Am J Respir Cell Mol Biol 37(2):222–231

Snider AJ, Orr Gandy KA, Obeid LM (2010) Sphingosine kinase: role in regulation of bioactive sphingolipid mediators in inflammation. Biochimie 92(6):707–715

Spiegel S, Milstien S (2003) Sphingosine-1-phosphate: an enigmatic signalling lipid. Nat Rev Mol Cell Biol 4(5):397–407

Sun X, Shikata Y, Wang L, Ohmori K, Watanabe N, Wada J, Shikata K, Birukov KG, Makino H, Jacobson JR, Dudek SM, Garcia JG (2009) Enhanced interaction between focal adhesion and adherens junction proteins: involvement in sphingosine 1-phosphate-induced endothelial barrier enhancement. Microvasc Res 77(3):304–313

Tani M, Ito M, Igarashi Y (2007) Ceramide/sphingosine/sphingosine 1-phosphate metabolism on the cell surface and in the extracellular space. Review. Cell Signal 19(2):229–237

Venkataraman K, Thangada S, Michaud J, Oo ML, Ai Y, Lee YM, Wu M, Parikh NS, Khan F, Proia RL, Hla T (2006) Extracellular export of sphingosine kinase-1a contributes to the vascular S1P gradient. Biochem J 397(3):461–471

Venkiteswaran K, Xiao K, Summers S, Calkins CC, Vincent PA, Pumiglia K, Kowalczyk AP (2002) Regulation of endothelial barrier function and growth by VE-cadherin, plakoglobin, and beta-catenin. Am J Physiol Cell Physiol 283(3):C811–C821

Vestweber D (2008) VE-cadherin: the major endothelial adhesion molecule controlling cellular junctions and blood vessel formation. Review. Arterioscler Thromb Vasc Biol 28(2):223–232

Vestweber D, Winderlich M, Cagna G, Nottebaum AF (2009) Cell adhesion dynamics at endothelial junctions: VE-cadherin as a major player. Trends Cell Biol 19(1):8–15

Vouret-Craviari V, Bourcier C, Boulter E, van Obberghen-Schilling E (2002) Distinct signals via Rho GTPases and Src drive shape changes by thrombin and sphingosine-1-phosphate in endothelial cells. J Cell Sci 115(Pt 12):2475–2484

Waeber C, Blondeau N, Salomone S (2004) Vascular sphingosine-1-phosphate S1P1 and S1P3 receptors. Review. Drug News Perspect 17(6):365–382

Wang L, Dudek SM (2009) Regulation of vascular permeability by sphingosine 1-phosphate. Microvasc Res 77(1):39–45

Wang Y, Zheng XR, Riddick N, Bryden M, Baur W, Zhang X, Surks HK (2009) ROCK isoform regulation of myosin phosphatase and contractility in vascular smooth muscle cells. Circ Res 104(4):531–540

Waterman-Storer CM, Worthylake RA, Liu BP, Burridge K, Salmon ED (1999) Microtubule growth activates Rac1 to promote lamellipodial protrusion in fibroblasts. Nat Cell Biol 1(1): 45–50

Wojciak-Stothard B, Tsang LY, Paleolog E, Hall SM, Haworth SG (2006) Rac1 and RhoA as regulators of endothelial phenotype and barrier function in hypoxia-induced neonatal pulmonary hypertension. Am J Physiol Lung Cell Mol Physiol 290(6):L1173–L1182

Worthylake RA, Lemoine S, Watson JM, Burridge K (2001) RhoA is required for monocyte tail retraction during transendothelial migration. J Cell Biol 154(1):147–160

Wu MH (2005) Endothelial focal adhesions and barrier function. Review. J Physiol 569(Pt 2): 359–366

Wysolmerski RB, Lagunoff D (1990) Involvement of myosin light-chain kinase in endothelial cell retraction. Proc Natl Acad Sci USA 87(1):16–20

Xu M, Waters CL, Hu C, Wysolmerski RB, Vincent PA, Minnear FL (2007) Sphingosine 1-phosphate rapidly increases endothelial barrier function independently of VE-cadherin but requires cell spreading and Rho kinase. Am J Physiol Cell Physiol 293(4):C1309–C1318

Yang L, Yatomi Y, Miura Y, Satoh K, Ozaki Y (1999) Metabolism and functional effects of sphingolipids in blood cells. Br J Haematol 107(2):282–293

Yatomi Y, Ruan F, Hakomori S, Igarashi Y (1995) Sphingosine-1-phosphate: a platelet-activating sphingolipid released from agonist-stimulated human platelets. Blood 86(1):193–202

Yuan CS, Doshan H, Charney MR, O'connor M, Karrison T, Maleckar SA, Israel RJ, Moss J (2005) Tolerability, gut effects, and pharmacokinetics of methylnaltrexone following repeated intravenous administration in humans. J Clin Pharmacol 45(5):538–546

Zhang Q, Magnusson MK, Mosher DF (1997) Lysophosphatidic acid and microtubule-destabilizing agents stimulate fibronectin matrix assembly through Rho-dependent actin stress fiber formation and cell contraction. Mol Biol Cell 8(8):1415–1425

Zhao Y, Kalari SK, Usatyuk PV, Gorshkova I, He D, Watkins T, Brindley DN, Sun C, Bittman R, Garcia JG, Berdyshev EV, Natarajan V (2007) Intracellular generation of sphingosine 1-phosphate in human lung endothelial cells: role of lipid phosphate phosphatase-1 and sphingosine kinase 1. J Biol Chem 282(19):14165–14177

Zondag GC, Postma FR, Etten IV, Verlaan I, Moolenaar WH (1998) Sphingosine 1-phosphate signalling through the G-protein-coupled receptor Edg-1. Biochem J 330(Pt 2):605–609

Sphingolipids in Acute Lung Injury

Stefan Uhlig and Yang Yang

Contents

Abstract Acute lung injury is a life-threatening disease that is characterized by pulmonary inflammation, loss of barrier functions, and hypoxemia. Sphingolipids are critically involved in the disease process that they can both expedite and extenuate: They expedite inflammation by promoting chemotaxis (neutral sphingomyelinase), increased endothelial permeability (acid sphingomyelinase, $S1P_3$-receptors), increased epithelial permeability ($S1P_2$- and $S1P_3$-receptors), and delaying neutrophil apoptosis (neutral sphingomyelinase, $S1P_1$-receptors). They extenuate inflammation by attenuating chemotaxis (S1P) and by stabilizing the endothelial and the epithelial barrier ($S1P_1$-receptor). This chapter discusses the multiple roles and therapeutic options that sphingolipids offer with respect to acute lung injury.

Keywords ARDS • PAF • caveolae • acid sphingomyelinase • sphingolipids • vascular permeability

---

S. Uhlig (✉) • Y. Yang
Institute of Pharmacology and Toxicology, RWTH Aachen University, Wendlingweg 2, 52074 Aachen, Germany
e-mail: suhlig@ukaachen.de

E. Gulbins and I. Petrache (eds.), *Sphingolipids in Disease*, Handbook of Experimental Pharmacology 216,
DOI 10.1007/978-3-7091-1511-4_11, © Springer-Verlag Wien 2013

1 Introduction

Acute respiratory distress syndrome (ARDS) represents an often-fatal disease with an acute onset that is characterized by severe hypoxemia ($PaO_2/FiO_2 < 300$), bilateral chest infiltrates, and respiratory failure that is not fully explained by cardiac failure or fluid overload (Ranieri et al. 2012). The latter point alludes to the common view that ARDS is the result of increased vascular permeability in the course of overshooting inflammation. The hyperinflammation hypothesis is based on abundant experimental evidence (Reiss et al. 2012b) such as the correlation of inflammatory biomarkers with disease severity and a plethora of preclinical studies (Cross and Matthay 2011); however, as of now, anti-inflammatory therapies have not been successful (Bosma et al. 2010; Frank and Thompson 2010; Adhikari et al. 2004).

In principle, the hyperinflammation hypothesis states that ARDS is caused by the over-activation of the innate immune system leading to tissue destruction by overexcited neutrophils, monocytes, and macrophages. The over-activation of the innate immune system can be triggered by typical causes of ARDS such as sepsis, pneumonia, or acid aspiration. In most experimental models of ARDS, neutrophils play a critical role, and hence, here we will largely focus on these cells; nonetheless, it should not be forgotten that ARDS can also occur in neutropenic patients (Maunder et al. 1986; Ognibene et al. 1986) and animals (Karsunky et al. 2002; Jin et al. 2006). The pathogenic sequence of events in the classical model where leukocytes are centrally implicated is depicted in Fig. 1, which also highlights the major roles played by sphingolipids in this model. In this chapter, we will review the multiple roles played by sphingolipids in acute lung injury.

2 Levels of Sphingolipids and Related Enzymes in ARDS

2.1 Clinical Data

Acid sphingomyelinase (ASMase) is rapidly upregulated during inflammation in the lungs and in other organs (Uhlig and Gulbins 2008). One of the first preparations of ASMase that yielded high amounts of the enzyme was from the urine of septic patients (Quintern et al. 1989). High ASMase levels have also been observed in the serum of three patients with hypercytokinemia (Takahashi et al. 2002). And most remarkably, in septic patient's serum ASMase activity and ceramide levels were not only increased but did even correlate with mortality (Claus et al. 2005; Delogu et al. 1999; Drobnik et al. 2003). For the lung itself, human data are lacking, except for the observation that alveolar neutrophils from ARDS patients contain increased intracellular ceramide levels (Lin et al. 2011).

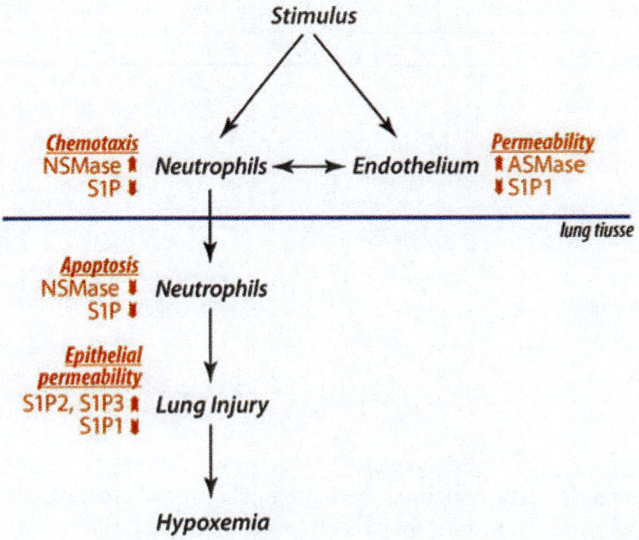

Fig. 1 Overview of the hyperinflammation hypothesis of acute lung injury. Shown in *red* are the contributions of sphingolipids

2.2 *Experimental Data*

Several pro-inflammatory and injurious stimuli such as PAF, TNF, LPS, extensive lavage, or sulfur mustard toxins will activate ASMase and cause an increase in pulmonary ceramide levels, independent of the species (Table 1). Sphingosine-1-phosphate (S1P) levels, on the other hand, are only moderately increased (von Bismarck et al. 2012) or remain unaltered in the lungs (Table 1) and also in the serum (Bachmaier et al. 2012). Other enzymes that have been found to be upregulated in response to i.t. LPS are SMS2 (Gowda et al. 2011), Sphk1 and Sphk2 (Wadgaonkar et al. 2009).

3 The Functional Role of Sphingolipids in ALI

3.1 *Neutrophils*

As depicted in Fig. 1, the activation of leukocytes and endothelial cells is a critical step in the pathophysiology of ALI. Drawn by chemotactic signals, the leukocytes leave the vasculature and sequester in the lungs. Once in the lungs they may attack surrounding microorganisms, structures and tissues by proteases, and reactive oxygen species, before they undergo apoptosis. Neutral sphingomyelinase (NSMase) and S1P

Table 1 Acid sphingomyelinase, ceramide, and S1P levels in models of acute lung injury

Stimulus	Species	ASMase	Ceramide	S1P	Reference
PAF i.v.	Mouse	▲	▲		Göggel et al. (2004), Yang et al. (2010b)
TNF (i.p., i.t.)	Rat	▲	▲		Ryan et al. (2003)
LPS (i.p., i.t.)	Mouse	▲	▲	◀▶/▲	von Bismarck et al. (2012), Bachmaier et al. (2012), Wadgaonkar et al. (2009), Haimovitz-Friedman et al. (1997), Zhao et al. (2011)
LPS antenatal	Sheep	▲	▲		Kunzmann et al. (2011)
Lavage model	Piglets	▲	▲		von Bismarck et al. (2008)

appear to be particularly important in the regulation of chemotaxis and apoptosis either in the same direction as in apoptosis or in an antithetic fashion as in chemotaxis.

3.1.1 Chemotaxis

Sphingolipids participate in the chemotactic response of neutrophils that is dependent on the polarized distribution of sphingolipid-enriched microdomains within the plasma membrane. Neutrophil chemotaxis is perturbed by disruption of these microdomains (Pierini et al. 2003; Bodin and Welch 2005). Crucial for their proper formation is the generation of ceramide by the NSMase, which is required for the asymmetrical distribution of Rac versus RhoA GTPases as well as for neutrophil elongation (Sitrin et al. 2011). Blocking NSMase caused loss of directional bias and decreased the chemotactic index of neutrophils, but had no effect on their motility or velocity (Sitrin et al. 2011). Intranasal application of NSMase resulted in pulmonary leukocyte sequestration and increased vascular permeability (Hayashida et al. 2009). The observation that intracellular ceramide levels were increased in neutrophils from the BAL of ARDS patients indicates that this mechanism may also be operative in these patients (Lin et al. 2011).

As so often, the actions of ceramide and S1P are contradictory, and S1P was shown to block neutrophil chemotaxis towards classical chemotaxins such as IL-8 (Kawa et al. 1997; Rahaman et al. 2006). Importantly, however, when given alone, S1P can act as a typical chemoattractant with a bell-shaped dose–response curve, and this effect became stronger in pneumonia patients where $S1P_3$-receptors were upregulated (Rahaman et al. 2006). Therefore, the effect of S1P on neutrophils in situ may depend on the plasticity of the neutrophils sequestered in the lungs.

3.1.2 Apoptosis

Neutrophils undergo spontaneous apoptosis unless external stimuli are provided (Martin et al. 2005). One important concept in the pathogenesis of ARDS is based on the longevity of neutrophils: The later the neutrophils go into apoptosis, the stronger the lung injury. Accordingly, neutrophil apoptosis is delayed in sepsis and in ARDS (Gilroy et al. 2004; Matute-Bello et al. 2000). Inflammatory signals such as LPS keep these cells alive by preventing neutrophil apoptosis.

S1P is well known for preventing apoptosis (Hannun and Obeid 2008), and this is also true for neutrophils (Chihab et al. 2003). In neutrophils, S1P can be formed from ceramide that is generated by the NSMase, the predominant sphingomyelinase isoenzyme in these cells (Hinkovska-Galcheva et al. 1998). Thus, inhibition of NSMase blocks the inhibitory effect of LPS on neutrophil apoptosis via S1P, at least in part by blocking the LPS-induced activation of p38 MAPK (Lin et al. 2011).

3.1.3 Neutrophils In Vivo: Chemotaxis Versus Apoptosis

Above, we have focused on chemotaxis and apoptosis; we have not discussed neutrophil effector functions such as degranulation, ROS production, or phagocytosis, because in these processes there is as yet no consistent role of sphingolipids apparent, although there are some interesting observations (Feldhaus et al. 2002; MacKinnon et al. 2002; Suchard et al. 1997).

With respect to ARDS, the findings above suggest that inhibition of NSMase would block the pathogenetic process on two levels: attenuation of pulmonary neutrophil sequestration and acceleration of neutrophil apoptosis. This hypothesis was confirmed by studies with the NSMase inhibitor sphingolactone-24 (Sph24) that reduced leukocyte counts in the lungs 24 h after i.t. administration of a lethal dose of LPS (Lin et al. 2011). In that study, Sph24 even reduced lung injury and mortality (Lin et al. 2011). ASMase, on the other hand, while critical for changes in vascular permeability (see below), appears to affect leukocyte counts only moderately as seen in studies with imipramine in the lavage model and in ASMase-deficient mice 3 days after inhalation of LPS (von Bismarck et al. 2008).

Like NSMase, S1P has two effects on neutrophils: an early one that attenuates chemotaxis and a late one that promotes their survival. The experimental findings support this hypothesis (Table 2), because S1P administration (i.t. or i.v.) reduced LPS-induced neutrophil sequestration in the alveolar space at 6 h but not at 12 h or 24 h (Wadgaonkar et al. 2009; Peng et al. 2004). The finding that tissue neutrophils—as opposed to BAL neutrophil counts—were decreased by S1P also at 18–24 h (Peng et al. 2004; Sammani et al. 2010) suggests that the anti-apoptotic effect of S1P affects mainly neutrophils in the alveolar but not in the interstitial space. These effects of S1P are mediated predominantly by the $S1P_1$-receptor, as anti-apoptotic actions have been described for the specific $S1P_1$-agonist SEW-2871 in vitro (Diab et al. 2010) and because in vivo that drug behaved similar to S1P in an LPS-based model of ARDS

Table 2 Sphingolipid-directed interventions in models of ALI

Stimulus	Intervention	PMN	Edema	PaO_2/ FiO_2	Reference
LPS i.t.	SMS2 ko	▼	▼		Gowda et al. (2011)
PAF	$ASMase^{-/-}$		▼		Göggel et al. (2004)
LPS i.p.	Imipramine		▼		Göggel et al. (2004), Yang et al. (2010a)
Lavage model	Imipramine	▼	▼	▲	von Bismarck et al. (2008)
LPS i.p.	D609		▼		Göggel et al. (2004)
LPS/zymosan i.v.	D609		▼		O'Dea et al. (2009)
Acid instillation	D609		▼	▲	Göggel et al. (2004)
LPS i.t.	D609	▼	▼		Sammani et al. (2010), Anjum et al. (2012)
Lavage/ ventilation/ LPS	Inositol-1,2,6-trisphosphate	◀▶	▼	▲	Preuss et al. (2012)
LPS instillation	Sph-24[a]	▼	▼		Lin et al. (2011)
LPS i.p.	SphK1 ko	▲			Bachmaier et al. (2012)
LPS i.t.	SphK1 ko	▲	▲		Wadgaonkar et al. (2009)
T/HS	SKI2[b]		▼		Lee et al. (2004)
LPS i.t.	S1P i.t.	▼/ ◀▶	▼	◀▶	Wadgaonkar et al. (2009), Szczepaniak et al. (2008)
LPS i.t.	S1P i.v.	▼/ ◀▶	▼	▲	Peng et al. (2004), McVerry et al. (2004)
LPS i.t.	SEW 2871[c] i.t.		▼		Sammani et al. (2010)
LPS i.t.	SEW 2871[c] i.v.	▼	▼		Sammani et al. (2010)
LPS i.t.	$S1P_1^{+/-}$	◀▶			Sammani et al. (2010)
LPS i.t.	$S1P_2^{-/-}$	◀▶	▼		Sammani et al. (2010)
LPS i.t.	$S1P_3$-knockdown		▼		Sammani et al. (2010)
LPS i.t.	S1P $lyase^{+/-}$, THI[d]	▼	▼		Zhao et al. (2011)
Bleomycin	AUY954[c]	▲	▲		Shea et al. (2010)
Pancreatitis	S1P	▼	▼		Liu et al. (2008)
Brain death	SEW2871[c]	▼	▼		Sammani et al. (2011)

[a]NSMase inhibitor
[b]Sphingosine kinase inhibitor
[c]$S1P_1$-receptor agonist
[d]S1P lyase inhibitor

(Sammani et al. 2010). Unfortunately, in the latter study, BAL cell counts are not available for all experimental groups.

S1P is synthesized by two sphingosine kinase isoenzymes: Sphk1 and Sphk2. SphK activity is stimulated by a variety of stimuli like histamine, growth factors, and various cytokines (Choi et al. 1996; Taha et al. 2006; Alemany et al. 2007; Oskeritzian et al. 2008) and contributes to the increased S1P levels in inflammation (Ledgerwood et al. 2008; Tauseef et al. 2008). In line with the notion that S1P blocks chemotaxis, in Sphk1-deficent mice, neutrophilic inflammation was increased with or without LPS challenge (Bachmaier et al. 2012; Wadgaonkar et al. 2009). In that latter study, the findings were also explained by the newly described S1P-dependent upregulation of the complement scavenger receptor C5L2 (Bachmaier et al. 2012).

In summary, the majority of the findings on the role of S1P and neutral sphingomyelinase in ARDS can be explained by taking into account their effects on chemotaxis and neutrophil apoptosis. It is noteworthy that in most studies the data on neutrophil sequestration correlate well with indices of vascular permeability. However, there are also differences as is illustrated by studies with S1P administration in the LPS model (Wadgaonkar et al. 2009; Peng et al. 2004) and with imipramine in the lavage model (von Bismarck et al. 2008), where there were pronounced effects on permeability but only small effects on BAL neutrophils. In the next chapter we will discuss the role of sphingolipids in regulation of vascular permeability in ARDS.

3.2 *Vascular Endothelial Permeability*

Increased vascular permeability is the hallmark of ARDS. It is frequently associated with leukocyte transmigration, but can also be regulated independently. Sphingolipids play a critical role in the regulation of endothelial permeability in the pulmonary vasculature, and again, S1P and ceramide act antithetically: ceramide weakens and S1P strengthens barrier functions in the lungs.

3.2.1 Acid Sphingomyelinase

A Role for Apoptosis?

In one of the first studies to implicate ASMase in acute lung injury, Haimovitz-Friedman and coworkers demonstrated increased numbers of TUNEL-positive endothelial cells in LPS-treated animals, a finding that was explained by the production of TNF and subsequent ASMase-dependent ceramide production (Haimovitz-Friedman et al. 1997). These findings suggested that endothelial cell apoptosis might be a mechanism of increased vascular permeability in acute lung injury. However, subsequent studies, which are summarized in Fig. 2, demonstrated that while ASMase is indeed involved in ARDS, this mechanism operates independently of endothelial cell apoptosis. Treatment with the pan-caspase inhibitor Z-VAD, at a concentration that was effective against liver apoptosis (Kunstle et al. 1997), had no effect against LPS-induced edema formation, while D609, an agent that blocks ceramide synthesis in this model, was protective (Fig. 2). In addition, the ceramide-induced hyperpermeability of endothelial cells in culture is also independent of apoptosis (Lindner et al. 2005; Uhlig et al. 2005). Furthermore, the finding that LPS-induced edema in vivo is dependent on PAF (Fig. 2) argues against a role of apoptosis in this process, because PAF acts within 10 min, which is too short for apoptosis. And finally, all

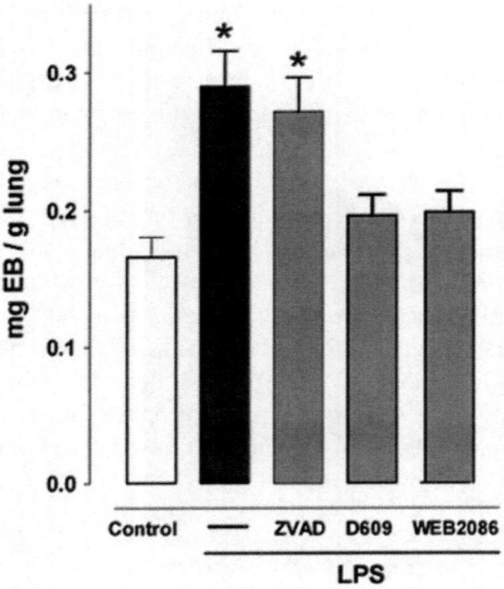

Fig. 2 LPS-induced edema formation is independent of apoptosis. Mice were treated with 10 mg/kg i.p.LPS. 30 min before LPS the animals were treated with 10 mg/kg of the pan-caspase inhibitor Z-VAD-FMK, with 5 mg/kg of the PAF-receptor antagonist WEB2086 or with 40 mg/kg D609. Because of its short half-life the animals received D609 also 30 min, 60 min and 120 min after the LPS challenge. Evans Blue extravasation was assessed as described (Goggel et al. 2004). The D609 data are from Goggel et al. (2004), the WEB2086 data from Falk et al. (1999). Controls, $n = 21$; LPS, $n = 15$; Z-VAD/LPS, $n = 6$; D609/LPS, $n = 5$. $^{*}p < 0.05$ vs. controls

these findings are reconciled by the observation that PAF increases vascular permeability by an ASMase-dependent process (Goggel et al. 2004).

Organization of Lipid Microdomains and Caveolae

Most of the mechanistic insights of how the ASMase promotes increased vascular permeability come from studies with PAF (Fig. 3). PAF causes pulmonary edema by a dual mechanism: one part of the edema depends on activation of cyclooxygenase and the other part on the activation of ASMase (Uhlig et al. 2005). The case for ASMase in PAF-induced edema formation is strong: ASMase deficiency (Goggel et al. 2004), inhibition of the ASMase activity by ARC039 (Yang et al. 2010b; Roth et al. 2009), promotion of ASMase degradation by imipramine (Goggel et al. 2004; Yang et al. 2010b), inhibition of ASMase activation by D609 (Goggel et al. 2004; Yang et al. 2010b), and antibodies against ceramide (Goggel et al. 2004) do all attenuate pulmonary edema induced by PAF. Within less than 10 min, PAF leads to increased ASMase activity in caveolae, which is prevented by imipramine (Yang et al. 2010b). The fact that imipramine simulates proteolysis (Hurwitz et al. 1994) raises the possibility that ASMase is transferred to the caveolae by exocytotic

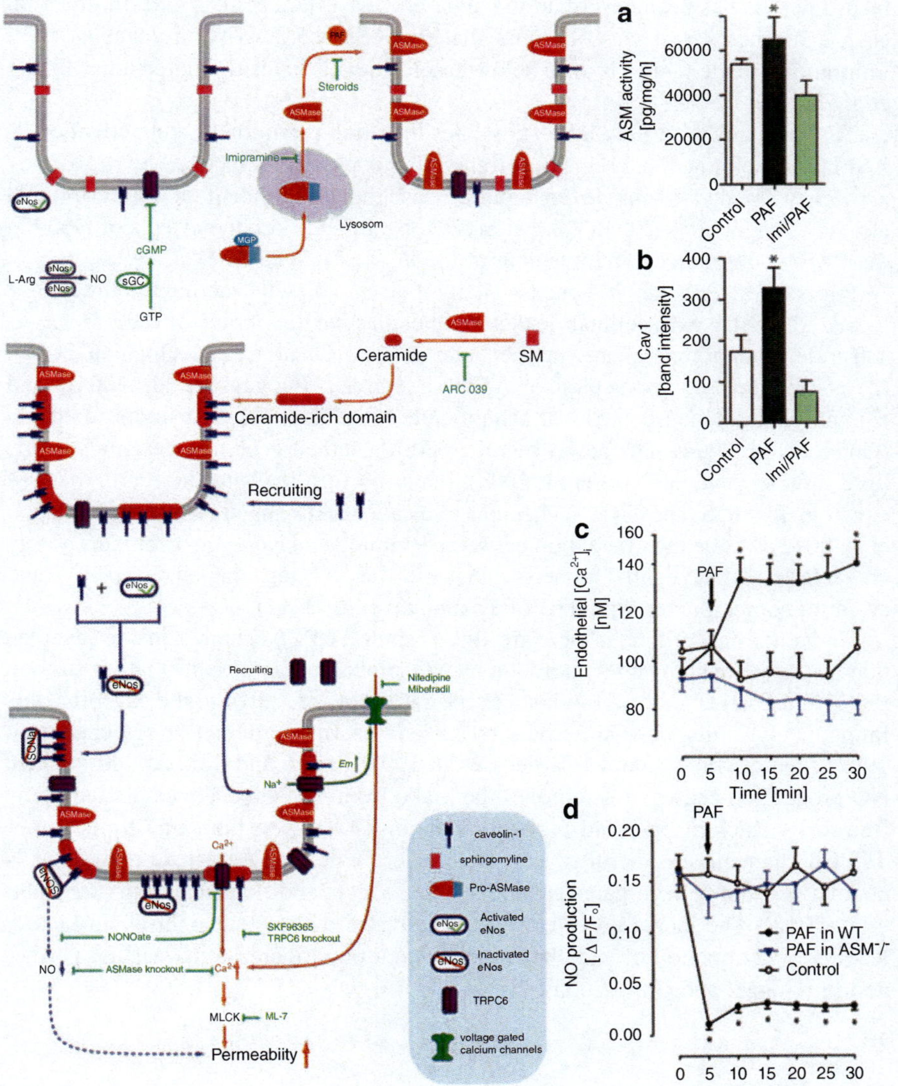

Fig. 3 The central role of ASMase in PAF-induced edema. PAF triggers rapid (<10 min) recruitment of ASMase, probably from the lysosomes to endothelial caveolae. ASMase produces ceramide which is responsible for the subsequent recruitment of caveolin-1, eNOS, and TRPC6 channels into caveolae. The inhibition of eNOS by caveolin-1 stops endothelial NO production that normally blocks TRPC6 channels. The recruitment and activation of TRPC6 increases endothelial $[Ca^{2+}]_i$ and $[Na^+]_i$ and stimulates additional $[Ca^{2+}]_i$ increase through voltage-gated calcium channels. The sudden drop in endothelial NO together with the rapid increase of endothelial $[Ca^{2+}]$ activates the myosin light chain kinase (MLCK) causing endothelial cell contraction and hyperpermeability. Interventions that blocked the various steps are shown in green. The cartoon is largely based on the references (Goggel et al. 2004; Yang et al. 2010b; Samapati et al. 2012). The panels **a–e** give

lysosomes, as has been described in the case of FasL (Jin et al. 2008). In line with such a mechanism, it was observed that the ASMase activity in caveolae from imipramine-treated lungs is even below baseline levels in unstimulated lungs (Yang et al. 2010b).

The mechanism of how PAF stimulates the rapid recruitment and activation of ASMase is unknown, but remarkably enough this process is blocked by steroids by a mechanism that owing to its rapidity must be independent of transcriptional effects (Goggel et al. 2004; Yang et al. 2010b). In addition, the effects of D609 in the PAF model indicate involvement of the PC-PLC (Schutze et al. 1992).

For several reasons it appears likely that the newly recruited ASMase is localized to the extracellular leaflet of the plasma membrane (1) 90 % of its substrate sphingomyelin are in the outer leaflet (Ohanian and Ohanian 2001); (2) PAF increases extracellullar ASMase activity (Goggel et al. 2004); and (3) anti-ceramide antibodies that cannot enter the cells block PAF-induced edema (Goggel et al. 2004). The newly recruited ASMase then generates ceramide inside lipid rafts or caveolae (Jin et al. 2008), which in turn leads to the recruitment of caveolin-1, eNOS, and TRPC6 channels into caveolae (Yang et al. 2010b; Samapati et al. 2012). While the interaction between ceramide and caveolin-1 remains poorly characterized, caveolin-1 itself binds eNOS (Yang et al. 2010b) and co-immunoprecipitates with TRPC6 (Samapati et al. 2012).

Under normal conditions there are always some TRPC6 channels inside caveolae that however are kept largely dormant by NO, probably by phosphorylation through PKG (Samapati et al. 2012). When caveolin-1 is now recruited to the caveolae, this immediately inhibits eNOS activity by a well-described interaction (Michel et al. 1997; Ju et al. 1997; Garcia-Cardena et al. 1997; Parton and Simons 2007). Once NO production ceases, this removes the brake from TRPC6 channels that are now free to let calcium pass and probably sodium as well. At the same time further TRPC6 channels are recruited and let in more calcium. As a result, the cell is depolarized and voltage-gated calcium channels let in additional calcium (Samapati et al. 2012). The increased calcium then activates the myosin light chain kinase, leading to contraction of endothelial cells and a rapid increase in vascular permeability (Goggel and Uhlig 2005).

Fig. 3 (continued) experimental evidence obtained in isolated perfused lungs for the critical role of ASMase. Panel **a**: Enzymatic activity of acid sphingomyelinase in caveolar fractions of endothelial cells (Yang et al. 2010b) from lungs treated with PAF or imipramine/PAF. (**b**) Immunoblot against caveolin-1 in endothelial caveolae from lungs treated with PAF or imipramine/PAF (Yang et al. 2010b). (**c**) Endothelial $[Ca^{2+}]_i$ in response to PAF in ASMase-deficient mice (Samapati et al. 2012). (**d**) Endothelial NO concentrations in response to PAF in ASMase-deficient mice (Kuebler et al. 2010). (**a** and **b**) Rat lungs were infused with a bolus of 5 nMol PAF and pretreated with imipramine 10 min before PAF administration. Data are presented as mean ± SD, $n = 4$. $*p < 0.05$ vs. control. (**c** and **d**) mouse lungs were infused with a bolus of 5 nMol PAF, the *arrow* indicating where PAF was added. Data are presented as mean ± SEM, $n = 5$. $*p < 0.05$ vs. ASMase-deficient group

In Vivo Studies

The mechanistic data above have been largely worked out by studying the rapid effects of PAF in the model of the isolated perfused lung. The relevance for the in vivo situation (see Table 2) was demonstrated in ASMase-deficient mice that were protected against pulmonary edema induced by intravenous injection of PAF (Goggel et al. 2004). Moreover, imipramine reduced pulmonary edema in endotoxemic mice (Yang et al. 2010a) and in piglets where ARDS was induced by repetitive lavage (von Bismarck et al. 2008). A remarkable finding in the latter study, where imipramine was given together with surfactant, was that the protective effect of imipramine was apparent after 24 h, thus indicating that the significance of ASM is sustained. Similarly, D-myo-Inositol-1,2,6-trisphosphate that was shown to reduce ASMase activity and subsequent ceramide production was protective in a clinically relevant 3-hit model that lasted for 3 days (Preuss et al. 2012).

In addition, the beneficial effects of D609, an inhibitor of sphingomyelin synthase and phosphatidylcholine-specific phospholipase C (Adibhatla et al. 2012), in four different ARDS models (Table 2), are at least partly explained by the effect of this drug on the ASMase-dependent ceramide production. This conclusion is corroborated by the findings that D609 prevented the ASM-dependent synthesis of ceramide in response to TNF (Schutze et al. 1992), Fas activation (Cifone et al. 1995), CD5 activation (Simarro et al. 1999), and PAF (Goggel et al. 2004). At present it cannot be excluded that the beneficial effects of D609 in LPS-induced lung injury are also related to its effects on sphingomyelin synthase (SMS2) (Anjum et al. 2012).

Of note, not only ASMase but also SMS2-deficient mice are protected against LPS-induced lung injury (Gowda et al. 2011). This finding would be in line with the notion that SMS2 is part of homeostatic de novo pathway that maintains the basal plasma sphingomyelin levels (Tafesse et al. 2007; Li et al. 2007), i.e., the substrate levels of the ASMase. Involvement of the de novo pathway of sphingolipids synthesis in the course of LPS-induced lung injury was also suggested in a recent study showing regulation of ASMase-dependent ceramide formation by NF-κB (von Bismarck et al. 2012), because this transcription factor regulates a subunit of serine palmitoyltransferase (Sptlc2) (Chang et al. 2011). Petrache and coworkers suggested that ceramides derived from the Spltc1 isoenzyme activity may contribute to the amplification of pulmonary vascular injury induced by excessive ceramides (Medler et al. 2008). Clearly, the contribution of the de novo pathway for the regulation of sphingolipids in ARDS remains to be further defined.

3.2.2 Sphingosine-1-Phosphate

Receptor Diversity

The divergent effects of S1P can be explained on the basis of the five cognate G-protein-coupled receptors ($S1P_{1–5}$). Sphingosine-1-phosphate protects the

Table 3 Bidirectional regulation of vascular permeability by Sphingosine-1-phosphate

Outcome	Receptors	G-protein	Pathways	References
▲Endothelial barrier	$S1P_1$	G_i	Rac/Tiam	Peng et al. (2004), McVerry et al. (2004), McVerry and Garcia (2004), Spiegel and Milstien (2002)
▼Endothelial barrier	$S1P_2$	$G_{12/12}$, G_q		Sanchez et al. (2005, 2007), Inoki et al. (2006)
▼Endothelial barrier	$S1P_3$	G_q	Ca, Rho-kinase	Sanchez et al. (2007), Singleton et al. (2007), Peters and Alewijnse (2007), Takuwa et al. (2008), Nofer et al. (2004)
▲NO via eNOS		G_i	Akt	

endothelial barrier by activation of Rac GTPase-dependent cytoskeletal reorganization and focal adhesion assembly via $S1P_1$-receptors (McVerry and Garcia 2004, 2005; Wang and Dudek 2009). On the other hand, S1P also impairs adherence junctions (Sanchez et al. 2007) and increases epithelial (Gon et al. 2005, 2009) and endothelial permeability (Singleton et al. 2007) via $S1P_2$- or $S1P_3$-receptors (Table 3). Accordingly, high doses of S1P are associated with increased vascular permeability and even death (Sammani et al. 2010) (Table 2).

Vascular endothelial cells primarily express $S1P_1$, $S1P_2$, and $S1P_3$ (Hla et al. 2001), and the endothelial barrier function is enhanced by $S1P_1$ but impaired by $S1P_2$ and $S1P_3$ (Peng et al. 2004; McVerry et al. 2004; McVerry and Garcia 2004; Spiegel and Milstien 2002). $S1P_2$ triggers the loosening of adherence junctions (Sanchez et al. 2007). At higher S1P concentrations (5 μM), $S1P_3$ (Shikata et al. 2003; Komarova et al. 2007) increases endothelial and probably also epithelial permeability (Sammani et al. 2010; Spiegel and Milstien 2002). The signaling mechanisms involved are discussed in detail in the chapter by Garcia and colleagues (Abbasi and Garcia 2013).

Studies In Vivo

The majority of studies with S1P in models of ARDS have focused on the remarkable barrier-stabilizing properties of S1P in the lungs that was first described by Garcia and colleagues (Garcia et al. 2001). This concept has been fully confirmed in all subsequent studies in a variety of animal species where ARDS was induced by stimuli such as LPS, pancreatitis, brain death, or hemorrhagic shock (Table 2). All these barrier protective actions appear to be mediated by $S1P_1$-receptors as is exemplified by the finding that the $S1P_1$-receptor agonist SEW-2871 reduced edema formation in experimental ARDS (Sammani et al. 2010). In addition, mutant mice that selectively lack S1P in the plasma show increased vascular leak in the lungs and are more susceptible to stimulation with PAF, a phenotype that can be reversed by S1P transfusion (Camerer et al. 2009).

$S1P_2$- and $S1P_3$-receptors, in contrast, augment permeability and hence their genetic deletion reduced edema formation following intratracheal administration of LPS (Sammani et al. 2010). It is also thought that these receptors are responsible for the increased permeability that is observed after administration of higher doses of S1P or SEW2871 (Sammani et al. 2010).

4 Summary

Sphingolipids assume multiple roles in the regulation of neutrophil chemotaxis, neutrophil apoptosis, as well as endothelial and epithelial barrier functions. Quite remarkably, each of these functions can be regulated by sphingolipids in an opposite manner. Therefore, pharmacological interventions that are directed against sphingolipids need to be highly specific. For instance, despite the fact that S1P is barrier protective does not only lack of S1P (Bachmaier et al. 2012; Wadgaonkar et al. 2009; Camerer et al. 2009), but also excessive amounts thereof cause or aggravate lung injury (Sammani et al. 2010). One way to achieve S1P concentrations in a beneficial and physiological range may be to inhibit S1P lyase, the S1P-degrading enzyme (Zhao et al. 2011). The insight that at least parts of the detrimental effects of S1P are mediated by $S1P_2$- and $S1P_3$-receptors has led to the development of specific $S1P_1$-receptor agonists such as AUY954 or SEW2871. The latter drug attenuated edema formation and acute neutrophil influx into the lungs (Sammani et al. 2010; McVerry et al. 2004). However, because in the alveolar space S1P is preventing the apoptosis of neutrophils and may thereby possibly aggravate lung injury, it will be important to study such drugs in long-term experiments. This need is emphasized by a recent study that showed a massive aggravation of acute lung injury including elevated neutrophil counts at day 7 in the bleomycin model by treatment with AUY954 (Shea et al. 2010). This study highlighted another critical aspect with regard to the therapeutic use of S1P, namely, the fact that upon engagement S1P-receptors are rapidly internalized thus causing functional antagonism (Shea et al. 2010).

Alternative targets to S1P-receptors are given by the sphingomyelinase enzymes. Inhibition of NSMase prevents both neutrophil chemotaxis and apoptosis, which makes it a more attractive strategy than exogenous S1P with respect to dampening neutrophil activities in the lungs. So far, however, in experimental ARDS there is only one study using an inhibitor with a relatively high IC_{50} value (Lin et al. 2011); further studies are needed to further clarify this interesting approach.

The role of ASMase in acute lung injury has been explored in greater detail. Inhibition of this pathway attenuated lung injury not only in response to PAF and LPS but also in the lavage model, in acid-induced lung injury, and in novel 3-hit model (Table 2). These data do also suggest that ASMase inhibition is not only protective in the acute situation but also in models that last for 1–3 days (von Bismarck et al. 2008;

Preuss et al. 2012). One noteworthy point here is that in all studies, the inhibition of ASMase attenuated edema formation and—when examined—also hypoxemia but that the effects on pulmonary neutrophil sequestration were usually moderate. This might indicate that neutrophil sequestration and edema formation can be separated, as has been observed before (Chignard and Balloy 2000; Davidson et al. 1999; Miotla et al. 1996; Ogawa et al. 2006; Wilson et al. 2007; Wosten-van Asperen et al. 2008).

In addition, increased ASMase levels are present in the circulation of septic patients and correlate with mortality (Claus et al. 2005). In keeping with this, recent experimental studies suggest that ASMase may have a role in the development of septic shock, because ASMase-deficient mice were protected against the TNF-induced alterations in blood pressure and heart rate (Reiss et al. 2012a). Above we discussed that ASMase is most likely acting at the outer leaflet of the plasma membrane. Thus, both the circulating as well as the pulmonary ASMase could be inhibited by drugs that need not enter the cells, which might be an advantage if one needs to minimize side effects. The extent of such side effects seems tolerable as may be surmised from patients with Niemann-Pick disease and other patients treated with imipramine (Uhlig and Gulbins 2008).

To conclude, sphingolipids are involved in nearly all steps of the pathogenesis of ARDS. These insights suggest a number of interesting pharmacological interventions. So far, the majority of such studies have focused either on giving exogenous $S1P_1$-receptor agonists or on inhibiting ASMase. Both approaches have been successful in preventing edema formation and to a lesser extent also pulmonary leukocyte sequestration. However, because the defining factor for ARDS is hypoxemia and because edema formation does not necessarily translate into hypoxemia [for discussion and examples, see Reiss et al. (2012b)], the significance of any experimental model is doubtful unless blood gas data are provided. Those few studies that have examined blood gases (Table 2) show that both exogenous S1P and ASMase inhibitors can improve oxygenation and strongly suggest that sphingolipid metabolism offers valuable novel therapeutic targets for the treatment of acute lung injury.

References

Abbasi T, Garcia JGN (2013) Sphingolipids in lung endothelial biology and regulation of vascular integrity. In: Gulbins E, Petrache I (eds) Sphingolipids in disease, vol 216, Handbook of experimental pharmacology. Springer, Heidelberg

Adhikari N, Burns KE, Meade MO (2004) Pharmacologic therapies for adults with acute lung injury and acute respiratory distress syndrome. Cochrane Database Syst Rev (4):CD004477

Adibhatla RM, Hatcher JF, Gusain A (2012) Tricyclodecan-9-yl-xanthogenate (D609) mechanism of actions: a mini-review of literature. Neurochem Res 37(4):671–679

Alemany R, van Koppen CJ, Danneberg K, Ter Braak M, Meyer Zu Heringdorf D (2007) Regulation and functional roles of sphingosine kinases. Naunyn Schmiedebergs Arch Pharmacol 374(5–6):413–428

Anjum F, Joshi K, Grinkina N, Gowda S, Cutaia M, Wadgaonkar R (2012) Role of sphingomyelin synthesis in pulmonary endothelial cells cytoskeletal activation and endotoxin induced lung injury. Am J Respir Cell Mol Biol 47(1):94–103

Bachmaier K, Guzman E, Kawamura T, Gao X, Malik AB (2012) Sphingosine kinase 1 mediation of expression of the anaphylatoxin receptor C5L2 dampens the inflammatory response to endotoxin. PLoS One 7(2):e30742

Bodin S, Welch MD (2005) Plasma membrane organization is essential for balancing competing pseudopod- and uropod-promoting signals during neutrophil polarization and migration. Mol Biol Cell 16(12):5773–5783

Bosma KJ, Taneja R, Lewis JF (2010) Pharmacotherapy for prevention and treatment of acute respiratory distress syndrome: current and experimental approaches. Drugs 70(10):1255–1282

Camerer E, Regard JB, Cornelissen I, Srinivasan Y, Duong DN, Palmer D, Pham TH, Wong JS, Pappu R, Coughlin SR (2009) Sphingosine-1-phosphate in the plasma compartment regulates basal and inflammation-induced vascular leak in mice. J Clin Invest 119(7):1871–1879

Chang ZQ, Lee SY, Kim HJ, Kim JR, Kim SJ, Hong IK, Oh BC, Choi CS, Goldberg IJ, Park TS (2011) Endotoxin activates de novo sphingolipid biosynthesis via nuclear factor kappa B-mediated upregulation of Sptlc2. Prostaglandins Other Lipid Mediat 94(1–2):44–52

Chignard M, Balloy V (2000) Neutrophil recruitment and increased permeability during acute lung injury induced by lipopolysaccharide. Am J Physiol Lung Cell Mol Physiol 279(6): L1083–L1090

Chihab R, Porn-Ares MI, Alvarado-Kristensson M, Andersson T (2003) Sphingosine 1-phosphate antagonizes human neutrophil apoptosis via p38 mitogen-activated protein kinase. Cell Mol Life Sci 60(4):776–785

Choi OH, Kim JH, Kinet JP (1996) Calcium mobilization via sphingosine kinase in signalling by the Fc epsilon RI antigen receptor. Nature 380(6575):634–636

Cifone MG, Roncaioli P, De Maria R, Camarda G, Santoni A, Ruberti G, Testi R (1995) Multiple pathways originate at the Fas/APO-1 (CD95) receptor: sequential involvement of phosphatidylcholine-specific phospholipase C and acidic sphingomyelinase in the propagation of the apoptotic signal. EMBO J 14(23):5859–5868

Claus RA, Bunck AC, Bockmeyer CL, Brunkhorst FM, Losche W, Kinscherf R, Deigner HP (2005) Role of increased sphingomyelinase activity in apoptosis and organ failure of patients with severe sepsis. FASEB J 19(12):1719–1721

Cross LJ, Matthay MA (2011) Biomarkers in acute lung injury: insights into the pathogenesis of acute lung injury. Crit Care Clin 27(2):355–377

Davidson BA, Knight PR, Helinski JD, Nader ND, Shanley TP, Johnson KJ (1999) The role of tumor necrosis factor-alpha in the pathogenesis of aspiration pneumonitis in rats. Anesthesiology 91(2):486–499

Delogu G, Famularo G, Amati F, Signore L, Antonucci A, Trinchieri V, Di Marzio L, Cifone MG (1999) Ceramide concentrations in septic patients: a possible marker of multiple organ dysfunction syndrome. Crit Care Med 27(11):2413–2417

Diab KJ, Adamowicz JJ, Kamocki K, Rush NI, Garrison J, Gu Y, Schweitzer KS, Skobeleva A, Rajashekhar G, Hubbard WC, Berdyshev EV, Petrache I (2010) Stimulation of sphingosine 1-phosphate signaling as an alveolar cell survival strategy in emphysema. Am J Respir Crit Care Med 181(4):344–352

Drobnik W, Liebisch G, Audebert FX, Frohlich D, Gluck T, Vogel P, Rothe G, Schmitz G (2003) Plasma ceramide and lysophosphatidylcholine inversely correlate with mortality in sepsis patients. J Lipid Res 44(4):754–761

Falk S, Göggel R, Heydasch U, Brasch F, Müller KM, Wendel A, Uhlig S (1999) Quinolines attenuate PAF-induced pulmonary pressor responses and edema formation. Am J Respir Crit Care Med 160(5 Pt 1):1734–1742

Feldhaus MJ, Weyrich AS, Zimmerman GA, McIntyre TM (2002) Ceramide generation in situ alters leukocyte cytoskeletal organization and beta 2-integrin function and causes complete degranulation. J Biol Chem 277(6):4285–4293

Frank AJ, Thompson BT (2010) Pharmacological treatments for acute respiratory distress syndrome. Curr Opin Crit Care 16(1):62–68

Garcia JG, Liu F, Verin AD, Birukova A, Dechert MA, Gerthoffer WT, Bamberg JR, English D (2001) Sphingosine 1-phosphate promotes endothelial cell barrier integrity by Edg-dependent cytoskeletal rearrangement. J Clin Invest 108(5):689–701

Garcia-Cardena G, Martasek P, Masters BS, Skidd PM, Couet J, Li S, Lisanti MP, Sessa WC (1997) Dissecting the interaction between nitric oxide synthase (NOS) and caveolin Functional significance of the nos caveolin binding domain in vivo. J Biol Chem 272(41):25437–25440

Gilroy DW, Lawrence T, Perretti M, Rossi AG (2004) Inflammatory resolution: new opportunities for drug discovery. Nat Rev Drug Discov 3(5):401–416

Göggel R, Uhlig S (2005) The inositol trisphosphate pathway mediates platelet-activating-factor-induced pulmonary oedema. Eur Respir J 25(5):849–857

Göggel R, Winoto-Morbach S, Vielhaber G, Imai Y, Lindner K, Brade L, Brade H, Ehlers S, Slutsky AS, Schutze S, Gulbins E, Uhlig S (2004) PAF-mediated pulmonary edema: a new role for acid sphingomyelinase and ceramide. Nat Med 10(2):155–160

Gon Y, Wood MR, Kiosses WB, Jo E, Sanna MG, Chun J, Rosen H (2005) S1P3 receptor-induced reorganization of epithelial tight junctions compromises lung barrier integrity and is potentiated by TNF. Proc Natl Acad Sci USA 102(26):9270–9275

Gon Y, Wood MR, Kiosses WB, Jo E, Sanna MG, Chun J, Rosen H (2009) Retraction for "S1P3 receptor-induced reorganization of epithelial tight junctions compromises lung barrier integrity and is potentiated by TNF". Proc Natl Acad Sci USA 106(30):12561

Gowda S, Yeang C, Wadgaonkar S, Anjum F, Grinkina N, Cutaia M, Jiang XC, Wadgaonkar R (2011) Sphingomyelin synthase 2 (SMS2) deficiency attenuates LPS-induced lung injury. Am J Physiol Lung Cell Mol Physiol 300(3):L430–L440

Haimovitz-Friedman A, Cordon-Cardo C, Bayoumy S, Garzotto M, McLoughlin M, Gallily R, Edwards CK 3rd, Schuchman EH, Fuks Z, Kolesnick R (1997) Lipopolysaccharide induces disseminated endothelial apoptosis requiring ceramide generation. J Exp Med 186(11):1831–1841

Hannun YA, Obeid LM (2008) Principles of bioactive lipid signalling: lessons from sphingolipids. Nat Rev Mol Cell Biol 9(2):139–150

Hayashida A, Bartlett AH, Foster TJ, Park PW (2009) Staphylococcus aureus beta-toxin induces lung injury through syndecan-1. Am J Pathol 174(2):509–518

Hinkovska-Galcheva V, Kjeldsen L, Mansfield PJ, Boxer LA, Shayman JA, Suchard SJ (1998) Activation of a plasma membrane-associated neutral sphingomyelinase and concomitant ceramide accumulation during IgG-dependent phagocytosis in human polymorphonuclear leukocytes. Blood 91(12):4761–4769

Hla T, Lee MJ, Ancellin N, Paik JH, Kluk MJ (2001) Lysophospholipids–receptor revelations. Science 294(5548):1875–1878

Hurwitz R, Ferlinz K, Sandhoff K (1994) The tricyclic antidepressant desipramine causes proteolytic degradation of lysosomal sphingomyelinase in human fibroblasts. Biol Chem Hoppe Seyler 375(7):447–450

Inoki I, Takuwa N, Sugimoto N, Yoshioka K, Takata S, Kaneko S, Takuwa Y (2006) Negative regulation of endothelial morphogenesis and angiogenesis by S1P2 receptor. Biochem Biophys Res Commun 346(1):293–300

Jin J, Zeng H, Schmid KW, Toetsch M, Uhlig S, Möröy T (2006) The zinc finger protein Gfi1 acts upstream of TNF to attenuate endotoxin-mediated inflammatory responses in the lung. Eur J Immunol 36(2):421–430

Jin S, Yi F, Zhang F, Poklis JL, Li PL (2008) Lysosomal targeting and trafficking of acid sphingomyelinase to lipid raft platforms in coronary endothelial cells. Arterioscler Thromb Vasc Biol 28(11):2056–2062

Ju H, Zou R, Venema VJ, Venema RC (1997) Direct interaction of endothelial nitric-oxide synthase and caveolin-1 inhibits synthase activity. J Biol Chem 272(30):18522–18525

Karsunky H, Zeng H, Schmidt T, Zevnik B, Kluge R, Schmid KW, Duhrsen U, Möröy T (2002) Inflammatory reactions and severe neutropenia in mice lacking the transcriptional repressor Gfi1. Nat Genet 30(3):295–300

Kawa S, Kimura S, Hakomori S, Igarashi Y (1997) Inhibition of chemotactic motility and trans-endothelial migration of human neutrophils by sphingosine 1-phosphate. FEBS Lett 420(2–3):196–200

Komarova YA, Mehta D, Malik AB (2007) Dual regulation of endothelial junctional permeability. Sci STKE 2007(412):re8

Kuebler WM, Yang Y, Samapati R, Uhlig S (2010) Vascular barrier regulation by PAF, ceramide, caveolae, and NO – an intricate signaling network with discrepant effects in the pulmonary and systemic vasculature. Cell Physiol Biochem 26(1):29–40

Künstle G, Leist M, Uhlig S, Revesz L, Feifel R, MacKenzie A, Wendel A (1997) ICE-protease inhibitors block murine liver injury and apoptosis caused by CD95 or by TNF-alpha. Immunol Lett 55(1):5–10

Kunzmann S, Collins JJ, Yang Y, Uhlig S, Kallapur SG, Speer CP, Jobe AH, Kramer BW (2011) Antenatal inflammation reduces expression of caveolin-1 and influences multiple signaling pathways in preterm fetal lungs. Am J Respir Cell Mol Biol 45(5):969–976

Ledgerwood LG, Lal G, Zhang N, Garin A, Esses SJ, Ginhoux F, Merad M, Peche H, Lira SA, Ding Y, Yang Y, He X, Schuchman EH, Allende ML, Ochando JC, Bromberg JS (2008) The sphingosine 1-phosphate receptor 1 causes tissue retention by inhibiting the entry of peripheral tissue T lymphocytes into afferent lymphatics. Nat Immunol 9(1):42–53

Lee C, Xu DZ, Feketeova E, Kannan KB, Yun JK, Deitch EA, Fekete Z, Livingston DH, Hauser CJ (2004) Attenuation of shock-induced acute lung injury by sphingosine kinase inhibition. J Trauma 57(5):955–960

Li Z, Hailemariam TK, Zhou H, Li Y, Duckworth DC, Peake DA, Zhang Y, Kuo MS, Cao G, Jiang XC (2007) Inhibition of sphingomyelin synthase (SMS) affects intracellular sphingomyelin accumulation and plasma membrane lipid organization. Biochim Biophys Acta 1771(9):1186–1194

Lin WC, Lin CF, Chen CL, Chen CW, Lin YS (2011) Inhibition of neutrophil apoptosis via sphingolipid signaling in acute lung injury. J Pharmacol Exp Ther 339(1):45–53

Lindner K, Uhlig U, Uhlig S (2005) Ceramide alters endothelial cell permeability by a nonapoptotic mechanism. Br J Pharmacol 145(1):132–140

Liu HB, Cui NQ, Wang Q, Li DH, Xue XP (2008) Sphingosine-1-phosphate and its analogue FTY720 diminish acute pulmonary injury in rats with acute necrotizing pancreatitis. Pancreas 36(3):e10–e15

MacKinnon AC, Buckley A, Chilvers ER, Rossi AG, Haslett C, Sethi T (2002) Sphingosine kinase: a point of convergence in the action of diverse neutrophil priming agents. J Immunol 169(11):6394–6400

Martin TR, Hagimoto N, Nakamura M, Matute-Bello G (2005) Apoptosis and epithelial injury in the lungs. Proc Am Thorac Soc 2(3):214–220

Matute-Bello G, Liles WC, Radella F 2nd, Steinberg KP, Ruzinski JT, Hudson LD, Martin TR (2000) Modulation of neutrophil apoptosis by granulocyte colony-stimulating factor and granulocyte/macrophage colony-stimulating factor during the course of acute respiratory distress syndrome. Crit Care Med 28(1):1–7

Maunder RJ, Hackman RC, Riff E, Albert RK, Springmeyer SC (1986) Occurrence of the adult respiratory distress syndrome in neutropenic patients. Am Rev Respir Dis 133(2):313–316

McVerry BJ, Garcia JG (2004) Endothelial cell barrier regulation by sphingosine 1-phosphate. J Cell Biochem 92(6):1075–1085

McVerry BJ, Garcia JG (2005) In vitro and in vivo modulation of vascular barrier integrity by sphingosine 1-phosphate: mechanistic insights. Cell Signal 17(2):131–139

McVerry BJ, Peng X, Hassoun PM, Sammani S, Simon BA, Garcia JG (2004) Sphingosine 1-phosphate reduces vascular leak in murine and canine models of acute lung injury. Am J Respir Crit Care Med 170(9):987–993

Medler TR, Petrusca DN, Lee PJ, Hubbard WC, Berdyshev EV, Skirball J, Kamocki K, Schuchman E, Tuder RM, Petrache I (2008) Apoptotic sphingolipid signaling by ceramides in lung endothelial cells. Am J Respir Cell Mol Biol 38(6):639–646

Michel JB, Feron O, Sacks D, Michel T (1997) Reciprocal regulation of endothelial nitric-oxide synthase by Ca2+–calmodulin and caveolin. J Biol Chem 272(25):15583–15586

Miotla JM, Williams TJ, Hellewell PG, Jeffery PK (1996) A role for the beta2 integrin CD11b in mediating experimental lung injury in mice. Am J Respir Cell Mol Biol 14(4):363–373

Nofer JR, van der Giet M, Tolle M, Wolinska I, von Wnuck LK, Baba HA, Tietge UJ, Godecke A, Ishii I, Kleuser B, Schafers M, Fobker M, Zidek W, Assmann G, Chun J, Levkau B (2004) HDL induces NO-dependent vasorelaxation via the lysophospholipid receptor S1P3. J Clin Invest 113(4):569–581

O'Dea KP, Wilson MR, Dokpesi JO, Wakabayashi K, Tatton L, van Rooijen N, Takata M (2009) Mobilization and margination of bone marrow Gr-1high monocytes during subclinical endotoxemia predisposes the lungs toward acute injury. J Immunol 182(2):1155–1166

Ogawa EN, Ishizaka A, Tasaka S, Koh H, Ueno H, Amaya F, Ebina M, Yamada S, Funakoshi Y, Soejima J, Moriyama K, Kotani T, Hashimoto S, Morisaki H, Abraham E, Takeda J (2006) Contribution of high-mobility group box-1 to the development of ventilator-induced lung injury. Am J Respir Crit Care Med 174(4):400–407

Ognibene FP, Martin SE, Parker MM, Schlesinger T, Roach P, Burch C, Shelhamer JH, Parrillo JE (1986) Adult respiratory distress syndrome in patients with severe neutropenia. N Engl J Med 315(9):547–551

Ohanian J, Ohanian V (2001) Sphingolipids in mammalian cell signalling. Cell Mol Life Sci 58(14):2053–2068

Oskeritzian CA, Alvarez SE, Hait NC, Price MM, Milstien S, Spiegel S (2008) Distinct roles of sphingosine kinases 1 and 2 in human mast-cell functions. Blood 111(8):4193–4200

Parton RG, Simons K (2007) The multiple faces of caveolae. Nat Rev Mol Cell Biol 8(3):185–194

Peng X, Hassoun PM, Sammani S, McVerry BJ, Burne MJ, Rabb H, Pearse D, Tuder RM, Garcia JG (2004) Protective effects of sphingosine 1-phosphate in murine endotoxin-induced inflammatory lung injury. Am J Respir Crit Care Med 169(11):1245–1251

Peters SL, Alewijnse AE (2007) Sphingosine-1-phosphate signaling in the cardiovascular system. Curr Opin Pharmacol 7(2):186–192

Pierini LM, Eddy RJ, Fuortes M, Seveau S, Casulo C, Maxfield FR (2003) Membrane lipid organization is critical for human neutrophil polarization. J Biol Chem 278(12):10831–10841

Preuss S, Stadelmann S, Omam FD, Scheiermann J, Winoto-Morbach S, von Bismarck P, Knerlich-Lukoschus F, Lex D, Adam-Klages S, Wesch D, Held-Feindt J, Uhlig S, Schutze S, Krause MF (2012) Inositol-trisphosphate reduces alveolar apoptosis and pulmonary edema in neonatal lung injury. Am J Respir Cell Mol Biol 47(2):158–169

Quintern LE, Zenk TS, Sandhoff K (1989) The urine from patients with peritonitis as a rich source for purifying human acid sphingomyelinase and other lysosomal enzymes. Biochim Biophys Acta 1003(2):121–124

Rahaman M, Costello RW, Belmonte KE, Gendy SS, Walsh MT (2006) Neutrophil sphingosine 1-phosphate and lysophosphatidic acid receptors in pneumonia. Am J Respir Cell Mol Biol 34(2):233–241

Ranieri VM, Rubenfeld GD, Thompson BT, Ferguson ND, Caldwell E, Fan E, Camporota L, Slutsky AS (2012) Acute respiratory distress syndrome: the Berlin Definition. JAMA 307(23):2526–2533

Reiss LK, Adam D, Uhlig S (2012a) Acid sphingomyelinase-deficient mice are protected from the lethal cardiovascular effects in TNF-induced septic shock. In: Paper presented at the Deutsche Gesellschaft für Experimentelle und Klinische Pharmakologie und Toxikologie e.V. Abstracts of the 78th annual meeting, Dresden, Germany, 19–22 Mar 2012

Reiss LK, Uhlig U, Uhlig S (2012b) Models and mechanisms of acute lung injury caused by direct insults. Eur J Cell Biol 91(6–7):590–601

Roth AG, Drescher D, Yang Y, Redmer S, Uhlig S, Arenz C (2009) Potent and selective inhibition of acid sphingomyelinase by bisphosphonates. Angew Chem Int Ed Engl 48(41):7560–7563

Ryan AJ, McCoy DM, McGowan SE, Salome RG, Mallampalli RK (2003) Alveolar sphingolipids generated in response to TNF-alpha modifies surfactant biophysical activity. J Appl Physiol 94 (1):253–258

Samapati R, Yang Y, Yin J, Stoerger C, Arenz C, Dietrich A, Gudermann T, Adam D, Wu S, Freichel M, Flockerzi V, Uhlig S, Kuebler WM (2012) Lung endothelial Ca^{2+} and permeability response to platelet-activating factor is mediated by acid sphingomyelinase and transient receptor potential classical 6. Am J Respir Crit Care Med 185(2):160–170

Sammani S, Moreno-Vinasco L, Mirzapoiazova T, Singleton PA, Chiang ET, Evenoski CL, Wang T, Mathew B, Husain A, Moitra J, Sun X, Nunez L, Jacobson JR, Dudek SM, Natarajan V, Garcia JG (2010) Differential effects of sphingosine 1-phosphate receptors on airway and vascular barrier function in the murine lung. Am J Respir Cell Mol Biol 43(4):394–402

Sammani S, Park KS, Zaidi SR, Mathew B, Wang T, Huang Y, Zhou T, Lussier YA, Husain AN, Moreno-Vinasco L, Vigneswaran WT, Garcia JG (2011) A sphingosine 1-phosphate 1 receptor agonist modulates brain death-induced neurogenic pulmonary injury. Am J Respir Cell Mol Biol 45(5):1022–1027

Sanchez T, Thangada S, Wu MT, Kontos CD, Wu D, Wu H, Hla T (2005) PTEN as an effector in the signaling of antimigratory G protein-coupled receptor. Proc Natl Acad Sci USA 102(12):4312–4317

Sanchez T, Skoura A, Wu MT, Casserly B, Harrington EO, Hla T (2007) Induction of vascular permeability by the sphingosine-1-phosphate receptor-2 (S1P2R) and its downstream effectors ROCK and PTEN. Arterioscler Thromb Vasc Biol 27(6):1312–1318

Schutze S, Potthoff K, Machleidt T, Berkovic D, Wiegmann K, Kronke M (1992) TNF activates NF-kappa B by phosphatidylcholine-specific phospholipase C-induced "acidic" sphingomyelin breakdown. Cell 71(5):765–776

Shea BS, Brooks SF, Fontaine BA, Chun J, Luster AD, Tager AM (2010) Prolonged exposure to sphingosine 1-phosphate receptor-1 agonists exacerbates vascular leak, fibrosis, and mortality after lung injury. Am J Respir Cell Mol Biol 43(6):662–673

Shikata Y, Birukov KG, Birukova AA, Verin A, Garcia JG (2003) Involvement of site-specific FAK phosphorylation in sphingosine-1 phosphate- and thrombin-induced focal adhesion remodeling: role of Src and GIT. FASEB J 17(15):2240–2249

Simarro M, Calvo J, Vila JM, Places L, Padilla O, Alberola-Ila J, Vives J, Lozano F (1999) Signaling through CD5 involves acidic sphingomyelinase, protein kinase C-zeta, mitogen-activated protein kinase kinase, and c-Jun NH2-terminal kinase. J Immunol 162(9):5149–5155

Singleton PA, Moreno-Vinasco L, Sammani S, Wanderling SL, Moss J, Garcia JG (2007) Attenuation of vascular permeability by methylnaltrexone: role of mOP-R and S1P3 transactivation. Am J Respir Cell Mol Biol 37(2):222–231

Sitrin RG, Sassanella TM, Petty HR (2011) An obligate role for membrane-associated neutral sphingomyelinase activity in orienting chemotactic migration of human neutrophils. Am J Respir Cell Mol Biol 44(2):205–212

Spiegel S, Milstien S (2002) Sphingosine 1-phosphate, a key cell signaling molecule. J Biol Chem 277(29):25851–25854

Suchard SJ, Hinkovska-Galcheva V, Mansfield PJ, Boxer LA, Shayman JA (1997) Ceramide inhibits IgG-dependent phagocytosis in human polymorphonuclear leukocytes. Blood 89(6):2139–2147

Szczepaniak WS, Zhang Y, Hagerty S, Crow MT, Kesari P, Garcia JG, Choi AM, Simon BA, McVerry BJ (2008) Sphingosine 1-phosphate rescues canine LPS-induced acute lung injury and alters systemic inflammatory cytokine production in vivo. Transl Res 152(5):213–224

Tafesse FG, Huitema K, Hermansson M, van der Poel S, van den Dikkenberg J, Uphoff A, Somerharju P, Holthuis JC (2007) Both sphingomyelin synthases SMS1 and SMS2 are required for sphingomyelin homeostasis and growth in human HeLa cells. J Biol Chem 282(24):17537–17547

Taha TA, Hannun YA, Obeid LM (2006) Sphingosine kinase: biochemical and cellular regulation and role in disease. J Biochem Mol Biol 39(2):113–131

Takahashi T, Abe T, Sato T, Miura K, Takahashi I, Yano M, Watanabe A, Imashuku S, Takada G (2002) Elevated sphingomyelinase and hypercytokinemia in hemophagocytic lymphohistiocytosis. J Pediatr Hematol Oncol 24(5):401–404

Takuwa Y, Okamoto Y, Yoshioka K, Takuwa N (2008) Sphingosine-1-phosphate signaling and biological activities in the cardiovascular system. Biochim Biophys Acta 1781(9):483–488

Tauseef M, Kini V, Knezevic N, Brannan M, Ramchandaran R, Fyrst H, Saba J, Vogel SM, Malik AB, Mehta D (2008) Activation of sphingosine kinase-1 reverses the increase in lung vascular permeability through sphingosine-1-phosphate receptor signaling in endothelial cells. Circ Res 103(10):1164–1172

Uhlig S, Gulbins E (2008) Sphingolipids in the lungs. Am J Respir Crit Care Med 178(11):1100–1114

Uhlig S, Göggel R, Engel S (2005) Mechanisms of platelet-activating factor (PAF)-mediated responses in the lung. Pharmacol Rep 57(Suppl):206–221

von Bismarck P, Wistadt CF, Klemm K, Winoto-Morbach S, Uhlig U, Schutze S, Adam D, Lachmann B, Uhlig S, Krause MF (2008) Improved pulmonary function by acid sphingomyelinase inhibition in a newborn piglet lavage model. Am J Respir Crit Care Med 177(11):1233–1241

von Bismarck P, Winoto-Mohrbach S, Herzberg M, Uhlig U, Schutze S, Lucius R, Krause MF (2012) IKK NBD peptide inhibits LPS induced pulmonary inflammation and alters sphingolipid metabolism in a murine model. Pulm Pharmacol Ther 25(3):228–235

Wadgaonkar R, Patel V, Grinkina N, Romano C, Liu J, Zhao Y, Sammani S, Garcia JG, Natarajan V (2009) Differential regulation of sphingosine kinases 1 and 2 in lung injury. Am J Physiol Lung Cell Mol Physiol 296(4):L603–L613

Wang L, Dudek SM (2009) Regulation of vascular permeability by sphingosine 1-phosphate. Microvasc Res 77(1):39–45

Wilson MR, Goddard ME, O'Dea KP, Choudhury S, Takata M (2007) Differential roles of p55 and p75 tumor necrosis factor receptors on stretch-induced pulmonary edema in mice. Am J Physiol Lung Cell Mol Physiol 293(1):L60–L68

Wosten-van Asperen RM, Lutter R, Haitsma JJ, Merkus MP, van Woensel JB, van der Loos CM, Florquin S, Lachmann B, Bos AP (2008) ACE mediates ventilator-induced lung injury in rats via angiotensin II but not bradykinin. Eur Respir J 31(2):363–371

Yang J, Qu JM, Summah H, Zhang J, Zhu YG, Jiang HN (2010a) Protective effects of imipramine in murine endotoxin-induced acute lung injury. Eur J Pharmacol 638(1–3):128–133

Yang Y, Yin J, Baumgartner W, Samapati R, Solymosi EA, Reppien E, Kuebler WM, Uhlig S (2010b) Platelet-activating factor reduces endothelial nitric oxide production: role of acid sphingomyelinase. Eur Respir J 36(2):417–427

Zhao Y, Gorshkova IA, Berdyshev E, He D, Fu P, Ma W, Su Y, Usatyuk PV, Pendyala S, Oskouian B, Saba JD, Garcia JG, Natarajan V (2011) Protection of LPS-induced murine acute lung injury by sphingosine-1-phosphate lyase suppression. Am J Respir Cell Mol Biol 45(2):426–435

The Involvement of Sphingolipids in Chronic Obstructive Pulmonary Diseases

Irina Petrache and Daniela N. Petrusca

Contents

Abstract Chronic obstructive pulmonary disease (COPD) includes a spectrum of conditions that have in common varying degrees of airflow obstruction, such as chronic bronchitis and emphysema. There is an increasing evidence of involvement of sphingolipids as key molecular mediators or biomarkers of disease in emphysema, chronic bronchitis, and more recently in asthma, another disease characterized by (reversible) airflow obstruction. Given the recognized central role of oxidative stress and inflammatory stimuli along with involvement of

I. Petrache (✉)
Division of Pulmonary and Critical Care Medicine, Department of Medicine,
Indiana University School of Medicine, Indianapolis, IN, USA

Division of Pulmonary, Allergy, Critical Care and Occupational Medicine, Department of Medicine, Indiana University School of Medicine, Walther Hall-R3 C400, 980 W. Walnut Street, Indianapolis, IN 46202-5120, USA
e-mail: ipetrach@iupui.edu

D.N. Petrusca
Division of Pulmonary and Critical Care Medicine, Department of Medicine,
Indiana University School of Medicine, Indianapolis, IN, USA

E. Gulbins and I. Petrache (eds.), *Sphingolipids in Disease*,
Handbook of Experimental Pharmacology 216,
DOI 10.1007/978-3-7091-1511-4_12,

immune responses, apoptosis, and tissue remodeling in the development of chronic obstructive lung diseases, it is not surprising that sphingolipids have been shown to play important role in their pathobiology. In particular the pro-apoptotic effects of ceramide were suspected as events in the lung destruction that occurs as a result of apoptotic loss of structural cells comprising the alveolar walls, such as microvascular endothelial cells and alveolar epithelial cells. In addition, the role of ceramide was investigated in models of larger airway epithelial cell stress responses to cigarette smoke, in the context of ensuing airway remodeling and inflammation. This chapter discusses current evidence of sphingolipid perturbations in experimental models of COPD and relevant links to human disease based on translational and epidemiological data.

Keywords COPD • Emphysema • Asthma • Chronic bronchitis • Airflow obstruction • Lung • CFTR • Sphingolipids • Ceramide • Sphingomyelinase • De novo sphingolipid synthesis • Inflammation • Human • Mouse • Cigarette smoking • Apoptosis • Autophagy • Senescence • Alveolar macrophage • Endothelial cells • Epithelial cells

Chronic obstructive pulmonary diseases encompass a spectrum of conditions that have in common varying degrees of obstruction of airflow in the lungs. The functional definition of chronic obstructive pulmonary diseases traditionally includes asthma, chronic bronchitis, and emphysema. While asthma is defined by reversible airflow obstruction, with normal lung function in-between episodes of obstructive ventilatory defects, chronic bronchitis and emphysema are considered irreversible conditions typically included in the clinical diagnosis of chronic obstructive pulmonary disease (COPD). Significant overlap exists among these three conditions, and furthermore, some patients with COPD clinical features do not manifest obstructive functional defects on pulmonary function tests. The clinical heterogeneity along with a lack of satisfactory therapeutic options for COPD led to ongoing research to better characterize and understand COPD. In this chapter, we will briefly introduce the clinical aspects of the lung diseases covered and highlight discoveries from published data revealing involvement of sphingolipids as key molecular mediators or biomarkers of disease in emphysema, chronic bronchitis, and asthma, highlighting the progress and potential applications but also the challenges facing their immediate implementation in clinical care.

1 Introduction into the Pathogenesis of Chronic Obstructive Diseases of the Lung

While asthma is a disease of all ages and may be associated with atopy, chronic bronchitis and emphysema are found almost exclusively in adults and are frequently associated with cigarette smoking or exposure to other sources of combustive smoke.

Both asthma and COPD are common conditions with increasing prevalence not only in developed countries but worldwide. Unfortunately, COPD has now become a societal burden, with significant morbidity, healthcare and productivity costs, and increasing mortality, surpassing stroke as the third cause of death in the USA.

The pathophysiology of chronic obstructive lung diseases includes distinct and also several overlapping mechanisms explaining the development of lung pathology as well as systemic comorbidities (Tuder and Petrache 2012b). A comprehensive review of these pathogenic mechanisms is beyond the scope of this chapter. Briefly, in susceptible individuals, chronic cigarette smoke exposure triggers oxidative and endoplasmic reticulum stress responses, loss of epithelial and endothelial barrier function, structural cell apoptosis, and impairment of apoptotic cell clearance by specialized phagocytes, autophagy, and inflammatory lung responses. In the case of chronic bronchitis, such cellular processes lead to increased mucus production, decreased airway clearance, airway wall inflammation, bronchoconstriction, and airway remodeling, which culminate in chronic cough, airflow obstruction, and increased propensity for exacerbations due to viral or bacterial respiratory infections. Patients with chronic bronchitis may go on to develop pulmonary hypertension and cor pulmonale, as well as impaired ventilation with carbon dioxide retention and respiratory acidosis. In those who develop emphysema, while the airway changes described above may be present in various degrees, there is also marked destruction of distal airspaces and terminal bronchioles, with matrix proteolysis and loss of alveolar surface available for gas exchange and a loss of elastic recoil of the lung with increased lung compliance and air trapping. These patients become hypoxic and may also develop cor pulmonale as well as chronic respiratory insufficiency, leading to death. COPD patients have important comorbidities in addition to cor pulmonale, including cachexia or severe weight loss and muscle atrophy; various degrees of bone marrow dysfunction, depression, anxiety; and an increased risk of lung cancer and cardiovascular diseases (Barnes and Celli 2009). Finally, in asthma, the airway inflammation, bronchoconstriction or bronchial hyperreactivity, and mucus hypersecretion lead to episodic airflow obstruction triggered by exposure to allergic stimuli, viral infections, or other airway irritants. In some, the disease progresses to irreversible phenotypes, with airway remodeling and persistence of functional and clinical symptoms (Wenzel 2012).

Given the recognized central role of oxidative stress and inflammatory stimuli along with involvement of immune responses, apoptosis, and tissue remodeling in the development of chronic obstructive lung diseases, it is not surprising sphingolipids have been shown to play important role in their pathobiology.

2 The Role of Sphingolipids in COPD

The central role of sphingolipids as second messengers and important mediators of oxidative stress, apoptosis, cell growth, and proliferation led to investigations into their involvement in cell stress responses to cigarette smoke exposure of the lungs. In particular, the pro-apoptotic effects of ceramide were suspected as events in the lung destruction that occurs as a result of apoptotic loss of structural cells comprising the alveolar walls. Most studies that investigated ceramide in this context addressed the apoptosis of capillary or microvascular endothelial cells and alveolar epithelial cells. In addition, the role of ceramide was investigated in models of larger airway epithelial cell stress responses to cigarette smoke, in the context of ensuing airway remodeling and inflammation. We will review the current evidence of sphingolipid perturbations in response to cigarette smoke in experimental models of COPD and relevant links to human disease based on translational and epidemiological data.

2.1 *Sphingolipid Involvement in Emphysema Development*

The disruption of the balance between apoptosis and replenishment of structural cells in the lung is considered a major mechanism of destruction of lung tissue in response to cigarette smoke (Henson et al. 2006; Kasahara et al. 2000; Tuder and Petrache 2012a). In a murine model of apoptosis-dependent emphysema induced by blockage of vascular endothelial growth factor (VEGF) signaling, the lung exhibited marked increases in ceramides (Petrache et al. 2005). Furthermore, direct exposure to CS increased ceramides in cultured lung cells (Petrache et al. 2005), mouse lungs (Petrache et al. 2008), and in the lung tissue and acellular bronchoalveolar fluid of individuals who actively smoke (Petrache et al. 2005; Petrusca et al. 2010). The notion that ceramides play an important role in the pathogenesis of COPD was suggested by evidence that ceramide upregulation in the lungs was responsible for both alveolar cell apoptosis and oxidative stress, therefore amplifying lung destruction and being causal elements in the development of airspace enlargement (Petrache et al. 2005). Direct intratracheal instillation of ceramide in the lungs recapitulated the phenotype of the VEGFR blockade and caused oxidative-stress-dependent apoptosis of lung structural cells in mice (Petrache et al. 2005, 2008). The mechanisms by which VEGFR blockade upregulated ceramide in the lung are not yet well established but are known to include activation of both the de novo and the ASMase ceramide synthesis pathways (Petrache et al. 2005, 2008), which may be activated as part of the stress response which includes both deprivation of lung endothelial cell survival factors (Chavakis and Dimmeler 2002) and oxidative stress. While the requirement of ceramide synthesis in cigarette smoke-induced emphysema remains to be demonstrated, ceramides are increased in the lungs of mice after 4 weeks of intermittent cigarette smoke exposure (Petrache et al. 2008), a timepoint which precedes the development of morphological changes of emphysema that typically occur after 4–6 months of exposure. Interestingly, dihydroceramides were also increased by cigarette smoke, suggesting activation of the de novo pathway of ceramide synthesis as a potential mechanism of initial ceramide upregulation following chronic smoke exposure in mice. Overall, the published evidence implicates abnormal increases in ceramides as a pathogenic link of airspace destruction in emphysema as well as potential hubs of amplification self-synthesis and, more importantly, of injury mechanisms such as oxidative stress, apoptosis, and inflammation. Such self-augmentation loops when fully engaged may overcome endogenous repair and may explain the relentless tissue destruction in emphysema, which progresses even after cessation of cigarette smoke exposure.

2.2 Ceramide-Induced Self-Amplification of Synthesis in the Lung

Two main lines of evidence support the notion of self-amplification of ceramide synthesis in the lung. First, intratracheal ceramide instillation induced a positive feedback of ceramide synthesis, allowing for amplification of apoptosis and lung destruction (Petrache et al. 2008). While the method of adding ceramides into the lung is not physiological, we measured increased extra- or paracellular ceramides in the lungs (Petrusca et al. 2010) and plasma of smokers, likely as part of microparticles or bound to proteins. Secondly, administration of ceramide antibodies attenuated VEGF blockade-induced apoptosis in the lung (Petrache et al. 2005), suggesting extra- or paracellular ceramide may be biologically active and may exert pro-apoptotic effects in the lung. Interestingly, the endogenous ceramide species synthesized following exogenous ceramide instillation were associated with significant activation of lung acid sphingomyelinase (ASMase) activity, suggesting sphingomyelin breakdown as a source of amplified endogenous ceramide production (Petrache et al. 2008) as previously indicated by experiments in ASMase-deficient human fibroblasts derived from individuals with Niemann-Pick disease (Petrache et al. 2005). These findings support the presence of active paracellular ceramide pools, potentially generated from plasma membrane sphingomyelin through a ceramide-triggered positive feedback mechanism.

2.3 Ceramide-Induced Oxidative Stress Amplification in the Lung

Oxidative stress plays a major role in emphysema development and pathophysiology (Rahman et al. 1996). Oxidative stress can be induced by cigarette smoke itself, since it contains abundant free radicals, or can result from activated or injured inflammatory and structural lung cells, respectively. There is a mutual connection between lung oxidative stress and apoptosis and alveolar destruction in the lung, as shown by studies using antioxidants in the VEGF receptor blockade (Tuder et al. 2003b), or in the chronic cigarette smoke exposure (Foronjy et al. 2006) models. At the cellular level, a mutual connection between oxidative stress and ceramide production in the induction of apoptosis has long been recognized (Pettus et al. 2002), where ceramides induced reactive oxygen species (ROS) directly by altering mitochondrial function (Andrieu-Abadie et al. 2001; Quillet-Mary et al. 1997) or indirectly by inactivating antioxidant enzymes such as catalase (Iwai et al. 2003). These findings indicated a potentially complex cross talk between ceramides and oxidative stress in the development of lung

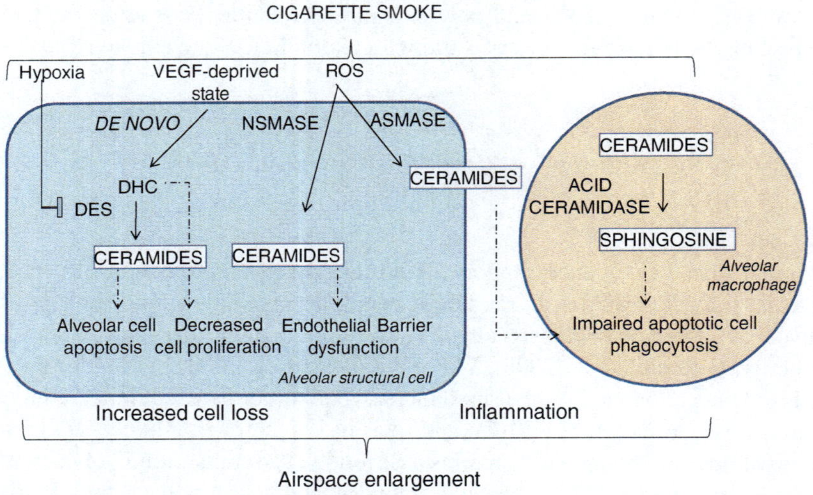

Fig. 1 Schematic of ceramide involvement in cigarette smoke-induced responses in the lung

injury emphysema. Such a mutual interaction was demonstrated using transgenic overexpression of the antioxidant enzyme Cu/Zn SOD which was protected against ceramide-initiated lung damage, suggesting that ceramide induces redox-dependent cell death and alveolar enlargement (Petrache et al. 2008). Furthermore, ceramide upregulation after airway instillation of bioactive synthetic ceramide inhibited SOD activity, subsequently causing increased oxidative stress in the lung. In turn, the oxidative stress directly contributed to caspase activation and alveolar space enlargement as documented by the protective effects of overexpressing Cu/Zn SOD in mice. This study also showed that ASMase activation in the lung is redox dependent, implicating oxidative stress both upstream of ceramide amplification and downstream, as a pathological mediator of alveolar enlargement caused by increased lung ceramides.

In the context of cigarette smoke exposure, an important interplay between oxidative stress and ceramide production has been noted during one of the earliest changes induced by cigarette smoke, that of lung alveolar barrier dysfunction. Previous reports have documented that CS increases epithelial cell permeability (Rusznak et al. 2000; Burns et al. 1989) in an oxidative stress-dependent manner (Li et al. 1996; Serikov et al. 2006). Whereas breaches of the epithelial barrier may induce wound-repair inflammatory responses, disruption of the endothelial barrier may directly increase the access of circulating proteins, plasma, and inflammatory cells to the interstitium and alveolar spaces. CS and its highly diffusible oxidative free radicals may also directly injure endothelial cells in an oxidative stress- and also ceramide-dependent manner (Schweitzer et al. 2011). Interestingly, the ceramides responsible for the breach induced by CS in the rat lung endothelial barrier were produced as a result of oxidative-stress-induced neutral SMase activation (Fig. 1). This mechanism is similar to that reported in large epithelial cells exposed to CS, in which the ceramide upregulation via neutral SMase was responsible for cell death (Levy et al. 2009). In lung endothelial cells, however, neutral SMase-generated ceramides activated the cytoskeleton and disrupted cell–cell junctions, in part via Rho kinase activation, independent of pro-apoptotic effects. At high ceramide concentrations, this effect culminated in edema formation in the intact animal, consistent with earlier reports implicating ASM-generated ceramide as a mediator of platelet-activating factor induced pulmonary edema in a pump-perfused mouse lung preparation (Goggel et al. 2004; Kuebler et al. 2010). However, at low concentrations, ceramide, similar to cigarette smoke exposure, induced subtle changes in the alveolo-capillary barrier which may render endothelial cells more susceptible to edemogenic effects of subsequent injuries associated with acute respiratory distress syndrome (ARDS) and may independently contribute, over time, to chronic inflammatory changes associated with COPD.

2.4 Death and Survival Signaling Induced by Sphingolipids in COPD

2.4.1 Ceramides and Lung Apoptosis

For more than a decade, it has been recognized that alveolar cell apoptosis is mechanistically involved in emphysema development (Kasahara et al. 2000; Rangasamy et al. 2004; Tuder et al. 2003a, b), accounting for the unique nature of alveolar septal destruction when compared with other lung diseases also characterized by inflammation and increased matrix proteolysis. Similar to caspase inhibition, interventions aimed at normalizing the lung ceramide content significantly ameliorated lung alveolar cell apoptosis (Petrache et al. 2005). In turn, direct intratracheal instillation of ceramides in mice increased total lung ceramide levels in a dose-dependent manner and was associated with increased lung alveolar cell apoptosis (Petrache et al. 2005). However, it is important to point out that excessive inhibition of ceramide synthesis such as with high doses of fumonisin B1 also increased lung cell apoptosis and even caused airspace enlargement, emphasizing the notion that it is the homeostasis of ceramides and/or their metabolites that is required for proper lung structure maintenance, and abnormally low ceramide levels may actually be detrimental.

In many experimental models, including cultured pulmonary cells and in the VEGF receptor blockade model of emphysema, ceramides were accumulated upstream in the apoptosis pathways, preceding caspase-3 activation (Petrache et al. 2005; Ravid et al. 2003; Sawada et al. 2002; Tepper et al. 1997). Both SMases and de novo pathways of ceramide synthesis have been implicated in augmentation of pro-apoptotic ceramides in lung cells in response to various stimuli relevant to COPD, such as oxidative stress, cigarette smoking, or TNF-α (Medler et al. 2008) (Fig. 1). In this context of increased inflammatory mediators such as TNF-α, ceramides facilitate pro-apoptotic TNF-α effects on its receptor (Tchikov et al. 2011). In addition, ceramides may activate caspase-independent apoptotic pathways (Thon et al. 2005), involving mitochondria plasma membrane depolarization and release of pro-apoptotic factors. It is becoming clear that the mechanism of apoptosis induction by ceramides appears to depend on the species, cell type, context, and the ceramide molecular species involved. While some of these pathways have been elucidated in murine models and human cell lines, more studies are needed to determine this specificity in the context of cigarette smoke-induced lung damage in primary human lung cells in culture and ideally in situ. It has been reported that endothelial cells are particularly susceptible to ceramide-dependent apoptosis (Haimovitz-Friedman et al. 1997) and themselves contain very high levels of ASMase and ceramides. These findings, together with indications that excessive lung microvascular endothelial cell apoptosis is sufficient to cause emphysema-like disease (Giordano et al. 2008; Taraseviciene-Stewart et al. 2005), explain why many studies in the field have been focused on endothelial cells. Nevertheless, it is not known in humans or even in mouse models what is the relative importance of cell type-specific apoptosis in emphysema development. It is likely that cell death of either epithelial or endothelial cells or their supporting fibroblasts in the alveolar septum could trigger events that culminate in airspace enlargement. Analyzing the cell type-specific sensitivity to sphingolipid-induced cell death, we found that ceramide instillation in mice lungs triggered apoptosis of both type II alveolar epithelial cells and endothelial cells (Kamocki et al., 2012, in press). Ongoing work in our lab indicates that both human primary lung epithelial and endothelial cells undergo apoptosis when exposed to ceramides, although to lesser degree than murine cells.

2.4.2 Ceramides and Lung Autophagy

Autophagy occurs in most cells at a basal rate where it acts as a cytoplasmic quality control (QC) mechanism to eliminate protein aggregates, damaged organelles, and other non-active structures such as ROS and stressed ER (Komatsu et al. 2007; Mehrpour et al. 2010). The role of autophagy in human diseases is less clear, as it has been associated with both protective and detrimental consequences. COPD lung specimens exhibit increased markers of autophagy, which are evoked in response to cigarette smoke exposure, in vitro and in vivo (Kim et al. 2008; Ryter et al. 2010), possibly as part of the stress response. Ceramides have been shown to induce a similar stress by mimicking a state of amino acid deprivation, which may explain the induction of autophagy upon ceramide treatment. The relevance of this mechanism to ceramides upregulated by cigarette smoking remains to be determined. However, an interesting link between CS exposure and autophagy/apoptosis of lung cells has been established via a direct correlation between decreased cystic fibrosis transmembrane receptor (CFTR) expression and emphysema through increased ceramide accumulation. CFTR is a chloride channel and ATP family transporter which is essential for the function of airway lung cells, as learned from the common genetic condition caused by CFTR dysfunction, cystic fibrosis. Interestingly, cigarette smoke can directly inhibit CFTR function which facilitates ceramide accumulation by regulating membrane ASMase activity and/or intracellular pH (Bodas et al. 2011; Noe et al. 2009; Teichgraber et al. 2008). In addition to ceramide-assisted formation of signaling platforms that are required for activation of the apoptotic cascade (Miyaji et al. 2005), membrane CFTR was required for proper autophagy responses to cigarette smoke. In the absence of CFTR, dysregulated autophagy was associated with accumulation of damaged proteins triggered by CS exposure, and increased apoptosis (Bodas et al. 2011).

2.4.3 Sphingolipids and Lung Cell Survival

Given the multiple pathways of ceramide upregulation in response to stress and the multiple metabolites that depend on homeostatic ceramide levels, it may be difficult to inhibit its production in a clinical setting. Given that one of the most studied ceramide metabolites, sphingosine-1-phosphate (S1P), has potent pro-survival effects, a strategy to effectively inhibit the pro-apoptotic effect of ceramide has been to rebalance the sphingolipid homeostasis (improving the sphingolipid rheostat) (Payne et al. 2002) by increasing S1P levels (Osawa et al. 2005). S1P has complex mechanisms of actions, many explained by ligation of G-protein-coupled S1P receptors (S1P1–S1P5), of which of particular interest is the S1P1 receptor that exerts anti-apoptotic actions in vascular cells (Bonnaud et al. 2007). Augmentation of S1P signaling in the VEGFR blockade model of emphysema attenuated lung cell apoptosis and airspace enlargement (Diab et al. 2010), supporting the notion that the effect of high levels of ceramides may be antagonized by increasing S1P signaling.

2.5 Role of Sphingolipids in Alveolar Macrophage Dysfunction in COPD

The clearance of apoptotic cells is normally promptly and efficiently executed via a highly dynamic and regulated engulfment process of central importance for tissue homeostasis (Henson et al. 2001). One mechanism postulated to link apoptosis and inflammation in lung emphysema pathology is a defect in clearance of apoptotic cells in the lung. An impairment of apoptotic cell phagocytosis (known as efferocytosis, to differentiate it from bacterial phagocytosis) primarily carried out by specialized phagocytes such as alveolar macrophages may explain the increased detection of apoptotic cells in emphysema lungs (Henson et al. 2001). Another important consequence of impaired efferocytosis is that non-engulfed apoptotic cells may undergo secondary necrosis with ensuing inflammation. Furthermore, efferocytosis differs from classical phagocytosis by being actively involved in a protective anti-inflammatory response (Ravichandran and Lorenz 2007). Smoking impairs efferocytosis and alveolar macrophages from COPD patients who exhibit defective efferocytosis (Hodge et al. 2003, 2007). Alveolar macrophages exposed to cigarette smoking exhibit increased ceramide levels, and ceramides or sphingosine, similar to cigarette smoke, dose-dependently impairs AM efferocytosis (Petrusca et al. 2010). Moreover, CS-induced efferocytosis impairment is in large part explained by an upregulation of acid ceramidase which causes endogenous sphingosine accumulation, which, in turn, affected the plasma membrane recruitment of an important effector of cytoskeletal actin remodeling, the Rac1 GTPase, a dynamic process required for engulfment. These data suggested that ceramides and their metabolite not only cause alveolar structural apoptosis but also impair the clearance of these apoptotic bodies by specialized phagocytes, which may contribute to lung inflammation despite an apoptotic rather than necrotic primary form of death of structural lung cells (Fig. 1).

2.6 *Sphingolipid Responses in Hypoxic Lungs*

Alveolar hypoxia and consequent hypoxemia are late effects of severe COPD. The key driver of lung hypoxia is the ventilation/perfusion mismatch resulting from progressive airflow limitation and loss of alveolar surface area. Uncorrected chronic hypoxemia is associated with the development of adverse sequels of COPD, including pulmonary hypertension, secondary polycythemia, systemic inflammation, and skeletal muscle dysfunction (Kent et al. 2011). Hypoxia elicits diverse and insufficiently understood effects on lipid metabolism, including modulation of lipid second messengers, such as ceramides. Both de novo ceramide synthesis (Kang et al. 2010) and sphingomyelinase-mediated ceramide production (Cogolludo et al. 2009) have been reported in the activation of ceramide responses to hypoxia in addition to an inhibition of the intracellular ceramide transport during oxygen deprivation (Kendler and Dawson 1992), depending on the cell type and the injury type (ischemia versus hypoxia). Cellular sensing of hypoxia involves a rapid upregulation of DHC in a non-transcriptional, dose-dependent but hypoxia inducible factor (HIF)-independent manner that may in turn regulate the cell proliferation response to low oxygen tension (Devlin et al. 2011). While this effect may diminish levels of intracellular ceramides with possible protective effects, the marked increases in DHCs may not be inconsequential, as they may exert biological effects, which are distinct from those of ceramides (Crowder 2009). Indirect augmentation of DHCs either by siRNA blockade of DEGS1 or DEGS2 or direct treatment with DHC decreased cell proliferation (Devlin et al. 2011) (Fig. 1). Additionally, DHCs may actively preserve cell viability by increasing autophagy (Signorelli et al. 2009) during conditions of low oxygen tension.

3 Chronic Bronchitis

Chronic bronchitis is a clinical diagnosis, primarily based on the presence of chronic cough. In this chapter we will focus on the chronic bronchitis syndrome associated with COPD or exposure to cigarette smoking, excluding chronic bronchitis due to cystic fibrosis, which is covered elsewhere. From a pathophysiology standpoint, however, there may be significant overlap, since, as discussed above, it has been shown that cigarette smoking impairs the expression of CFTR, a protein whose genetically determined abnormalities are the cause of cystic fibrosis. It is therefore conceivable that the paradigms of sphingolipid involvement in CF apply to a certain extent to the pathogenesis of chronic bronchitis induced by cigarette smoking or other pollutants that may impair CFTR expression and/or function. A recent study compared the ceramide content of lower airway epithelium in CF with that in emphysema, pulmonary hypertension, or unused lungs from transplant donors, free of pathology (Brodlie et al. 2010). Both emphysema and CF lungs exhibited increased ceramide immunostaining in the epithelium of the lower airway compared to non-diseased lungs, but relative to CF airways, emphysema airways had less striking ceramide elevations in these regions of the lung. In all lungs, there was a correlation between epithelial ceramide expression and markers of neutrophil inflammation. While it is tempting to speculate that the ceramide content of airway epithelium may correlate with the degree of CFTR dysfunction, it is notable that all CF lungs had *Pseudomonas aeruginosa* colonization, whereas only 50 % of emphysema lungs did; however, airways from chronic bronchitis forms of COPD were not included. In addition to findings implicating increased ceramides in its pathogenesis, the S1P signaling may also have a role in the inflammatory manifestations of COPD such as chronic bronchitis. In a recent report based on lung tissue from COPD patients who presented for lung cancer biopsies, the authors measured significantly decreased S1PR5 expression in these lungs and a potentially important cross talk between sphingosine kinase activity producing S1P and S1P receptors that respond to S1P in the lung (Cordts et al. 2011). However, correlations with COPD phenotype or with topographical differences in expression (large airway versus parenchyma) remain to be established.

Overall, these studies encourage future investigations in the role of ceramides and its metabolites not only in airway remodeling and aberrant mucous production, both central components of chronic bronchitis pathology, but also in the risk for bacterial colonization, an important determinant of COPD exacerbations which have an enormous impact on disease morbidity.

4 Sphingolipid Involvement in Asthma

Asthma is a Th2-driven inflammatory lung disease that leads to bronchial narrowing from increased mucus production, airway smooth muscle hyperreactivity and contraction, inflammatory cell accumulation, and airway wall remodeling (Jackson et al. 2011; Szefler and Dakhama 2011). Asthma has complex pathogenesis that underlies several clinical phenotypes, and despite major advances, it remains incompletely understood. Sphingolipids have been implicated in several aspects of asthma pathogenesis, ranging from modulation of inflammatory cell function, smooth muscle contractility, and airway remodeling to susceptibility of disease. Glycoceramides have been shown to be ligand activators of invariant natural killer T (CD1d-restricted NKT) cells, being presented via the CD1d molecule on antigen-presenting cells. Their importance in asthmatic responses has been recently studied by several groups. In naïve (non-sensitized) mice, alpha-galactosylceramide instillation via the nose but not via intravenous administration induced severe airway hyperreactivity in an NKT-dependent manner (Akbari et al. 2010; Chuang et al. 2011). Interestingly, repeated administration of alpha-galactosylceramide attenuated the subsequent capability of NKT cells to induce airway hyperreactivity and led to reduced levels of airway inflammation and airway hyperreactivity, suggesting a titration effect on NKT activation which is directly related to subsequent functional responses (Meyer et al. 2006). However, there was augmentation of eosinophilic inflammation noted in allergic mice injected with alpha-galactosylceramide before immunization. These findings raise concern that the use of alpha-galactosylceramide as a vaccine adjuvant may have untoward effect of increased allergic inflammation in predisposed individuals. When alpha-galactosylceramide was administered following sensitization, on the first day of allergen challenge, it attenuated the airway allergic inflammation (Matsuda et al. 2005). In a similar murine model of ovalbumin-induced bronchoreactivity, intranasal treatment with beta-glucosylceramide decreased bronchial reactivity to ovalbumin challenge and decreased inflammatory airway responses to a similar extent as steroids and was associated with T cells trapping in the liver (Horani et al. 2011). Intriguingly, Chuang et al. (2011) administered intranasally alpha-galactosylceramide in a similar model of ovalbumin-induced allergic asthma with distinct results (Chuang et al. 2011). It has been indeed noted that the fashion and timing of alpha-galactosylceramide are important determinants of its ultimate effects during murine allergic responses (Iwamura and Nakayama 2007). This study reveals that the role of glycoceramides as a therapeutic intervention in asthma deserves further investigations, including precise dose responses and expansion to other asthma models. In addition, while suggesting an interesting cross talk between the liver and lung T cell trafficking and opportunities of NKT manipulations in the asthma treatment, these studies did not address the role of endogenous glycoceramides and other sphingolipid metabolites in the pathogenesis of asthma. Their complex involvement is suggested by several studies in which inhibition of enzymes leading to multiple glycosphingolipid metabolites was associated with a protective effect in the ovalbumin asthma model, primarily linked to inhibition of mast cell degranulation (Karman et al. 2010). Similarly, inhibition of ceramide synthase with fumonisin B1 inhibited allergic airway responses and airway inflammation which was associated with epithelial cell apoptosis and nitro-oxidative stress (Masini et al. 2008). Important downstream metabolites of ceramide in addition to glycosphingolipids have been indeed long been recognized as important asthma mediators. Of these, S1P received particular attention, since it has been measured in increased levels in the bronchoalveolar lavage fluid of asthma patients and animal models (Ammit et al. 2001; Nishiuma et al. 2008). S1P has been involved in regulating the airway smooth muscle function (Ammit et al. 2001), and its effect on S1P2 receptors has been implicated as a major cause of mast cell degranulation in allergic asthma (Oskeritzian et al. 2007, 2010; Price et al. 2009). Interestingly, activation of mast cells causes itself activation of sphingosine kinase further increasing the S1P, leading to calcium-dependent activation of these cells (Melendez 2008). More recently, other ceramide metabolites such as Cer1P have been suggested as potential mediators of asthmatic pathogenic mechanisms in the lung (Niwa et al. 2010).

In addition to roles in asthma pathogenesis, an interesting potential role of endogenous sphingolipids in an individual's risk for asthma development has been discovered through genome-wide association studies. A single-nucleotide polymorphism (SNP) at chromosome 17q21 near the ORMDL3 gene (Moffatt et al. 2007) has been discovered and since then, confirmed in several cohorts, was associated with an increased risk of childhood asthma. ORMDL 1/2/ and 3 are human genes homologs of the yeast ORM1/2 that encode transmembrane proteins in endoplasmic reticulum. Subsequent work demonstrated that the Orm proteins modulate sphingolipid production by inhibition of the first step in the de novo sphingolipid synthesis, via a tightly regulated inhibitory interaction with serine palmitoyltransferase (Breslow et al. 2010), thus controlling sphingolipid homeostasis.

In conclusion, ceramides and their sphingolipid metabolites are involved in the pathogenesis of chronic obstructive lung diseases, but their precise regulation and importance are awaiting further investigations. In emphysema, they may act as a proximal hub of amplification for apoptosis, oxidative stress, and own synthesis. Cigarette smoke and possibly other pollutant exposures disrupt sphingolipid homeostasis in the lung, generating distinct acute and chronic ceramide responses, responsible for the death of structural alveolar epithelial and endothelial cells, inhibition of clearance of apoptotic cells by alveolar macrophages, and potentially impairment of cell repair that sustain the irreversible lung destruction in emphysema. Human cells exhibit an adaptive response to increased ceramide concentrations in the lung that include induction of autophagy and decreased cell proliferation. Understanding the mechanisms by which the cells respond to complex sphingolipid changes in the lungs exposed to environmental insults may enhance our ability to diagnose early and even intervene therapeutically at various stages of these chronic debilitating lung diseases.

Acknowledgments We are thanking the Petrache Lab technicians and trainees, the Reba and Floyd Smith Chair in Respiratory Research, and the Calvin English Chair for their support of this work. We would also like to acknowledge the funding support from the National Institutes of Health (NHLBI-HL077328, NIH-R21-DA029249), BSF (to I.P.), and the AHA (to D.P.).

References

Akbari O, Stock P, Singh AK, Lombardi V, Lee WL, Freeman GJ, Sharpe AH, Umetsu DT, Dekruyff RH (2010) PD-L1 and PD-L2 modulate airway inflammation and iNKT-cell-dependent airway hyperreactivity in opposing directions. Mucosal Immunol 3:81–91

Ammit AJ, Hastie AT, Edsall LC, Hoffman RK, Amrani Y, Krymskaya VP, Kane SA, Peters SP, Penn RB, Spiegel S, Panettieri RA Jr (2001) Sphingosine 1-phosphate modulates human airway smooth muscle cell functions that promote inflammation and airway remodeling in asthma. FASEB J 15:1212–1214

Andrieu-Abadie N, Gouaze V, Salvayre R, Levade T (2001) Ceramide in apoptosis signaling: relationship with oxidative stress. Free Radic Biol Med 31:717–728

Barnes PJ, Celli BR (2009) Systemic manifestations and comorbidities of COPD. Eur Respir J 33: 1165–1185

Bodas M, Min T, Vij N (2011) Critical role of CFTR dependent lipid-rafts in cigarette smoke induced lung epithelial injury. Am J Physiol Lung Cell Mol Physiol 300(6):L811–L820

Bonnaud S, Niaudet C, Pottier G, Gaugler MH, Millour J, Barbet J, Sabatier L, Paris F (2007) Sphingosine-1-phosphate protects proliferating endothelial cells from ceramide-induced apoptosis but not from DNA damage-induced mitotic death. Cancer Res 67:1803–1811

Breslow DK, Collins SR, Bodenmiller B, Aebersold R, Simons K, Shevchenko A, Ejsing CS, Weissman JS (2010) Orm family proteins mediate sphingolipid homeostasis. Nature 463: 1048–1053

Brodlie M, McKean MC, Johnson GE, Gray J, Fisher AJ, Corris PA, Lordan JL, Ward C (2010) Ceramide is increased in the lower airway epithelium of people with advanced cystic fibrosis lung disease. Am J Respir Crit Care Med 182:369–375

Burns AR, Hosford SP, Dunn LA, Walker DC, Hogg JC (1989) Respiratory epithelial permeability after cigarette smoke exposure in guinea pigs. J Appl Physiol 66:2109–2116

Chavakis E, Dimmeler S (2002) Regulation of endothelial cell survival and apoptosis during angiogenesis. Arterioscler Thromb Vasc Biol 22:887–893

Chuang YH, Wang TC, Jen HY, Yu AL, Chiang BL (2011) alpha-Galactosylceramide-induced airway eosinophilia is mediated through the activation of NKT cells. J Immunol 186: 4687–4692

Cogolludo A, Moreno L, Frazziano G, Moral-Sanz J, Menendez C, Castaneda J, Gonzalez C, Villamor E, Perez-Vizcaino F (2009) Activation of neutral sphingomyelinase is involved in acute hypoxic pulmonary vasoconstriction. Cardiovasc Res 82:296–302

Cordts F, Pitson S, Tabeling C, Gibbins I, Moffat DF, Jersmann H, Hodge S, Haberberger RV (2011) Expression profile of the sphingosine kinase signalling system in the lung of patients with chronic obstructive pulmonary disease. Life Sci 89:806–811

Crowder CM (2009) Cell biology. Ceramides–friend or foe in hypoxia? Science 324:343–344

Devlin CM, Lahm T, Hubbard WC, Van Demark M, Wang KC, Wu X, Bielawska A, Obeid LM, Ivan M, Petrache I (2011) Dihydroceramide-based response to hypoxia. J Biol Chem 286: 38069–38078

Diab KJ, Adamowicz JJ, Kamocki K, Rush NI, Garrison J, Gu Y, Schweitzer KS, Skobeleva A, Rajashekhar G, Hubbard WC, Berdyshev EV, Petrache I (2010) Stimulation of sphingosine 1-phosphate signaling as an alveolar cell survival strategy in emphysema. Am J Respir Crit Care Med 181:344–352

Foronjy RF, Mirochnitchenko O, Propokenko O, Lemaitre V, Jia Y, Inouye M, Okada Y, D'Armiento JM (2006) Superoxide dismutase expression attenuates cigarette smoke- or elastase-generated emphysema in mice. Am J Respir Crit Care Med 173:623–631

Giordano RJ, Lahdenranta J, Zhen L, Chukwueke U, Petrache I, Langley RR, Fidler IJ, Pasqualini R, Tuder RM, Arap W (2008) Targeted induction of lung endothelial cell apoptosis causes emphysema-like changes in the mouse. J Biol Chem 283:29447–29460

Goggel R, Winoto-Morbach S, Vielhaber G, Imai Y, Lindner K, Brade L, Brade H, Ehlers S, Slutsky AS, Schutze S, Gulbins E, Uhlig S (2004) PAF-mediated pulmonary edema: a new role for acid sphingomyelinase and ceramide. Nat Med 10:155–160

Haimovitz-Friedman A, Cordon-Cardo C, Bayoumy S, Garzotto M, McLoughlin M, Gallily R, Edwards CK 3rd, Schuchman EH, Fuks Z, Kolesnick R (1997) Lipopolysaccharide induces disseminated endothelial apoptosis requiring ceramide generation. J Exp Med 186:1831–1841
Henson PM, Bratton DL, Fadok VA (2001) The phosphatidylserine receptor: a crucial molecular switch? Nat Rev Mol Cell Biol 2:627–633
Henson PM, Vandivier RW, Douglas IS (2006) Cell death, remodeling, and repair in chronic obstructive pulmonary disease? Proc Am Thorac Soc 3:713–717
Hodge S, Hodge G, Scicchitano R, Reynolds PN, Holmes M (2003) Alveolar macrophages from subjects with chronic obstructive pulmonary disease are deficient in their ability to phagocytose apoptotic airway epithelial cells. Immunol Cell Biol 81:289–296
Hodge S, Hodge G, Ahern J, Jersmann H, Holmes M, Reynolds PN (2007) Smoking alters alveolar macrophage recognition and phagocytic ability: implications in chronic obstructive pulmonary disease. Am J Respir Cell Mol Biol 37:748–755
Horani A, Shoseyov D, Doron S, Mruwat R, Amer J, Kerem E, Safadi R (2011) Immune modulation of ovalbumin-induced lung injury in mice using beta-glucosylceramide and a potential role of the liver. Immunobiology 216:548–557
Iwai K, Kondo T, Watanabe M, Yabu T, Kitano T, Taguchi Y, Umehara H, Takahashi A, Uchiyama T, Okazaki T (2003) Ceramide increases oxidative damage due to inhibition of catalase by caspase-3-dependent proteolysis in HL-60 cell apoptosis. J Biol Chem 278: 9813–9822
Iwamura C, Nakayama T (2007) Role of alpha-galactosylceramide-activated Valpha14 natural killer T cells in the regulation of allergic diseases. Allergol Int 56:1–6
Jackson DJ, Sykes A, Mallia P, Johnston SL (2011) Asthma exacerbations: origin, effect, and prevention. J Allergy Clin Immunol 128:1165–1174
Kamocki K, Van Demark M, Fisher A, Rush NI, Presson RG Jr, Hubbard W, Berdyshev EV, Adamsky S, Feinstein E, Gandjeva A, Tuder RM, Petrache I (2012) RTP801 is required for ceramide-induced cell-specific death in the murine lung. Am J Respir Cell Mol Biol 48:87–93
Kang MS, Ahn KH, Kim SK, Jeon HJ, Ji JE, Choi JM, Jung KM, Jung SY, Kim DK (2010) Hypoxia-induced neuronal apoptosis is mediated by de novo synthesis of ceramide through activation of serine palmitoyltransferase. Cell Signal 22:610–618
Karman J, Tedstone JL, Gumlaw NK, Zhu Y, Yew N, Siegel C, Guo S, Siwkowski A, Ruzek M, Jiang C, Cheng SH (2010) Reducing glycosphingolipid biosynthesis in airway cells partially ameliorates disease manifestations in a mouse model of asthma. Int Immunol 22:593–603
Kasahara Y, Tuder RM, Taraseviciene-Stewart L, Le Cras TD, Abman S, Hirth PK, Waltenberger J, Voelkel NF (2000) Inhibition of VEGF receptors causes lung cell apoptosis and emphysema. J Clin Invest 106:1311–1319
Kendler A, Dawson G (1992) Hypoxic injury to oligodendrocytes: reversible inhibition of ATP-dependent transport of ceramide from the endoplasmic reticulum to the Golgi. J Neurosci Res 31:205–211
Kent BD, Mitchell PD, McNicholas WT (2011) Hypoxemia in patients with COPD: cause, effects, and disease progression. Int J Chron Obstruct Pulmon Dis 6:199–208
Kim HP, Wang X, Chen ZH, Lee SJ, Huang MH, Wang Y, Ryter SW, Choi AM (2008) Autophagic proteins regulate cigarette smoke-induced apoptosis: protective role of heme oxygenase-1. Autophagy 4:887–895
Komatsu M, Ueno T, Waguri S, Uchiyama Y, Kominami E, Tanaka K (2007) Constitutive autophagy: vital role in clearance of unfavorable proteins in neurons. Cell Death Differ 14: 887–894
Kuebler WM, Yang Y, Samapati R, Uhlig S (2010) Vascular barrier regulation by PAF, ceramide, caveolae, and NO—an intricate signaling network with discrepant effects in the pulmonary and systemic vasculature. Cell Physiol Biochem 26:29–40
Levy M, Khan E, Careaga M, Goldkorn T (2009) Neutral sphingomyelinase 2 is activated by cigarette smoke to augment ceramide-induced apoptosis in lung cell death. Am J Physiol Lung Cell Mol Physiol 297:L125–L133
Li XY, Rahman I, Donaldson K, MacNee W (1996) Mechanisms of cigarette smoke induced increased airspace permeability. Thorax 51:465–471

Masini E, Giannini L, Nistri S, Cinci L, Mastroianni R, Xu W, Comhair SA, Li D, Cuzzocrea S, Matuschak GM, Salvemini D (2008) Ceramide: a key signaling molecule in a Guinea pig model of allergic asthmatic response and airway inflammation. J Pharmacol Exp Ther 324: 548–557

Matsuda H, Suda T, Sato J, Nagata T, Koide Y, Chida K, Nakamura H (2005) alpha-Galactosylceramide, a ligand of natural killer T cells, inhibits allergic airway inflammation. Am J Respir Cell Mol Biol 33:22–31

Medler TR, Petrusca DN, Lee PJ, Hubbard WC, Berdyshev EV, Skirball J, Kamocki K, Schuchman E, Tuder RM, Petrache I (2008) Apoptotic sphingolipid signaling by ceramides in lung endothelial cells. Am J Respir Cell Mol Biol 38:639–646

Mehrpour M, Esclatine A, Beau I, Codogno P (2010) Autophagy in health and disease. 1. Regulation and significance of autophagy: an overview. Am J Physiol Cell Physiol 298: C776–785

Melendez AJ (2008) Allergy therapy: the therapeutic potential of targeting sphingosine kinase signalling in mast cells. Eur J Immunol 38:2969–2974

Meyer EH, Goya S, Akbari O, Berry GJ, Savage PB, Kronenberg M, Nakayama T, DeKruyff RH, Umetsu DT (2006) Glycolipid activation of invariant T cell receptor + NK T cells is sufficient to induce airway hyperreactivity independent of conventional CD4+ T cells. Proc Natl Acad Sci USA 103:2782–2787

Miyaji M, Jin ZX, Yamaoka S, Amakawa R, Fukuhara S, Sato SB, Kobayashi T, Domae N, Mimori T, Bloom ET, Okazaki T, Umehara H (2005) Role of membrane sphingomyelin and ceramide in platform formation for Fas-mediated apoptosis. J Exp Med 202:249–259

Moffatt MF, Kabesch M, Liang L, Dixon AL, Strachan D, Heath S, Depner M, von Berg A, Bufe A, Rietschel E, Heinzmann A, Simma B, Frischer T, Willis-Owen SA, Wong KC, Illig T, Vogelberg C, Weiland SK, von Mutius E, Abecasis GR, Farrall M, Gut IG, Lathrop GM, Cookson WO (2007) Genetic variants regulating ORMDL3 expression contribute to the risk of childhood asthma. Nature 448:470–473

Nishiuma T, Nishimura Y, Okada T, Kuramoto E, Kotani Y, Jahangeer S, Nakamura S (2008) Inhalation of sphingosine kinase inhibitor attenuates airway inflammation in asthmatic mouse model. Am J Physiol Lung Cell Mol Physiol 294:L1085–L1093

Niwa S, Urtz N, Baumruker T, Billich A, Bornancin F (2010) Ovalbumin-induced plasma interleukin-4 levels are reduced in ceramide kinase-deficient DO11.10 RAG1−/− mice. Lipids Health Dis 9:1

Noe J, Petrusca D, Rush N, Deng P, VanDemark M, Berdyshev E, Gu Y, Smith P, Schweitzer K, Pilewsky J, Natarajan V, Xu Z, Obukhov AG, Petrache I (2009) CFTR regulation of intracellular pH and ceramides is required for lung endothelial cell apoptosis. Am J Respir Cell Mol Biol 41:314–323

Osawa Y, Uchinami H, Bielawski J, Schwabe RF, Hannun YA, Brenner DA (2005) Roles for C16-ceramide and sphingosine 1-phosphate in regulating hepatocyte apoptosis in response to tumor necrosis factor-alpha. J Biol Chem 280:27879–27887

Oskeritzian CA, Milstien S, Spiegel S (2007) Sphingosine-1-phosphate in allergic responses, asthma and anaphylaxis. Pharmacol Ther 115:390–399

Oskeritzian CA, Price MM, Hait NC, Kapitonov D, Falanga YT, Morales JK, Ryan JJ, Milstien S, Spiegel S (2010) Essential roles of sphingosine-1-phosphate receptor 2 in human mast cell activation, anaphylaxis, and pulmonary edema. J Exp Med 207:465–474

Payne SG, Milstien S, Spiegel S (2002) Sphingosine-1-phosphate: dual messenger functions. FEBS Lett 531:54–57

Petrache I, Natarajan V, Zhen L, Medler TR, Richter AT, Cho C, Hubbard WC, Berdyshev EV, Tuder RM (2005) Ceramide upregulation causes pulmonary cell apoptosis and emphysema-like disease in mice. Nat Med 11:491–498

Petrache I, Medler TR, Richter AT, Kamocki K, Chukwueke U, Zhen L, Gu Y, Adamowicz J, Schweitzer KS, Hubbard WC, Berdyshev EV, Lungarella G, Tuder RM (2008) Superoxide dismutase protects against apoptosis and alveolar enlargement induced by ceramide. Am J Physiol Lung Cell Mol Physiol 295:L44–L53

Petrusca DN, Gu Y, Adamowicz JJ, Rush NI, Hubbard WC, Smith PA, Berdyshev EV, Birukov KG, Lee CH, Tuder RM, Twigg HL III, Vandivier RW, Petrache I (2010)

Sphingolipid-mediated inhibition of apoptotic cell clearance by alveolar macrophages. J Biol Chem 285(51):40322–40332
Pettus BJ, Chalfant CE, Hannun YA (2002) Ceramide in apoptosis: an overview and current perspectives. Biochim Biophys Acta 1585:114–125
Price MM, Kapitonov D, Allegood J, Milstien S, Oskeritzian CA, Spiegel S (2009) Sphingosine-1-phosphate induces development of functionally mature chymase-expressing human mast cells from hematopoietic progenitors. FASEB J 23:3506–3515
Quillet-Mary A, Jaffrezou JP, Mansat V, Bordier C, Naval J, Laurent G (1997) Implication of mitochondrial hydrogen peroxide generation in ceramide-induced apoptosis. J Biol Chem 272: 21388–21395
Rahman I, Morrison D, Donaldson K, MacNee W (1996) Systemic oxidative stress in asthma, COPD, and smokers. Am J Respir Crit Care Med 154:1055–1060
Rangasamy T, Cho CY, Thimmulappa RK, Zhen L, Srisuma SS, Kensler TW, Yamamoto M, Petrache I, Tuder RM, Biswal S (2004) Genetic ablation of Nrf2 enhances susceptibility to cigarette smoke-induced emphysema in mice. J Clin Invest 114:1248–1259
Ravichandran KS, Lorenz U (2007) Engulfment of apoptotic cells: signals for a good meal. Nat Rev Immunol 7:964–974
Ravid T, Tsaba A, Gee P, Rasooly R, Medina EA, Goldkorn T (2003) Ceramide accumulation precedes caspase-3 activation during apoptosis of A549 human lung adenocarcinoma cells. Am J Physiol Lung Cell Mol Physiol 284:L1082–L1092
Rusznak C, Mills PR, Devalia JL, Sapsford RJ, Davies RJ, Lozewicz S (2000) Effect of cigarette smoke on the permeability and IL-1beta and sICAM-1 release from cultured human bronchial epithelial cells of never-smokers, smokers, and patients with chronic obstructive pulmonary disease. Am J Respir Cell Mol Biol 23:530–536
Ryter SW, Lee SJ, Choi AM (2010) Autophagy in cigarette smoke-induced chronic obstructive pulmonary disease. Expert Rev Respir Med 4:573–584
Sawada M, Nakashima S, Kiyono T, Yamada J, Hara S, Nakagawa M, Shinoda J, Sakai N (2002) Acid sphingomyelinase activation requires caspase-8 but not p53 nor reactive oxygen species during Fas-induced apoptosis in human glioma cells. Exp Cell Res 273:157–168
Schweitzer KS, Hatoum H, Brown MB, Gupta M, Justice MJ, Beteck B, Van Demark MJ, Gu Y, Presson RG Jr, Hubbard WC, Petrache I (2011) Mechanisms of lung endothelial barrier disruption induced by cigarette smoke: role of oxidative stress and ceramides. Am J Physiol Lung Cell Mol Physiol 301(6):L836–L846
Serikov VB, Leutenegger C, Krutilina R, Kropotov A, Pleskach N, Suh JH, Tomilin NV (2006) Cigarette smoke extract inhibits expression of peroxiredoxin V and increases airway epithelial permeability. Inhal Toxicol 18:79–92
Signorelli P, Munoz-Olaya JM, Gagliostro V, Casas J, Ghidoni R, Fabrias G (2009) Dihydroceramide intracellular increase in response to resveratrol treatment mediates autophagy in gastric cancer cells. Cancer Lett 282:238–243
Szefler SJ, Dakhama A (2011) New insights into asthma pathogenesis and treatment. Curr Opin Immunol 23:801–807
Taraseviciene-Stewart L, Scerbavicius R, Choe KH, Moore M, Sullivan A, Nicolls MR, Fontenot AP, Tuder RM, Voelkel NF (2005) An animal model of autoimmune emphysema. Am J Respir Crit Care Med 171:734–742
Tchikov V, Bertsch U, Fritsch J, Edelmann B, Schutze S (2011) Subcellular compartmentalization of TNF receptor-1 and CD95 signaling pathways. Eur J Cell Biol 90:467–475
Teichgraber V, Ulrich M, Endlich N, Riethmuller J, Wilker B, De Oliveira-Munding CC, van Heeckeren AM, Barr ML, von Kurthy G, Schmid KW, Weller M, Tummler B, Lang F, Grassme H, Doring G, Gulbins E (2008) Ceramide accumulation mediates inflammation, cell death and infection susceptibility in cystic fibrosis. Nat Med 14:382–391
Tepper AD, Cock JG, de Vries E, Borst J, van Blitterswijk WJ (1997) CD95/Fas-induced ceramide formation proceeds with slow kinetics and is not blocked by caspase-3/CPP32 inhibition. J Biol Chem 272:24308–24312

Thon L, Mohlig H, Mathieu S, Lange A, Bulanova E, Winoto-Morbach S, Schutze S, Bulfone-Paus S, Adam D (2005) Ceramide mediates caspase-independent programmed cell death. FASEB J 19:1945–1956

Tuder R, Petrache I (2012) Pathogenesis of chronic obstructive pulmonary disease. J Clin Invest 122(8):2749–2755

Tuder RM, Petrache I, Elias JA, Voelkel NF, Henson PM (2003a) Apoptosis and emphysema: the missing link. Am J Respir Cell Mol Biol 28:551–554

Tuder RM, Zhen L, Cho CY, Taraseviciene-Stewart L, Kasahara Y, Salvemini D, Voelkel NF, Flores SC (2003b) Oxidative stress and apoptosis interact and cause emphysema due to vascular endothelial growth factor receptor blockade. Am J Respir Cell Mol Biol 29:88–97

Wenzel S (2012) Severe asthma: from characteristics to phenotypes to endotypes. Clin Exp Allergy 42:650–658

Ceramide in Cystic Fibrosis

Heike Grassmé, Joachim Riethmüller, and Erich Gulbins

Contents

Abstract Cystic fibrosis is caused by mutations in the cystic fibrosis transmembrane conductance regulator (*CFTR*) molecule; these mutations result in a defect in chloride secretion in epithelial cell layers. The disease is characterized by severe gastrointestinal and pulmonary symptoms, but it is the pulmonary symptoms that dominate the clinical course of the disease and determine patients' life expectancy. These pulmonary symptoms include reduced mucociliary clearance, chronic inflammation, and recurrent and chronic pulmonary infections with *Pseudomonas aeruginosa*, *Staphylococcus aureus*, *Burkholderia cepacia*, and *Haemophilus influenzae*. Recent studies have shown that sphingolipids, especially ceramide, play a crucial role in the pathogenesis of cystic fibrosis. These studies have demonstrated that ceramide accumulates in the lungs of cystic fibrosis patients and mice, causing inflammation and high susceptibility to bacterial infections. The results of initial clinical studies suggest that interfering with sphingolipids may be a novel treatment strategy for cystic fibrosis.

Keywords Cystic fibrosis • Cftr • Ceramide • Pulmonary infections • *P. aeruginosa*

H. Grassmé • J. Riethmüller • E. Gulbins (✉)
Department of Molecular Biology, University of Duisburg-Essen, Hufelandstrasse 55, 45122 Essen, Germany

Children's Clinic, University of Tuebingen, Hoppe-Seyler Str. 1, 72076 Tuebingen, Germany
e-mail: erich.gulbins@uni-due.de

E. Gulbins and I. Petrache (eds.), *Sphingolipids in Disease*,
Handbook of Experimental Pharmacology 216,
DOI 10.1007/978-3-7091-1511-4_13,

1 Cystic Fibrosis: Introduction

Cystic fibrosis (CF) is the most common autosomal recessive disorder in Western countries: approximately 1 child per 2,500 births will be affected, and approximately 90,000 patients in the European Union and the United States have the disease (Cystic Fibrosis Registry of the USA). Cystic fibrosis is caused by mutations in the cystic fibrosis transmembrane conductance regulator (*CFTR*) (Kerem et al. 1989; Rommens et al. 1989). Many genetic mutations in *CFTR* have been identified, and most of them result either in decreased expression of mRNA or in decreased expression, instability, or inactivity of the CFTR protein (Kerem et al. 1992). CFTR is a chloride channel that mediates chloride secretion in epithelial cell layers (Welsh et al. 1994). However, it has also been suggested that the channel is involved in bicarbonate transport (Poulsen et al. 1994). Furthermore, the channel is expressed in intracellular vesicles, such as lysosomes, secretory lysosomes, and phagosomes, and it has been shown to be involved in acidifying these organelles, although this finding is controversial (Barasch et al. 1991; Di et al. 2006; Haggie and Verkman 2007).

Clinically, *CFTR* mutations cause severe gastrointestinal and pulmonary symptoms. The gastrointestinal symptoms, which include pancreatic insufficiency and associated maldigestion, meconium ileus in newborns, and biliary liver cirrhosis, can usually be controlled. However, the pulmonary symptoms are currently most important because they determine the quality of life and the life expectancy of CF patients. These pulmonary problems include reduced mucociliary clearance and accumulation of mucus in the bronchi, chronic inflammation, lung fibrosis, and, in particular, recurrent and chronic infections with *Staphylococcus aureus*, *Pseudomonas aeruginosa*, and *Burkholderia cepacia*, as well as *Haemophilus influenza*e.

Most of the studies described below used *Cftr*-deficient mice. When bred to a C57BL/6 background, these mice spontaneously exhibit pulmonary changes that are typical of CF: reduced mucociliary clearance, alveolar enlargement, lung fibrosis and inflammation, and a mucus-like coating of the proximal and distal airways (Kent et al. 1996, 1997; Durie et al. 2004; Charizopoulou et al. 2004, 2006; Teichgräber et al. 2008). The mice also show chronic obstruction and extension of pancreatic ducts, acinar atrophy of the pancreas, cholangitis, fibrosis and biliary cirrhosis, abdominal distention, mucoid substances in the ileus lumen, and increased numbers of goblet cells (Durie et al. 2004). Thus, these mouse models are appropriate for studies of the pathophysiology of CF.

2 Cystic Fibrosis and Inflammation or Infection

Studies using animal or human cells and tissues have demonstrated that CF airways are chronically inflamed because of an imbalance between proinflammatory and anti-inflammatory cytokines in the lungs or the lung cells (Inoue et al. 1994;

Khan et al. 1995; Bonfield et al. 1995; Tirouvanziam et al. 2000; Venkatakrishnan et al. 2000; Tabary et al. 2001; Oceandy et al. 2002; Verhaeghe et al. 2007). At present, it is unclear whether the chronic inflammation predisposes to pulmonary infections (see below) or is caused by them. However, studies involving newborns and, in particular, aborted fetuses carrying *Cftr* mutations suggest that chronic inflammation in the lungs of CF patients precedes infection (Khan et al. 1995; Verhaeghe et al. 2007). The studies using aborted fetuses demonstrated increased concentrations of inflammatory markers in the lungs, a finding strongly suggesting that aseptic inflammation occurs before infection in CF. However, studies using CF pigs did not detect airway inflammation at birth (Stoltz et al. 2010), leaving the issue still unresolved.

Experiments using *Cftr*-deficient mice demonstrated that the levels of keratinocyte chemoattractant (KC; mouse homolog of interleukin-8) are increased even before infection (Inoue et al. 1994; Tabary et al. 2001) and that the levels of anti-inflammatory cytokines, in particular IL-10, are concomitantly decreased (Bonfield et al. 1995). Furthermore, *Cftr*-deficient epithelial cells displayed a constitutive activation of nuclear factor kappaB (NF-κB), a finding that also suggests a proinflammatory phenotype (Venkatakrishnan et al. 2000). The pathogenetic role of these inflammatory mediators in the lungs of CF patients still requires definition. The lungs of CF patients contain large numbers of neutrophils (Goldstein and Döring 1986), although these cells do not seem to express *Cftr*. However, the increased death of neutrophils in CF lungs may result in the release of DNA and neutrophil proteases that contribute to increased viscosity of mucus and increased inflammation (Matsui et al. 2005; Whitchurch et al. 2002; Worlitzsch et al. 2002). Studies by Teichgräber et al. (2008) indicated that sphingolipids are critical for the induction of aseptic inflammation: the increased pulmonary concentrations of IL-8/KC and IL-1 and the peribronchial accumulation of macrophages and neutrophils in CF mice were corrected by normalization of ceramide levels in these mice (see also below).

The second hallmark of the pulmonary problems associated with CF is patients' increased susceptibility to infection with pathogens such as *P. aeruginosa*, *S. aureus*, *B. cepacia*, and *H. influenzae*. More than 80 % of adult CF patients are infected with *P. aeruginosa* (Cystic Fibrosis Registry of the USA); therefore, it is very important to identify the molecular mechanisms mediating *P. aeruginosa* infections in CF and to develop novel treatment options for CF patients. This chapter focuses on the role of sphingolipids in *P. aeruginosa* infections of the CF lung.

Several studies have found that, unlike healthy control samples (donor lungs) or mice, lung samples from CF patients or *Cftr*-deficient mice exhibit an accumulation of ceramide in bronchial epithelial cells and alveolar macrophages in the lung and also in tracheal and intestinal epithelial cells (Teichgräber et al. 2008; Zhang et al. 2009; Becker et al. 2010a, b; Brodlie et al. 2010; Bodas et al. 2011a, b; Ulrich et al. 2010). These in vivo findings were confirmed by in vitro studies using cultured cells, which also demonstrated an accumulation of ceramide in *Cftr*-deficient cells (Bodas et al. 2011a, b). Interestingly, these in vitro studies also found that CFTR is downregulated in emphysema patients, and this downregulation may contribute to the accumulation of ceramide in the lungs of patients with chronic obstructive pulmonary disease (COPD) (Bodas et al. 2011b). Studies using human and mouse tissues employed several techniques to demonstrate the accumulation of ceramide: biochemical assay, fluorescence microscopy, electron microscopy, and mass spectrometry. Mass spectrometry studies by Brodlie et al. (2010) demonstrated a substantial accumulation of C16, C18, and C20 ceramides in the lungs of CF patients, whereas the level of C22 ceramide was not significantly changed. It is important to note that the murine studies found an age-dependent accumulation of ceramide in the lungs: the increase in ceramide concentrations was seen only in mice older than 12 weeks (Teichgräber et al. 2008) consistent with the slow progression of the disease in young children.

What are the effects of increased ceramide concentrations in CF lungs? Among other pathways, ceramide is formed when acid sphingomyelinase hydrolyzes sphingomyelin to ceramide. Thus, one option for reducing ceramide in the lungs of CF mice is the genetic or pharmacological inhibition of acid sphingomyelinase. To this end, CF mice were crossed with acid sphingomyelinase-deficient mice to obtain *Cftr*-deficient, acid sphingomyelinase-heterozygous mice ($Cftr^{-/-}/Smpd1^{+/-}$, *Smpd1* is the gene symbol for the acid sphingomyelinase) that, in fact, exhibited a normalization of ceramide concentrations in the lungs (Teichgräber et al. 2008). This normalization was also achieved by treating CF mice with pharmacological functional inhibitors of acid sphingomyelinase, such as amitriptyline, trimipramine, desipramine, chlorprothixene, fluoxetine, amlodipine, and sertraline (Teichgräber et al. 2008; Becker et al. 2010a). These drugs are weak bases that are trapped in lysosomes or secretory lysosomes after protonation, bind to the inner membrane, and displace acid sphingomyelinase from the membrane; this sequence of activity results in proteolytic degradation of acid sphingomyelinase in the lysosome (Hurwitz et al. 1994; Kornhuber et al. 2008).

The hallmarks of CF (inflammation, reduced mucociliary clearance, and increased susceptibility to infection) were then measured in these genetically modified or pharmacologically treated mice.

1. Inflammation: *Cftr*-deficient mice exhibit high numbers of peribronchial macrophages and neutrophils, which are absent in wild-type mice (Teichgräber et al. 2008). These numbers of inflammatory cells are similar to the high numbers of neutrophils and macrophages in the peribronchial tissue of CF patients. Pharmacological or genetic inhibition of Asm and the concomitant normalization of ceramide concentrations normalized the numbers of neutrophils and macrophages in the peribronchial tissue of CF mice (Teichgräber et al. 2008). Furthermore, the lungs of CF mice exhibited increased concentrations of proinflammatory mediators such as IL-1 and IL-8. The application of acid sphingomyelinase inhibitors or genetic heterozygosity of acid sphingomyelinase in CF mice normalized the concentrations of these proinflammatory mediators (Teichgräber et al. 2008). The increased release of cytokines in CF mice seems to be mediated, at least in part, by a constitutive activation of CD95 in the bronchial epithelial cells of these mice (Becker et al. 2012). Mice lacking *Cftr* exhibit an upregulation and

activation of CD95, and this upregulation is caused by plasma membrane ceramide. CD95 clusters in ceramide-enriched domains of CF bronchial epithelial cells, a clustering that results in the activation of CD95 and the subsequent release of proinflammatory cytokines. The activation of CD95 also further increases pulmonary ceramide concentrations.

2. Mucociliary clearance: Histological studies of tissue from the lungs of CF mice indicated that the death of epithelial cells is increased in these mice (Maiuri et al. 1997; Durie et al. 1999; Teichgräber et al. 2008) and that this cell death results in increased DNA deposits in CF airways and in mucus or DNA plugs in the lower airways. When ceramide concentrations in CF mice were normalized by the application of amitriptyline or by genetic heterozygosity of Asm, epithelial cell death and DNA deposits in the airways of these mice were abrogated (Teichgräber et al. 2008).
3. Infection susceptibility: Mice lacking *Cftr* are very susceptible to pulmonary infections with *P. aeruginosa*, as is the case with CF patients. While intranasal application of *P. aeruginosa* into wild-type mice results in only a very mild pneumonia or even in no symptoms, CF mice cannot eliminate and control the bacteria; thus, severe pneumonia develops in CF mice after intranasal challenge with the bacteria (Pier et al. 1996; Teichgräber et al. 2008). In contrast, intratracheal injection of the bacteria results in a similar pneumonia in wild-type and CF mice, (Munder et al. 2011) a finding suggesting that CF mice have a severe impairment in mucociliary clearance. The central role of ceramide in the susceptibility to infection is indicated by the finding that normalizing pulmonary ceramide concentrations in CF mice by treating them with Asm inhibitors or by heterozygosity of Asm prevents pulmonary infection with *P. aeruginosa* after intranasal application (Teichgräber et al. 2008). Ceramide and CD95 seem to play a crucial although not exclusive role in this process: the genetic suppression of CD95 in CF mice reduces cell death and at least partially susceptibility to infection (Becker et al. 2012). Furthermore, hydrolyzing DNA deposits in the bronchi of CF mice by the inhalation of DNAase rescued the mice from pulmonary *P. aeruginosa* infection (Teichgräber et al. 2008). At present, it is unknown whether the DNA in bronchi serves to bind bacteria or whether it primarily reduces mucociliary clearance, facilitating *P. aeruginosa* infection, or both. A role of CD95 in CF is also indicated by genetic studies involving CF twins; these studies found that CF symptoms are associated with allelic variants in intron 2 of the CD95 receptor gene (Kumar et al. 2008). However, it is unknown whether such changes in CF patients result in changes in cell death in CF airways or whether other factors, such as proinflammatory cytokines, are primarily regulated by CD95.

It is important to note that ceramide concentrations are increased in CF airways even without infection and that ceramide-induced inflammation and cell death in CF airways occur without *P. aeruginosa* infection.

In summary, ceramide is a crucial mediator of important pathophysiological alterations in *Cftr*-deficient lungs.

Several additional studies supported the finding that sphingolipids play an important role in CF. These studies identified severe alterations of rafts that contain predominantly sphingolipids in cells lacking *Cftr* (Kowalski and Pier 2004; Gadjeva et al. 2010; Zaas et al. 2005; Teichgräber et al. 2008; Bajmoczi et al. 2009) and demonstrated that the destruction of rafts impairs the internalization of *P. aeruginosa* into epithelial cells and also impairs host cell apoptosis in epithelial cells after infection with *P. aeruginosa* (Grassmé et al. 2003). Likewise, caveolae, which are small, invaginated membrane organelles enriched with sphingolipids, are reorganized after infection with *P. aeruginosa* and recruit caveolin-1 and Cftr during the infection process. Cells lacking caveolin-1 or -2 exhibited reduced internalization of *P. aeruginosa*, and mice deficient in caveolin-1 were much more susceptible than wild-type mice to pulmonary infection with *P. aeruginosa* (Gadjeva et al. 2010; Bajmoczi et al. 2009).

These findings indicate that rafts and caveolae play an important role in the functions of CFTR.

3 Novel Therapies for Cystic Fibrosis: Manipulating Sphingolipids

As described above, treating CF mice with pharmacological inhibitors of the acid sphingomyelinase normalizes ceramide concentrations in the lungs, reduces inflammation in the lungs, and prevents infection with *P. aeruginosa* (Teichgräber et al. 2008). In these initial studies, acid sphingomyelinase inhibitors were applied by intraperitoneal injection. Additional studies demonstrated that inhalation of the acid sphingomyelinase inhibitors amitriptyline, trimipramine, desipramine, chlorprothixene, fluoxetine, amlodipine, and sertraline also reduces pulmonary inflammation in and prevents infection of CF mice (Becker et al. 2010a).

Previous studies in our laboratory demonstrated that mice completely lacking acid sphingomyelinase cannot activate nicotinamide adenine dinucleotide phosphate (NADPH) oxidases, release reactive oxygen species, or kill *P. aeruginosa* (Zhang et al. 2007). Thus, it may seem somewhat paradoxical to propose that CF patients should be treated with acid sphingomyelinase inhibitors. However, CF cells exhibit increased ceramide concentrations, and heterozygosity of acid sphingomyelinase or treatment with acid sphingomyelinase inhibitors results in a *normalization* of ceramide concentrations, not in a *reduction* of ceramide concentrations. This normalization of ceramide concentrations reduces inflammation, normalizes mucociliary clearance, and prevents pulmonary *P. aeruginosa* infections, but still permits NADPH-oxidase activation and ROS release. Thus, for future CF treatments based on the inhibition of ceramide production, it will be important to normalize ceramide concentrations but not to suppress the production of ceramide completely, because complete suppression would result in a Niemann-Pick type A/B phenotype and reduced killing of *P. aeruginosa* in the airways.

Drugs inhibiting acid sphingomyelinase activity are commonly used antidepressants with only very mild adverse effects, at least at low doses (Kornhuber et al. 2008). Thus, the findings from animal studies were transferred into a clinical study (Riethmüller et al. 2009). An initial randomized, double-blind, placebo-controlled, crossover pilot study involving only four adult CF patients demonstrated that treatment with amitriptyline was safe and exerted a beneficial effect on lung function. A phase IIa study treated 19 adult CF patients with 25, 50, or 75 mg amitriptyline per day (Riethmüller et al. 2009). Except for patients receiving the highest dose who suffered from dry mucosa and tiredness, only mild and transient adverse effects, if any, were observed. These studies also indicate that doses of amitriptyline as low as 25 or 50 mg per day significantly improved lung functions, as determined by measurements of forced expiratory volume in the first second (FEV_1). Finally, long-term observation of 20 patients who had taken amitriptyline for more than 1 year showed an impressive increase in lung function (Riethmüller and Gulbins, unpublished observations). Although the sizes of study groups and the study times were limited and thus permitted only limited conclusions about the effect of ASM inhibitors on lung function in CF patients, the results of these early trials are encouraging and should be confirmed in a larger clinical study.

References

Bajmoczi M, Gadjeva M, Alper SL, Pier GB, Golan DE (2009) Cystic fibrosis transmembrane conductance regulator and caveolin-1 regulate epithelial cell internalization of *Pseudomonas aeruginosa*. Am J Physiol Cell Physiol 297:C263–C277

Barasch J, Kiss B, Prince A, Saiman L, Gruenert D, al-Awqati Q (1991) Defective acidification of intracellular organelles in cystic fibrosis. Nature 352:70–73

Becker KA, Riethmüller J, Lüth A, Döring G, Kleuser B, Gulbins E (2010a) Acid sphingomyelinase inhibitors normalize pulmonary ceramide and inflammation in cystic fibrosis. Am J Respir Cell Mol Biol 42:716–724

Becker KA, Tümmler B, Gulbins E, Grassmé H (2010b) Accumulation of ceramide in the trachea and intestine of cystic fibrosis mice causes inflammation and cell death. Biochem Biophys Res Commun 403:368–374

Becker KA, Henry B, Ziobro R, Tümmler B, Gulbins E, Grassmé H (2012) Role of CD95 in pulmonary inflammation and infection in cystic fibrosis. J Mol Med 90:1011–1023

Bodas M, Min T, Mazur S, Vij N (2011a) Critical modifier role of membrane-cystic fibrosis transmembrane conductance regulator-dependent ceramide signaling in lung injury and emphysema. J Immunol 186:602–613

Bodas M, Min T, Vij N (2011b) Critical role of CFTR-dependent lipid rafts in cigarette smoke-induced lung epithelial injury. Am J Physiol Lung Cell Mol Physiol 300:L811–L820

Bonfield TL, Konstan MW, Burfeind P, Panuska JR, Hilliard JB, Berger M (1995) Normal bronchial epithelial cells constitutively produce the anti-inflammatory cytokine interleukin-10, which is downregulated in cystic fibrosis. Am J Respir Cell Mol Biol 13:257–261

Brodlie M, McKean MC, Johnson GE, Gray J, Fisher AJ, Corris PA, Lordan JL, Ward C (2010) Ceramide is increased in the lower airway epithelium of people with advanced cystic fibrosis lung disease. Am J Respir Crit Care Med 182:369–375

Charizopoulou N, Jansen S, Dorsch M, Stanke F, Dorin JR, Hedrich H-J, Tümmler B (2004) Instability of the insertional mutation in CftrTgH(neoim)Hgu cystic fibrosis mouse model. BMC Genet 5:6

Charizopoulou N, Wilke M, Dorsch M, Bot A, Jorna H, Jansen S, Stanke F, Hedrich H-J, de Jonge HR, Tümmler B (2006) Spontaneous rescue from cystic fibrosis in a mouse model. BMC Genet 7:18

Cystic fibrosis registry of the USA. http://www.cff.org

Di A, Brown ME, Deriy LV, Li C, Szeto FL, Chen Y, Huang P, Tong J, Naren AP, Bindokas V, Palfrey HC, Nelson DJ (2006) CFTR regulates phagosome acidification in macrophages and alters bactericidal activity. Nat Cell Biol 8:933–944

Durie I, Amsellem C, Paulin C, Chambe MT, Bienvenu J, Bellon G, Pacheco Y (1999) Fas and Fas ligand expression in cystic fibrosis airway epithelium. Thorax 54:1093–1098

Durie PR, Kent G, Phillips MJ, Ackerley CA (2004) Characteristic multiorgan pathology of cystic fibrosis in a long-living cystic fibrosis transmembrane regulator knockout murine model. Am J Pathol 164:1481–1493

Gadjeva M, Paradis-Bleau C, Priebe GP, Fichorova R, Pier GB (2010) Caveolin-1 modifies the immunity to *Pseudomonas aeruginosa*. J Immunol 184:296–302

Goldstein W, Döring G (1986) Lysosomal enzymes from polymorphonuclear leukocytes and protease inhibitors in patients with cystic fibrosis. Am Rev Respir Dis 134:49–56

Grassmé H, Jendrossek V, Riehle A, von Kurthy G, Berger J, Schwarz H, Weller M, Kolesnick R, Gulbins E (2003) Host defense against *Pseudomonas aeruginosa* requires ceramide-rich membrane rafts. Nat Med 9:322–330

Haggie PM, Verkman AS (2007) Cystic fibrosis transmembrane conductance regulator-independent phagosomal acidification in macrophages. J Biol Chem 282:31422–31428

Hurwitz R, Ferlinz K, Sandhoff K (1994) The tricyclic antidepressant desipramine causes proteolytic degradation of lysosomal sphingomyelinase in human fibroblasts. Biol Chem Hoppe Seyler 375:447–450

Inoue H, Massion PP, Ueki IF, Grattan KM, Hara M, Dohrman AF, Chan B, Lausier JA, Golden JA, Nadel JA (1994) Pseudomonas stimulates interleukin-8 mRNA expression selectively in airway epithelium, in gland ducts, and in recruited neutrophils. Am J Respir Cell Mol Biol 11:651–663

Kent G, Oliver M, Foskett JK, Frndova H, Durie P, Forstner J, Forstner GG, Riordan JR, Percy D, Buchwald M (1996) Phenotypic abnormalities in long-term surviving cystic fibrosis mice. Pediatr Res 40:233–241

Kent G, Iles R, Bear CE, Huan LJ, Griesenbach U, McKerlie C, Frndova H, Ackerley C, Gosselin D, Radzioch D, O'Brodovich H, Tsui LC, Buchwald M, Transwell AK (1997) Lung disease in mice with cystic fibrosis. J Clin Invest 100:3060–3069

Kerem B, Rommens JM, Buchanan JA, Markiewicz D, Cox TK, Chakravarti A, Buchwald M, Tsui LC (1989) Identification of the cystic fibrosis gene: genetic analysis. Science 245: 1073–1080

Kerem E, Reisman J, Corey M, Canny GJ, Levison H (1992) Prediction of mortality in patients with cystic fibrosis. N Engl J Med 326:1187–1191

Khan TZ, Wagener JS, Bost T, Martinez J, Accurso FJ, Riches DW (1995) Early pulmonary inflammation in infants with cystic fibrosis. Am J Respir Crit Care Med 151:1075–1082

Kornhuber J, Tripal P, Reichel M, Terfloth L, Bleich S, Wiltfang J, Gulbins E (2008) Identification of new functional inhibitors of acid sphingomyelinase using a structure–property-activity relation model. J Med Chem 51:219–237

Kowalski MP, Pier GB (2004) Localization of cystic fibrosis transmembrane conductance regulator to lipid rafts of epithelial cells is required for *Pseudomonas aeruginosa*-induced cellular activation. J Immunol 172:418–425

Kumar V, Becker T, Jansen S, van Barneveld A, Boztug K, Wolfl S, Tümmler B, Stanke F (2008) Expression levels of FAS are regulated through an evolutionary conserved element in intron 2, which modulates cystic fibrosis disease severity. Genes Immun 9:689–696

Maiuri L, Raia V, De Marco G, Coletta S, de Ritis G, Londei M, Auricchio S (1997) DNA fragmentation is a feature of cystic fibrosis epithelial cells: a disease with inappropriate apoptosis? FEBS Lett 408:225–231

Matsui H, Verghese MW, Kesimer M, Schwab UE, Randell SH, Sheehan JK, Grubb BR, Boucher RC (2005) Reduced three-dimensional motility in dehydrated airway mucus prevents neutrophil capture and killing bacteria on airway epithelial surfaces. J Immunol 175:1090–1099

Munder A, Wölbeling F, Kerber-Momot T, Wedekind D, Baumann U, Gulbins E, Tümmler B (2011) Acute intratracheal *Pseudomonas aeruginosa* infection in cystic fibrosis mice is age-independent. Respir Res 12:148

Oceandy D, McMorran BJ, Smith SN, Schreiber R, Kunzelmann K, Alton EW, Hume DA, Wainwright BJ (2002) Gene complementation of airway epithelium in the cystic fibrosis mouse is necessary and sufficient to correct the pathogen clearance and inflammatory abnormalities. Hum Mol Genet 11:1059–1067

Pier GB, Grout M, Zaidi TS, Olsen JC, Johnson LG, Yankaskas JR, Goldberg JB (1996) Role of mutant CFTR in hypersusceptibility of cystic fibrosis patients to lung infections. Science 271: 64–67

Poulsen JH, Fischer H, Illek B, Machen TE (1994) Bicarbonate conductance and pH regulatory capability of cystic fibrosis transmembrane conductance regulator. Proc Natl Acad Sci USA 91:5340–5344

Riethmüller J, Anthonysamy J, Emilio Serra E, Schwab M, Döring G, Gulbins E (2009) Therapeutic efficacy and safety of amitriptyline in patients with cystic fibrosis. Cell Physiol Biochem 24:65–72

Rommens JM, Iannuzzi MC, Kerem B, Drumm ML, Melmer G, Dean M, Rozmahel R, Cole JL, Kennedy D, Hidaka N et al (1989) Identification of the cystic fibrosis gene: chromosome walking and jumping. Science 245:1059–1065

Stoltz DA, Meyerholz DK, Pezzulo AA, Ramachandran S, Rogan MP, Davis GJ, Hanfland RA, Wohlford-Lenane C, Dohrn CL, Bartlett JA, Nelson GA IV, Chang EH, Taft PJ, Ludwig PS, Estin M, Hornick EE, Launspach JL, Samuel M, Rokhlina T, Karp PH, Ostedgaard LS, Uc A, Starner TD, Horswill AR, Brogden KA, Prather RS, Richter SS, Shilyansky J, McCray PB Jr, Zabner J, Welsh MJ (2010) Cystic fibrosis pigs develop lung disease and exhibit defective bacterial eradication at birth. Sci Transl Med 2:29ra31

Tabary O, Escotte S, Couetil JP, Hubert D, Dusser D, Puchelle E, Jacquot J (2001) Relationship between IkappaBalpha deficiency, NFkappaB activity and interleukin-8 production in CF human airway epithelial cells. Pflugers Arch 443(Suppl 1):S40–S44

Teichgräber V, Ulrich M, Endlich N, Riethmüller J, Wilker B, De Oliveira-Munding CC, van Heeckeren AM, Barr ML, von Kürthy G, Schmid KW, Weller M, Tümmler B, Lang F, Grassmé H, Döring G, Gulbins E (2008) Ceramide accumulation mediates inflammation, cell death and infection susceptibility in cystic fibrosis. Nat Med 14:382–391

Tirouvanziam R, de Bentzmann S, Hubeau C, Hinnrasky J, Jacquot J, Peault B, Puchelle E (2000) Inflammation and infection in naive human cystic fibrosis airway grafts. Am J Respir Cell Mol Biol 23:121–127

Ulrich M, Worlitzsch D, Viglio S, Siegmann N, Iadarola P, Shute JK, Geiser M, Pier GB, Friedel G, Barr ML, Schuster A, Meyer KC, Ratjen F, Bjarnsholt T, Gulbins E, Döring G (2010) Alveolar inflammation in cystic fibrosis. J Cyst Fibros 9:217–227

Venkatakrishnan A, Stecenko AA, King G, Blackwell TR, Brigham KL, Christman JW, Blackwell TS (2000) Exaggerated activation of nuclear factor-kappaB and altered IkappaB-beta processing in cystic fibrosis bronchial epithelial cells. Am J Respir Cell Mol Biol 23: 396–403

Verhaeghe C, Delbecque K, de Leval L, Oury C, Bours V (2007) Early inflammation in the airways of a cystic fibrosis foetus. J Cyst Fibros 6:304–308

Welsh MJ, Anderson MP, Rich DP, Berger HA, Sheppard DN (1994) The CFTR chloride channel. In: Guggino WB (ed) Chloride channels. Academic, San Diego, pp 153–171

Whitchurch CB, Tolker-Nielsen T, Ragas PC, Mattick JS (2002) Extracellular DNA required for bacterial biofilm formation. Science 295:148

Worlitzsch D, Tarran R, Ulrich M, Schwab U, Cekici A, Meyer KC, Birrer P, Bellon G, Berger J, Weiss T, Botzenhart K, Yankaskas JR, Randell S, Boucher RC, Döring G (2002) Reduced oxygen concentrations in airway mucus contribute to the early and late pathogenesis of *Pseudomonas aeruginosa* CF airways infection. J Clin Invest 109:317–325

Zaas DW, Duncan MJ, Li G, Wright JR, Abraham SN (2005) Pseudomonas invasion of type I pneumocytes is dependent on the expression and phosphorylation of caveolin-2. J Biol Chem 280:4864–4872

Zhang AY, AY, Yi F, Jin S, Xia M, Chen QZ, Gulbins E, Li PL (2007) Acid sphingomyelinase and its redox amplification in formation of lipid raft redox signaling platforms in endothelial cells. Antioxid Redox Signal 9:817–828

Zhang Y, Li X, Grassmé H, Döring G, Gulbins E (2009) Alterations in ceramide concentration and pH determine the release of reactive oxygen species by Cftr-deficient macrophages on infection. J Immunol 184:5104–5111.

Regulation of the Sphingosine Kinase/Sphingosine 1-Phosphate Pathway

K. Alexa Orr Gandy and Lina M. Obeid

Contents

K.A. Orr Gandy
The Department of Molecular and Cellular Biology and Pathobiology,
The Medical University of South Carolina, Charleston, SC 29425, USA

L.M. Obeid (✉)
The Department of Medicine, Health Science Center, Stony Brook University, L-4, 179,
Stony Brook, NY 11794-8430, USA

The Northport VA Medical Center, Northport, NY 11768, USA
e-mail: lina.obeid@stonybrookmedicine.edu

E. Gulbins and I. Petrache (eds.), *Sphingolipids in Disease*,
Handbook of Experimental Pharmacology 216,
DOI 10.1007/978-3-7091-1511-4_14,

Abstract Sphingolipids have emerged as pleiotropic signaling molecules with roles in numerous cellular and biological functions. Defining the regulatory mechanisms governing sphingolipid metabolism is crucial in order to develop a complete understanding of the biological functions of sphingolipid metabolites. The sphingosine kinase/ sphingosine 1-phosphate pathway was originally thought to function in the irreversible breakdown of sphingoid bases; however, in the last few decades it has materialized as an extremely important signaling pathway involved in a plethora of cellular events contributing to both normal and pathophysiological events. Recognition of the SK/S1P pathway as a second messaging system has aided in the identification of many mechanisms of its regulation; however, a cohesive, global understanding of the regulatory mechanisms controlling the SK/S1P pathway is lacking. In this chapter, the role of the SK/S1P pathway as a second messenger is discussed, and its role in mediating TNF-α- and EGF-induced biologies is examined. This work provides a comprehensive look into the roles and regulation of the sphingosine kinase/ sphingosine 1-phosphate pathway and highlights the potential of the pathway as a therapeutic target.

Keywords sphingolipids • growth factors • epidermal growth factor (EGF) • platelet derived growth factor (PDGF) • tumor necrosis-α (TNFα) • cytokines • interleukin • tumor growth factor-β (TGFβ) • hormones

1 Introduction

Lipid molecules defined by long-chain or sphingoid base backbones are referred to as sphingolipids and over the span of several decades have been found to carry out functions as diverse as the thousands of sphingolipid subspecies known to exist (Merrill 2011). In mammals, the prominent sphingoid base is sphingosine, an amino alcohol compound with multiple fates (Merrill 2011). Sphingosine can be phosphorylated at the primary hydroxyl group to form the bioactive sphingolipid metabolite sphingosine 1-phosphate; alternatively, sphingosine can be acylated at the secondary amine group to form numerous ceramide species (Merrill 2011). Ceramide can then be further metabolized into various complex sphingolipids, discussed below. A fine balance exists between the bioactive sphingolipid metabolites sphingosine, sphingosine 1-phosphate, and ceramide, and the enzymes responsible for the interconversion of these molecules are tightly regulated. Because sphingolipids are involved in a plethora of physiological and pathophysiological processes, it is important to understand the mechanisms governing the generation and degradation of these important lipid molecules.

2 Sphingolipid Metabolism

Originally, sphingolipids were thought to function only as structural components of the plasma membrane; however, sphingosine, ceramide, and sphingosine 1-phosphate, along with other sphingolipid metabolites, have been shown to possess important cell signaling capabilities. Ceramide is considered the epicenter of sphingolipid metabolism, as the different arms of sphingolipid metabolism converge at this point (Gault et al. 2010) (Fig. 1). Three pathways of ceramide generation exist: the de novo pathway, the sphingomyelinase pathway, and the salvage pathway (Fig. 1). The initial reaction in the de novo pathway of ceramide synthesis is the condensation of serine and palmitoyl-CoA by serine palmitoyl transferase (SPT) to form 3-keto-dihydrosphingosine (KDHS) (Gault et al. 2010). KDHS is then reduced via the action of KDHS reductase to form dihydrosphingosine (DHS). Addition of varying lengths of acyl chains, by the family of (dihydro) ceramide synthases, to the amine group of DHS results in the precursor to ceramide, dihydroceramide (DHC). Through the addition of a double bond at C4 of the DHC backbone, carried out by the enzyme dihydroceramide desaturase, DHC becomes ceramide (Fig. 1). The de novo pathway occurs in the endoplasmic reticulum, and ceramide travels vesicularly or via ceramide transport protein (CERT) through the Golgi network while being further metabolized to more complex sphingolipids such as sphingomyelin and glycosphingolipids. Also, ceramide can be generated from the sphingomyelinase pathway (Fig. 1). Sphingomyelinase is the enzyme responsible for the hydrolysis of the major plasma membrane constituent, sphingomyelin, by removal of a phosphocholine head group from phosphatidylcholine to form ceramide and diacylglycerol (DAG). Sphingomyelin hydrolysis occurs at varying subcellular localizations including the plasma membrane, the lysosome, and the Golgi (Bartke and Hannun 2009). Lastly, ceramide can be generated in the cell through the salvage pathway (Fig. 1). In the salvage pathway of ceramide generation, long-chain sphingoid bases derived from catabolism of more complex sphingolipids are re-acylated by ceramide synthases to yield ceramide (Mullen et al. 2012). Sphingolipid metabolism and ceramide generation are important in maintaining membrane homeostasis, as well as in generating bioactive sphingolipid metabolites for the purpose of cell signaling.

Following its generation, ceramide has numerous fates. As mentioned briefly above, addition of a phosphocholine head group to ceramide species, by action of the sphingomyelin synthases (SMS), leads to production of sphingomyelin (Fig. 1) which is incorporated into cellular membranes or acts as a signaling molecule. Also mentioned previously, ceramide can be used to generate more complex sphingolipid species such as glycosphingolipids (GSL). GSL consist of ceramide with varying types and/or numbers of sugars attached as a head group. When a single glucose molecule is attached as the head group, glucosylceramide species are formed; furthermore, when a single galactose molecule is attached to ceramide,

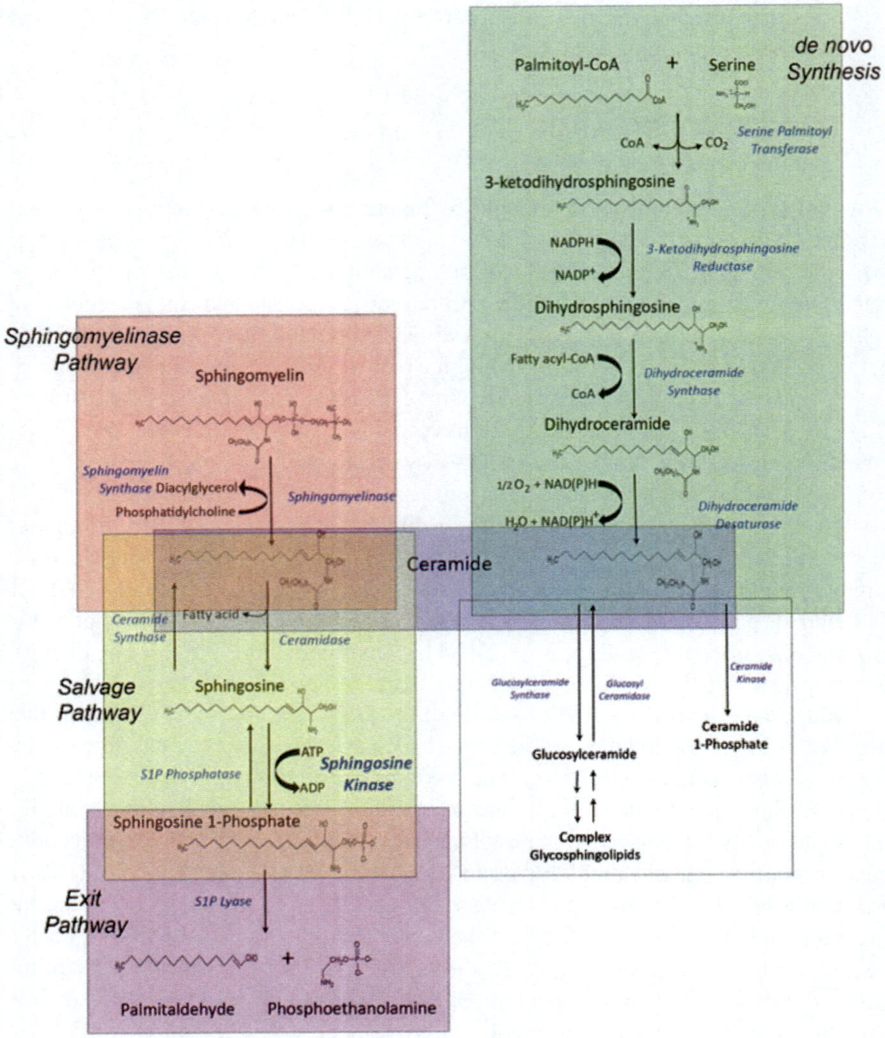

Fig. 1 *Sphingolipid metabolism*. In the de novo pathway of ceramide synthesis, serine and palmitoyl-CoA are condensed via action of the enzyme serine palmitoyltransferase (SPT) to form 3-keto-dihydrosphingosine. 3-Keto-dihydrosphingosine is then reduced to dihydrosphingosine by 3-keto-dihydrosphingosine reductase. Dihydroceramide is formed following addition of a fatty acid chain to dihydrosphingosine via the action of (dihydro)ceramide synthase. Dihydroceramide is then desaturated by dihydroceramide desaturase to form ceramide. In the sphingomyelinase pathway, ceramide is formed from membrane sphingomyelin by removal of a choline head group by the enzyme sphingomyelinase. In the reverse reaction, ceramide, by the addition of a choline head group from phosphatidylcholine, is converted back to sphingomyelin by sphingomyelin synthase. The salvage pathway of ceramide formation occurs following addition of a fatty acyl chain to already formed sphingosine by ceramide synthase. Sphingosine can be formed by the dephosphorylation of sphingosine 1-phosphate by sphingosine 1-phosphate phosphatase. In the exit pathway, sphingosine 1-phosphate is broken down irreversibly by sphingosine 1-phosphate lyase into phosphoethanolamine and palmitaldehyde

galactosylceramide species are formed (Gault et al. 2010). These sphingolipids are quite complex for two reasons: (1) ceramide subspecies are numerous due to varying lengths of acyl chains and different levels of saturation, and (2) head groups of glycosphingolipids can vary by type, linkage, order, and number of sugar molecules attached. These complex sphingolipids are outside the scope of this chapter, and readers are referred to a recent review (Merrill 2011). Ceramide can be phosphorylated to form ceramide 1-phosphate (C1P) which functions as a signaling molecule; however, effects of C1P signaling are less well characterized than other bioactive sphingolipid molecules (Gault et al. 2010). C1P may play a role in inflammation via its activation of cytosolic phospholipase A2 (Lamour and Chalfant 2005). Lastly, ceramide can enter the breakdown pathway whereby ceramidase-mediated removal of the acyl chains leads to the production of sphingosine (Fig. 1). Sphingosine is then phosphorylated to sphingosine 1-phosphate by the sphingosine kinases (SK), of which there are two isoforms: SK1 and SK2. S1P can ultimately be broken down into hexadecenal and phosphoethanolamine by the enzyme S1P lyase; alternatively, through the action of S1P-specific or nonspecific lipid phosphatases, S1P can undergo dephosphorylation, reforming sphingosine (Fig. 1). On the other hand, and central to this chapter, S1P can function as an important bioactive signaling molecule leading to numerous important cellular functions. S1P and its synthesizing enzymes, SK1 and SK2, as well as their regulatory mechanisms will be discussed.

3 Sphingosine 1-Phosphate

3.1 S1P Metabolism and Function

S1P is an 18-carbon amino alcohol that was originally thought to exist strictly as an intermediate in the sole, permanent degradative pathway of long-chain sphingoid bases. All sphingolipids are broken down through the sphingolipid exit pathway by conversion to S1P and subsequent S1P lyase-mediated formation of hexadecenal and phosphoethanolamine. S1P was first found to have important cell signaling functions only many years following its initial detection in the early 1970s (Stoffel 1970). It was not until the early 1990s when S1P was implicated in intracellular calcium release (Ghosh et al. 1990). Next was the discovery of a phorbol-myristic-acetate (PMA)-inducible message that led to the formation of capillary networks and was therefore termed the endothelial differentiation gene (EDG)-1 (Hla and Maciag 1990a, b). EDG-1 was determined to be an orphan G protein-coupled receptor (GPCR) by cDNA sequence analysis (Hla and Maciag 1990a, b). From the mid-to-late 1990s, numerous other EDG receptors were cloned (Hla 2001), and S1P was determined to be a bona fide high-affinity ligand for the EDG family of receptors. Since the discovery of sphingosine 1-phosphate receptors (S1PRs), or EDG receptors, there has been an eruption of literary evidence supporting S1P's role as a multifaceted, bioactive signaling lipid molecule (Lamour and Chalfant 2005).

At concentrations of ~0.5 μM, and even higher at local areas of inflammation (Ikeda et al. 2010), S1P is a major component of human plasma (Ikeda et al. 2010). Thought to prevent dephosphorylation of S1P by nonspecific lipid ecto-phosphatases (Pyne et al. 2009) or other degradative enzymes, the majority of plasma S1P is carried by lipoproteins (LDL, VLDL, HDL), although a portion is known to be carried on albumin (Bode et al. 2010). Interestingly, S1P displays differential functions depending on the carrier molecule (Hammad 2011). S1P in the blood functions to maintain vascular integrity under normal and pathological conditions (Camerer et al. 2009). Over the last several decades, S1P has been implicated in a plethora of cellular functions as well as in normal physiological and pathophysiological events. At the cellular level, S1P has been shown to modulate cell survival, proliferation, cytoskeletal structure, and adhesive properties. Meanwhile, S1P is known to play vital functions in many physiological processes including vascular development, cardiovascular function, and immunity (Hla and Brinkmann 2011; Siow and Wattenberg 2011; Siow et al. 2011). On the other hand, S1P is known to mediate numerous pathologies such as atherosclerosis, cancer and metastasis, and multiple sclerosis (Johnson et al. 2005; Kawamori et al. 2006, 2009; Hla and Brinkmann 2011; Pyne and Pyne 2011). Given the wide variety of processes in which the SK/S1P pathway is involved, the importance of understanding the mechanisms regulating this pathway becomes immediately obvious.

3.2 S1P Signaling

S1P has been implicated as both an extracellular signaling ligand by binding to cell surface S1P receptors (S1PRs), as well as an intracellular second messenger; however, the exact intracellular targets are only just being discovered. First of all, the absence of S1PRs in lower classes of organisms, such as *Saccharomyces cerevisiae* (Lanterman and Saba 1998), indicates an intracellular function for S1P. Also, S1P has been shown to promote growth and survival independent of its G protein-coupled receptors (GPCR) in mouse embryonic fibroblasts devoid of S1P receptors (Olivera et al. 2003). There is also evidence supporting a role for intracellular S1P in calcium mobilization (van Koppen et al. 2001). More recently, nuclear S1P produced specifically by the action of SK2 was shown to bind directly to histone deacetylases (HDACs), inhibiting their function, thus playing a role in epigenetic-mediated gene expression (Alvarez et al. 2010). Additionally, in response to TNF-α treatment, intracellular S1P produced explicitly by the action of SK1 was shown to act as a cofactor for tumor necrosis factor (TNF-α) receptor-associated factor 2 (TRAF2) E3 ubiquitin ligase activity, which is required for many TNF-α-mediated functions (Alvarez et al. 2010). Despite the identified intracellular roles of S1P, as of yet, most S1P functions have been shown to be receptor mediated. There have been five specific S1P G protein-coupled receptors (S1P1R-S1P5R) discovered; however, there have been a number of putative S1PRs described in the literature (Niedernberg et al. 2003), for example, Gpr12 and Gpr63 (Uhlenbrock et al. 2002, 2003). The S1P family of receptors signal by coupling to different Gα subunits including $G\alpha_i$, $G\alpha_q$, and $G\alpha_{12/13}$ (Hla and Brinkmann 2011), and this, along with receptor expression patterns, results in activation of various effector molecules such as Rac, ERK, PI_3K, adenylyl cyclase, phospholipase C, Rho, and JNK, leading to different downstream responses (Spiegel and Milstien 2003).

Intracellular produced S1P is capable of being exported from the cell via transporters and acting on cell surface receptors on the cell from which it was produced or on neighboring cells (Pyne and Pyne 2000). Ligand-mediated activation of SK1 typically drives S1P "inside-out" signaling. Many ligands are known to induce activation of the SK/S1P pathway upon binding to their respective receptors, ultimately leading to S1P receptor transactivation. Not surprisingly, the subcellular localization of S1P production is extremely important for "inside-out" S1P signaling. Most ligand-induced activation of the SK/S1P pathway occurs via activation of SK1, presumably by ligand-induced ERK, PKC, or other undefined downstream effector molecule activation, and this is thought to localize SK1 to the plasma membrane, bringing it into contact with its membrane-localized substrate, sphingosine (Siow and Wattenberg 2011). S1P formed at the inner leaflet of the plasma membrane can then be immediately exported from the cell to act as a ligand for its own receptors (Nieuwenhuis et al. 2010). Alternatively, S1P produced by endoplasmic reticulum (ER)-localized SK2 is subject to degradation by S1P lyase and S1P phosphatase, both of which are also localized to the ER (Gault et al. 2010), limiting the ability of S1P to be exported from the cell and act as an extracellular ligand (Siow and Wattenberg 2011). S1PR transactivation by receptor tyrosine kinases, like those responsible for growth factor-mediated signaling, further confounds S1P signaling. Activation of the SK/S1P pathway by growth factors and cytokines is a well-documented phenomenon, yet it is only one of the ways SK/S1P signaling is regulated. The crossing of S1PR pathways with other GPCR and RTK signaling pathways, along with the activation of SK by numerous agonists or stimuli and "inside-out" signaling capabilities of S1P, clearly demonstrates the complexity involved in the SK/S1P signaling pathway. Teasing out various S1P-mediated signals and the manner in which they are regulated is imperative for revealing potential therapeutic targets.

4 Sphingosine Kinase

SK1 and SK2 are located on separate genes and are highly homologous, with the exception of an extended N-terminal tail possessed by SK2 (Alemany et al. 2007; Hait et al. 2009). In contrast to SK1 which is localized to both the cytoplasmic and membrane portions of the cell, SK2 is thought to be localized to the ER, nucleus, and mitochondria (Igarashi et al. 2003; Strub et al. 2011). There is some functional redundancy between SK1 and SK2, evidenced by a lack of major phenotypes in SK1 or SK2 single knockout mice (Michaud et al. 2006); however, SK1/SK2 double knockouts are embryonic lethal due to improper formation of blood vessels and neural tube closure, providing compelling evidence for the necessity of SK in development (Mizugishi et al. 2005).

4.1 Sphingosine Kinase 1

A key step in the sphingolipid pathway is the formation of bioactive signaling molecule, S1P. Two known isoforms of sphingosine kinase (SK), SK1 and SK2, are responsible for phosphorylation of the primary hydroxyl group of sphingosine to form S1P. Cloning of SK from yeast dramatically facilitated the study of the enzyme (Lanterman and Saba 1998). In humans, three splice variants of SK1 (SK1a, SK1b, and SK1c) have been identified (Venkataraman et al. 2006), and a high degree of homology exists between mammalian SK1 and SK1 from other organisms (Kohama et al. 1998). These evolutionarily conserved regions include a diacylglycerol kinase (DGK) catalytic domain, glycine 82, which is imperative for SK1 catalytic activity (Pitson et al. 2000), and aspartate 278, which is necessary for sphingosine binding (Yokota et al. 2004) (Fig. 2). ERK phosphorylation sites (Pitson et al. 2003), phosphatidylserine binding residues (Stahelin et al. 2005), ATP binding sites (Pitson et al. 2002), Ca^{2+}/calmodulin (Sutherland et al. 2006), and TNF receptor-associated factor 2 (TRAF2) binding sites (Xia et al. 2002) are among other homologous regions of the SK1 enzyme. SK1 resides mostly in the cytosol but can translocate to the plasma membrane upon activation by various stimuli (Johnson et al. 2002), and this is thought to be one of the major mechanisms by which SK/S1P signaling is regulated (Siow et al. 2011).

Unlike many well-known enzymes, there is no concrete single or set of regulatory mechanisms known for SK1; however, a small portion of the literature is dedicated to deciphering the direct mechanisms of SK1 regulation. Touched upon earlier, SK1 localization is pivotal in determining the outcome of S1P signaling, and SK1 phosphorylation is the mechanism thought to regulate subcellular localization. Studies examining SK1 activation by extracellular ligands have revealed a serine phosphorylation site at residue 225 (Pitson et al. 2003). There have been reports that ligand-induced phosphorylation of SK1 at serine 225 causes conformational or electrostatic changes that allow SK1 to localize to and remain at the plasma membrane, enhancing the chance of contacting its substrate (Stahelin et al. 2005). SK1 has been shown to be activated by acidic phospholipids in vitro and selectively bind phosphatidylserine at the plasma membrane (Gao et al. 2012). Furthermore, calmodulin, PKC, and CIB1 (calcium- and integrin-binding protein 1) have been shown to facilitate SK1 translocation to the plasma membrane (Johnson et al. 2002; Stahelin et al. 2005; Sutherland et al. 2006). On the other hand, protein phosphatase 2A (PP2A), via interaction of its B'α subunit with the C-terminal region of SK1, has been implicated in the dephosphorylation and deactivation of SK1 (Barr et al. 2008; Pitman et al. 2011). Other, less well characterized, regulators of SK1 localization exist. For example, the mutated form of the Ras oncogene, K-RasG12V, was recently shown to induce SK1 translocation to the plasma membrane, as well as increase

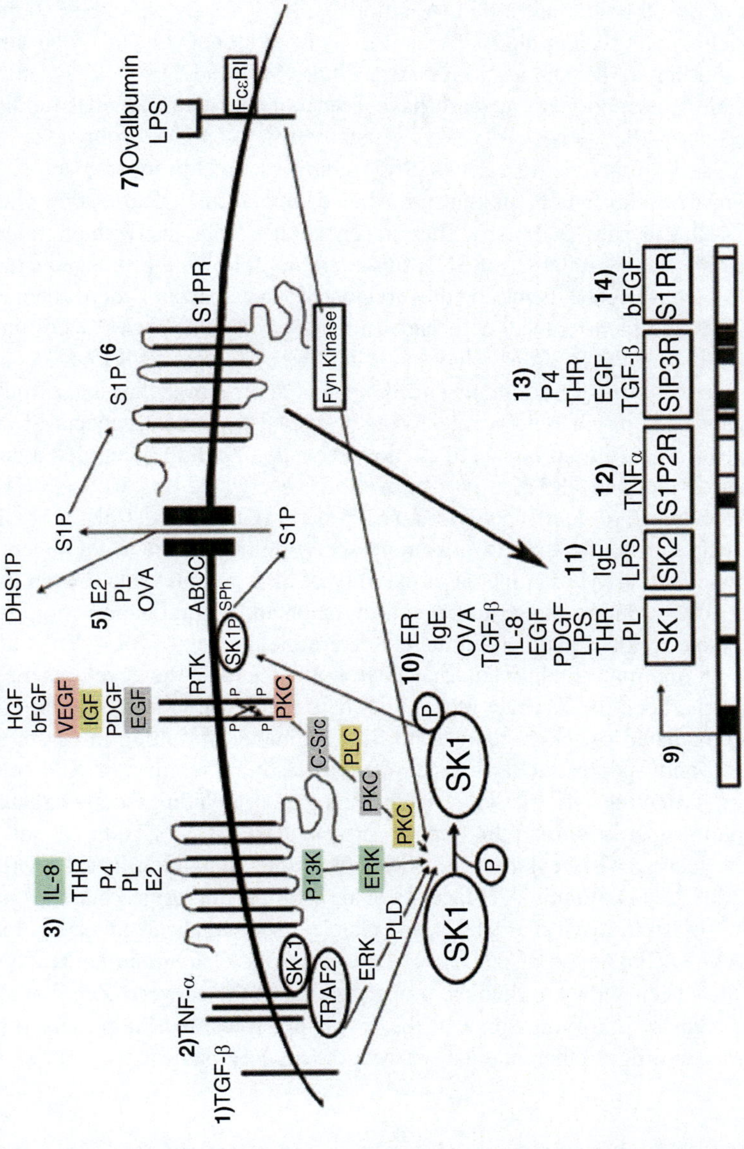

Fig. 2 *Regulation of the SK/S1P Pathway*. Schematic representation of ligands and types of receptors involved in activation of the SK/S1P pathway and the mechanisms involved. (1) TGF-β contributes to SK/S1P signaling by acute activation of SK1 and transcriptional up regulation of (10) SK1 and (13) S1P3R. (2) TNF-α activation of SK/S1P pathway by TRAF2-meditaed recruitment to the membrane; TNF-α activation of SK1 through ERK and PLD; TNF-α-induced transcriptional up regulation of (12) S1P2R. (3) G protein-coupled receptor regulation of the SK/S1P pathway, as an example in green: IL-8 binds its cognate GPCR and activates SK1 through PI3K and ERK and induces transcriptional up regulation of (10) SK1. (4) Growth factor/ Receptor Tyrosine Kinase regulation of the SK/S1P pathway. For example, highlighted in *purple*: EGF has been shown to regulate the SK/S1P pathway via c-src and PKC-mediated activation of SK1 and transcriptional up

levels of S1P (Gault et al. 2012). Additionally, SK1 has shown to be exported from endothelial cells (Hla et al. 2008), as well as from monocytic cells (Hammad et al. 2006), further solidifying a role for extracellular SK1 and S1P.

Besides phosphorylation, there have been other posttranslational modifications that regulate the activity of SK1. A study by the Pyne group examined the mechanisms of action of various SK1 inhibitors and discovered an unforeseen mechanism of inhibition: ubiquitin-mediated proteasomal degradation (Loveridge et al. 2010; Lim et al. 2011a, b). These recent studies are in line with previous studies implicating the degradation of SK1. For example, following prolonged exposure to TNF, SK1 has been shown to be downregulated via cathepsin B-dependent degradation in MCF-7 breast cancer cells and via reactive oxygen species (ROS) in cardiac cells (Taha et al. 2005, 2006; Pchejetski et al. 2007). Moreover, SK1 has recently been shown to be proteolyzed in response to genotoxic stress through a mechanism involving p53, implicating a role for SK1 in p53-dependent cancers (Taha et al. 2004; Heffernan-Stroud et al. 2012). Very recently, Yu et al. identified a conserved acetylation motif in SK1 that was shown to be acetylated by p300/cAMP-response element-binding protein (CREB)-binding protein (CBP) (Yu et al. 2012). Interestingly, using an SK1 lysine to glutamine acetyl-mimetic mutant, they were able to determine that acetylation increases stability of SK1 and prevents it from ubiquitin-mediated degradation, suggesting that acetylation and ubiquitination compete for the same lysine residues and can ultimately determine the fate of SK1 (Yu et al. 2012). This is an important finding and may have implications in the development of SK1-targeted therapeutics. Another recent finding in neurons implicated NADPH oxidase (NOX)-mediated oxidative damage to SK1, ultimately resulting in ceramide accumulation and apoptosis (Barth et al. 2011). Lastly, a number of SK1-interacting proteins affecting its activity have emerged, including delta-catenin/neural plakophilin-related armadillo repeat protein (NPRAP) (Fujita et al. 2004), aminoacyclase 1 (Maceyka et al. 2004), eukaryotic elongation factor 1A (Leclercq et al. 2008, 2011), filamin A (Maceyka et al. 2008), sphingosine kinase 1-interacting protein (SKIP) (Lacana et al. 2002), and platelet endothelial cell adhesion molecule-1 (PECAM-1) (Fukuda et al. 2004), whereas cytosolic chaperonin containing TCP-1 (CCT) has been shown to mediate proper folding of the enzyme (Zebol et al. 2009). Future studies solidifying some of the regulatory roles of SK1 presented here are necessary in order to therapeutically target the SK/S1P pathway.

Fig. 2 (continued) regulation of (10) SK1 and (13) S1P3R. (5) E2, prolactin and ovalbumin challenge have all been shown to increase S1P or DHS1P export from the cell; E2 for example also induces the transcriptional up regulation of SK1 through activation of ER. (6) Mechanism of ligand-induced S1P inside-out signaling. Ligand activation of SK, production and export of S1P, (6) activation of S1P receptors and subsequent (9) transcriptional up regulation of various genes. (7) FcεR1 receptor activation leads to fyn-kinase-mediated activation of SK1 and in the case of (7) LPS, induces transcriptional up regulation of SK1 and SK2

4.2 *Sphingosine Kinase 2*

While SK1 has been the subject of intense investigation, until recently SK2 has remained the less well-characterized enzyme isoform. In early studies, SK2 was shown to possess functions that are contradictory to those of SK1, including induction of apoptosis via its role as a putative BH3-only protein (Liu et al. 2003; Maceyka et al. 2005; Okada et al. 2005, 2009). In an important study that identified the first direct intracellular target of S1P, SK2 was reported to play a role in the epigenetic regulation of gene expression via modulation of histone acetylation, whereby SK2-generated nuclear S1P was shown to prevent the removal of acetyl groups contained within histone tails via direct interaction with and inhibition of histone deacetylases 1 and 2 (HDAC1 and HDAC2) (Hait et al. 2009). Despite the absence of detailed characterization of SK2 regulation and function, the development of SK2-specific enzyme inhibitors has exposed many novel functions of SK2. In contrast to its initially identified functions, SK2 has been implicated in promotion of cancer such that pharmacological inhibition of SK2 has been shown to sensitize cells to apoptotic stimuli (Gao and Smith 2011). Moreover, siRNA-mediated downregulation of SK2 was shown to inhibit proliferation and migration in tumor cells (Gao and Smith 2011). Another recently identified role for SK2 is in precondition-mediated protection from cerebral ischemic injury (Shi et al. 2012). SK2 inhibition prevented precondition-induced tolerance to ischemic injury in a rodent stroke model (Garofalo et al. 2012). Shown to be in the ER (Siow and Wattenberg 2011), nucleus (Hait et al. 2009; Riccio 2010), and mitochondria (Strub et al. 2011), SK2 is localized to subcellular regions different from those of SK1 (Siow et al. 2011). Similarly to SK1, SK2 has been shown to be phosphorylated by ERK in response to EGF treatment, leading to MDA-MB-453 migration (Hait et al. 2005, 2007). Besides subcellular localization and phosphorylation, additional direct regulatory mechanisms remain to be determined for SK2 (Hait et al. 2005, 2007). Expansion on the already known functions of SK2 is essential for the development of drugs targeting the SK/S1P pathway.

5 Regulation of the SK/S1P Pathway

As mentioned above, the SK/S1P pathway is activated by numerous extracellular ligands, and the resultant "inside-out" signaling has emerged as a major mechanism by which SK/S1P signaling is controlled (Fig. 2). Extracellular stimuli shown to activate the SK/S1P pathway are numerous and diverse in nature; however, cytokines and growth factors are perhaps the best characterized. In addition to cytokines and growth factors, other factors known to regulate S1P signaling include nuclear receptors, inducers of immune responses such as endotoxins, as well as the export machinery responsible for S1P release from the cell. Ligand-mediated regulation of SK/S1P signaling highlights the role of S1P as a secondary messenger and provides additional targets for the treatment of cytokine and/or growth factor-mediated diseases.

6 Cytokine Regulation of the SK/S1P Pathway

6.1 *Transforming Growth Factor-β (TGF-β)*

Sphingolipid metabolism has long been known to be influenced by immunomodulating cytokines with activation of the sphingomyelinase pathway leading to apoptosis and activation of the SK/S1P pathway leading to inflammation (Jenkins et al. 2011). Evidenced by the immunomodulatory drug FTY720, the SK/S1P pathway has been shown to be involved in inflammation and cellular processes brought about by numerous cytokines (Zhu and Da 2007). Transforming growth factor-β is one such cytokine shown to regulate SK/S1P signaling. TGF-β is best known for its role in the development of tissue fibrosis, such as occurring in cardiac remodeling and scleroderma (Silver 1996; Derrett-Smith et al. 2010). In myofibroblasts, TGF-β treatment led to Smad-induced SK1 upregulation and subsequent S1PR modulation, indicating a role for TGF-β in regulation of the SK/S1P pathway (Cencetti et al. 2010). Also, in dermal fibroblasts, TGF-β treatment increased SK1 mRNA and protein levels resulting in prolonged SK1 activation and subsequent increases in S1P with corresponding decreases in ceramide and sphingosine, as well as inhibition of S1P phosphatase activity (Yamanaka et al. 2004) (Fig. 2). In a mouse model of lung fibrosis, TGF-β treatment induced activation of SK1 and transactivation of S1P receptors (Kono et al. 2007). Another manner in which TGF-β regulates SK/S1P signaling is through organization of signaling platforms. In renal proximal tubules, Song et al. demonstrated that isoflurane induced caveolae enriched in SK/S1P signaling components, including SK1, S1P, and downstream effectors of SK/S1P signaling, MAPK, and ERK (Song et al. 2010). Caveolae formed following isoflurane treatment were also enriched in TGF-β1 receptor, and inhibition of TGF-β1 signaling prevented caveolae formation, suggesting a role for TGF-β in organizing SK/S1P signaling (Song et al. 2010). This organization may also occur at particular subcellular localizations, which has been shown to mediate the effect of S1P signaling, as mentioned above. In the case of TGF-β, SK1-specific inhibitors may serve as potential therapeutic treatment modalities in pathologies mediated by TGF-β.

6.2 *Interleukins*

Interleukins (ILs) are cytokines derived mostly from immune cells including B cells, T cells, monocytes, and macrophages (Chen et al. 2006). Numerous families of ILs exist and exhibit many functions including differentiation and proliferation (Chen et al. 2006). IL-8 is one IL known to regulate the SK/S1P pathway as it was shown to dose-dependently increase SK activity in multiple myeloma cells through a mechanism involving PI3K and ERK/MAPK signaling (Li et al. 2007) (Fig. 2). Also known to regulate SK/S1P is IL-1β. SK1 activity was acutely and transiently activated in A549 lung epithelial cells treated with IL-1β (Billich et al. 2005). IL-1β was also shown to regulate SK/S1P signaling in glioblastoma cells via transcriptional upregulation (Fig. 2) (Paugh et al. 2008). Further research into IL regulation of SK/S1P signaling has the potential to reveal novel therapeutic targets.

6.3 *Tumor Necrosis-Alpha (TNF-α)*

TNF-α is a pleiotropic cytokine involved in mediating numerous inflammatory and immune responses. It has long been known that TNF-α elicits many of its signals via the sphingolipid metabolic pathway (Candela et al. 1991; Yang et al. 1993; Muller et al. 1995). TNF-α is known to generate ceramide via activation of sphingomyelinase (Higuchi et al. 1996). Moreover, TNF-α has been shown to regulate SK/S1P in many ways and is perhaps the most well-characterized activator of SK1. A major mechanism of SK activation, as mentioned prior, is phosphorylation. Chow et al. proposed that TNF-α induces the ERK-mediated phosphorylation of SK1 at residue 225 which increases the chances of substrate-enzyme interaction (Fig. 2) (Stahelin et al. 2005). Furthermore, Pitson et al. also implicated MAPK in the phosphorylation of serine residue 225 and activation of SK1 in response to TNF-α (Fig. 2) (Pitson et al. 2003; Chandru and Boggaram 2007). Interestingly, a TRAF2-binding motif in SK1 has been identified and has been shown to be necessary for the TNF-α-induced activation of SK1 (Fig. 2 and Table 2) (Xia et al. 2002). In line with the previous findings, TNF-α was shown to elicit acute increases in S1P in L929 fibroblasts (Table 2) (Pettus et al. 2003). More recently, TNF-α was shown to modulate SK/S1P signaling independent of SK1; in endothelial cells, TNF-α treatment was shown to induce the expression of S1P2R mRNA and protein, thus shifting S1P signaling towards S1P2R-mediated events (Fig. 2 and Table 1) (Du et al. 2012). Lastly, TNF-α was shown to negatively regulate SK/S1P signaling via promotion of NADPH oxidase (Nox)-mediated oxidative damage to SK1 in neuronal cells (Barth et al. 2012). Discussed in the next chapter, the SK/S1P pathway has been shown to mediate a large portion of TNF-α-mediated events, supporting TNF-α induced “inside-out” S1P signaling. With the overwhelming evidence for TNF-α-mediated activation of SK1, establishing the mechanisms by which it does so can help reveal possible therapeutic targets for the many diseases caused by and associated with inflammation.

7 Growth Factor Regulation of the SK/S1P Pathway

7.1 *Miscellaneous Growth Factors*

S1P receptors can undergo sequential activation by ligands other than S1P, such as growth factors. This transactivation of receptors is also known as “receptor cross talk” (Spiegel and Milstien 2003). For example, platelet-derived growth factor (PDGF) has been shown to activate SK1, ultimately leading to activation of S1P receptors (Fig. 2) (Olivera and Spiegel 1993; Rosenfeldt et al. 2001a, b). The recently coined term “integrative signaling” is used to describe functional signaling complexes, which in

Table 1 Role of the SK/S1P pathway in growth factor mediated events

Cell type	Response	Mechanism	References
Platelet-derived growth factor (PDGF)			
HEK-293, ASMC, and fibroblasts	Cell motility	S1P1R-mediated Rac activation	Hobson et al. (2001), Olivera and Spiegel (1993), and Rosenfeldt et al. (2001a, b)
Fibroblasts	Cell motility	S1P1R-mediated Src and FAK activation	Olivera and Spiegel (1993) and Rosenfeldt et al. (2001a, b)
Swiss 3T3 fibroblasts	Cell proliferation	ND	Olivera and Spiegel (1993)
Oligodendroglial progenitors	Expression of delayed rectifier current (I(K))	Activation of Src family kinases	Soliven et al. (2003)
Mesangial cells	Cell proliferation	Increased SK1 mRNA	Katsuma et al. (2002)
Myoblasts	Inhibition of cell proliferation	Upregulation of SK1 and S1P1R signaling	Nincheri et al. (2010)
Epidermal growth factor (EGF)			
HEK293	Calcium signaling	ND	Meyer zu Heringdorf et al. (1999)
MDA-MB-453	Cellular migration	ND	Hait et al. (2005)
MDA-MB-453	Cellular migration	ERK-mediated SK2 phosphorylation	Hait et al. (2007)
Cytotrophoblasts	Inhibits TNFα-induced apoptosis	PI3K-dependent SK1 activation	Johnstone et al. (2005)
MCF-7	Cell growth and migration	ND	Doll et al. (2005) and Sarkar et al. (2005)
Glioblastoma	Cell growth and survival	ND	Estrada-Bernal et al. (2011)

ND not determined

this case consists of PDGFR and S1P1R (Pyne and Pyne 2010). Upon binding its receptor, PDGF activates SK1 resulting in the transactivation of the S1P1R receptor leading to Rac-mediated cell motility; additionally, it was shown that PDGF-mediated S1P1 transactivation results in activation of Src and FAK leading to subsequent lamellipodia formation and migration (Table 1) (Hobson et al. 2001). Furthermore, PDGF was shown to elevate SK1 mRNA, protein levels, and activity in human coronary artery smooth muscle cells (HCASM) through a mechanism involving the PI3K/AKT2/mTOR (mammalian target of rapamycin) signaling pathway (Table 1) (Francy et al. 2007). Also via upregulation of SK1, PDGF is known to promote cellular proliferation in mesangial cells (Katsuma et al. 2002); however, PDGF inhibits cellular proliferation through upregulation of SK1 in myoblasts

Table 2 Role of the SK/S1P pathway in TNF-α-mediated events

Cell type	Response	Mechanism	References
HUVEC	Expression of E-selectin and VCAM; ERK and NFkB activation	ND	Xia et al. (1998)
HUVEC	eNOS activation	S1PR1 and S1PR3 activation	De Palma et al. (2006)
HUVEC	ICAM expression	Akt, ERK, and NFkB activation	Kang et al. (2006)
HAEC	MCP-1 expression and secretion, VCAM expression	p38	Chen et al. (2004)
Human neutrophils	Neutrophil priming	ND	MacKinnon et al. (2002)
RAW 264.7 macrophages	COX-2 expression, PGE2 production	ND	Hammad et al. (2008)
U937 human monocytic cells	ERK and NFkB activation	PLD activation	Sethu et al. (2008)
1321N1 glioblastoma	Proliferation	Akt and cyclin D activation	Radeff-Huang et al. (2007)
SKNBE human neuroblastoma	eNOS activation	SK/S1P/S1PR and Akt activation	De Palma et al. (2008)
H441 lung epithelial	IL-8 expression	ERK, p38, and AP-1 activation	Chandru and Boggaram (2007)
A549	Cox-2 expression and PGE2 production, IL-6, RANTES, MCP-1, and VCAM expression	NFkB activation	Billich et al. (2005)
A549 and L929	COX-2 and PGE2 production	ND	Pettus et al. (2003)
Fibroblasts	ERK activation and MMP1 expression	Dihydro-S1P via $G\alpha_i$ signaling	Bu et al. (2006)
C2C12 myoblasts	Myogenesis	S1P2R	Donati et al. (2007)
MC3T3-E1 osteoblast-like	IL-6 production	PLC and PKC	Wang et al. (2002)
HEK 293T	NFkB activation	SK:TRAF2 binding	Xia et al. (2002)
Synoviocytes	MMP1a and IL-6 production	ND	Baker et al. (2010)
HepG2	Downregulation of Na+/K + ATPase	JNK and NFκB inhibition	Dakroub and Kreydiyyeh (2012)

ND not determined

(Nincheri et al. 2010). PDGF has been shown to induce the expression of delayed rectifier current (I(K)) in oligodendroglial precursor cells, also through a mechanism involving Src activation and dependence on the SK/S1P pathway (Soliven et al. 2003). Many growth factors other than PDGF have also been shown to activate SK1, including vascular endothelial growth factor (VEGF), nerve growth factor (NGF), and insulin-like growth factor (IGF) (Fig. 2) (Johnson et al. 2002; Billich et al. 2005; El-Shewy et al. 2006; Kusner et al. 2007). A study by Shu et al. demonstrates the mechanism by which VEGF mediates activation of SK1 (Shu et al. 2002). VEGF binds its receptor, activating PKC, which stimulates SK1 to produce S1P, and this was shown to occur in a dose-dependent manner (Shu et al. 2002). IGF has also been shown to activate SK1 and cause S1P production followed by GFP-S1P1R receptor internalization in HEK-293 cells (El-Shewy et al. 2006). More recently, this IGF-mediated activation of SK1 was shown to occur through IGFRII-activation of PLC and PKC-βII (Fig. 2) (El-Shewy et al. 2011). Basic fibroblast growth factor (bFGF) is known to regulate SK/S1P signaling via upregulation of S1P1R expression in smooth muscle cells (SMCs) and activation of SK1 in neuronal cells (Fig. 2) (Rius 1997). Lastly, hepatocyte growth factor (HGF) has been shown to differentially regulate SK activity, such that at high doses it inhibits SK, while at low doses SK is activated (Liu et al. 2011); results from this study highlight the importance of examining dose-dependent effects of ligand regulation of the SK/S1P signaling pathway. Given the large number of molecules capable of activating SK1, resulting in S1P production and ultimately S1P receptor transactivation, this may be a common mechanism by which G protein-dependent signals are elicited by non-G protein-coupled receptors; in fact, "GPCR-jacking" is a term used to represent this phenomenon (Delcourt et al. 2007).

7.2 Epidermal Growth Factor

The effects of epidermal growth factor (EGF) on cell growth, proliferation, and differentiation are very well studied, and EGF has also been extensively studied as an activator of the SK/S1P pathway, especially in diseases associated with growth and proliferation such as in cancer. EGF has been shown to induce acute activation and late transcriptional upregulation of SK1 (Paugh et al. 2008), as well as induce SK1 translocation to the plasma membrane in MCF-7 breast cancer cells (Fig. 2 and Table 1) (Sarkar et al. 2005). EGF has also been shown to activate the SK/S1P pathway via ERK-mediated phosphorylation and subsequent activation of SK2 (Hait et al. 2007). Through a mechanism termed "oncogene tolerance," EGFR increased the expression of SK1; however, this occurred only in estrogen receptor-positive MCF-7 cells, whereby SK1 then functions in negative feedback inhibition of EGF signaling (Long et al. 2010). Not only has EGF been shown to regulate S1P signaling in breast cancers, but cross talk between EGFR and S1PRs has been extensively studied in glioblastoma multiforme, the most common and aggressive brain tumors occurring in humans (Mao et al. 2012). EGF, through a mechanism involving c-Src and PKCδ, induced SK1 translocation to the plasma membrane (Table 1) (Paugh et al. 2008), while EGF treatment or overexpression of constitutively active EGFRVIII resulted in increased expression and activity of SK1 in glioblastoma cells (Fig. 2) (Estrada-Bernal et al. 2011). Interestingly, lysophosphatidic acid (LPA) upregulated SK1 and S1P3R expression in gastric cancer cells; however, this occurred downstream of LPA-induced activation of EGFR (Shida et al. 2008). Determining the mechanisms by which EGF activates SK1 can prove useful in the treatment of EGF-driven diseases, such as in various malignancies.

8 Other Regulators of the SK/S1P Pathway

8.1 Hormones

A wide variety of stimuli, which do not fall into the cytokine or growth factor category, are also known to regulate SK/S1P signaling. Such types of signaling molecules are hormones. Perhaps the best characterized hormonal regulation of SK/S1P signaling emerges in 17β-estradiol (E2) signaling. E2 has been shown to regulate SK/S1P signaling on many levels. First, it has been demonstrated that SK1 undergoes acute activation by E2 (Sukocheva et al. 2003, 2006; Doll et al. 2005), as well as by STAT5/PKC/MAPK-mediated SK1 transcriptional upregulation at a later phase in MCF-7 breast cancer cells (Fig. 2) (Sukocheva et al. 2003; Doll et al. 2005). Another manner in which E2 regulates SK/S1P signaling is through stimulating the rapid release of S1P and dihydro-S1P from cells; interestingly, E2 also results in late phase activation of SK1 through transcriptional upregulation of estrogen receptors which, when activated, further induce SK1 activity (Fig. 2) (Sukocheva et al. 2003). Not only does E2 regulate SK/S1P signaling, but prolactin has been shown to do so as well. Similar to E2, prolactin biphasically activates SK1 with acute activation of SK1 and late phase activation via SK1 gene transcription (Fig. 2) (Doll et al. 2007). Lastly, progesterone has been shown to regulate SK1. Progesterone increases SK1 mRNA, and as gestational age increases, so does SK1 protein and activity, along with S1P lyase and S1P3R protein expression, which tracks with E2 and progesterone levels in late pregnancy (Fig. 2) (Yamamoto et al. 2010). It is well demonstrated that estrogen and other related sex hormones regulate the SK/S1P pathway; therefore, it will be important to examine the role of nonsex hormones in its regulation as well.

8.2 Immune Activators

Immune activators encompass any ligand that induces an immune response. The SK/S1P is intimately involved in inflammation and the immune response; therefore, it is not unusual that many activators of the immune response also directly affect SK/S1P function. Antigen challenge has been shown to regulate the SK/S1P pathway in numerous accounts. Mast cell activation, via IgE binding to its receptor FcεR1, is known to lead to SK activation and subsequent S1P formation (Melendez

2008); for example, IgE receptor activation has been shown to stimulate the expression and activation of both sphingosine kinase isoforms (Fig. 2) (Olivera et al. 2006). Activation of FcεRI in bone marrow-derived mast cells led to S1P production through a Fyn-kinase-dependent mechanism (Fig. 2) (Olivera et al. 2006). Antigen challenge in the form of ovalbumin administration is also known to affect the SK/S1P signaling pathway. In a mouse model of lung inflammation, ovalbumin challenge led to increased expression of SK1 in areas near bronchial epithelial walls and in areas of inflammation (Fig. 2) (Nishiuma et al. 2008). Ovalbumin also caused release of S1P into lavage fluid, thus activating S1P signaling in the lung (Fig. 2) (Nishiuma et al. 2008). Endotoxins are also known to induce robust immune responses and have been studied in the regulation of SK/S1P pathway. Using microarray technology and RTPCR, Wadgaonkar et al. found that LPS treatment significantly increased SK1 expression within 6 h which returned to basal by 24 h; unexpectedly, SK2 expression also increased, although more slowly and more sustained than that of SK1 (Fig. 2) (Wadgaonkar et al. 2009). This suggests that LPS differentially regulates SK/S1P signaling through early and late upregulation of SK1 and SK2, respectively (Wadgaonkar et al. 2009). Also implicating LPS in SK/S1P regulation, treatment of glial cells with LPS resulted in increased SK1 message in one study (Fig. 2) (Lin et al. 2011) and upregulation of both SK1 and S1P3R in another (Fig. 2) (Fischer et al. 2011).

8.3 *Miscellaneous*

Other activators, which do not fall into the aforementioned categories, are known to mediate SK/S1P signaling and deserve recognition, as well. First, agonists of peroxisome-proliferator-activated receptor gamma (PPARγ), such as those used to treat type 2 diabetes (Salvado et al. 2012), have been shown to induce SK1 expression and activity leading to increased S1P levels (Chao et al. 2010). Also implicated in regulation of the pathway are thrombin and its receptor, PAR-1. Thrombin activation of PAR-1 was shown to regulate SK/S1P signaling such that treatment of A549 with thrombin led to elevated levels of SK1 mRNA and subsequent increases in S1P (Fig. 2) (Billich et al. 2009). Thrombin also contributes to regulation of SK/S1P signaling through "integrative signaling," by PAR-1-mediated activation of S1P3R (Fig. 2) (Niessen et al. 2008). Moreover, under hypoxic conditions, Hif-2α has been shown to bind the SK1 promoter and induce transcriptional upregulation in U87MG glioblastoma cells (Anelli et al. 2008). An additional, very important mechanism that controls SK/S1P signaling is S1P export from cells. The ABC family of transporters and the spinster 2 (Spns2) transporter are known to be responsible for shuttling S1P from the cell to the extracellular space whereby it can act on its receptors (Fig. 2) (Kawahara et al. 2009; Kobayashi et al. 2009; Nieuwenhuis et al. 2010; Hisano et al. 2011, 2012). Importantly, expression levels and localization of transporters can affect the amount of S1P exported and, in the case of cancers, can serve to pump out drugs as well as pro-survival S1P, perhaps promoting the progression of disease (Fletcher et al. 2010). While many ligands are known to induce the activation of the SK/S1P pathway, the roles the SK/S1P pathway play in mediating ligand-induced cellular events is of great interest, potentially allowing exploitation of the pathway as a secondary target for therapeutic intervention.

9 Role of the SK/S1P Pathway in Mediating EGF and TNF-α Biologies

Having previously focused on cytokine and growth factor mechanisms of SK/S1P regulation, we will now explore the role of the SK/S1P pathway in mediating growth factor- and cytokine-induced biologies. Not only is the SK/S1P signaling pathway activated by other cell signaling molecules, such as TNF and EGF, but it also functions as a mediator of TNF- and EGF-induced biologies. Many of the ligands known to activate and utilize the SK/S1P pathway to carry out their biological functions are also known to contribute to disease when deregulated: TNF-α, EGF, TGF-β, and PDGF, for example. Development of TNF-α- and EGF-targeted drugs has led to successful cancer and immunomodulatory therapies; however, not everyone responds to these therapies and those who do may acquire resistance during treatment. The role of the SK/S1P pathway in mediating many TNF-α and EGF-induced biologies makes it an additional drug target in diseases resulting from aberrant TNF-α or EGF signaling. The role of the SK/S1P pathway in mediating EGF and TNF-α biologies and its capacity to serve as a therapeutic target will be discussed.

9.1 Role of SK/S1P in EGF-Mediated Biologies

The SK/S1P pathway is used by numerous ligands to carry out vast cellular functions, and intact SK/S1P signaling has been shown to be especially required for mediating the biologies of growth factors (Table 1). It is not by coincidence that EGF and S1P share overlapping functions, as SK/S1P signaling has been shown to induce growth (Meng et al. 2011), survival (Meng et al. 2011), and motility (Berdyshev et al. 2011) itself, as well as mediate EGF-stimulated calcium signaling (Meyer zu Heringdorf et al. 1999), growth, survival, and motility (Doll et al. 2005). Mentioned previously, EGF is known to induce activation, transcription, and translocation of SK1 in numerous cell types. Activation of the SK/S1P pathway is not a mere side effect of EGF treatment such that EGF activation of the SK/S1P pathway has been shown to mediate MCF-7 and glioma cell growth and survival (Doll et al. 2005); however, the best characterized role of the SK/S1P pathway in EGF-mediated events stems from cell motility and invasion studies (Table 1). For example, EGF activation and upregulation of the SK/S1P pathway have been shown to be required for motility in MDA-MB-453, MCF10A, and MCF-7 breast cancer cells (Table 1) (Hait et al. 2005; Sarkar et al. 2005; Martin et al. 2009).

Also, EGF has been shown to promote invasion of glioblastoma cells through a mechanism involving S1P3R (Hsu et al. 2012) and of MNK1 gastric cancer cells via SK1-mediated c-Jun phosphorylation and upregulation of plasminogen activator inhibitor 1 (PAI-1) (Table 1) (Paugh et al. 2008). Similar to other ligands, EGF employs the SK/S1P pathway as a mechanism for cellular evasion of apoptosis, and this is supported by many studies. For example, inhibition of SK1 sensitized MCF-7 cells to chemotherapeutic agents (Sarkar et al. 2005). In addition, in glioma cells, constitutively active EGFRVIII protects from apoptosis via upregulation of SK1 message and activity (Table 1) (Estrada-Bernal et al. 2011). Based on these studies, it seems that the SK/S1P pathway may be an alternative or additional treatment target for proliferative diseases, especially breast cancers and glioblastomas that do not respond to currently available treatments.

EGF not only regulates the SK/S1P pathway in pathological processes but in normal physiological functions as well, for example, regulation of trophoblast function. Trophoblasts are specialized cells that function in embryo implantation as well as in nutrient delivery to developing embryos (Singh et al. 2012); furthermore, in addition to JNK and p38 activation in trophoblasts, EGF stimulated SK1 activity preventing apoptosis (Singh et al. 2012). Also, both EGF and S1P have been shown to activate Na^+/H^+ exchanger activity in syncytiotrophoblasts, which is essential for embryonic invasion of the uterine wall in order to establish a nutrient connection (Johnstone et al. 2005); however, it remains to be determined if EGF regulates this function in an SK/S1P-dependent manner. Given the role of EGF and S1P in invasion (Hsu et al. 2012), and the connections between Na^+/H^+ exchangers and ezrin, radixin, and moesin (ERM) proteins (Denker et al. 2000), it is possible that EGF induces the invasion of syncytiotrophoblasts via SK/S1P-mediated activation of ERM proteins. Indeed, we have found that EGF induces the invasion of HeLa cells via SK/S1P activation of ERM proteins (unpublished results). While EGF has been known to regulate ERM proteins for some time (Bretscher 1989; Krieg and Hunter 1992), our lab has shown that it is through activation of the SK/S1P pathway (unpublished results); therefore, a novel EGF-induced biology mediated by the SK/S1P pathway is regulation of ERM proteins. The SK/S1P pathway has been shown to mediate EGF signals in normal and disease processes and is a prime example of growth factor regulation of the SK/S1P pathway.

9.2 Role of SK/S1P in TNF-α-Mediated Biologies

While many different ligands activate the SK/S1P pathway, TNF-α activation of SK1 is quite well studied, and the signaling events and the biologies elicited by TNF-α are perhaps the most intimately linked with the SK/S1P pathway. As mentioned in the previous chapter, TNF-α activates the SK/S1P pathway on numerous levels, and similar to EGF, the activation of SK and production of S1P serve as major mediators of TNF-α-induced cellular events. TNF-α regulates SK/S1P in many different cell lines (Table 2), with much of the work being done in endothelial cells and cells of the immune system. Using HUVEC and HAEC,

SK/S1P has been shown to mediate the TNF-stimulated expression of adhesion molecules including E-selectin (Xia et al. 1998), VCAM (Xia et al. 1998), and ICAM (Chen et al. 2004; Kang et al. 2006) (Table 2). Also in endothelial cells, in response to TNF, SK/S1P has been shown to mediate transcription, activation, or secretion of other inflammatory signaling molecules (Table 2) including MCP-1 (Chen et al. 2004), PGE2 (Pettus et al. 2003), IL-8 (Chandru and Boggaram 2007), and eNOS (De Palma et al. 2006, 2008). Contributing to perpetuation of the inflammatory response, TNF-α is also known to induce Cox2 expression, PGE2 production, and NFκB activation in an SK/S1P-dependent manner, although the mechanisms by which this occurs have yet to be clearly defined (Table 2) (Pettus et al. 2003; Billich et al. 2005). One study has identified phospholipase D (PLD) as the mechanism by which TNF activates the SK/S1P pathway in order to induce ERK and NFκB activation in monocytes (Sethu et al. 2008). Myogenesis of C2C12 myoblasts (Donati et al. 2007), proliferation of 1321N1 glioblastoma cells (Radeff-Huang et al. 2007), priming of neutrophils (MacKinnon et al. 2002), MMP1α (Bu et al. 2006; Baker et al. 2011) and IL-6 production (Wang et al. 2002; Baker et al. 2010), and downregulation of the Na^+/K^+ ATPase (Dakroub and Kreydiyyeh 2012) are among other functions dictated by TNF-α and carried out by SK/S1P signaling (Table 2).

Similar to that of EGF, TNF-α activates the SK/S1P pathway in order for cellular evasion of apoptosis. Xia et al. first detected a role for the SK/S1P in evasion from TNF-mediated apoptosis in HUVEC cells (Xia et al. 1998): addition of exogenous S1P or phorbol ester-induced increases in S1P protected cells from TNF-mediated apoptosis (Xia et al. 1998). In addition, RAW 264.7 cells were protected from TNF-induced apoptosis in an SK/S1P manner such that knockdown of SK1 abrogated increases in PGE2 and sensitized cells to apoptosis (Hammad et al. 2008). The role of the SK/S1P pathway in mediating TNF-induced biologies emphasizes the SK/S1P pathway as an attractive target for the development of drugs used to treat diseases with important inflammatory components such as atherosclerosis, arthritis, and diabetes.

10 Conclusions

While the SK/S1P pathway is involved in numerous cellular events, the mechanisms of its regulation have yet to be fully elucidated. A better understanding of the manner in which this pathway is regulated may potentially reveal druggable targets for the treatment of diseases known to possess important SK/S1P signaling components.

Funding This manuscript is based upon work supported in part by a MERIT Award [BX000156-01A1] (LMO) by the Office of Research and Development, Department of Veterans Affairs, Northport VA Medical Center, Northport, NY. The content of this material does not represent the views of the Department of Veterans Affairs or the United States Government, US Department of Education Graduate Assistance in Areas of National Need (GAANN) in Lipid Biology and

New Technologies [P200A070596] (KAOG), and National Cancer Institute [PO1-CA97132 (LMO)] and National Institutes of Health National Institute of General Medical Sciences [R01-GM062887 (LMO)].

References

Alemany R, van Koppen CJ et al (2007) Regulation and functional roles of sphingosine kinases. Naunyn Schmiedebergs Arch Pharmacol 374(5–6):413–428

Alvarez SE, Harikumar KB et al (2010) Sphingosine-1-phosphate is a missing cofactor for the E3 ubiquitin ligase TRAF2. Nature 465(7301):1084–1088

Anelli V, Gault CR et al (2008) Sphingosine kinase 1 is up-regulated during hypoxia in U87MG glioma cells. Role of hypoxia-inducible factors 1 and 2. J Biol Chem 283(6):3365–3375

Baker DA, Barth J et al (2010) Genetic sphingosine kinase 1 deficiency significantly decreases synovial inflammation and joint erosions in murine TNF-alpha-induced arthritis. J Immunol 185(4):2570–2579

Baker DA, Obeid LM et al (2011) Impact of sphingosine kinase on inflammatory pathways in fibroblast-like synoviocytes. Inflamm Allergy Drug Targets 10(6):464–471

Barr RK, Lynn HE et al (2008) Deactivation of sphingosine kinase 1 by protein phosphatase 2A. J Biol Chem 283(50):34994–35002

Barth BM, Gustafson SJ et al (2011) Ceramide kinase regulates TNFalpha-stimulated NADPH oxidase activity and eicosanoid biosynthesis in neuroblastoma cells. Cell Signal 24(6): 1126–1133

Barth BM, Gustafson SJ et al (2012) Neutral sphingomyelinase activation precedes NADPH oxidase-dependent damage in neurons exposed to the proinflammatory cytokine tumor necrosis factor-alpha. J Neurosci Res 90(1):229–242

Bartke N, Hannun YA (2009) Bioactive sphingolipids: metabolism and function. J Lipid Res 50(Suppl): S91–S96

Berdyshev EV, Gorshkova I et al (2011) Intracellular S1P generation is essential for S1P-induced motility of human lung endothelial cells: role of sphingosine kinase 1 and S1P lyase. PLoS One 6(1):e16571

Billich A, Bornancin F et al (2005) Basal and induced sphingosine kinase 1 activity in A549 carcinoma cells: function in cell survival and IL-1beta and TNF-alpha induced production of inflammatory mediators. Cell Signal 17(10):1203–1217

Billich A, Urtz N et al (2009) Sphingosine kinase 1 is essential for proteinase-activated receptor-1 signalling in epithelial and endothelial cells. Int J Biochem Cell Biol 41(7):1547–1555

Bode C, Sensken SC et al (2010) Erythrocytes serve as a reservoir for cellular and extracellular sphingosine 1-phosphate. J Cell Biochem 109(6):1232–1243

Bretscher A (1989) Rapid phosphorylation and reorganization of ezrin and spectrin accompany morphological changes induced in A-431 cells by epidermal growth factor. J Cell Biol 108(3): 921–930

Bu S, Yamanaka M et al (2006) Dihydrosphingosine 1-phosphate stimulates MMP1 gene expression via activation of ERK1/2-Ets1 pathway in human fibroblasts. FASEB J 20(1):184–186

Camerer E, Regard JB et al (2009) Sphingosine-1-phosphate in the plasma compartment regulates basal and inflammation-induced vascular leak in mice. J Clin Invest 119(7):1871–1879

Candela M, Barker SC et al (1991) Sphingosine synergistically stimulates tumor necrosis factor alpha-induced prostaglandin E2 production in human fibroblasts. J Exp Med 174(6): 1363–1369

Cencetti F, Bernacchioni C et al (2010) Transforming growth factor-beta1 induces transdifferentiation of myoblasts into myofibroblasts via up-regulation of sphingosine kinase-1/S1P3 axis. Mol Biol Cell 21(6):1111–1124

Chandru H, Boggaram V (2007) The role of sphingosine 1-phosphate in the TNF-alpha induction of IL-8 gene expression in lung epithelial cells. Gene 391(1–2):150–160

Chao HH, Chen CH et al (2010) L-Carnitine attenuates angiotensin II-induced proliferation of cardiac fibroblasts: role of NADPH oxidase inhibition and decreased sphingosine-1-phosphate generation. J Nutr Biochem 21(7):580–588

Chen XL, Grey JY et al (2004) Sphingosine kinase-1 mediates TNF-alpha-induced MCP-1 gene expression in endothelial cells: upregulation by oscillatory flow. Am J Physiol Heart Circ Physiol 287(4):H1452–H1458

Chen Q, Carroll HP et al (2006) The newest interleukins: recent additions to the ever-growing cytokine family. Vitam Horm 74:207–228

Dakroub Z, Kreydiyyeh SI (2012) Sphingosine-1-phosphate is a mediator of TNF-alpha action on the Na+/K + ATPase in HepG2 cells. J Cell Biochem 113(6):2077–2085

De Palma C, Meacci E et al (2006) Endothelial nitric oxide synthase activation by tumor necrosis factor alpha through neutral sphingomyelinase 2, sphingosine kinase 1, and sphingosine 1 phosphate receptors: a novel pathway relevant to the pathophysiology of endothelium. Arterioscler Thromb Vasc Biol 26(1):99–105

De Palma C, Falcone S et al (2008) Endothelial nitric oxide synthase overexpression by neuronal cells in neurodegeneration: a link between inflammation and neuroprotection. J Neurochem 106(1):193–204

Delcourt N, Bockaert J et al (2007) GPCR-jacking: from a new route in RTK signalling to a new concept in GPCR activation. Trends Pharmacol Sci 28(12):602–607

Denker SP, Huang DC et al (2000) Direct binding of the Na–H exchanger NHE1 to ERM proteins regulates the cortical cytoskeleton and cell shape independently of H(+) translocation. Mol Cell 6(6):1425–1436

Derrett-Smith EC, Dooley A et al (2010) Systemic vasculopathy with altered vasoreactivity in a transgenic mouse model of scleroderma. Arthritis Res Ther 12(2):R69

Doll F, Pfeilschifter J et al (2005) The epidermal growth factor stimulates sphingosine kinase-1 expression and activity in the human mammary carcinoma cell line MCF7. Biochim Biophys Acta 1738(1–3):72–81

Doll F, Pfeilschifter J et al (2007) Prolactin upregulates sphingosine kinase-1 expression and activity in the human breast cancer cell line MCF7 and triggers enhanced proliferation and migration. Endocr Relat Cancer 14(2):325–335

Donati C, Nincheri P et al (2007) Tumor necrosis factor-alpha exerts pro-myogenic action in C2C12 myoblasts via sphingosine kinase/S1P(2) signaling. FEBS Lett 581(23):4384–4388

Du J, Zeng C et al (2012) LPS and TNF-alpha induce expression of sphingosine-1-phosphate receptor-2 in human microvascular endothelial cells. Pathol Res Pract 208(2):82–88

El-Shewy HM, Johnson KR et al (2006) Insulin-like growth factors mediate heterotrimeric G protein-dependent ERK1/2 activation by transactivating sphingosine-1-phosphate receptors. J Biol Chem 281(42):31399–31407

El-Shewy HM, Abdel-Samie SA et al (2011) Phospholipase C and protein kinase C-beta 2 mediate insulin-like growth factor II-dependent sphingosine kinase 1 activation. Mol Endocrinol 25(12): 2144–2156

Estrada-Bernal A, Lawler SE et al (2011) The role of sphingosine kinase-1 in EGFRvIII-regulated growth and survival of glioblastoma cells. J Neurooncol 102(3):353–366

Fischer I, Alliod C et al (2011) Sphingosine kinase 1 and sphingosine 1-phosphate receptor 3 are functionally upregulated on astrocytes under pro-inflammatory conditions. PLoS One 6(8): e23905

Fletcher JI, Haber M et al (2010) ABC transporters in cancer: more than just drug efflux pumps. Nat Rev Cancer 10(2):147–156

Francy JM, Nag A et al (2007) Sphingosine kinase 1 expression is regulated by signaling through PI3K, AKT2, and mTOR in human coronary artery smooth muscle cells. Biochim Biophys Acta 1769(4):253–265

Fujita T, Okada T et al (2004) Delta-catenin/NPRAP (neural plakophilin-related armadillo repeat protein) interacts with and activates sphingosine kinase 1. Biochem J 382(Pt 2):717–723

Fukuda Y, Aoyama Y et al (2004) Identification of PECAM-1 association with sphingosine kinase 1 and its regulation by agonist-induced phosphorylation. Biochim Biophys Acta 1636(1):12–21
Gao P, Smith CD (2011) Ablation of sphingosine kinase-2 inhibits tumor cell proliferation and migration. Mol Cancer Res 9(11):1509–1519
Gao H, Fan Y et al (2012) Optimization on preparation condition of epimedium polysaccharide liposome and evaluation of its adjuvant activity. Int J Biol Macromol 50(1):207–213
Garofalo M, Romano G et al (2012) EGFR and MET receptor tyrosine kinase-altered microRNA expression induces tumorigenesis and gefitinib resistance in lung cancers. Nat Med 18(1): 74–82
Gault CR, Obeid LM et al (2010) An overview of sphingolipid metabolism: from synthesis to breakdown. Adv Exp Med Biol 688:1–23
Gault CR, Eblen ST et al (2012) Oncogenic K-Ras regulates bioactive sphingolipids in a sphingosine kinase 1 dependent manner. J Biol Chem 287(38):31794–31803
Ghosh TK, Bian J et al (1990) Intracellular calcium release mediated by sphingosine derivatives generated in cells. Science 248(4963):1653–1656
Hait NC, Sarkar S et al (2005) Role of sphingosine kinase 2 in cell migration toward epidermal growth factor. J Biol Chem 280(33):29462–29469
Hait NC, Bellamy A et al (2007) Sphingosine kinase type 2 activation by ERK-mediated phosphorylation. J Biol Chem 282(16):12058–12065
Hait NC, Allegood J et al (2009) Regulation of histone acetylation in the nucleus by sphingosine-1-phosphate. Science 325(5945):1254–1257
Hammad SM (2011) Blood sphingolipids in homeostasis and pathobiology. Adv Exp Med Biol 721:57–66
Hammad SM, Taha TA et al (2006) Oxidized LDL immune complexes induce release of sphingosine kinase in human U937 monocytic cells. Prostaglandins Other Lipid Mediat 79(1–2):126–140
Hammad SM, Crellin HG et al (2008) Dual and distinct roles for sphingosine kinase 1 and sphingosine 1 phoshate in the response to inflammatory stimuli in RAW macrophages. Prostaglandins Other Lipid Mediat 85(3–4):107–114
Heffernan-Stroud LA, Helke KL et al (2012) Defining a role for sphingosine kinase 1 in p53-dependent tumors. Oncogene 31(9):1166–1175
Higuchi M, Singh S et al (1996) Acidic sphingomyelinase-generated ceramide is needed but not sufficient for TNF-induced apoptosis and nuclear factor-kappa B activation. J Immunol 157(1): 297–304
Hisano Y, Kobayashi N et al (2011) The sphingosine 1-phosphate transporter, SPNS2, functions as a transporter of the phosphorylated form of the immunomodulating agent FTY720. J Biol Chem 286(3):1758–1766
Hisano Y, Kobayashi N et al (2012) Mouse SPNS2 functions as a sphingosine-1-phosphate transporter in vascular endothelial cells. PLoS One 7(6):e38941
Hla T (2001) Sphingosine 1-phosphate receptors. Prostaglandins Other Lipid Mediat 64(1–4): 135–142
Hla T, Brinkmann V (2011) Sphingosine 1-phosphate (S1P): physiology and the effects of S1P receptor modulation. Neurology 76(8 Suppl 3):S3–S8
Hla T, Maciag T (1990a) An abundant transcript induced in differentiating human endothelial cells encodes a polypeptide with structural similarities to G-protein-coupled receptors. J Biol Chem 265(16):9308–9313
Hla T, Maciag T (1990b) Isolation of immediate-early differentiation mRNAs by enzymatic amplification of subtracted cDNA from human endothelial cells. Biochem Biophys Res Commun 167(2):637–643
Hla T, Venkataraman K et al (2008) The vascular S1P gradient-cellular sources and biological significance. Biochim Biophys Acta 1781(9):477–482
Hobson JP, Rosenfeldt HM et al (2001) Role of the sphingosine-1-phosphate receptor EDG-1 in PDGF-induced cell motility. Science 291(5509):1800–1803

Hsu A, Zhang W et al (2012) Sphingosine-1-phosphate receptor-3 signaling up-regulates epidermal growth factor receptor and enhances epidermal growth factor receptor-mediated carcinogenic activities in cultured lung adenocarcinoma cells. Int J Oncol 40(5):1619–1626

Igarashi N, Okada T et al (2003) Sphingosine kinase 2 is a nuclear protein and inhibits DNA synthesis. J Biol Chem 278(47):46832–46839

Ikeda H, Ohkawa R et al (2010) Plasma concentration of bioactive lipid mediator sphingosine 1-phosphate is reduced in patients with chronic hepatitis C. Int J Clin Chim Acta 411 (9–10):765–770

Jenkins RW, Clarke CJ et al (2011) Regulation of CC ligand 5/RANTES by acid sphingomyelinase and acid ceramidase. J Biol Chem 286(15):13292–13303

Johnson KR, Becker KP et al (2002) PKC-dependent activation of sphingosine kinase 1 and translocation to the plasma membrane. Extracellular release of sphingosine-1-phosphate induced by phorbol 12-myristate 13-acetate (PMA). J Biol Chem 277(38):35257–35262

Johnson KR, Johnson KY et al (2005) Immunohistochemical distribution of sphingosine kinase 1 in normal and tumor lung tissue. J Histochem Cytochem 53(9):1159–1166

Johnstone ED, Mackova M et al (2005) Multiple anti-apoptotic pathways stimulated by EGF in cytotrophoblasts. Placenta 26(7):548–555

Kang JS, Yoon YD et al (2006) Glabridin suppresses intercellular adhesion molecule-1 expression in tumor necrosis factor-alpha-stimulated human umbilical vein endothelial cells by blocking sphingosine kinase pathway: implications of Akt, extracellular signal-regulated kinase, and nuclear factor-kappaB/Rel signaling pathways. Mol Pharmacol 69(3):941–949

Katsuma S, Hada Y et al (2002) Signalling mechanisms in sphingosine 1-phosphate-promoted mesangial cell proliferation. Genes Cells 7(12):1217–1230

Kawahara A, Nishi T et al (2009) The sphingolipid transporter spns2 functions in migration of zebrafish myocardial precursors. Science 323(5913):524–527

Kawamori T, Osta W et al (2006) Sphingosine kinase 1 is up-regulated in colon carcinogenesis. FASEB J 20(2):386–388

Kawamori T, Kaneshiro T et al (2009) Role for sphingosine kinase 1 in colon carcinogenesis. FASEB J 23(2):405–414

Kobayashi N, Yamaguchi A et al (2009) Characterization of the ATP-dependent sphingosine 1-phosphate transporter in rat erythrocytes. J Biol Chem 284(32):21192–21200

Kohama T, Olivera A et al (1998) Molecular cloning and functional characterization of murine sphingosine kinase. J Biol Chem 273(37):23722–23728

Kono Y, Nishiuma T et al (2007) Sphingosine kinase 1 regulates differentiation of human and mouse lung fibroblasts mediated by TGF-beta1. Am J Respir Cell Mol Biol 37(4):395–404

Krieg J, Hunter T (1992) Identification of the two major epidermal growth factor-induced tyrosine phosphorylation sites in the microvillar core protein ezrin. J Biol Chem 267 (27):19258–19265

Kusner DJ, Thompson CR et al (2007) The localization and activity of sphingosine kinase 1 are coordinately regulated with actin cytoskeletal dynamics in macrophages. J Biol Chem 282(32): 23147–23162

Lacana E, Maceyka M et al (2002) Cloning and characterization of a protein kinase A anchoring protein (AKAP)-related protein that interacts with and regulates sphingosine kinase 1 activity. J Biol Chem 277(36):32947–32953

Lamour NF, Chalfant CE (2005) Ceramide-1-phosphate: the "missing" link in eicosanoid biosynthesis and inflammation. Mol Interv 5(6):358–367

Lanterman MM, Saba JD (1998) Characterization of sphingosine kinase (SK) activity in Saccharomyces cerevisiae and isolation of SK-deficient mutants. Biochem J 332(Pt 2):525–531

Leclercq TM, Moretti PA et al (2008) Eukaryotic elongation factor 1A interacts with sphingosine kinase and directly enhances its catalytic activity. J Biol Chem 283(15):9606–9614

Leclercq TM, Moretti PA et al (2011) Guanine nucleotides regulate sphingosine kinase 1 activation by eukaryotic elongation factor 1A and provide a mechanism for eEF1A-associated oncogenesis. Oncogene 30(3):372–378

Li QF, Wu CT et al (2007) Activation of sphingosine kinase mediates suppressive effect of interleukin-6 on human multiple myeloma cell apoptosis. Br J Haematol 138(5):632–639

Lim KG, Sun C et al (2011a) (R)-FTY720 methyl ether is a specific sphingosine kinase 2 inhibitor: effect on sphingosine kinase 2 expression in HEK 293 cells and actin rearrangement and survival of MCF-7 breast cancer cells. Cell Signal 23(10):1590–1595

Lim KG, Tonelli F et al (2011b) FTY720 analogues as sphingosine kinase 1 inhibitors: enzyme inhibition kinetics, allosterism, proteasomal degradation, and actin rearrangement in MCF-7 breast cancer cells. J Biol Chem 286(21):18633–18640

Lin H, Baby N et al (2011) Expression of sphingosine kinase 1 in amoeboid microglial cells in the corpus callosum of postnatal rats. J Neuroinflammation 8:13

Liu H, Toman RE et al (2003) Sphingosine kinase type 2 is a putative BH3-only protein that induces apoptosis. J Biol Chem 278(41):40330–40336

Liu X, Guo XZ et al (2011) KAI1 inhibits HGF-induced invasion of pancreatic cancer by sphingosine kinase activity. Hepatobiliary Pancreat Dis Int 10(2):201–208

Long JS, Edwards J et al (2010) Sphingosine kinase 1 induces tolerance to human epidermal growth factor receptor 2 and prevents formation of a migratory phenotype in response to sphingosine 1-phosphate in estrogen receptor-positive breast cancer cells. Mol Cell Biol 30 (15):3827–3841

Loveridge C, Tonelli F et al (2010) The sphingosine kinase 1 inhibitor 2-(p-hydroxyanilino)-4-(p-chlorophenyl)thiazole induces proteasomal degradation of sphingosine kinase 1 in mammalian cells. J Biol Chem 285(50):38841–38852

Maceyka M, Nava VE et al (2004) Aminoacylase 1 is a sphingosine kinase 1-interacting protein. FEBS Lett 568(1–3):30–34

Maceyka M, Sankala H et al (2005) SphK1 and SphK2, sphingosine kinase isoenzymes with opposing functions in sphingolipid metabolism. J Biol Chem 280(44):37118–37129

Maceyka M, Alvarez SE et al (2008) Filamin A links sphingosine kinase 1 and sphingosine-1-phosphate receptor 1 at lamellipodia to orchestrate cell migration. Mol Cell Biol 28(18): 5687–5697

MacKinnon AC, Buckley A et al (2002) Sphingosine kinase: a point of convergence in the action of diverse neutrophil priming agents. J Immunol 169(11):6394–6400

Mao H, Lebrun DG et al (2012) Deregulated signaling pathways in glioblastoma multiforme: molecular mechanisms and therapeutic targets. Cancer Invest 30(1):48–56

Martin JL, Lin MZ et al (2009) Potentiation of growth factor signaling by insulin-like growth factor-binding protein-3 in breast epithelial cells requires sphingosine kinase activity. J Biol Chem 284(38):25542–25552

Melendez AJ (2008) Sphingosine kinase signalling in immune cells: potential as novel therapeutic targets. Biochim Biophys Acta 1784(1):66–75

Meng H, Yuan Y et al (2011) Loss of Sphingosine kinase 1/S1P signaling impairs cell growth and survival of neurons and progenitor cells in the developing sensory ganglia. PLoS One 6(11): e27150

Merrill AH Jr (2011) Sphingolipid and glycosphingolipid metabolic pathways in the era of sphingolipidomics. Chem Rev 111(10):6387–6422

Meyer zu Heringdorf D, Lass H et al (1999) Role of sphingosine kinase in Ca(2+) signalling by epidermal growth factor receptor. FEBS Lett 461(3):217–222

Michaud J, Kohno M et al (2006) Normal acute and chronic inflammatory responses in sphingosine kinase 1 knockout mice. FEBS Lett 580(19):4607–4612

Mizugishi K, Yamashita T et al (2005) Essential role for sphingosine kinases in neural and vascular development. Mol Cell Biol 25(24):11113–11121

Mullen TD, Hannun YA et al (2012) Ceramide synthases at the centre of sphingolipid metabolism and biology. Biochem J 441(3):789–802

Muller G, Ayoub M et al (1995) PKC zeta is a molecular switch in signal transduction of TNF-alpha, bifunctionally regulated by ceramide and arachidonic acid. EMBO J 14(9):1961–1969

Niedernberg A, Tunaru S et al (2003) Comparative analysis of functional assays for characterization of agonist ligands at G protein-coupled receptors. J Biomol Screen 8(5):500–510

Niessen F, Schaffner F et al (2008) Dendritic cell PAR1–P3 signalling couples coagulation and inflammation. Nature 452(7187):654–658

Nieuwenhuis B, Luth A et al (2010) Dexamethasone protects human fibroblasts from apoptosis via an S1P3-receptor subtype dependent activation of PKB/Akt and Bcl XL. Pharm Res 61(5): 449–459

Nincheri P, Bernacchioni C et al (2010) Sphingosine kinase-1/S1P1 signalling axis negatively regulates mitogenic response elicited by PDGF in mouse myoblasts. Cell Signal 22(11): 1688–1699

Nishiuma T, Nishimura Y et al (2008) Inhalation of sphingosine kinase inhibitor attenuates airway inflammation in asthmatic mouse model. Am J Physiol Lung Cell Mol Physiol 294(6): L1085–L1093

Okada T, Ding G et al (2005) Involvement of N-terminal-extended form of sphingosine kinase 2 in serum-dependent regulation of cell proliferation and apoptosis. J Biol Chem 280(43): 36318–36325

Okada T, Kajimoto T et al (2009) Sphingosine kinase/sphingosine 1-phosphate signalling in central nervous system. Cell Signal 21(1):7–13

Olivera A, Spiegel S (1993) Sphingosine-1-phosphate as second messenger in cell proliferation induced by PDGF and FCS mitogens. Nature 365(6446):557–560

Olivera A, Rosenfeldt HM et al (2003) Sphingosine kinase type 1 induces G12/13-mediated stress fiber formation, yet promotes growth and survival independent of G protein-coupled receptors. J Biol Chem 278(47):46452–46460

Olivera A, Urtz N et al (2006) IgE-dependent activation of sphingosine kinases 1 and 2 and secretion of sphingosine 1-phosphate requires Fyn kinase and contributes to mast cell responses. J Biol Chem 281(5):2515–2525

Paugh BS, Paugh SW et al (2008) EGF regulates plasminogen activator inhibitor-1 (PAI-1) by a pathway involving c-Src, PKCdelta, and sphingosine kinase 1 in glioblastoma cells. FASEB J 22(2):455–465

Pchejetski D, Kunduzova O et al (2007) Oxidative stress-dependent sphingosine kinase-1 inhibition mediates monoamine oxidase A-associated cardiac cell apoptosis. Circ Res 100(1):41–49

Pettus BJ, Bielawski J et al (2003) The sphingosine kinase 1/sphingosine-1-phosphate pathway mediates COX-2 induction and PGE2 production in response to TNF-alpha. FASEB J 17(11): 1411–1421

Pitman MR, Barr RK et al (2011) A critical role for the protein phosphatase 2A B'alpha regulatory subunit in dephosphorylation of sphingosine kinase 1. Int J Biochem Cell Biol 43(3):342–347

Pitson SM, D'Andrea RJ et al (2000) Human sphingosine kinase: purification, molecular cloning and characterization of the native and recombinant enzymes. Biochem J 350(Pt 2):429–441

Pitson SM, Moretti PA et al (2002) The nucleotide-binding site of human sphingosine kinase 1. J Biol Chem 277(51):49545–49553

Pitson SM, Moretti PA et al (2003) Activation of sphingosine kinase 1 by ERK1/2-mediated phosphorylation. EMBO J 22(20):5491–5500

Pyne S, Pyne N (2000) Sphingosine 1-phosphate signalling via the endothelial differentiation gene family of G-protein-coupled receptors. Pharmacol Ther 88(2):115–131

Pyne NJ, Pyne S (2010) Sphingosine 1-phosphate and cancer. Nat Rev Cancer 10(7):489–503

Pyne S, Pyne NJ (2011) Translational aspects of sphingosine 1-phosphate biology. Trends Mol Med 17(8):463–472

Pyne S, Lee SC et al (2009) Role of sphingosine kinases and lipid phosphate phosphatases in regulating spatial sphingosine 1-phosphate signalling in health and disease. Cell Signal 21(1): 14–21

Radeff-Huang J, Seasholtz TM et al (2007) Tumor necrosis factor-alpha-stimulated cell proliferation is mediated through sphingosine kinase-dependent Akt activation and cyclin D expression. J Biol Chem 282(2):863–870

Riccio A (2010) New endogenous regulators of class I histone deacetylases. Sci Signal 3(103):pe1

Rius J (1997) A new probability distribution of the triplet from Patterson function arguments V. Acta Crystallogr D Biol Crystallogr 53(Pt 5):535–539

Rosenfeldt HM, Hobson JP et al (2001a) EDG-1 links the PDGF receptor to Src and focal adhesion kinase activation leading to lamellipodia formation and cell migration. FASEB J 15(14): 2649–2659

Rosenfeldt HM, Hobson JP et al (2001b) The sphingosine-1-phosphate receptor EDG-1 is essential for platelet-derived growth factor-induced cell motility. Biochem Soc Trans 29(Pt 6):836–839

Salvado L, Serrano-Marco L et al (2012) Targeting PPARbeta/delta for the treatment of type 2 diabetes mellitus. Expert Opin Ther Targets 16(2):209–223

Sarkar S, Maceyka M et al (2005) Sphingosine kinase 1 is required for migration, proliferation and survival of MCF-7 human breast cancer cells. FEBS Lett 579(24):5313–5317

Sethu S, Mendez-Corao G et al (2008) Phospholipase D1 plays a key role in TNF-alpha signaling. J Immunol 180(9):6027–6034

Shi Y, Rehman H et al (2012) Sphingosine kinase-2 inhibition improves mitochondrial function and survival after hepatic ischemia-reperfusion. J Hepatol 56(1):137–145

Shida D, Fang X et al (2008) Cross-talk between LPA1 and epidermal growth factor receptors mediates up-regulation of sphingosine kinase 1 to promote gastric cancer cell motility and invasion. Cancer Res 68(16):6569–6577

Shu X, Wu W et al (2002) Sphingosine kinase mediates vascular endothelial growth factor-induced activation of ras and mitogen-activated protein kinases. Mol Cell Biol 22(22): 7758–7768

Silver RM (1996) Scleroderma. Clinical problems. The lungs. Rheum Dis Clin North Am 22(4): 825–840

Singh AT, Dharmarajan A et al (2012) Sphingosine-sphingosine-1-phosphate pathway regulates trophoblast differentiation and syncytialization. Reprod Biomed Online 24(2):224–234

Siow D, Wattenberg B (2011) The compartmentalization and translocation of the sphingosine kinases: mechanisms and functions in cell signaling and sphingolipid metabolism. Crit Rev Biochem Mol Biol 46(5):365–375

Siow DL, Anderson CD et al (2011) Sphingosine kinase localization in the control of sphingolipid metabolism. Adv Enzyme Regul 51(1):229–244

Soliven B, Ma L et al (2003) PDGF upregulates delayed rectifier via Src family kinases and sphingosine kinase in oligodendroglial progenitors. Am J Physiol Cell Physiol 284(1): C85–C93

Song JH, Kim M et al (2010) Isoflurane via TGF-beta1 release increases caveolae formation and organizes sphingosine kinase signaling in renal proximal tubules. Am J Physiol Renal Physiol 298(4):F1041–F1050

Spiegel S, Milstien S (2003) Sphingosine-1-phosphate: an enigmatic signalling lipid. Nat Rev Mol Cell Biol 4(5):397–407

Stahelin RV, Hwang JH et al (2005) The mechanism of membrane targeting of human sphingosine kinase 1. J Biol Chem 280(52):43030–43038

Stoffel W (1970) Studies on the biosynthesis and degradation of sphingosine bases. Chem Phys Lipids 5(1):139–158

Strub GM, Paillard M et al (2011) Sphingosine-1-phosphate produced by sphingosine kinase 2 in mitochondria interacts with prohibitin 2 to regulate complex IV assembly and respiration. FASEB J 25(2):600–612

Sukocheva OA, Wang L et al (2003) Sphingosine kinase transmits estrogen signaling in human breast cancer cells. Mol Endocrinol 17(10):2002–2012

Sukocheva O, Wadham C et al (2006) Estrogen transactivates EGFR via the sphingosine 1-phosphate receptor Edg-3: the role of sphingosine kinase-1. J Cell Biol 173(2):301–310

Sutherland CM, Moretti PA et al (2006) The calmodulin-binding site of sphingosine kinase and its role in agonist-dependent translocation of sphingosine kinase 1 to the plasma membrane. J Biol Chem 281(17):11693–11701

Taha TA, Osta W et al (2004) Down-regulation of sphingosine kinase-1 by DNA damage: dependence on proteases and p53. J Biol Chem 279(19):20546–20554

Taha TA, Kitatani K et al (2005) Tumor necrosis factor induces the loss of sphingosine kinase-1 by a cathepsin B-dependent mechanism. J Biol Chem 280(17):17196–17202

Taha TA, El-Alwani M et al (2006) Sphingosine kinase-1 is cleaved by cathepsin B in vitro: identification of the initial cleavage sites for the protease. FEBS Lett 580(26):6047–6054

Uhlenbrock K, Gassenhuber H et al (2002) Sphingosine 1-phosphate is a ligand of the human gpr3, gpr6 and gpr12 family of constitutively active G protein-coupled receptors. Cell Signal 14(11): 941–953

Uhlenbrock K, Huber J et al (2003) Fluid shear stress differentially regulates gpr3, gpr6, and gpr12 expression in human umbilical vein endothelial cells. Cell Physiol Biochem 13 (2):75–84

van Koppen CJ, Meyer zu Heringdorf D et al (2001) Sphingosine kinase-mediated calcium signaling by muscarinic acetylcholine receptors. Life Sci 68(22–23):2535–2540

Venkataraman K, Thangada S et al (2006) Extracellular export of sphingosine kinase-1a contributes to the vascular S1P gradient. Biochem J 397(3):461–471

Wadgaonkar R, Patel V et al (2009) Differential regulation of sphingosine kinases 1 and 2 in lung injury. Am J Physiol Lung Cell Mol Physiol 296(4):L603–L613

Wang LS, Chow KC et al (2002) Effects of platelet activating factor, butyrate and interleukin-6 on cyclooxygenase-2 expression in human esophageal cancer cells. Scand J Gastroenterol 37(4): 467–475

Xia P, Gamble JR et al (1998) Tumor necrosis factor-alpha induces adhesion molecule expression through the sphingosine kinase pathway. PNAS 95(24):14196–14201

Xia P, Wang L et al (2002) Sphingosine kinase interacts with TRAF2 and dissects tumor necrosis factor-alpha signaling. J Biol Chem 277(10):7996–8003

Yamamoto Y, Olson DM et al (2010) Increased expression of enzymes for sphingosine 1-phosphate turnover and signaling in human decidua during late pregnancy. Biol Reprod 82 (3):628–635

Yamanaka M, Shegogue D et al (2004) Sphingosine kinase 1 (SPHK1) is induced by transforming growth factor-beta and mediates TIMP-1 up-regulation. J Biol Chem 279(52):53994–54001

Yang Z, Costanzo M et al (1993) Tumor necrosis factor activation of the sphingomyelin pathway signals nuclear factor kappa B translocation in intact HL-60 cells. J Biol Chem 268(27): 20520–20523

Yokota S, Taniguchi Y et al (2004) Asp177 in C4 domain of mouse sphingosine kinase 1a is important for the sphingosine recognition. FEBS Lett 578(1–2):106–110

Yu H, Shao Y et al (2012) Acetylation of sphingosine kinase 1 regulates cell growth and cell-cycle progression. Biochem Biophys Res Commun 417(4):1242–1247

Zebol JR, Hewitt NM et al (2009) The CCT/TRiC chaperonin is required for maturation of sphingosine kinase 1. Int J Biochem Cell Biol 41(4):822–827

Zhu HY, Da WM (2007) Regulation of immunity by sphingosine 1-phosphate and its G protein-coupled receptors–review. J Exp Hematol 5(6):1317–1324

Bacterial Infections and Ceramide

Heike Grassmé and Katrin Anne Becker

Contents

Abstract Ceramide is released from sphingomyelin primarily by the activity of acid, neutral, or alkaline sphingomyelinases or is synthesized de novo. Several bacteria, viruses, and even parasites infect mammalian cells by exploiting the acid sphingomyelinase or the neutral sphingomyelinase-ceramide system, or both. Sphingomyelinases and ceramide have been shown to be crucially involved in the internalization of pathogens, the induction of apoptosis in infected cells, the intracellular activation of signaling pathways, and the release of cytokines. The diverse functions of ceramide in infections suggest that the sphingomyelinase-ceramide system is a key player in the host response to many pathogens.

Keywords Pathogens • Ceramide • Acid sphingomyelinase • Infection

H. Grassmé (✉) • K.A. Becker
Department of Molecular Biology, University of Duisburg-Essen, Hufelandstr. 55, 45122 Essen, Germany
e-mail: heike.gulbins@uni-due.de

E. Gulbins and I. Petrache (eds.), *Sphingolipids in Disease*,
Handbook of Experimental Pharmacology 216,
DOI 10.1007/978-3-7091-1511-4_15,

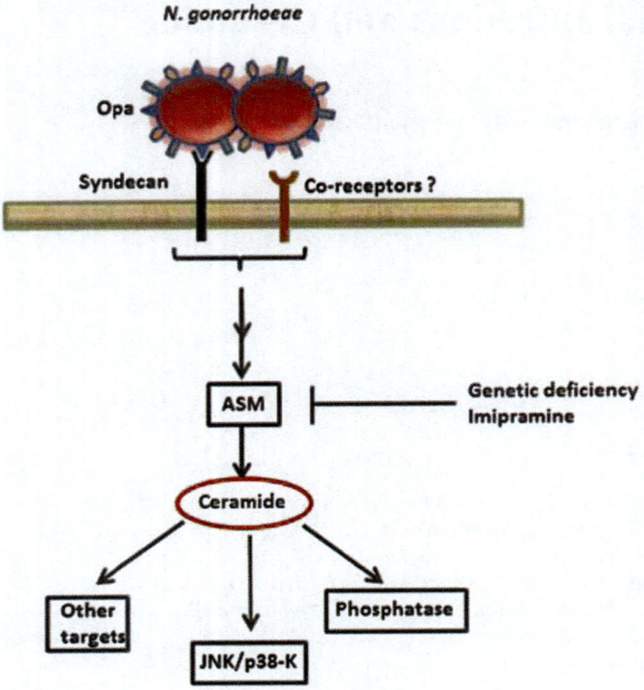

Fig. 1 Diagram of the signaling events involved in the internalization of *N. gonorrhoeae* into nonphagocytic cells. Invasive opacity-associated (Opa) protein variants bind to receptors belonging to the syndecan family and to unknown co-receptors. This binding leads to the activation of acid sphingomyelinase (ASM), which in turn generates ceramide from sphingomyelin. Finally, ceramide affects various target molecules, such as JNK/p38 kinase. The inhibition of acid sphingomyelinase, either inherently by genetic deficiency or pharmacologically by imipramine, prevents infection of the cells with *N. gonorrhoeae* (Modified from Grassmé et al. (1997), with permission from Cell Press)

1 *Neisseria gonorrhoeae* (*N. gonorrhoeae*)

The first studies on the function of the acid sphingomyelinase-ceramide system in bacterial infections investigated the role of this system in the invasion of *N. gonorrhoeae* into human epithelial cells (Grassmé et al. 1997). *N. gonorrhoeae* cause the human venereal disease gonorrhea. During the course of infection, these bacteria interact with a variety of human cell types, including epithelial cells and phagocytes. Actin filament-dependent entry of *N. gonorrhoeae* into human mucosal epithelial cells is considered to be a crucial event during infection (McGee et al. 1983; Apicella et al. 1996; Grassmé et al. 1996). This event is mediated by outer membrane proteins, the so-called opacity-associated (Opa) proteins (Lambden et al. 1979; Stern et al. 1986; Makino et al. 1991). Specific Opa variants mediate the binding of the microorganisms to heparan sulfate proteoglycan (HSPG) receptors on epithelial cells, and this binding leads to the uptake of the pathogen (Chen et al.

1995; van Putten and Paul 1995). Initial studies aimed at identifying the cellular mechanisms involved in this process showed that stimulation of acid sphingomyelinase is crucial for gonococcal entry into epithelial cells and fibroblasts. Pharmacological inhibition of acid sphingomyelinase by imipramine prevents the invasion of *N. gonorrhoeae* into epithelial cells. Additionally, acid sphingomyelinase-deficient fibroblasts obtained from patients with Niemann–Pick disease type A (NPDA), who lack acid sphingomyelinase because of an inborn genetic defect, did not internalize Opa-expressing HSPG-specific gonococci. The uptake of *N. gonorrhoeae* by these fibroblasts was restored by transfection with acid sphingomyelinase (Grassmé et al. 1997). These studies demonstrated that acid sphingomyelinase and ceramide play a crucial role in the infection of human epithelial cells with *N. gonorrhoeae*.

Additional studies with human neutrophils demonstrated that the acid sphingomyelinase-ceramide system is also necessary for the opsonin-independent uptake of *N. gonorrhoeae* into human phagocytes via carcinoembryonic antigen-related cellular adhesion molecule (CEACAM) receptors (Hauck et al. 2000). Infecting these cells with invasive Opa variants of *N. gonorrhoeae* results in the rapid activation of acid sphingomyelinase and various kinases, such as Src-like tyrosine kinases and Jun N-terminal kinases (JNK). Pharmacological inhibition of acid sphingomyelinase by imipramine prevents both the internalization of the bacteria into macrophages and the activation of Src-like tyrosine kinases and JNK (Hauck et al. 2000).

In summary, these findings showed for the first time that acid sphingomyelinase and ceramide play a central role in the infection of human cells with a bacterial pathogen (Fig. 1).

2 *Staphylococcus aureus* (*S. aureus*)

Although acid sphingomyelinase and ceramide mediate the internalization of *N. gonorrhoeae* into epithelial cells and macrophages, the enzyme has been shown to be involved in the induction of cell death after endothelial cells, and fibroblasts have been infected with *S. aureus* (Esen et al. 2001). *S. aureus* very often causes nosocomial infections, such as pneumonia and wound infections. These diseases can result in sepsis and toxigenic diseases, including toxic shock syndrome (Lowy 1998; Chesney 1989). The cocci are not considered to be typical intracellular pathogens, but their uptake into mammalian cells results in the induction of apoptosis in epithelial and endothelial cells (Menzies and Kourteva 1998; Bayles et al. 1998). Apoptosis, or programmed cell death, is an important active cellular mechanism that is induced by a wide variety of stimuli, including stress stimuli, death receptors such as the endogenous CD95/CD95 ligand system, and infections with microbes (Thompson 1995; Gulbins et al. 1995), such as *S. aureus* (Menzies and Kourteva 1998), *Shigella flexneri* (Zychlinsky et al. 1994), *Yersinia enterocolitica* (Monack et al. 1997; Ruckdeschel et al. 1997), *Salmonella* spp.

(Hersh et al. 1999), enteropathogenic *Escherichia coli* (Crane et al. 1999), and *Pseudomonas aeruginosa* (Hauser and Engel 1999; Grassmé et al. 2000). Apoptosis involves the activation of several cellular molecules, including caspases and stress-activated protein kinases, and the release of cytochrome c from the mitochondria into the cytosol (Takahashi 1999; Kroemer et al. 1998; Green and Reed 1998). Acid sphingomyelinase and ceramide have been demonstrated to be important in many forms of programmed cell death (Basu and Kolesnick 1998). It has been shown that ceramide functions in apoptosis by forming ceramide-enriched membrane domains that cluster receptor molecules and thereby facilitate the activation of downstream signals (Grassmé et al. 2001). For *S. aureus*, Esen et al. (2001) showed, for the first time, that the induction of endothelial cell apoptosis involves the activation of acid sphingomyelinase and the resultant release of ceramide, which mediates the stimulation of cellular caspases and JNK and triggers the release of cytochrome c from mitochondria to the cytosol. The central role of acid sphingomyelinase in this process is indicated by the finding that *S. aureus*-triggered apoptosis is prevented in acid sphingomyelinase-deficient fibroblasts obtained from NPDA patients and in cells in which JNK and caspases have been pharmacologically inhibited.

3 *Pseudomonas aeruginosa* (*P. aeruginosa*)

Several studies have demonstrated that acid sphingomyelinase and ceramide play a central role in the infection of mammalian cells with *P. aeruginosa*. Many investigations have shown that ceramide-enriched membrane platforms are crucial for the infection of epithelial cells with *P. aeruginosa*. *P. aeruginosa* infections are most important in patients with cystic fibrosis but are also serious for immunocompromised patients, patients with burn wounds, and patients with sepsis or ventilator-associated pneumonia (CF Patient Registry 2010; Crouch Brewer et al. 1996; Keen et al. 2010; McManus et al. 1985; Sadikot et al. 2005; Vidal et al. 1996). In addition, many patients with chronic obstructive pulmonary disease also experience pulmonary *P. aeruginosa* infections.

The infection of mammalian cells by this pathogen triggers rapid activation of acid sphingomyelinase, and this activation is correlated with the translocation of the enzyme to the bacterial infection site on the extracellular leaflet of the cell membrane (Grassmé et al. 2003) (Fig. 2).

Surface sphingomyelinase converts sphingomyelin to ceramide, which spontaneously fuses to small and, subsequently, to large ceramide-enriched membrane domains. Ceramide-enriched membrane domains have been shown to be crucial for the internalization of *P. aeruginosa* into epithelial cells and fibroblasts, the induction of death of infected cells, and the controlled release of cytokines (Grassmé et al. 2003). At present, it is not known how ceramide-enriched membrane platforms may regulate these events, but it is possible that such regulation is due to the clustering of cystic fibrosis conductance regulator (Cftr) and CD95 (Grassmé et al. 2003). The notion that ceramide-enriched membrane platforms cluster Cftr is consistent with findings demonstrating that Cftr concentrates in the raft fraction

Fig. 2 The infection of epithelial cells by *P. aeruginosa* results in the activation of acid sphingomyelinase and in its translocation onto the extracellular leaflet of the membrane. In this figure, acid sphingomyelinase was stained with an anti-acid sphingomyelinase antibody coupled with nanogold and was visualized by scanning electron microscopy

after infection (Kowalski and Pier 2004). Furthermore, pharmacological destruction of rafts after *P. aeruginosa* infection prevents both the internalization of the bacteria into pulmonary epithelial cells and the nuclear translocation of nuclear factor-kappaB (NF-κB) in respiratory epithelial cells (Kowalski and Pier 2004). Because CD95 triggers the death of epithelial cells after infection with *P. aeruginosa* (Grassmé et al. 2000), it is very likely that the clustering of CD95 in ceramide-enriched membrane platforms induces apoptosis. Consistent with this notion is the finding that a genetic deficiency of acid sphingomyelinase prevents cell death in vitro and in vivo after infection with *P. aeruginosa* (Grassmé et al. 2003). The role of ceramide-enriched membrane platforms in the control of cytokine release (Grassmé et al. 2003) is much less defined, and the mechanisms involved remain to be elucidated.

Studies using mice genetically deficient in acid sphingomyelinase demonstrated that the acid sphingomyelinase-ceramide system plays an important role in vitro and in vivo after pulmonary infection with *P. aeruginosa* (Grassmé et al. 2003). The in vivo studies showed that acid sphingomyelinase-deficient mice are highly susceptible to pulmonary *P. aeruginosa* infections. These mice were unable to clear

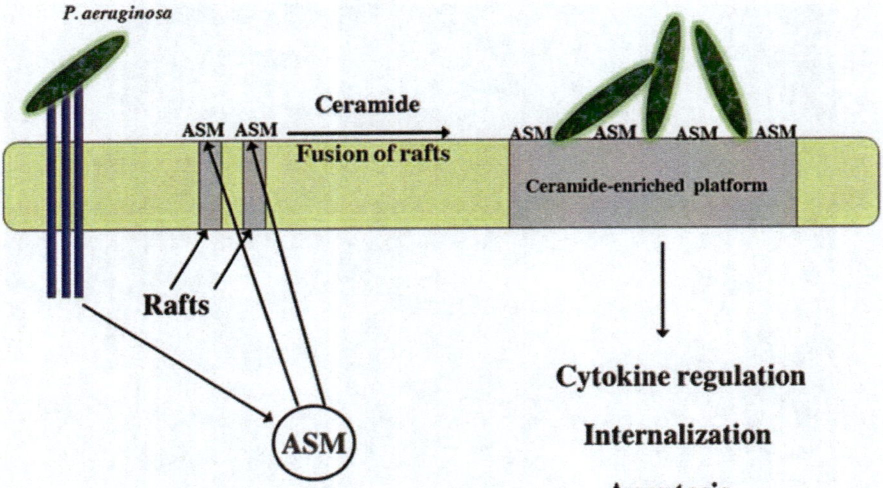

Fig. 3 Model of the signaling cascade induced by infecting nonphagocytic host cells with *P. aeruginosa*. The binding of the bacteria to their putative receptor leads to the activation of acid sphingomyelinase (ASM), the release of ceramide, and the formation of ceramide-enriched membrane platforms that enable the pathogen to interact with host cells and determine the response of the host cell to the pathogen

the bacteria after an acute infection; thus, the infection became generalized, sepsis developed, and the mice died (Grassmé et al. 2003). Studies using freshly isolated alveolar macrophages demonstrated that acid sphingomyelinase and ceramide are required for the clustering and activation of nicotinamide adenine dinucleotide phosphate-oxidase (NADPH) oxidases and the production of reactive oxygen species (ROS) that kill *P. aeruginosa* (Zhang et al. 2008). Whether the acid sphingomyelinase-ceramide system is also required for the formation of reactive oxygen species in epithelial cells of the lung is presently unknown. However, the failure of acid sphingomyelinase-deficient cells to cluster and activate NADPH oxidases may be the reason for the high susceptibility of these mice to pulmonary infections with *P. aeruginosa*. This high susceptibility may also be linked to the inability of acid sphingomyelinase-deficient cells to internalize the bacteria and to undergo apoptosis (Grassmé et al. 2003). Most of the events described above, e.g., internalization of the bacteria and induction of cell death, require expression of the type III secretion system in *P. aeruginosa* (Galán and Collmer 1999). Mutants of *P. aeruginosa* that lack the type III secretion system fail to trigger death and also fail to invade epithelial cells (Jendrossek et al. 2003). Although it is unknown whether ceramide-enriched membrane platforms play a role in the transfer of bacterial proteins into mammalian host cells via the type III secretion system, an interesting speculation suggests that the specific environment of ceramide-enriched membrane platforms permits the adhesion, penetration, or both of the multi-protein complex that composes the bacterial type III secretion system (Fig. 3).

4 *Listeria monocytogenes* (*L. monocytogenes*)

L. monocytogenes are facultative intracellular bacteria that cause listeriosis. This disease affects, in particular, pregnant women, neonates, and immunocompromised patients and is characterized by several clinical features, such as mild gastrointestinal symptoms, and by spontaneous abortion, sepsis, perinatal infections, and meningoencephalitis. These differing aspects of the disease are due to the ability of the pathogen to cross several barriers in the host and to enter a wide variety of mammalian cells (Vázquez-Boland et al. 2001). After being taken up into host macrophages, *L. monocytogenes* are rapidly killed in preactivated cells; however, in quiescent cells, the bacteria can lyse the phagosomal membrane. This lysis is mediated by a pore-forming toxin, i.e., listeriolysin O, and by the concerted action of cellular phospholipases that ultimately allow the pathogens to escape into the cytoplasm, where they replicate inside the cell. By avoiding phagolysosomal fusion and thereby preventing degradation by lysosomes, *L. monocytogenes* spread from one cell to the next and escape the humoral immune response (Cossart et al. 1989; Goldfine and Wadsworth 2002; Portnoy et al. 1988; Schnupf and Portnoy 2007). Acid sphingomyelinase-deficient mice are highly susceptible to infection with *L. monocytogenes*, failing to kill the intracellular bacteria in macrophages (Utermöhlen et al. 2003). Further studies show that the acid sphingomyelinase-ceramide system is necessary if bacteria-containing late phagosomes are to fuse properly with lysosomes. Phagosome maturation occurs stepwise by sequential fusion of the phagosome with endosomes and finally with lysosomes; this fusion results in the degradation of pathogens. Several lysosomal proteases and hydrolases are responsible for effective control of the pathogens (Haas 2007). In acid sphingomyelinase-deficient macrophages, the transfer of lysosomal markers such as Lamp1 from the lysosome to the phagosome is delayed, and the activities of listericidal proteases such as cathepsin D, B, and L are decreased because of the reduced ability of the membranes to fuse with each other (Utermöhlen et al. 2008; Schramm et al. 2008). In wild-type macrophages, acid sphingomyelinase-derived ceramide enhances the fusion of biomembranes (e.g., the fusion of lysosomal and phagosomal membranes) because of its cone-shaped structure (Utermöhlen et al. 2008). In summary, acid sphingomyelinase-derived ceramide mediates the efficient fusion of *L. monocytogenes*-containing phagosomes with lysosomes by enhancing the activation of bactericidal cathepsins, such as cathepsin D (Heinrich et al. 1999), and the transfer of lysosomal hydrolases that lead to efficient killing of the bacteria. In contrast, acid sphingomyelinase is not necessary for the uptake of *L. monocytogenes* in vivo or in vitro (Utermöhlen et al. 2003).

5 *Salmonella enterica* serovar typhimurium (*S. typhimurium*)

Human infections with *S. typhimurium* are manifested as nontyphoidal gastroenteritis that resolves spontaneously without antibiotic treatment. In immunocompromised patients, however, the disease may lead to severe consequences. In mice, *S. typhimurium* causes a serious infection that resembles typhoid fever in humans, and the mouse model is preferentially used as a model for systemic *Salmonella* infections in humans. If bacteria are ingested with contaminated water or food, they enter Peyer's patches and undergo rapid phagocytosis. An important virulence factor of *S. typhimurium* is the ability to reside in macrophages by preventing the lysosomal maturation of the phagosomes (Oh et al. 1996; McCollister et al. 2005). In this way, *S. typhimurium* differs from *L. monocytogenes*, which rapidly escape from the acidified phagosome into the cytoplasm (Cossart et al. 1989). The survival of *S. typhimurium* within the lysosomal compartment of macrophages is closely associated with the *Salmonella* pathogenicity island 2 (SPI2) type III secretion system, which translocates bacterial proteins across the phagosomal membrane into the host cell cytoplasm (Shea et al. 1996; Kuhle and Hensel 2002; Miao and Miller 2000). This effective system allows the pathogen to avoid contact with lysosomal enzymes and prevents killing of the pathogen by the active NADPH phagocyte oxidase enzymatic complex (Chakravortty et al. 2002; Uchiya et al. 1999; Vázquez-Torres et al. 2000). Studies of molecular mechanisms in the infection process of macrophages by *S. typhimurium* have shown that acid sphingomyelinase is associated with the host defense against these bacteria: Investigations by McCollister et al. (2007) found that constitutive expression of acid sphingomyelinase is important for the killing of *S. typhimurium* by macrophages, as indicated by the finding that acid sphingomyelinase-deficient macrophages were much less able than wild-type macrophages to kill *S. typhimurium*, which was mediated by a defect NADPH-mediated release of ROS in acid sphingomyelinase-deficient cells.

Acid sphingomyelinase is found in cellular and secreted forms in mammalian cells. These variations appear to reflect posttranslational processing of the primary gene product (Ferlinz et al. 1994). The secreted form of acid sphingomyelinase is stimulated by Zn^{2+}, whereas the lysosomal form is not (Schissel et al. 1996). Human macrophages, fibroblasts, and endothelial cells can secrete substantial amounts of acid sphingomyelinase (Marathe et al. 1998). Like *Escherichia coli*, another enterobacterial species, *S. typhimurium*, induces the secretion of acid sphingomyelinase during phagocytosis (McCollister et al. 2007). In contrast, the infection of host cells with bacteria such as *N.* gonorrhoeae does not induce acid sphingomyelinase secretion (Grassmé et al. 1997; Hauck et al. 2000). At present it is not known how secreted acid sphingomyelinase is involved in the host defense system.

Studies by Utermöhlen et al. (2003) showed that acid sphingomyelinase not only participates in the killing of *S. typhimurium* by phagocytes in vitro but also mediates the resistance of mice to *S. typhimurium* infection in vivo, as indicated by the finding that acid sphingomyelinase-deficient mice are highly susceptible to infection with *S. typhimurium.*

6 *Escherichia coli* (*E. coli*)

E. coli are ubiquitous bacteria that are widespread in various environments, including the gastrointestinal tract of humans. Although these bacteria are opportunistic, *E. coli* can cause a variety of infectious diseases, including urinary tract infections, meningitis, wound infections, and sepsis (Gyles 1992). Sepsis is a life-threatening disease and is often followed by endotoxic shock, one of the most common causes of death in the industrialized world (Angus et al. 2001). This disorder is due primarily to systemic infection with Gram-negative bacteria such as *E. coli*. Lipopolysaccharide (LPS), a potent inflammatory glycolipid in the outer membrane molecule of *E. coli* and many other Gram-negative bacteria, triggers both sepsis and endotoxic shock (Morrison and Ryan 1987; Bone 1991). LPS acts by activating a variety of messenger molecules, including protein kinases, cytokines, and acid sphingomyelinase-derived ceramide (Narumi et al. 1992; Hambleton et al. 1996; Herrera-Velit and Reiner 1996; Barber et al. 1996). One of the hallmarks of sepsis and endotoxic shock is the destruction of dendritic cells and endothelial cells by apoptosis (Haimowitz-Friedman et al. 1997; Kanto et al. 2001). Ceramide has been shown to trigger apoptosis in many cell types, thereby contributing to the effects induced by death-receptor activation or physical stimuli (Grassmé et al. 2001; Kolesnick and Krönke 1998; Levade and Jaffrézou 1999; Pettus et al. 2002). Recent studies have shown that immature dendritic cells undergo apoptosis when treated with high doses of *E. coli* or LPS, a process that contributes to the development of sepsis (Falcone et al. 2004). The induction of apoptosis is dependent on the activity of acid sphingomyelinase, as demonstrated by the finding that pharmacological inhibition by imipramine prevents apoptosis in these cells and that apoptosis is in turn restored by the administration of exogenous ceramide. Acid sphingomyelinase-dependent apoptosis is inhibited by nitric oxide (NO), which is generated by inducible NO synthase (iNOS). This pathway involves the formation of cyclic guanosine monophosphate (cGMP) and the stimulation of the cGMP-dependent protein kinase (G-kinase). Falcone et al. (2004) showed that the mechanism by which NO acts is the inhibition of acid sphingomyelinase. This finding was confirmed in an in vivo model using iNOS-deficient mice. Dendritic cells from these mice were very sensitive to apoptosis but were protected by exposure to exogenous NO/cGMP or imipramine. Thus, the acid sphingomyelinase-ceramide system plays a central role in *E. coli*/LPS-induced apoptosis of dendritic cells and is regulated by NO/cGMP.

Besides damaging dendritic cells, endothelial cell apoptosis also plays a crucial role in the pathogenesis of endotoxic shock syndrome. Studies have shown that this LPS-inducible process is mediated by tumor necrosis factor alpha (TNFα), a putative effector of LPS, and by the generation of ceramide via acid sphingomyelinase (Haimowitz-Friedman et al. 1997). In contrast to wild-type mice, acid sphingomyelinase-deficient mice were protected against LPS-induced endothelial cell apoptosis, exhibited a reduced increase in TNFα, and survived endotoxic shock syndrome.

Toll-like receptor 4 (TLR4) is a well-known receptor of LPS from Gram-negative bacteria (Takeda and Akira 2004). TLR4 signaling leads to the production of various cytokines and NF-κB, thereby inducing the activation of the adaptive immune system (Martin and Wesche 2002). However, the role of ceramide metabolites in LPS signaling still requires definition (Cuschieri et al. 2007; Pfeiffer et al. 2001; MacKichan and DeFranco 1999). The results of several studies have suggested that ceramide promotes CD14/TLR4-mediated signaling, probably by forming ceramide-enriched domains that serve to reorganize the receptors. The role of endogenous ceramide-1-P in LPS signaling remains to be clarified (Hankins et al. 2011).

7 Pathogenic Mycobacteria

Mycobacterial species include pathogenic bacteria, such as *Mycobacterium tuberculosis* (*M. tuberculosis*) and *Mycobacterium avium* (*M. avium*), and nonpathogenic bacteria, such as *Mycobacterium smegmatis* (*M. smegmatis*). The bacteria that reach the lung via aerosols are ingested by alveolar macrophages. A characteristic feature of pathogenic Mycobacteria is that they reside in macrophages within phagosomes. The bacteria can prevent the fusion of this phagosome with late endosomes and lysosomes. As a consequence, they survive and replicate in macrophages (Russel 2001; Oh et al. 1996). Infected macrophages, together with epithelioid macrophages and multinucleated giant cells, become organized in granulomas in infected tissues, such as liver, spleen, and lung (Russel 2007). It was recently shown that for *M. avium*, the formation of multinucleated giant cells in the granuloma requires acid sphingomyelinase expression. Acid sphingomyelinase-deficient mice infected with *M. avium* fail to form giant cells. Instead, infected tissues contain only small, clearly delimited granulomas. Acid sphingomyelinase-deficient mice are more resistant to lethal infections with *M. avium* than wild-type mice (Utermöhlen et al. 2008).

Human tuberculosis (TB), caused by *M. tuberculosis*, is still one of the deadliest infectious diseases worldwide (Dye et al. 2008). This disease progresses from latent TB, in which granulomas are controlled by the immune system, to the active form of TB, in which caseation of the granuloma is increased. The granulomas become necrotic because of the death of infected macrophages, and this necrosis results in rupture into the lung airway and the release of many free infectious rods (Dannenberg and Sugimoto 1976; Dannenberg 1994; Russel et al. 2009). The increased caseation of the granulomas seems to be associated with a shift in the host's lipid metabolism and the accumulation of specific lipids within the granulomas. Biochemical analysis revealed high concentrations of distinct neutral lipids, as well as an accumulation of cholesterol ester, cholesterol, triacylglycerol, and lactosylceramide (Kim et al. 2010). Lactosylceramide is an intermediate of the glycosphingolipid metabolism and is generated either by increased activity of ceramide synthase or by a decrease in complex ceramides (Kolter and Sandhoff 2005; Chatterjee and Pandey 2008). Lactosylceramide has been found to play a role in the degradation of cholesterol and in cell death (Chatterjee et al. 1997; Garner et al. 2002). In vitro, *M. tuberculosis* seems to trigger a similar dysregulation of host lipid metabolism in macrophages.

Anes et al. (2003) showed that several lipids are responsible for killing *M. avium* and *M. smegmatis* in macrophages by modulating the macrophage's innate immune response against the bacteria. Among other lipids, sphingomyelin and ceramide can trigger actin nucleation on phagosomes and probably also on other membranes, such as the plasma membrane, thereby promoting phagolysosomal fusion and enhancing mycobacterial killing. In contrast to those of several other tested lipids, the mycobactericidal properties of ceramide do not require the activation of NF-κB, one of the key transcription factors stimulated during the inflammatory process.

In conclusion, these findings suggest that lipids such as sphingomyelin and ceramide can modulate both the proinflammatory response and the state of Mycobacteria in macrophages (Gutierrez et al. 2009).

8 Summary

Lipid-mediated signaling initiated by sphingomyelinases, which hydrolyze sphingomyelin to ceramide, is recognized as a crucial component in diverse cellular processes. In particular, acid sphingomyelinase has been shown to be crucially involved in the host response to *N. gonorrhoeae, S. aureus, L. monocytogenes, P. aeruginosa*, *S. typhimurium*, *E. coli*, and *M. avium*. Acid sphingomyelinase deficiency in the target cells often leads to increased susceptibility to these pathogens. Acid sphingomyelinase is involved in killing bacteria predominantly by facilitating the activation of NADPH oxidases, the internalization of pathogens, the intracellular processing of phagosomes, in particular phagolysosomal fusion, and the induction of apoptosis in infected host cells. Ceramide seems to trigger these diverse biological effects by modifying the biophysical properties of membranes, thereby resulting in the formation of ceramide-enriched membrane platforms that serve to trap and cluster receptor molecules and in the fusion of membranes from various vesicles, such as phagosomes and lysosomes. Whether ceramide directly acts on signaling molecules that are important for the pathogen–host interaction is presently unknown. Because of the central role of the acid sphingomyelinase-ceramide system in the infection of mammalian cells by pathogens, this system may serve as a future target for the development of anti-infective drugs.

References

Anes E, Kühnel MP, Bos E, Moniz-Pereira J, Habermann A, Griffith G (2003) Selected lipids activate phagosome actin assembly and maturation resulting in killing of pathogenic mycobacteria. Nat Cell Biol 5:793–802

Angus DC, Linde-Zwirble WT, Lidicker J, Clermont G, Carcillo J, Pinsky MR (2001) Epidemiology of severe sepsis in the United States: analysis of incidence, outcome, and associated costs of care. Crit Care Med 29:1303–1310

Apicella MA, Ketterer M, Lee FK, Zhou D, Rice PA, Blake MS (1996) The pathogenesis of gonococcal urethritis in men; confocal and immunoelectron microscopic analysis of urethral exudates from men infected with *Neisseria gonorrhoeae*. J Infect Dis 173:636–646

Barber SA, Detore G, McNally R, Vogel SN (1996) Stimulation of the ceramide pathway partially mimics lipopolysaccharide-induced responses in murine peritoneal macrophages. Infect Immun 64:3397–3400

Basu S, Kolesnick RN (1998) Stress signals for apoptosis: ceramide and c-Jun kinase. Oncogene 17:3277–3285

Bayles KW, Wesson CA, Liou LE, Fox LK, Bohach GA, Trumble WR (1998) Intracellular *Staphylococcus aureus* escapes the endosome and induces apoptosis in epithelial cells. Infect Immun 66:336–342

Bone RC (1991) Sepsis, the sepsis syndrome, multi-organ failure: a plea for comparable definitions. Ann Intern Med 114:332–333

CF Patient Registry (2010) CF foundation, patient registry annual report. Bethesda, MD, USA. http://www.cff.org/livingwithcf/carecenternetwork/patientregistry/

Chakravortty D, Hansen-Wester I, Hensel M (2002) *Salmonella* pathogenicity island 2 mediates protection of intracellular *Salmonella* from reactive nitrogen intermediates. J Exp Med 195:1155–1166

Chatterjee SB, Pandey A (2008) The Yin and Yang of lactosylceramide metabolism: implications in cell function. Biochem Biophys Acta 1780:370–382

Chatterjee SB, Dey S, Shi WY, Thomas K, Hutchins GM (1997) Accumulation of glycosphingolipids in human atherosclerotic plaque and unaffected aorta tissues. Glycobiology 7:57–65

Chen T, Belland RJ, Wilson J, Swanson J (1995) Adherence of pilus$^-$ Opa$^+$ gonococci to epithelial cells in vitro involves heparin sulfate. J Exp Med 182:511–517

Chesney PJ (1989) Clinical aspects and spectrum of illness of toxic shock syndrome: overview. Rev Infect Dis 11:1–7

Cossart P, Vicente MF, Mengaud J, Baquero F, Perez-Diaz JC, Berche P (1989) Listeriolysin O is essential for virulence of *Listeria monocytogenes*: direct evidence obtained by gene complementation. Infect Immun 57:3629–3636

Crane JK, Majumdar S, Pickhardt DF 3rd (1999) Host cell death due to enteropathogenic *Escherichia coli* has features of apoptosis. Infect Immun 67:2575–2584

Crouch Brewer S, Wunderink RG, Jones CB, Leeper KV Jr (1996) Ventilator-associated pneumonia due to *Pseudomonas aeruginosa*. Chest 109:1019–1029

Cuschieri J, Bulger E, Billgrin J, Garcia I, Maier RV (2007) Acid sphingomyelinase is required for lipid raft TLR4 complex formation. Surg Infect (Larchmt) 8:91–106

Dannenberg AM Jr (1994) Roles of cytotoxic delayed-type hypersensitivity and macrophage-activating cell-mediated immunity in the pathogenesis of tuberculosis. Immunobiology 191:461–473

Dannenberg AM Jr, Sugimoto M (1976) Liquefaction of caseous foci in tuberculosis. Am Rev Respir Dis 113:257–259

Dye C, Bassili A, Bierenbach AL, Broekmans JF, Chada VK, Glaziou P, Gopi PG, Hosseini M, Kim SJ, Manissero D, Onozaki I, Rieder HL, Scheele S, van Leth F, van der Werf M, Wiliams BG (2008) Measuring tuberculosis burden, trends, and the impact of control programmes. Lancet Infect Dis 8:233–243

Esen M, Schreiner B, Jendrossek V, Lang F, Fassbender K, Grassmé H, Gulbins E (2001) Mechanism of *Staphylococcus aureus* induced apoptosis of human endothelial cells. Apoptosis 6:431–439

Falcone S, Perrotta C, De Palma C, Pisconti A, Sciorati C, Capobianco A, Rovere-Querini P, Manfredi AA, Clementi E (2004) Activation of acid sphingomyelinase and its inhibition by the nitric oxide/cyclic guanosine 3′,5′-monophosphate pathway: key events in *Escherichia coli*-elicited apoptosis of dendritic cells. J Immunol 173:4452–4463

Ferlinz K, Hurwitz R, Vielhaber G, Suzuki K, Sandhoff K (1994) Occurrence of two molecular forms of human acid sphingomyelinase. Biochem J 301:855–862

Galán JE, Collmer A (1999) Type III secretion machines: bacterial devices for protein delivery into host cells. Science 284:1322–1328

Garner B, Mellor HR, Butters TD, Dwek RA, Platt FM (2002) Modulation of THP-1 macrophage and cholesterol-loaded foam cell apolipoprotein E levels by glycosphingolipids. Biochem Biophys Res Commun 290:1361–1367

Goldfine H, Wadsworth SJ (2002) Macrophage intracellular signaling induced by *Listeria monocytogenes*. Microbes Infect 4:1335–1343

Grassmé HU, Ireland RM, van Putten JP (1996) Gonococcal opacity protein promotes bacterial entry-associated rearrangements of epithelial cell actin-cytoskeleton. Infect Immun 64:1621–1630

Grassmé H, Gulbins E, Brenner B, Ferlinz K, Sandhoff K, Harzer K, Lang F, Meyer TF (1997) Acidic sphingomyelinase mediates entry of *N. gonorrhoeae* into nonphagocytic cells. Cell 91:605–615

Grassmé H, Kirschnek S, Riethmüller J, Riehle A, von Kürthy G, Lang F, Weller M, Gulbins E (2000) CD95/CD95 ligand interactions on epithelial cells in host defense to *Pseudomonas aeruginosa*. Science 290:527–530

Grassmé H, Jekle A, Riehle A, Schwarz H, Berger J, Sandhoff K, Kolesnick R, Gulbins E (2001) CD95 Signaling via ceramide-rich membrane rafts. J Biol Chem 276:20589–20596

Grassmé H, Jendrossek V, Riehle A, von Kürthy G, Berger J, Schwarz H, Weller M, Kolesnick R, Gulbins E (2003) Host defense against *Pseudomonas aeruginosa* requires ceramide-rich membrane rafts. Nat Med 9:322–330

Green DR, Reed JC (1998) Mitochondria and apoptosis. Science 281:1309–1312

Gulbins E, Bissonette R, Mahboubi A, Martin S, Nishioka W, Brunner T, Baier G, Baier-Bitterlich G, Byrd C, Lang F et al (1995) FAS-induced apoptosis is mediated via a ceramide-initiated RAS signaling pathway. Immunity 2:341–351

Gutierrez MG, Gonzales AP, Anes E, Griffith G (2009) Role of lipids in killing mycobacteria by macrophages: evidence for NF-κB-dependent and -independent killing induced by different lipids. Cell Microbiol 11:406–420

Gyles CL (1992) *Escherichia coli* cytotoxins and enterotoxins. Can J Microbiol 38:734–746

Haas A (2007) The phagosome: compartment with a license to kill. Traffic 8:311–330

Haimowitz-Friedman A, Cordon-Cardo C, Bayoumy S, Garzotto M, McLoughlin M, Gallily R, Edwards CK 3rd, Schuchman EH, Fuks Z, Kolesnick R (1997) Lipopolysaccharide induces disseminated endothelial apoptosis requiring ceramide generation. J Exp Med 186:1831–1841

Hambleton J, Weinstein SL, Lem L, DeFranco AL (1996) Activation of c-Jun N-terminal kinase in bacterial lipopolysaccharide-stimulated macrophages. Proc Natl Acad Sci USA 93:2774–2778

Hankins JL, Fox TE, Barth BM, Unrath KA, Kester M (2011) Exogenous ceramide-1-phosphate reduces lipopolysaccharide (LPS)-mediated cytokine expression. J Biol Chem 286:44357–44366

Hauck CR, Grassmé H, Bock J, Jendrossek V, Ferlinz K, Meyer TF, Gulbins E (2000) Acid sphingomyelinase is involved in CEACAM receptor-mediated phagocytosis of *Neisseria gonorrhoeae*. FEBS Lett 478:260–266

Hauser AR, Engel JN (1999) *Pseudomonas aeruginosa* induces type-III-secretion-mediated apoptosis of macrophages and epithelial cells. Infect Immun 67:5530–5537

Heinrich M, Wickel M, Schneider-Brachert W, Sandberg C, Gahr J, Schwander R, Weber T, Saftig P, Peters C, Brunner J, Krönke M, Schütze S (1999) Cathepsin D targeted by acid sphingomyelinase-derived ceramide. EMBO J 18:5252–5263

Herrera-Velit P, Reiner NE (1996) Bacterial lipopolysaccharide induces the association and coordinate activation of p53/561yn and phosphatidylinositol 3-kinase in human monocytes. J Immunol 156:1157–1165

Hersh D, Monack DM, Smith MR, Ghori N, Falkow S, Zychlinsky A (1999) The *Salmonella* invasion SipB induces macrophage apoptosis by binding to caspase-1. Proc Natl Acad Sci USA 96:2396–2401

Jendrossek V, Fillon S, Belka C, Müller I, Puttkammer B, Lang F (2003) Apoptotic response of Chang cells to infection with *Pseudomonas aeruginosa* strains PAK and PAO-I: molecular ordering of the apoptosis signaling cascade and role of type IV pili. Infect Immun 71:2665–2673

Kanto T, Kalinski P, Hunter OC, Lotze MT, Amoscato AA (2001) Ceramide mediates tumor-induced dendritic cell apoptosis. J Immunol 167:3773–3784

Keen EF 3rd, Robinson BJ, Hospenthal DR, Aldous WK, Wolf SE, Chung KK, Murray CK (2010) Incidence and bacteriology of burn infections at a military burn center. Burns 36:461–468

Kim MJ, Wainwright HC, Lochetz M, Bekker LG, Walther GB, Dittrich C, Visser A, Wang W, Hsu FF, Wiehart U, Tsenova L, Kaplan G, Russel DG (2010) Caseation of human tuberculosis granulomas correlates with elevated host lipid metabolism. EMBO Mol Med 2: 258–274

Kolesnick R, Krönke M (1998) Regulation of ceramide production and apoptosis. Annu Rev Physiol 60:643–665

Kolter T, Sandhoff K (2005) Principles of lysosomal membrane digestion: stimulation of sphingolipid degradation by sphingolipid activator proteins and anionic lysosomal lipids. Annu Rev Cell Dev Biol 21:81–103

Kowalski MP, Pier GB (2004) Localization of cystic fibrosis transmembrane conductance regulator to lipid rafts of epithelial cells is required for Pseudomonas aeruginosa-induced cellular activation. J Immunol 172:418–425

Kroemer G, Dallaporta B, Resche-Rigon M (1998) The mitochondrial death/life regulator in apoptosis and necrosis. Annu Rev Physiol 60:619–642

Kuhle V, Hensel M (2002) SseF and SseG are translocated effectors of the type III secretion system of *Salmonella* pathogenicity island 2 that modulates aggregation of endosomal compartments. Cell Microbiol 4:813–824

Lambden PR, Heckels JE, James LT, Watt PJ (1979) Variations in surface protein composition associated with virulence properties in opacity types of *Neisseria gonorrhoeae*. J Gen Microbiol 114:305–312

Levade T, Jaffrézou JP (1999) Signaling sphingomyelinases: which, where, how and why? Biochem Biophys Acta 1438:1–17

Lowy FD (1998) *Staphylococcus aureus* infections. N Engl J Med 339:520–532

MacKichan ML, DeFranco AL (1999) Role of ceramide in lipopolysaccharide (LPS)-induced signaling. LPS increases ceramide rather than acting as a structural homolog. J Biol Chem 274:1767–1775

Makino S, van Putten JP, Meyer TF (1991) Phase variation of the opacity outer membrane protein controls invasion by *Neisseria gonorrhoeae* into human epithelial cells. EMBO J 10:1307–1315

Marathe S, Schissel SL, Yellin MJ, Beatini N, Mintzer R, Williams KJ, Tabas I (1998) Human vascular endothelial cells are a rich and regulatable source of secretory sphingomyelinase. Implications for early atherogenesis and ceramide-mediated cell signaling. J Biol Chem 273:4081–4088

Martin MU, Wesche H (2002) Summary and comparison of the signaling mechanisms of the Toll/interleukin-1 receptor family. Biochim Biophys Acta 1592:265–280

McCollister BD, Bouret TJ, Gill R, Jones-Carson J, Vázquez-Torres A (2005) Repression of SPI2 transcription by nitric oxide-producing, IFNgamma-activated macrophages promotes maturation of Salmonella phagosomes. J Exp Med 5:625–635

McCollister BD, Myers JT, Jones-Carson J, Voelker DR, Vázquez-Torres A (2007) Constitutive acid sphingomyelinase enhances early and late macrophage killing of *Salmonella enterica* serovar Typhimurium. Infect Immun 75:5346–5352

McGee ZA, Stephens DS, Hoffman LH, Schlech WF 3rd, Horn RG (1983) Mechanisms of mucosal invasion by pathogenic Neisseria. Rev Infect Dis 5(suppl 4):S708–S714

McManus AT, Mason AD Jr, McManus WF, Pruitt BA Jr (1985) Twenty-five year review of *Pseudomonas aeruginosa* bacteremia in a burn center. Eur J Clin Microbiol 4:219–223

Menzies BE, Kourteva I (1998) Internalization of *Staphylococcus aureus* by endothelial cells induces apoptosis. Infect Immun 66:5994–5998

Miao EA, Miller SI (2000) A conserved amino acid sequence directing intracellular type III secretion by *Salmonella typhimurium*. Proc Natl Acad Sci USA 97:7539–7544

Monack DM, Mecsas J, Ghori N, Falkow S (1997) *Yersinia* signals macrophages to undergo apoptosis and YopJ is necessary for this cell death. Proc Natl Acad Sci USA 94:10385–10390

Morrison DC, Ryan JL (1987) Endotoxins and disease mechanisms. Annu Rev Med 38:417–432

Narumi S, Tebo JM, Finke JH, Hamilton TA (1992) IFN-gamma and IL-2 cooperatively activate NF kappa B in murine peritoneal macrophages. J Immunol 149:529–534

Oh YK, Alpuche-Aranda C, Berthiaume E, Jinks T, Miller SI, Swanson JA (1996) Rapid and complete fusion of macrophage lysosomes with phagosomes containing *Salmonella typhimurium*. Infect Immun 64:3877–3883

Pettus BJ, Chalfant CE, Hannun YA (2002) Ceramide in apoptosis: an overview and current perspectives. Biochim Biophys Acta 1585:114–125

Pfeiffer A, Böttcher A, Orsó E, Kapinsky M, Nagy P, Bodnár A, Spreizer I, Liebisch G, Drobnik W, Gempel K, Horn M, Holmer S, Hartung T, Multhoff G, Schütz G, Schindler H, Ulmer AJ, Heine H, Stelter F, Schütt C, Rothe G, Szöllôsi J, Damjanovich S, Schmitz G (2001) Lipopolysaccharide and ceramide docking to CD14 provokes ligand-specific receptor clustering in rafts. Eur J Immunol 31:3153–3164

Portnoy DA, Jacks PS, Hinrichs DJ (1988) Role of hemolysin for the intracellular growth of *Listeria monocytogenes*. J Exp Med 167:1459–1471

Ruckdeschel K, Roggenkamp A, Lafont V, Mangeat P, Heeseman J, Rouot B (1997) Interaction of *Yersinia enterocolitica* with macrophages leads to macrophage cell death through apoptosis. Infect Immun 65:4813–4821

Russel DG (2001) *Mycobacterium tuberculosis*: here today, and here tomorrow. Nat Rev Mol Cell Biol 2:569–577

Russel DG (2007) Who puts the tubercle in tuberculosis? Nat Rev Microbiol 5:39–47

Russel DG, Cardona PJ, Kim MJ, Allain S, Altare F (2009) Foamy macrophages and the progression of the human tuberculosis granuloma. Nat Immunol 10:943–948

Sadikot RT, Blackwell TS, Christman JW, Prince AS (2005) Pathogen-host interactions in *Pseudomonas aeruginosa* pneumonia. Am J Respir Crit Care Med 171:1209–1223

Schissel SL, Schuchman EH, Williams KJ, Tabas I (1996) Zn^{2+}-stimulated sphingomyelinase is secreted by many cell types and is a product of the acid sphingomyelinase gene. J Biol Chem 271:18431–18436

Schnupf P, Portnoy DA (2007) Listeriolysin O: a phagosome-specific lysin. Microbes Infect 9:1176–1187

Schramm M, Herz J, Haas A, Krönke M, Utermöhlen O (2008) Acid sphingomyelinase is required for efficient phago-lysosomal fusion. Cell Microbiol 10:1839–1853

Shea JE, Hensel M, Gleeson C, Holden DW (1996) Identification of a virulence locus encoding a second type III secretion system in *Salmonella typhimurium*. Proc Natl Acad Sci USA 93:2593–2597

Stern A, Brown M, Nickel P, Meyer TF (1986) Opacity genes in *Neisseria gonorrhoeae*: control of phase and antigenic variation. Cell 47:61–71

Takahashi A (1999) Caspase: executioner and undertaker of apoptosis. Int J Hematol 70:226–232

Takeda K, Akira S (2004) Microbial recognition by Toll-like receptors. J Dermatol Sci 34:73–82

Thompson CB (1995) Apoptosis in the pathogenesis and treatment of disease. Science 267:1456–1462

Uchiya K, Barbieri MA, Funato K, Shah AH, Stahl PD, Groisman EA (1999) A *Salmonella* virulence protein that inhibits cellular trafficking. EMBO J 18:3924–3933

Utermöhlen O, Karow U, Löhler J, Krönke M (2003) Severe impairment in early host defense against *Listeria monocytogenes* in mice deficient in acid sphingomyelinase. J Immunol 170:2621–2628

Utermöhlen O, Herz J, Schramm M, Krönke M (2008) Fusogenicity of membranes: the impact of acidic sphingomyelinase on innate immune responses. Immunobiology 213:307–314

van Putten JP, Paul SM (1995) Binding of syndecan-like cell surface proteoglycan receptors is required for *Neisseria gonorrhoeae* entry into human mucosal cells. EMBO J 14:2144–2154

Vázquez-Boland JA, Kuhn M, Berche P, Chakraborty T, Domínguez-Bernal B, Goebel W, González-Zorn B, Wehland J, Kreft J (2001) Listeria pathogenesis and molecular virulence determinants. Clin Microbiol Rev 14:584–640

Vázquez-Torres A, Jones-Carson J, Mastroeni P, Ischiropoulos H, Fang FC (2000) Antimicrobial actions of the NADPH phagocyte oxidase and inducible nitric oxide synthase in experimental salmonellosis. I. Effects on microbial killing by activated peritoneal macrophages in vitro. J Exp Med 192:227–236

Vidal F, Mensa J, Almela M, Martínez JA, Marco F, Casals C, Gatell JM, Soriano E, Jimenez de Anta MT (1996) Epidemiology and outcome of *Pseudomonas aeruginosa* bacteremia, with special emphasis on the influence of antibiotic treatment. Analysis of 189 episodes. Arch Intern Med 156:2121–2126

Zhang Y, Li X, Carpinteiro A, Gulbins E (2008) Acid sphingomyelinase amplifies redox signaling in *Pseudomonas aeruginosa-induced* macrophage apoptosis. J Immunol 181:4247–4254

Zychlinsky A, Kenny B, Ménard R, Prévost MC, Holland IB, Sansonetti PJ (1994) IpaB mediates macrophage apoptosis induced by *Shigella flexneri*. Mol Microbiol 11:619–627

Viral Infections and Sphingolipids

Jürgen Schneider-Schaulies and Sibylle Schneider-Schaulies

Contents

Abstract Besides their essential role in the immune system, sphingolipids and their metabolites are potential key regulators in the life cycle of obligatory intracellular pathogens such as viruses. They are involved in lateral and vertical segregation of receptors required for attachment, membrane fusion and endocytosis, as well as in the intracellular replication, assembly and release of viruses. Glycosphingolipids may themselves act as receptors for viruses, such as Galactosylceramide for human immunodeficiency virus (HIV). In addition, sphingolipids and their metabolites are inseparably interwoven in signal transduction processes, dynamic alterations of the cytoskeleton, and the regulation of innate and intrinsic responses of infected target cells. Depending on the nature of the intracellular pathogen, they may support or inhibit infections. Understanding of the underlying mechanisms depending on the specific virus, immune control, and type of disease may open new avenues for therapeutic interventions.

J. Schneider-Schaulies • S. Schneider-Schaulies (✉)
Institute for Virology and Immunobiology, University of Wuerzburg, Versbacher Str. 7, 97078 Wuerzburg, Germany
e-mail: s-s-s@vim.uni-wuerzburg.de

E. Gulbins and I. Petrache (eds.), *Sphingolipids in Disease*,
Handbook of Experimental Pharmacology 216,
DOI 10.1007/978-3-7091-1511-4_16, © Springer-Verlag Wien 2013

Keywords Viral infection • Virus uptake • Viral replication • Viral receptor • Human immunodeficiency virus • Measles virus • DC-SIGN • CD150 • Cytolytic granule • Lymphocytic choriomeningitis virus • Sindbis virus

Abbreviations

ASMase	Acidic sphingomyelinase
bSMase	Bacterial sphingomyelinase
BVDV	Bovine viral diarrhea virus
CRAC	Ca^{2+} release-activated Ca^{2+}
DC	Dendritic cell
DC-SIGN	DC-specific intercellular adhesion molecule-3-grabbing non-integrin
DRM	Detergent-resistant membrane domain
Gal-Cer	Galactosylceramide
Gb3	Gal1α1-4Gal1β1-4 glucosyl ceramide or globotriaosylceramide, also referred to as p^k blood group antigen or CD77
Gb4Cer	Globotetraosylceramide
GSL	Glycosphingolipid
HCV	Hepatitis C virus
HCMV	Human cytomegalovirus
HIV	Human immunodeficiency virus
JEV	Japanese encephalitis virus
LCMV	Lmphocytic choriomeningitis virus
MV	Measles virus
NEM	*N*-ethylmaleimide
NSMase	Neutral sphingomyelinase
PPMP	1-Phenyl-2-hexadecanoylamino-3-morpholino-1-propanol
PBMC	Peripheral blood mononuclear cell
RV	Rhinovirus
RSV	Respiratory syncytial virus
S1P	Sphingosine 1-phosphate
SINV	Sindbis virus
SMase	Sphingomyelinase

1 Introduction

In addition to protein sorting, the dynamic sorting of lipids and mutual interdependence of both led—and continues to do so—to a refined vision on how membrane dynamics impacts on cellular processes such as vesicular transport and biogenesis, as especially important in endo- and exocytosis. The mechanistic understanding of lateral and vertical sorting of proteins associated with that of lipids has proven to be highly instrumental in the regulation of assembly and activity of signaling macromolecular complexes conveying extracellular signals, thereby essentially regulating key cellular processes such as activation, differentiation, and survival. Sphingolipids are major membrane lipid constituents, and as such, their biogenesis, modifications, and turnover (the biology of which is reviewed elsewhere in this book) are tightly linked to these processes all involving membrane dynamics. Moreover, the local segregation of sphingolipids and their metabolites directly affects biophysical properties of the membrane at a local basis regulating membrane deformation, vesiculation, and fusogenicity (Goni and Alonso 2002; Simons and Gerl 2010; Trajkovic et al. 2008; Utermohlen et al. 2008; Zhang et al. 2008).

Given these properties, it is obvious that sphingolipids and their metabolites are potential key regulators in the life cycle of obligatory intracellular pathogens such as viruses. Most evidently, this will extend to lateral or vertical segregation of receptors required for attachment, fusion, and entry. The latter processes can be directly influenced at the entry site by alterations of membrane fluidity or polarity, which may or may not provide an environment for fusogenic or endocytic uptake and subsequent sorting to specific subcellular compartments. Though formally proven for a limited number of viruses yet, it is conceivable that membrane-associated viral replication and assembly of viral components rely on defined lipid compositions, and this also refers to the selection of sites of particle formation.

It will be within the first part of this chapter where we review the current knowledge of the role of sphingolipids on viral replication at a molecular level. The second part will address issues relating their role in viral pathogenesis, which will focus on the impact of regulated sphingolipid breakdown on virus–host cell interactions and antiviral defense mechanism with particular focus on immunologically relevant targets. Finally, we will briefly extend on the therapeutic potential of strategies targeting sphingolipids yet will not extend on their role as vaccine adjuvants, which will be reviewed elsewhere in this issue.

2 Sphingolipids in Viral Uptake and Replication

The vast majority of findings relating to the role of sphingolipids in the viral life cycle deal with their property to mediate or regulate viral entry into target cells. For this, they may themselves serve as (co-)receptors as detailed for glycosphingolipids (GSLs) below and regulate lateral or vertical segregation of receptor proteins by altering biophysical properties of the membrane, by providing an environment supporting endocytosis and/or fusion, or by modulating actin dynamics as often required in viral entry processes.

2.1 Role of Globo Receptors in Viral Uptake

As well-established, the entry of HIV into T cells, monocytes, or dendritic cells (DCs) relies on binding of its envelope protein gp120 to CD4, thereby promoting a conformational change within the gp120 variable loop 3 (V3 loop) that is essential for interaction with its co-receptors CXCR4 or CCR5, respectively, and subsequent membrane fusion [recently reviewed in Caffrey (2011) and Cosset and Lavillette (2011)]. V3 loops, however, also harbor a motif interacting with the carbohydrate moieties of GSLs, of which Gb3 (Gal1α1-4Gal1β1-4 glucosylceramide or globotriaosylceramide, also referred to as p^k blood group antigen or CD77) and galactosylceramide (Gal-Cer, binding to both the gp120 and the gp41 subunit of the HIV env protein, see below) have mainly been investigated for their role in modulating HIV infection. In contrast to the binding to chemokine receptors, gp120 binding to GSLs does not require prior CD4 interaction, and it has been suggested that the smaller size of GSLs might render gp120 interaction independent of conformational changes within the V3 loop (Hammache et al. 1998b; Lingwood et al. 2010b). Interestingly, all GSLs bind within the V3 loop close to the center of the chemokine binding site, and thus might differentially regulate HIV fusion and uptake. Under certain conditions, Gb3 may either inhibit or promote fusion between viral and host cell membranes (see below). In line with the importance of GSLs in HIV pathogenesis, they have been implicated in HIV uptake into CD4 negative cells such as mucosal epithelial cells, in transcytosis, and HIV transmission to T cells or DCs [(Cook et al. 1994; Hammache et al. 1998a), and see below]. Moreover, HIV entry into target cells is highly sensitive to compounds affecting GSL biosynthesis such as 1-phenyl-2-hexadecanoylamino-3-morpholino-1-propanol (PPMP), which inhibits conversion of ceramide into glucosylceramide by glucosyltransferase (Puri et al. 2004), and to variations in cellular GSL contents (Rawat et al. 2004).

In mammalian cells, GSLs differ with regard to the length of their fatty acid chains, which has always even numbers and predominantly ranges between C_{16} and C_{24}, sometimes longer. They usually reside outside classical rafts, and the bioavailability of their carbohydrate region for ligand binding is strongly dependent on the surrounding lipid environment, which has been referred to as "aglycone modulation" (Lingwood et al. 2011). This has, for instance, been illustrated for the shiga toxins (verotoxins 1 and 2, VT1 and VT2), which bind to Gb3 with differential affinity that is sensitive to high cholesterol conditions (Khan et al. 2009; Lingwood et al. 2011). VT toxicity relies on retrograde transport to the Golgi after binding to Gb3 yet is completely dependent on Gb3 raft association, while Gb3 interaction outside rafts results in VT internalization followed by degradation and inactivation.

In vitro studies involving model membranes such as detergent-resistant membrane domain (DRM) vesicles or Gb3 monolayers revealed that gp120 mirrored binding characteristics with VT1 with regard to the fatty acid moiety in that C_{16}, C_{22}, and C_{24} Gb3 isoforms efficiently bound, while C_{18} and C_{20} were not recognized and acted as dominant negatives in lipid mixtures. Lipid mixture experiments also clearly supported the aglycone concept revealing modulation of ligand binding by lateral interaction of GSL fatty isoforms within the bilayers (Mahfoud et al. 2009). Inasmuch the similarities found for gp120 and VT1 binding in vitro translate into in vivo conditions are as yet unclear, because in tissues, monomeric gp120 preferentially binds Gb3 in non-raft fractions (Khan et al. 2009), and binding characteristics may differ for the biologically active trimeric gp120 (Lingwood et al. 2010a, b).

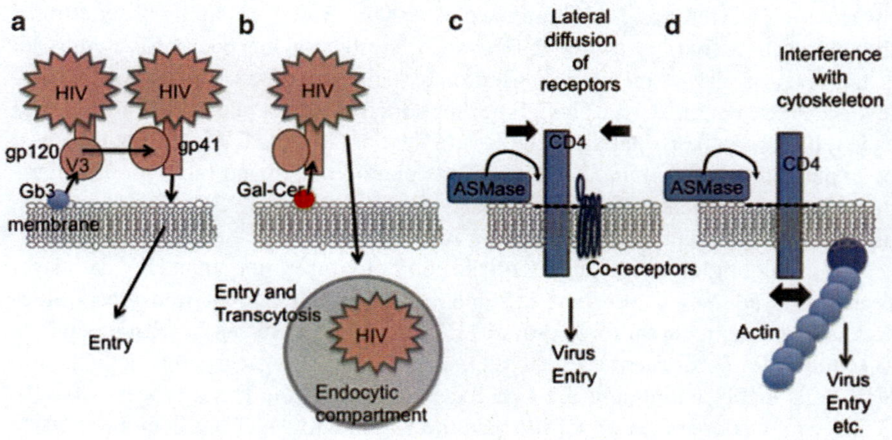

Fig. 1 *Sphingolipids in regulating HIV attachment and entry*. (**a**) At the level of the plasma membrane, globotriaosylceramide (Gb3) can either enhance (by trapping the V3 loop of gp120, thereby concentrating viral particles under conditions where receptor expression is low) or inhibit HIV entry [under high Gb3 conditions by preventing insertion of the fusion moiety (gp41)]. (**b**) Similar as for Gb3, Gal-Cer can support HIV entry into CD4 dim or negative epithelial cells yet also act in viral transcytosis through epithelial barriers or in viral trapping by DCs, both followed by transmission to T cells. (**c**) Accumulation of plasma membrane ceramides interferes with lateral motility of HIV attachment (CD4) and fusion co-receptors (chemokine receptors) and thus prevents fusion-mediated uptake and favors endocytosis. (**d**) SMase activation can effectively prevent actin cytoskeletal integrity and dynamics, both of which are essential in entry of most, if not all, viruses in supporting, for instance, endocytic uptake, membrane extensions, receptor clustering, and formation of virological synapses

The enhancing effect of Gb3 for HIV infection has been shown in cells which had remained insensitive to infection in spite of CD4 expression or after Gb3 ablation in CD4-expressing cells (Hug et al. 2000; Puri et al. 1998) (Fig. 1a). In line with a role of Gb3 in HIV entry into target cells, adamantly-Gb3 (a soluble Gb3-mimic, in which the ceramide moiety is replaced by a rigid adamantane hydrocarbon frame) efficiently prevented infection of primary lymphoid cells by both CXCR4 and CCR5 tropic HIV strains in vitro including those resistant to other fusion inhibitory drugs such as T20 (Lund et al. 2006). Surprisingly, Gb3 can also act as resistance factor for HIV infection. Thus, peripheral blood mononuclear cells (PBMCs) accumulating high levels of Gb3 due to an α-galactosidase defect (in Fabry disease) are resistant rather than hypersensitive to in vitro infection with CCR5 tropic HIV strains, while the susceptibility to CXCR4 tropic strains is equivalent to that of healthy controls (Lund et al. 2005). A study involving PBMCs isolated from healthy controls based on their association to the P blood group system (with p lacking, Pk having normal, and P1K having high Gb3 levels) confirmed the inverse correlation between Gb3 levels and susceptibility to HIV infection, and this did not differ for CXCR4 and CCR5 tropic strains (Lund et al. 2009). Finally, interference of high Gb3 levels on HIV infection was also observed in cell lines where levels of Gb3 were constitutively high exogenously modified by genetic interference (Ramkumar et al. 2009). Based on these observations, the inhibitory activity of synthetic Gb3 on HIV infection has been proposed to act at the level of plasma membrane insertion associated with an increase of the Gb3 level (Harrison et al. 2010). Though not supported by direct experimental evidence as yet, it is most likely that the differential effect of Gb3 on HIV infection may best be explained by a combination of overall expression level and turnover, compartmentalization to membrane microdomains, and accessibility, all of which may be highly dependent on type and activation status of the respective host cell analyzed.

The effect of globosylceramides on the entry of viruses other than HIV has as yet been barely addressed, yet there is evident that Gb4Cer (globotetraosylceramide) acts as a primary receptor for parvovirus B19 triggering viral capsid rearrangements as required for subsequent steps in internalization into megacaryoblastic leukemia UT7 cells stably expressing the Epo receptor. During the entry process of B19, Gb4Cer may cooperate with CD49e and the Ku-80 antigen (Bonsch et al. 2010).

In addition to globosylceramides, gangliosides (glycosphingolipids with one or more sialic acid residues linked to the sugar moiety) such as GD1a and GT1b or GM1 are used as entry receptors by murine polyoma virus and SV40, respectively [recently reviewed in Burckhardt and Greber (2009)].

2.2 The Role of Ceramides in Viral Receptor Secretion, Uptake, and Pathogen Sorting

2.2.1 Gal-Ceramides as HIV-1 Attachment Receptors

In polarized epithelial cells, GSLs are differentially sorted in apical vs. basolateral membranes resulting from lipid sorting within the secretory pathway (Harrison et al. 2010; Lingwood et al. 2010a, 2011). It is especially the GSL galactosylceramide (Gal-Cer) which has been proposed to act as mucosal epithelial cell receptor for both HIV gp120 and gp41, and at least for the latter, a structural dependence has been revealed by loss of interaction on modulation of Ca^{2+} levels. HIV binding to Gal-Cer promotes internalization and, rather than degradation, transcytosis followed by release from the basolateral membrane in an infectious form (Alfsen and Bomsel 2002; Magerus-Chatinet et al. 2007; Yu et al. 2008) (Fig. 1b). Gal-Cer trapping was, however, also found to efficiently promote HIV transmission to T cells by mammary epithelial cells (Dorosko and Connor 2010). Mechanistically, this resembles virus acquisition by T cells via DCs with HIV bound to DC-SIGN (trans-infection) in the absence of detectable DC infection (Geijtenbeek and van Kooyk 2003). There is evidence that Gal-Cer does also support T-cell trans-infection by immature DCs. Enhancement of HIV transmission by Gal-Cer trapping and raft-dependent endocytosis was proposed to act as an alternative route to the clathrin-mediated uptake following CD4 interaction. The potential importance of this particular mode of transmission is supported by the finding that gp41 was found to attach to Gal-Cer expressed on DCs isolated from human blood and mucosal tissue and in situ on mucosal tissues (Magerus-Chatinet et al. 2007).

2.2.2 Ceramides in Viral Uptake and Replication

As evidenced by the Gal-Cer example, ceramides may not only serve as attachment factors via their carbohydrate moiety yet also regulate further routing of viruses, and this may relate to their ability to alter membrane organization in terms of fluidity and segregation of host cell proteins including viral receptors. Membrane domains enriched in ceramides were suggested to support vesicular fusion (Utermohlen et al. 2008). Thus, conditions favoring generation of these domains, e.g., by activating acidic sphingomyelinase (ASMase) or inhibiting ceramidase, would be predicted to create an environment enhancing membrane fusion and entry of particularly enveloped viruses. This may, however, not generally apply as revealed for HIV, the entry of which proved to be highly sensitive to compounds elevating levels of ceramides (Finnegan and Blumenthal 2006; Finnegan et al. 2004, 2007; Puri et al. 2004). Thus, preexposure to fenretinide, bSMAse, or long-chain ceramide (C_{16}) dose-dependently prevented HIV uptake into T cells, monocytes, or macrophages (Finnegan et al. 2004). The compounds did not interfere with HIV binding nor with overall surface levels of CD4 and chemokine receptors, or their association with detergent-resistant membrane domains, but rather—as confirmed by fluorescence recovery after photobleaching (FRAP) experiments—with lateral diffusion of CD4 towards the co-receptors as required for viral entry (Finnegan et al. 2004, 2007; Rawat et al. 2008) (Fig. 1c). Supporting the hypothesis that availability of CD4 to the fusion receptor complex might be limited by entrapment in ceramide-enriched platforms, the inhibitory effect of bacterial sphingomyelinase (bSMase) was at least partially counteracted by increasing CD4 surface expression levels (Finnegan et al. 2007). In addition, the sphingolipid environment can directly regulate HIV fusion, because elevation of dihydroxyceramide was found to efficiently interfere with HIV uptake in tissue culture. This was linked to enhanced rigidity of the membrane which did not allow for insertion of the fusogenic gp41 (Vieira et al. 2010). Secondary to inhibition of fusion-mediated uptake and subsequent access to the cytosol, HIV was found to be endocytosed and targeted to lysosomal compartments upon fenretinide preexposure of target cells indicating that sphingolipid-dependent receptor segregation may directly impact on the mode of viral uptake into host cells (Finnegan and Blumenthal 2006; Marechal et al. 1998).

Similar as for HIV, overall elevation of ceramides by bSMase exposure interfered with uptake of hepatitis C virus (HCV) at the level of receptor segregation. There, one of the major entry factors, CD81, was partially internalized, while recruitment of this factor and the other components required for entry, scavenger receptor B1 and claudin-1, into detergent-resistant microdomains was abolished (Voisset et al. 2008). Whether this would also occur under more physiological conditions (i.e., an inflammatory environment where ceramides are produced as a result of receptor-dependent SMase activation) is unknown.

Based on their biophysical properties, ceramide-enriched membrane domains, and especially platforms formed on receptor-mediated SMase activation, are sites of endocytosis and uptake of pathogens (Bollinger et al. 2005; Grassme et al. 2007; Gulbins et al. 2004; Gulbins and Grassme 2002; Holopainen et al. 2000; Zha et al. 1998). Thus, ceramide induction would be predicted to have a positive impact on uptake of viruses relying on endocytosis followed by vesicular acidification. This has been verified for influenza A virus, the uptake of which depends on endocytosis, and on activation of the fusogenic activity of hemagglutinin, targeting to low pH compartments. It is thus not surprising that uptake of influenza A virus is enhanced rather than inhibited (as for HIV) by fenretinide (Finnegan and Blumenthal 2006). In line with this, SMase activation supports pH and clathrin-dependent entry and replication of the flavivirus Japanese encephalitis virus (JEV) in tissue culture, though the role of this enzyme in either process and its relevance in vivo have not been further analyzed (Tani et al. 2010).

Certain viruses are able to promote formation of ceramide-enriched platforms supporting their entry into target cells themselves. This has first been revealed for major and minor subgroup picornavirus rhinovirus (RV) attachment to epithelial cells (Dreschers et al. 2007; Grassme et al. 2005). Attachment of RV caused microtubule- and microfilament-dependent ASMase cell-surface translocation and subsequent formation of large membrane platforms enriched in ceramides or GSLs, both of which interacted with and were important for virus uptake (Dreschers et al. 2007; Grassme et al. 2005). Indicating that intracellular trafficking of RV involved GSLs rather than ceramide, GSLs were co-detected with endocytosed virus in membrane proximal and perinuclear vesicles. Though both types of sphingolipids promote RV binding and uptake, it is as yet unknown how they communicate with or impact on recruitment or concentration of the RV receptors, ICAM-1, or the low-density lipoprotein family members. Interestingly, RV raft interaction causes biphasic activation p38 MAPK in a RhoA-dependent manner, with late activation relying on viral replication, yet not integrity of raft domains (Dumitru et al. 2006).

Two independent large screening studies provided convincing evidence for the crucial role of the endo/lysosomal cholesterol transporter Niemann-Pick C protein 1 (NPC1) in Ebola virus uptake. Following attachment to host cells, Ebola virus particles are delivered to endosomes. There, fusion with the viral and endosomal membranes, which is absolutely required for subsequent replication, depends on NPC1 and recruitment of the homotypic fusion and vacuole protein-sorting (HOPS) complex (Carette et al. 2011; Cote et al. 2011). Interestingly, viral escape from this compartment was independent of the NPC1 cholesterol transporter activity (Cote et al. 2011). NPC1 was found to bind proteolytically activated viral gp and thus serves as bona fide entry receptor for Ebola virus which resides in an intracellular compartment rather than at the plasma membrane and might be a perfect target for antiviral intervention (Miller et al. 2012a). In addition to NPC1, both SM and ASMase activity were implicated in early steps of Ebola virus infection. Viral attachment appeared to be SM dependent, and viral particles strongly associated with surface-displayed ASMase indicating that viral interaction may occur in SM-enriched membrane domains followed by ASMase activation (Miller et al.

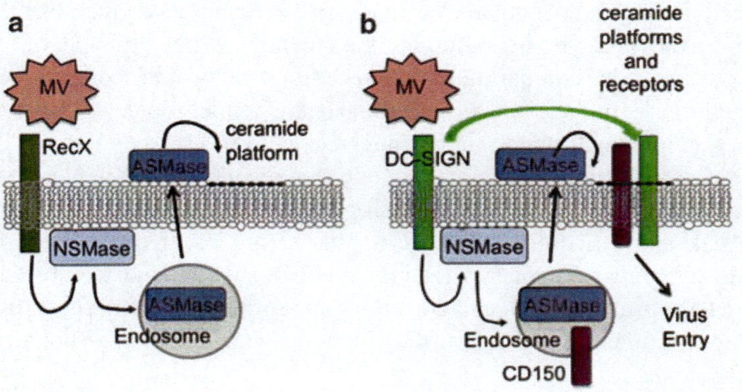

Fig. 2 *Functional consequences of MV-induced SMase activation in immune cells*. MV interaction with an as yet unknown receptor on T lymphocytes (recX) (**a**) or the C-type lectin receptor DC-SIGN on DCs (**b**) sequentially activates NSMase and ASMase followed by formation of ceramide in enriched membrane domains. SMase activation in T cells accounts for MV-induced loss of actin-based membrane protrusions and interference with cytoskeletal dynamics (**a**). In DCs, CD150, the MV entry receptor is co-recruited and displayed at the plasma membrane with ASMase on DC-SIGN-induced NSMase activation, and both MV attachment (DC-SIGN) and entry receptor (CD150) co-cluster in microdomains also enriched in ceramides (**b**)

2012b). Whether the latter is important in rendering NPC1 accessible to the viral particle (similar to what has been observed for measles virus, see below) as suggested has not yet been proven.

More recently, the ability of measles virus (MV) to activate both NSMase and ASMase upon ligation of an unknown receptor on T cells [(Gassert et al. 2009) (Fig. 2a), see below] and of DC-SIGN on immature DCs has been documented (Avota et al. 2011). On DCs, SMase activation and subsequent accumulation of outer membrane ceramides were required for an *N*-ethylmaleimide (NEM)-sensitive surface recruitment of the MV entry receptor CD150 from intracellular ASMase-containing compartments. CD150 was transiently co-displayed with DC-SIGN in ceramide-enriched platforms, thus explaining enhancement of MV infection of DCs by recruitment of the entry receptor as a consequence of loading the attachment receptor, DC-SIGN (Fig. 2b). As a C-type lectin receptor on DCs, DC-SIGN binds and signals in response to a variety of pathogens, and the ability of mannan and DC-SIGN-specific antibodies (after crosslink) to mediate SMAse activation shows that DC-SIGN clearly adds into the list of surface receptors generally involved in sphingolipid turnover (Avota et al. 2011). Because it also binds to this molecule, HIV gp120 would be expected to cause formation of ceramide-enriched surface platforms as well, which would cause inhibition of lateral CD4 mobility and thereby inhibit uptake by fusion and favor endocytosis (see above). This hypothesis is in line with HIV being predominantly internalized into nondegradative invaginated compartments in DCs where it is stored for subsequent transmission to T cells (Blanchet et al. 2011; Izquierdo-Useros et al. 2009, 2010). How DC-SIGN induced

SMase activation might regulate the mode and efficiency of entry of other viral ligands is unknown as yet. In addition to the potential association with or exclusion from SMase/ceramide compartments, the receptor density at the cell surface may be decisive, because high levels of the latter may render lipid sorting-dependent redistribution no longer important (Finnegan et al. 2007).

In addition to serving as pathogen recognition receptor, DC-SIGN is a signaling molecule which can modify TLR signaling and thereby shape adaptive immunity (Geijtenbeek and Gringhuis 2009; Gringhuis et al. 2009). Activation of major DC-SIGN signaling components (Raf-1 and ERK) was found to rely on SMase activation (Avota et al. 2011), and thus, the sphingolipid turnover especially in DCs is likely to essentially contribute to DC functions and, eventually, polarization of T-cell responses.

2.3 Ceramide-Dependent Modulation of Actin Cytoskeletal Dynamics

Actin dynamics is central to the uptake of viruses into host cells. This includes retrograde actin flow-dependent drifting and surfing of receptors (also including glycolipids) engaged by viruses along filopodia or on the cell body surface (Ewers et al. 2005; Lehmann et al. 2005; Schelhaas et al. 2008), receptor clustering, and formation of and transmission by defined structures such as virological synapses or filopodial bridges (Mothes et al. 2010; Sherer et al. 2007), to name only examples. Moreover, actin-mediated membrane ruffling and blebbing is essential for macropinocytic uptake of a number of viruses such as vaccinia, picorna, and adenoviruses (Mercer and Helenius 2008, 2009). Thus, interference with actin dynamics would be expected to have a significant impact on uptake efficiency, which has as yet not been investigated in relation to sphingolipid turnover (Fig. 1d). However, breakdown of actin cytoskeletal protrusions after ASMase activation and subsequent ceramide accumulation have been shown to occur in MCF-7 breast cancer cells (Zeidan et al. 2008) and, NSMase- and ASMase- dependently, after MV interaction with an unknown surface receptor on T cells, where actin cytoskeletal rearrangements in response to TCR or β1-integrin ligation were abrogated and filopodial protrusions were lost, referred to as physical T-cell paralysis (Gassert et al. 2009) (Fig. 2a). Inasmuch SMase activation might account for loss of stimulated T-cell expansion, which is efficiently induced by MV in vivo and in vitro and, at a cellular level, targets activation of the PI3/Akt kinase pathway (Avota et al. 2001; Gassert et al. 2009; Muller et al. 2006), is as yet not fully established. It might, however, be highly interesting to evaluate as to what extent SMase activation induced upon receptor ligation by viruses or inflammatory mediators might be able to affect viral entry, trafficking, and eventually assembly and release at the level of actin remodeling.

2.4 *Sphingolipids in Viral Maturation and Budding*

For many enveloped viruses, rafts have been implicated as sites of particle production, which was often based on co-floatation of raft-associated cellular proteins with viral components, their co-localization with raft components such as GM1, or lack of particle production on extraction of cholesterol, which is a major component of membrane rafts. As a direct evidence for the role of defined lipids in the production of enveloped viral particles, virion lipidomes are very informative yet are only available for hepatitis C virus (HCV), HIV, and human cytomegalovirus (HCMV) as yet (Brugger et al. 2006; Liu et al. 2011; Merz et al. 2011). Comparative analyses with the respective host cell lipidomes confirmed that first, viral infection modifies the latter and second, that lipid composition of viral budding sites does not only differ from that of their host cells but also between the viruses studied. At least for HIV, the importance of sphingolipids, as especially reflected by selective enrichment of SM and dihydroSM as virion constituents, in the viral replication cycle has been clearly revealed (Brugger et al. 2006). While, common to HCV, ceramides represented a very minor virion sphingolipid component, glycosphingolipids are apparently contained in virions from certain producer cells as revealed by the importance of GSLs in HIV infectivity in T-cell trans-infection in cocultures with DCs (Hatch et al. 2009). In addition to defined budding sites, lipid structures (and their biogenesis) are intimately coupled to viral replication and/or assembly for certain viruses. This has been especially analyzed for HCV and Dengue virus, where lipid droplets and their biogenesis have a crucial role in initiation of viral assembly (Alvisi et al. 2011; Fischl and Bartenschlager 2011). Though its specific role in this process is as yet unknown, inhibition of ceramide transfer protein (CERT) was found to inhibit HCV maturation pointing to an important contribution of the sphingolipid pathway in flavivirus replication (Amako et al. 2011). As referred to above, actin remodeling has been related to particle formation by many viruses as reflected by its incorporation into virions. Therefore, SMase-dependent alterations of actin dynamics as eventually provided in inflammatory conditions might as well substantially impact the viral maturation process.

3 Sphingolipids as Modulators of Viral Pathogenesis

3.1 *Modulation of Cell Autonomous Defense Mechanisms*

Sphingolipids may not only play an important role in essential steps of viral replication but may also influence the interaction of viruses with their host cells and organisms. Examples of those particularly include viral modulations of host cell apoptosis by targeting sphingolipids and their metabolites.

The alphavirus sindbis virus (SINV) accesses the cytosol of neurons in an endosomal fusion-dependent process, which relies on the presence of sphingomyelin in the endosomal target membrane and acidification of the endosomal compartment to catalyze conformational changes within the E1–E2 heterodimer of viral envelope proteins. The viral fusion process induced host cell apoptosis, which is linked to SINV-dependent activation of ASMase, later followed by NSMase and subsequent ceramide accumulation and was abolished on overexpression of ASMase in tissue culture (Jan et al. 2000). In spite of the importance for SMases in viral entry and induction of apoptosis, ASMase-deficient animals were more susceptible to SINV-induced encephalitis which was associated with enhancement of viral replication and spread, as well as frequency of apoptotic neurons in this compartment (Ng and Griffin 2006). In line with these observations, ASMase deficiency in human fibroblast cell lines enhanced SINV replication, cell loss, and infectivity of released virions, which revealed a highly condensed structure as compared to those released from wild-type ASMase-expressing cells (Ng et al. 2008a). Thus, though SMase activation positively affects SINV entry, it appears to interfere with intracellular replication and assembly of infectious particles explaining the moderate course of SINV encephalitis on ablation of the enzyme. It is, however, still unclear why the absence of SMase would enhance neuronal apoptosis and what role the NSMase, also activated upon SINV infection in tissue culture, might have in SINV pathogenesis.

The ability to modulate host cell apoptosis may directly influence the type of infection established with a given virus. Thus, the highly cytopathogenic bovine viral diarrhea virus (BVDV), a close relative of hepatitis C virus (HCV), encodes for a nonstructural protein (NS3), which binds to and inhibits SphK1 (while the HCV orthologue does not target this enzyme). This was found to be important for efficient viral replication yet also cytopathogenicity and induction of apoptosis (Yamane et al. 2009). Thus, BVDV obviously does not care for but rather counteracts host cell integrity, which may directly be associated with its particular mode of replication and spread. In contrast, respiratory syncytial virus (RSV) apparently actively struggles to retain viability of its host lung epithelial cells by infection-mediated activation of neutral ceramidase and sphingosine kinase and, due to S1P generation, Akt and ERK within minutes. Underlying mechanisms are as yet unknown. However, given the time frame, they may relate to receptor proximal signaling or the activity of components of the virion rather than viral replication (Monick et al. 2004).

Host cell survival is particularly important in persistent infections, and a plethora of mechanisms counteracting host cell apoptosis has been described. It is as yet largely unknown as to whether this also includes modulation of the sphingolipid metabolism. Whether induction of de novo sphingolipid synthesis or accumulation of sphingosine 1 phosphate (S1P) as a consequence of sphingosine kinase activation by human cytomegalovirus (HCMV) acts to support viral replication or to promote host cell survival and establishment of persistency has not been resolved (Machesky et al. 2008). A general role for S1P in persistent viral infections needs to be established, though FTY720 or other S1P receptor agonists did not restore exhausted T-cell responses in mice chronically infected with lymphocytic choriomeningitis virus (LCMV) (Walsh et al. 2010). Sphingosine analogs were recently found to have a protective role in mice experimentally infected with pathogenic influenza virus by preventing cytokine storm induced immunopathology (Walsh et al. 2011). Whether the enhancing effect of S1P on influenza virus cytopathogenicity and virus-induced apoptosis seen in vitro also contributes to viral pathogenicity in vivo, e.g., by rapid depletion of infected cells, needs to be established (Seo et al. 2010).

3.2 Modulation of Antiviral Immune Responses

As detailed elsewhere in this chapter, SMase deficiency per se did not affect steady-state parameters of the immune system in mice, such as the architecture of secondary immune tissues and the composition of the peripheral blood compartment. However, as established for ASMase deficiency, for a number of bacteria, the efficiency of pathogen clearance is affected (Gulbins et al. 2004; Gulbins and Kolesnick 2003; Teichgraber et al. 2008).

As detailed above, ASMase deficiency increased the sensitivity to fatal SINV-induced experimental encephalitis, yet this was predominantly analyzed with regard to viral and histological parameters so that the contribution of potential SMAse relating alterations in immune control remains unknown as yet (Ng et al. 2008a; Ng and Griffin 2006). More recently, a role of ASMase in the CD8 T-cell-mediated control of the non-cytolytic LCMV in mice has been directly documented (Herz et al. 2009). Thus, viral clearance from the spleen was significantly delayed, while the footpad swelling response was reduced. ASMase deficiency did affect perforin-mediated cytotoxicity, but not that depending on CD95 ligation, and neither frequencies of virus-specific CD8+ T cells nor their content of cytolytic granule was compromised. As revealed by single-cell analyses, ASMase-deficient LCMV-specific CD8+ T cells were perfectly able to conjugate to their targets and polarize the microtubule-organizing centre MTOC and the cytolytic granule towards the interface where they fused equally efficient in a SNARE-dependent manner (Pattu et al. 2011; Qu et al. 2011) with the membrane as seen for controls. Though not affecting TCR-stimulated Ca^{2+} mobilization as required for degranulation and target cell lysis of LCMV-specific CD8+ T cells (Maul-Pavicic et al. 2011), ASMase deficiency efficiently abrogated shrinking of granules as required for expulsion of effector molecules into the synaptic cleft (Herz et al. 2009). These data clearly revealed that ASMase affects a very late step in CD8+ T-cell effector functions and, at the same time, imply that primary activation and differentiation of these cells may not substantially be affected.

These studies were performed in a fully ASMase-deficient background (also including CD4+ and antigen-presenting cells) and thus are not informative with regard to cell type-specific SMase requirement in T-cell activation (e.g., by wild-type APCs), nor do they rule out an inhibitory effect of inappropriate SMase activation on T-cell activation as described in tissue culture. There, SMase activation in T cells (by MV, exogenous addition of bSMase or C_{16}, or ligation of CD95 or TNF-R) caused loss of actin-based protrusions and interfered with Ca^{2+} mobilization stimulated by phorbol ester or TCR signaling (Church et al. 2005; Gassert et al. 2009; Lepple-Wienhues et al. 1999). The latter observations were linked to lack of CRAC channel activation and thus inhibition of store-operated Ca^{2+} release (Lepple-Wienhues et al. 1999), and though not addressed in this study, activation and/or subcellular redistribution of the ER Ca^{2+} sensor STIM1 towards the Orai1 subunit of the channel relocating to the APC/T-cell interface might be targeted (Lioudyno et al. 2008). Interestingly, co-ligation of CD95 (which is an efficient SMase activator) or anti-CD3/CD28 stimulation of naive T cells inhibited TCR signaling at the level of ZAP-70, PLC-γ, and PKCθ redistribution as well as Ca^{2+} mobilization, nuclear translocation of NFAT, AP1, and NF-kB, and activation of caspase-3, caspase-8, and caspase-10 (Strauss et al. 2009). This study did, however, not attribute these alterations to SMase activation, and it might be interesting to dissect this, especially because CD95L was found to be highly expressed on HIV-infected antigen-presenting cells (Strauss et al. 2009).

In addition to altering TCR signaling, SMase activation also abrogated T-cell polarization, redistribution of chemokine receptors, and T-cell chemotaxis in vitro (Gassert et al. 2009). If this also applies in vivo, T cell homing to secondary tissues or extravasation through endothelial barriers might also be targeted, which, however, still needs to be investigated to be fully informative using ubiquitous and cell type-specific ablation/overexpression systems. On the other hand, SMase activation was found to be an important component of CD40 clustering and signaling in B cells in vitro. Though not analyzed in DCs as yet, ablation of SMase in antigen-presenting cells would be expected to impact on ensuing humoral and cellular responses in vivo (Grassme et al. 2002a, b).

4 Outlook and Perspectives

Sorting of lipids, especially that of sphingolipids, which are major components of the plasma membrane, continues to emerge as key to the understanding of sorting of membrane-associated proteins and biologically relevant macromolecular complexes as well as membrane dynamics in endo- and exocytosis. Sphingolipids may, however, also directly act as effectors, e.g., by acting as soluble mediators (S1P), or as receptors. Biological responses in that context have, however, to be explicitly appreciated at the level of particular host cells. Interestingly, the well-known adjuvant properties of compounds such as alum or uric acid crystals have recently been attributed to their ability to cause lipid sorting on DCs (Flach et al. 2011; Ng et al. 2008b) or, as detailed above, lateral or vertical recruitment or segregation of receptors. Alterations of the membrane sphingolipid composition may only be seen under conditions where viral receptors are not artificially abundant as often the case when standard tissue culture or overexpression systems are used.

In addition to these basic considerations, there are numerous issues that need to be further addressed in order to understand the role of sphingolipids in viral infections. Obviously, positional effects need to be analyzed in more detail because, for instance, overall modulations of sphingolipids (e.g., by bSMase or exogenous addition of ceramides) may differ from those seen in a specific membrane compartment, e.g., proximal to SMase-inducing receptors or viral entry sites. In this context, the impact of SMase activation by receptors not supporting viral entry per se (e.g., TNF-R) would be certainly interesting, because these would expectedly be engaged in inflammatory conditions and as some of them also act as membrane-associated proteins at a local basis.

Also just touched upon as yet, the understanding of regulation of viral fusion by the local sphingolipid environment, which may affect not only the availability and clustering of fusion active viral protein complexes but also biophysical parameters such as their ease of insertion and conformational stability, will be very important. Moreover, as well established, virus-membrane and cell–cell fusion are known to mechanistically differ to a certain extent, and available data as yet were mainly confined to entry processes rather than cell-to-cell spread which, for many viruses, would be the favorite way of propagation in vivo in an infected organism.

As outlined above, the understanding of the role of sphingolipids at the level of viral pathogenicity is still in its infancy and might, depending on the virus and type of disease (for instance, tissue, acute versus chronic, virally or immune mediated), vary considerably, ranging from control of viral entry and replication to balancing immunity. In that context, findings obtained on the impact of SMase activation on specific immune cells need to be elaborated as for instance the involvement of SMase activation in DC-SIGN signaling and subsequent shaping of immune responses. Similarly, the regulation of ceramide metabolites such as SP1 in immune cells in viral interactions and their role in immune control or immunopathology has barely been addressed as yet.

Acknowledgment The authors thank the Deutsche Forschungsgemeinschaft for financial support of their laboratory research.

References

Alfsen A, Bomsel M (2002) HIV-1 gp41 envelope residues 650-685 exposed on native virus act as a lectin to bind epithelial cell galactosyl ceramide. J Biol Chem 277:25649–25659

Alvisi G, Madan V, Bartenschlager R (2011) Hepatitis C virus and host cell lipids: an intimate connection. RNA Biol 8:258–269

Amako Y, Syed GH, Siddiqui A (2011) Protein kinase D negatively regulates hepatitis C virus secretion through phosphorylation of oxysterol-binding protein and ceramide transfer protein. J Biol Chem 286:11265–11274

Avota E, Avots A, Niewiesk S, Kane LP, Bommhardt U, ter Meulen V, Schneider-Schaulies S (2001) Disruption of Akt kinase activation is important for immunosuppression induced by measles virus. Nat Med 7:725–731

Avota E, Gulbins E, Schneider-Schaulies S (2011) DC-SIGN mediated sphingomyelinase-activation and ceramide generation is essential for enhancement of viral uptake in dendritic cells. PLoS Pathog 7:e1001290

Blanchet F, Moris A, Mitchell JP, Piguet V (2011) A look at HIV journey: from dendritic cells to infection spread in CD4+ T cells. Curr Opin HIV AIDS 6:391–397

Bollinger CR, Teichgraber V, Gulbins E (2005) Ceramide-enriched membrane domains. Biochim Biophys Acta 1746:284–294

Bonsch C, Zuercher C, Lieby P, Kempf C, Ros C (2010) The globoside receptor triggers structural changes in the B19 virus capsid that facilitate virus internalization. J Virol 84:11737–11746

Brugger B, Glass B, Haberkant P, Leibrecht I, Wieland FT, Krausslich HG (2006) The HIV lipidome: a raft with an unusual composition. Proc Natl Acad Sci USA 103:2641–2646

Burckhardt CJ, Greber UF (2009) Virus movements on the plasma membrane support infection and transmission between cells. PLoS Pathog 5:e1000621

Caffrey M (2011) HIV envelope: challenges and opportunities for development of entry inhibitors. Trends Microbiol 19:191–197

Carette JE, Raaben M, Wong AC, Herbert AS, Obernosterer G, Mulherkar N, Kuehne AI, Kranzusch PJ, Griffin AM, Ruthel G, Dal Cin P, Dye JM, Whelan SP, Chandran K, Brummelkamp TR (2011) Ebola virus entry requires the cholesterol transporter Niemann-Pick C1. Nature 477:340–343

Church LD, Hessler G, Goodall JE, Rider DA, Workman CJ, Vignali DA, Bacon PA, Gulbins E, Young SP (2005) TNFR1-induced sphingomyelinase activation modulates TCR signaling by impairing store-operated Ca^{2+} influx. J Leukoc Biol 78:266–278

Cook DG, Fantini J, Spitalnik SL, Gonzalez-Scarano F (1994) Binding of human immunodeficiency virus type I (HIV-1) gp120 to galactosylceramide (GalCer): relationship to the V3 loop. Virology 201:206–214

Cosset FL, Lavillette D (2011) Cell entry of enveloped viruses. Adv Genet 73:121–183

Cote M, Misasi J, Ren T, Bruchez A, Lee K, Filone CM, Hensley L, Li Q, Ory D, Chandran K, Cunningham J (2011) Small molecule inhibitors reveal Niemann-Pick C1 is essential for Ebola virus infection. Nature 477:344–348

Dorosko SM, Connor RI (2010) Primary human mammary epithelial cells endocytose HIV-1 and facilitate viral infection of CD4+ T lymphocytes. J Virol 84:10533–10542

Dreschers S, Franz P, Dumitru C, Wilker B, Jahnke K, Gulbins E (2007) Infections with human rhinovirus induce the formation of distinct functional membrane domains. Cell Physiol Biochem 20:241–254

Dumitru CA, Dreschers S, Gulbins E (2006) Rhinoviral infections activate p38MAP-kinases via membrane rafts and RhoA. Cell Physiol Biochem 17:159–166

Ewers H, Smith AE, Sbalzarini IF, Lilie H, Koumoutsakos P, Helenius A (2005) Single-particle tracking of murine polyoma virus-like particles on live cells and artificial membranes. Proc Natl Acad Sci USA 102:15110–15115

Finnegan CM, Blumenthal R (2006) Fenretinide inhibits HIV infection by promoting viral endocytosis. Antiviral Res 69:116–123

Finnegan CM, Rawat SS, Puri A, Wang JM, Ruscetti FW, Blumenthal R (2004) Ceramide, a target for antiretroviral therapy. Proc Natl Acad Sci USA 101:15452–15457

Finnegan CM, Rawat SS, Cho EH, Guiffre DL, Lockett S, Merrill AH Jr, Blumenthal R (2007) Sphingomyelinase restricts the lateral diffusion of CD4 and inhibits human immunodeficiency virus fusion. J Virol 81:5294–5304

Fischl W, Bartenschlager R (2011) Exploitation of cellular pathways by Dengue virus. Curr Opin Microbiol 14(4):470–475

Flach TL, Ng G, Hari A, Desrosiers MD, Zhang P, Ward SM, Seamone ME, Vilaysane A, Mucsi AD, Fong Y, Prenner E, Ling CC, Tschopp J, Muruve DA, Amrein MW, Shi Y (2011) Alum interaction with dendritic cell membrane lipids is essential for its adjuvanticity. Nat Med 17:479–487

Gassert E, Avota E, Harms H, Krohne G, Gulbins E, Schneider-Schaulies S (2009) Induction of membrane ceramides: a novel strategy to interfere with T lymphocyte cytoskeletal reorganisation in viral immunosuppression. PLoS Pathog 5:e1000623

Geijtenbeek TB, Gringhuis SI (2009) Signalling through C-type lectin receptors: shaping immune responses. Nat Rev Immunol 9:465–479

Geijtenbeek TB, van Kooyk Y (2003) DC-SIGN: a novel HIV receptor on DCs that mediates HIV-1 transmission. Curr Top Microbiol Immunol 276:31–54

Goni FM, Alonso A (2002) Sphingomyelinases: enzymology and membrane activity. FEBS Lett 531:38–46

Grassme H, Bock J, Kun J, Gulbins E (2002a) Clustering of CD40 ligand is required to form a functional contact with CD40. J Biol Chem 277:30289–30299
Grassme H, Jendrossek V, Bock J, Riehle A, Gulbins E (2002b) Ceramide-rich membrane rafts mediate CD40 clustering. J Immunol 168:298–307
Grassme H, Riehle A, Wilker B, Gulbins E (2005) Rhinoviruses infect human epithelial cells via ceramide-enriched membrane platforms. J Biol Chem 280:26256–26262
Grassme H, Riethmuller J, Gulbins E (2007) Biological aspects of ceramide-enriched membrane domains. Prog Lipid Res 46:161–170
Gringhuis SI, den Dunnen J, Litjens M, van der Vlist M, Geijtenbeek TB (2009) Carbohydrate-specific signaling through the DC-SIGN signalosome tailors immunity to Mycobacterium tuberculosis, HIV-1 and Helicobacter pylori. Nat Immunol 10:1081–1088
Gulbins E, Grassme H (2002) Ceramide and cell death receptor clustering. Biochim Biophys Acta 1585:139–145
Gulbins E, Kolesnick R (2003) Raft ceramide in molecular medicine. Oncogene 22:7070–7077
Gulbins E, Dreschers S, Wilker B, Grassme H (2004) Ceramide, membrane rafts and infections. J Mol Med 82:357–363
Hammache D, Pieroni G, Yahi N, Delezay O, Koch N, Lafont H, Tamalet C, Fantini J (1998a) Specific interaction of HIV-1 and HIV-2 surface envelope glycoproteins with monolayers of galactosylceramide and ganglioside GM3. J Biol Chem 273:7967–7971
Hammache D, Yahi N, Pieroni G, Ariasi F, Tamalet C, Fantini J (1998b) Sequential interaction of CD4 and HIV-1 gp120 with a reconstituted membrane patch of ganglioside GM3: implications for the role of glycolipids as potential HIV-1 fusion cofactors. Biochem Biophys Res Commun 246:117–122
Harrison AL, Olsson ML, Jones RB, Ramkumar S, Sakac D, Binnington B, Henry S, Lingwood CA, Branch DR (2010) A synthetic globotriaosylceramide analogue inhibits HIV-1 infection in vitro by two mechanisms. Glycoconj J 27:515–524
Hatch SC, Archer J, Gummuluru S (2009) Glycosphingolipid composition of human immunodeficiency virus type 1 (HIV-1) particles is a crucial determinant for dendritic cell-mediated HIV-1 trans-infection. J Virol 83:3496–3506
Herz J, Pardo J, Kashkar H, Schramm M, Kuzmenkina E, Bos E, Wiegmann K, Wallich R, Peters PJ, Herzig S, Schmelzer E, Kronke M, Simon MM, Utermohlen O (2009) Acid sphingomyelinase is a key regulator of cytotoxic granule secretion by primary T lymphocytes. Nat Immunol 10:761–768
Holopainen JM, Angelova MI, Kinnunen PK (2000) Vectorial budding of vesicles by asymmetrical enzymatic formation of ceramide in giant liposomes. Biophys J 78:830–838
Hug P, Lin HM, Korte T, Xiao X, Dimitrov DS, Wang JM, Puri A, Blumenthal R (2000) Glycosphingolipids promote entry of a broad range of human immunodeficiency virus type 1 isolates into cell lines expressing CD4, CXCR4, and/or CCR5. J Virol 74:6377–6385
Izquierdo-Useros N, Naranjo-Gomez M, Archer J, Hatch SC, Erkizia I, Blanco J, Borras FE, Puertas MC, Connor JH, Fernandez-Figueras MT, Moore L, Clotet B, Gummuluru S, Martinez-Picado J (2009) Capture and transfer of HIV-1 particles by mature dendritic cells converges with the exosome-dissemination pathway. Blood 113:2732–2741
Izquierdo-Useros N, Naranjo-Gomez M, Erkizia I, Puertas MC, Borras FE, Blanco J, Martinez-Picado J (2010) HIV and mature dendritic cells: Trojan exosomes riding the Trojan horse? PLoS Pathog 6:e1000740
Jan JT, Chatterjee S, Griffin DE (2000) Sindbis virus entry into cells triggers apoptosis by activating sphingomyelinase, leading to the release of ceramide. J Virol 74:6425–6432
Khan F, Proulx F, Lingwood CA (2009) Detergent-resistant globotriaosyl ceramide may define verotoxin/glomeruli-restricted hemolytic uremic syndrome pathology. Kidney Int 75:1209–1216
Lehmann MJ, Sherer NM, Marks CB, Pypaert M, Mothes W (2005) Actin- and myosin-driven movement of viruses along filopodia precedes their entry into cells. J Cell Biol 170:317–325
Lepple-Wienhues A, Belka C, Laun T, Jekle A, Walter B, Wieland U, Welz M, Heil L, Kun J, Busch G, Weller M, Bamberg M, Gulbins E, Lang F (1999) Stimulation of CD95 (Fas) blocks T lymphocyte calcium channels through sphingomyelinase and sphingolipids. Proc Natl Acad Sci USA 96:13795–13800

Lingwood CA, Binnington B, Manis A, Branch DR (2010a) Globotriaosyl ceramide receptor function—where membrane structure and pathology intersect. FEBS Lett 584:1879–1886

Lingwood CA, Manis A, Mahfoud R, Khan F, Binnington B, Mylvaganam M (2010b) New aspects of the regulation of glycosphingolipid receptor function. Chem Phys Lipids 163:27–35

Lingwood D, Binnington B, Rog T, Vattulainen I, Grzybek M, Coskun U, Lingwood CA, Simons K (2011) Cholesterol modulates glycolipid conformation and receptor activity. Nat Chem Biol 7(5):260–262

Lioudyno MI, Kozak JA, Penna A, Safrina O, Zhang SL, Sen D, Roos J, Stauderman KA, Cahalan MD (2008) Orai1 and STIM1 move to the immunological synapse and are up-regulated during T cell activation. Proc Natl Acad Sci USA 105:2011–2016

Liu ST, Sharon-Friling R, Ivanova P, Milne SB, Myers DS, Rabinowitz JD, Brown HA, Shenk T (2011) Synaptic vesicle-like lipidome of human cytomegalovirus virions reveals a role for SNARE machinery in virion egress. Proc Natl Acad Sci USA 108:12869–12874

Lund N, Branch DR, Sakac D, Lingwood CA, Siatskas C, Robinson CJ, Brady RO, Medin JA (2005) Lack of susceptibility of cells from patients with Fabry disease to productive infection with R5 human immunodeficiency virus. AIDS 19:1543–1546

Lund N, Branch DR, Mylvaganam M, Chark D, Ma XZ, Sakac D, Binnington B, Fantini J, Puri A, Blumenthal R, Lingwood CA (2006) A novel soluble mimic of the glycolipid, globotriaosyl ceramide inhibits HIV infection. AIDS 20:333–343

Lund N, Olsson ML, Ramkumar S, Sakac D, Yahalom V, Levene C, Hellberg A, Ma XZ, Binnington B, Jung D, Lingwood CA, Branch DR (2009) The human P(k) histo-blood group antigen provides protection against HIV-1 infection. Blood 113:4980–4991

Machesky NJ, Zhang G, Raghavan B, Zimmerman P, Kelly SL, Merrill AH Jr, Waldman WJ, Van Brocklyn JR, Trgovcich J (2008) Human cytomegalovirus regulates bioactive sphingolipids. J Biol Chem 283:26148–26160

Magerus-Chatinet A, Yu H, Garcia S, Ducloux E, Terris B, Bomsel M (2007) Galactosyl ceramide expressed on dendritic cells can mediate HIV-1 transfer from monocyte derived dendritic cells to autologous T cells. Virology 362:67–74

Mahfoud R, Manis A, Lingwood CA (2009) Fatty acid-dependent globotriaosyl ceramide receptor function in detergent resistant model membranes. J Lipid Res 50:1744–1755

Marechal V, Clavel F, Heard JM, Schwartz O (1998) Cytosolic Gag p24 as an index of productive entry of human immunodeficiency virus type 1. J Virol 72:2208–2212

Maul-Pavicic A, Chiang SC, Rensing-Ehl A, Jessen B, Fauriat C, Wood SM, Sjoqvist S, Hufnagel M, Schulze I, Bass T, Schamel WW, Fuchs S, Pircher H, McCarl CA, Mikoshiba K, Schwarz K, Feske S, Bryceson YT, Ehl S (2011) ORAI1-mediated calcium influx is required for human cytotoxic lymphocyte degranulation and target cell lysis. Proc Natl Acad Sci USA 108:3324–3329

Mercer J, Helenius A (2008) Vaccinia virus uses macropinocytosis and apoptotic mimicry to enter host cells. Science 320:531–535

Mercer J, Helenius A (2009) Virus entry by macropinocytosis. Nat Cell Biol 11:510–520

Merz A, Long G, Hiet MS, Brugger B, Chlanda P, Andre P, Wieland F, Krijnse-Locker J, Bartenschlager R (2011) Biochemical and morphological properties of hepatitis C virus particles and determination of their lipidome. J Biol Chem 286:3018–3032

Miller EH, Obernosterer G, Raaben M, Herbert AS, Deffieu MS, Krishnan A, Ndungo E, Sandesara RG, Carette JE, Kuehne AI, Ruthel G, Pfeffer SR, Dye JM, Whelan SP, Brummelkamp TR, Chandran K (2012a) Ebola virus entry requires the host-programmed recognition of an intracellular receptor. EMBO J 31:1947–1960

Miller ME, Adhikary S, Kolokoltsov AA, Davey RA (2012b) Ebola virus requires acid sphingomyelinase activity and plasma membrane sphingomyelin for infection. J Virol 86(14): 7473–7483

Monick MM, Cameron K, Powers LS, Butler NS, McCoy D, Mallampalli RK, Hunninghake GW (2004) Sphingosine kinase mediates activation of extracellular signal-related kinase and Akt by respiratory syncytial virus. Am J Respir Cell Mol Biol 30:844–852

Mothes W, Sherer NM, Jin J, Zhong P (2010) Virus cell-to-cell transmission. J Virol 84: 8360–8368

Muller N, Avota E, Schneider-Schaulies J, Harms H, Krohne G, Schneider-Schaulies S (2006) Measles virus contact with T cells impedes cytoskeletal remodeling associated with spreading, polarization, and CD3 clustering. Traffic 7:849–858

Ng CG, Griffin DE (2006) Acid sphingomyelinase deficiency increases susceptibility to fatal alphavirus encephalomyelitis. J Virol 80:10989–10999

Ng CG, Coppens I, Govindarajan D, Pisciotta J, Shulaev V, Griffin DE (2008a) Effect of host cell lipid metabolism on alphavirus replication, virion morphogenesis, and infectivity. Proc Natl Acad Sci USA 105:16326–16331

Ng G, Sharma K, Ward SM, Desrosiers MD, Stephens LA, Schoel WM, Li T, Lowell CA, Ling CC, Amrein MW, Shi Y (2008b) Receptor-independent, direct membrane binding leads to cell-surface lipid sorting and Syk kinase activation in dendritic cells. Immunity 29:807–818

Pattu V, Qu B, Marshall M, Becherer U, Junker C, Matti U, Schwarz EC, Krause E, Hoth M, Rettig J (2011) Syntaxin7 is required for lytic granule release from cytotoxic T lymphocytes. Traffic 12:890–901

Puri A, Hug P, Jernigan K, Barchi J, Kim HY, Hamilton J, Wiels J, Murray GJ, Brady RO, Blumenthal R (1998) The neutral glycosphingolipid globotriaosylceramide promotes fusion mediated by a CD4-dependent CXCR4-utilizing HIV type 1 envelope glycoprotein. Proc Natl Acad Sci USA 95:14435–14440

Puri A, Rawat SS, Lin HM, Finnegan CM, Mikovits J, Ruscetti FW, Blumenthal R (2004) An inhibitor of glycosphingolipid metabolism blocks HIV-1 infection of primary T-cells. AIDS 18:849–858

Qu B, Pattu V, Junker C, Schwarz EC, Bhat SS, Kummerow C, Marshall M, Matti U, Neumann F, Pfreundschuh M, Becherer U, Rieger H, Rettig J, Hoth M (2011) Docking of lytic granules at the immunological synapse in human CTL requires Vti1b-dependent pairing with CD3 endosomes. J Immunol 186:6894–6904

Ramkumar S, Sakac D, Binnington B, Branch DR, Lingwood CA (2009) Induction of HIV-1 resistance: cell susceptibility to infection is an inverse function of globotriaosyl ceramide levels. Glycobiology 19:76–82

Rawat SS, Gallo SA, Eaton J, Martin TD, Ablan S, KewalRamani VN, Wang JM, Blumenthal R, Puri A (2004) Elevated expression of GM3 in receptor-bearing targets confers resistance to human immunodeficiency virus type 1 fusion. J Virol 78:7360–7368

Rawat SS, Zimmerman C, Johnson BT, Cho E, Lockett SJ, Blumenthal R, Puri A (2008) Restricted lateral mobility of plasma membrane CD4 impairs HIV-1 envelope glycoprotein mediated fusion. Mol Membr Biol 25:83–94

Schelhaas M, Ewers H, Rajamaki ML, Day PM, Schiller JT, Helenius A (2008) Human papillomavirus type 16 entry: retrograde cell surface transport along actin-rich protrusions. PLoS Pathog 4:e1000148

Seo YJ, Blake C, Alexander S, Hahm B (2010) Sphingosine 1-phosphate-metabolizing enzymes control influenza virus propagation and viral cytopathogenicity. J Virol 84:8124–8131

Sherer NM, Lehmann MJ, Jimenez-Soto LF, Horensavitz C, Pypaert M, Mothes W (2007) Retroviruses can establish filopodial bridges for efficient cell-to-cell transmission. Nat Cell Biol 9:310–315

Simons K, Gerl MJ (2010) Revitalizing membrane rafts: new tools and insights. Nat Rev Mol Cell Biol 11:688–699

Strauss G, Lindquist JA, Arhel N, Felder E, Karl S, Haas TL, Fulda S, Walczak H, Kirchhoff F, Debatin KM (2009) CD95 co-stimulation blocks activation of naive T cells by inhibiting T cell receptor signaling. J Exp Med 206:1379–1393

Tani H, Shiokawa M, Kaname Y, Kambara H, Mori Y, Abe T, Moriishi K, Matsuura Y (2010) Involvement of ceramide in the propagation of Japanese encephalitis virus. J Virol 84: 2798–2807

Teichgraber V, Ulrich M, Endlich N, Riethmuller J, Wilker B, De Oliveira-Munding CC, van Heeckeren AM, Barr ML, von Kurthy G, Schmid KW, Weller M, Tummler B, Lang F, Grassme H, Doring G, Gulbins E (2008) Ceramide accumulation mediates inflammation, cell death and infection susceptibility in cystic fibrosis. Nat Med 14:382–391

Trajkovic K, Hsu C, Chiantia S, Rajendran L, Wenzel D, Wieland F, Schwille P, Brugger B, Simons M (2008) Ceramide triggers budding of exosome vesicles into multivesicular endosomes. Science 319:1244–1247

Utermohlen O, Herz J, Schramm M, Kronke M (2008) Fusogenicity of membranes: the impact of acid sphingomyelinase on innate immune responses. Immunobiology 213:307–314

Vieira CR, Munoz-Olaya JM, Sot J, Jimenez-Baranda S, Izquierdo-Useros N, Abad JL, Apellaniz B, Delgado R, Martinez-Picado J, Alonso A, Casas J, Nieva JL, Fabrias G, Manes S, Goni FM (2010) Dihydrosphingomyelin impairs HIV-1 infection by rigidifying liquid-ordered membrane domains. Chem Biol 17:766–775

Voisset C, Lavie M, Helle F, Op De Beeck A, Bilheu A, Bertrand-Michel J, Terce F, Cocquerel L, Wychowski C, Vu-Dac N, Dubuisson J (2008) Ceramide enrichment of the plasma membrane induces CD81 internalization and inhibits hepatitis C virus entry. Cell Microbiol 10:606–617

Walsh KB, Marsolais D, Welch MJ, Rosen H, Oldstone MB (2010) Treatment with a sphingosine analog does not alter the outcome of a persistent virus infection. Virology 397:260–269

Walsh KB, Teijaro JR, Wilker PR, Jatzek A, Fremgen DM, Das SC, Watanabe T, Hatta M, Shinya K, Suresh M, Kawaoka Y, Rosen H, Oldstone MB (2011) Suppression of cytokine storm with a sphingosine analog provides protection against pathogenic influenza virus. Proc Natl Acad Sci USA 108:12018–12023

Yamane D, Zahoor MA, Mohamed YM, Azab W, Kato K, Tohya Y, Akashi H (2009) Inhibition of sphingosine kinase by bovine viral diarrhea virus NS3 is crucial for efficient viral replication and cytopathogenesis. J Biol Chem 284:13648–13659

Yu H, Alfsen A, Tudor D, Bomsel M (2008) The binding of HIV-1 gp41 membrane proximal domain to its mucosal receptor, galactosyl ceramide, is structure-dependent. Cell Calcium 43:73–82

Zeidan YH, Jenkins RW, Hannun YA (2008) Remodeling of cellular cytoskeleton by the acid sphingomyelinase/ceramide pathway. J Cell Biol 181:335–350

Zha X, Pierini LM, Leopold PL, Skiba PJ, Tabas I, Maxfield FR (1998) Sphingomyelinase treatment induces ATP-independent endocytosis. J Cell Biol 140:39–47

Zhang Y, Li X, Becker KA, Gulbins E (2008) Ceramide-enriched membrane domains-Structure and function. Biochim Biophys Acta 1788(1):178–183

Ceramide in Plasma Membrane Repair

Annette Draeger and Eduard B. Babiychuk

Contents

Abstract The perforation of the plasmalemma by pore-forming toxins causes an influx of Ca^{2+} and an efflux of cytoplasmic proteins. In order to ensure cellular survival, lesions have to be identified, plugged and removed from the membrane. The Ca^{2+}-driven fusion of lysosomes with the plasma membrane leads to hydrolysis of sphingomyelin by acid sphingomyelinase and a formation of ceramide platforms in the outer leaflet of the lipid bilayer. We propose that the negative curvature, promoted by tighter packing of lipids in the outer layer, leads to an inward vesiculation of the damaged area for its endocytotic uptake and internal degradation. In contrast, the activation of neutral sphingomyelinase triggers the production of ceramide within the inner leaflet of the lipid bilayer, thereby promoting an outward curvature, which enables the cell to shed the membrane-containing toxin pore into the extracellular space. In this process, ceramide is supported by members of the annexin protein family which act as Ca^{2+} sensors and as membrane fusion agents.

Keywords annexins • sphingomyelinases • calcium

A. Draeger (✉) • E.B. Babiychuk
Department of Cell Biology, Institute of Anatomy, University of Bern, Baltzerstr. 2, 3012 Bern, Switzerland
e-mail: draeger@ana.unibe.ch

E. Gulbins and I. Petrache (eds.), *Sphingolipids in Disease*,
Handbook of Experimental Pharmacology 216,
DOI 10.1007/978-3-7091-1511-4_17,

The well-being of a cell is critically dependent on the integrity of its plasma membrane, which maintains the essential differences between the cytosol and the extracellular environment. However, episodes of plasmalemmal injury are a normal occurrence in the life of most cells and often lead to their premature, necrotic death. Physical disruption of the plasma membrane is common in cells that operate under conditions of mechanical stress. The permeability barrier can also be breached by chemical means: pathogens gain access to host cells by secreting pore-forming toxins, and the host's own immune system employs pore-forming proteins to eliminate both pathogens and the pathogen-invaded cells.

Whereas mechanical lesions leave irregularly shaped, gaping holes, injuries created by pore-forming toxins or complement are of defined structure and size. Moreover, dependent on the extent of the injury, repaired cells either return to their fully functional state or, in case of massive injury accompanied by irreversible changes in cellular homeostasis, are eliminated by apoptosis or necrosis.

Given the variety in the nature and intensity of cellular injury, it is not surprising that the mechanisms of membrane resealing differ. Hence, two distinct repair pathways, which excise patches of the damaged lipid bilayer from the plasma membrane and move them in opposing directions—either into or out of the cell—can be distinguished.

Extensive changes in the structure of the plasma membrane, which occur upon the formation of ceramide, are a prerequisite for the repair of plasmalemmal injuries. Thus, ceramide, which is produced from sphingomyelin in the outer lipid bilayer of the plasma membrane, appears to be of critical importance in the management of a correct internalisation programme during plasmalemmal repair; in addition, it is highly likely that ceramide, produced in the inner lipid bilayer of the plasma membrane, plays an important role in the survival processes, which lead to shedding of the toxin-induced pores in the extracellular milieu. It appears that the fine balance between ceramide produced either within the outer or inner leaflet of the plasma membrane may be critical in defining the fate of an injured cell.

1 Plasmalemmal Repair

Nucleated cells survive the disruption of their plasma membrane by a process of resealing (Chambers 1917; McNeil and Steinhardt 2003). If resealing fails, an abrupt elevation in the intracellular concentration of Ca^{2+} ($[Ca^{2+}]_i$) activates intracellular proteases and phospholipases, thereby causing generalised damage to the phospholipid bilayer and the cytoskeleton. It is thus hardly surprising that an elevation in $[Ca^{2+}]_i$ is sensed as an "immediate danger" signal by an injured cell (Morgan et al. 1986; Babiychuk et al. 2009). A localised, pore-induced $[Ca^{2+}]_i$ elevation designates the lesion, indicating the site where the patch has to be applied and sets the repair mechanisms in motion (McNeil and Steinhardt 2003). In addition, Ca^{2+} activates the transcriptional machinery, releases inflammatory molecules and stimulates the recovery process (Morgan 1989; Lewis 2003; McNeil and Kirchhausen 2005;

Pilzer et al. 2005). Yet at the same time, an unrestrained influx of Ca^{2+} is responsible for the death of injured cells (Babiychuk et al. 2008). Thus, the chance for a cell to recover from an injury critically depends on its ability to control the influx of Ca^{2+}. Repair mechanisms operate in the range of 5–20 μM $[Ca^{2+}]_i$ and lead to the recovery of the injured cell (Babiychuk et al. 2009, 2010; Potez et al. 2011); however, if $[Ca^{2+}]_i$ exceeds 20 μM, persistent changes in cellular homeostasis cause irreversible damage and death even in resealed cells (Babiychuk et al. 2008, 2009; Potez et al. 2011).

The process of plasmalemmal repair can be roughly divided into three stages: initially, the lesion must be sensed by the injured cell, then it must be plugged without delay, in order to prevent a massive influx of extracellular Ca^{2+} and a parallel efflux of cytoplasmic constituents, and finally the damaged region of the lipid bilayer, which carries not only the physical lesion but also protein and lipid molecules, which have been irreversibly modified in the wake of injury, must be removed from the membrane to minimise the detrimental effects of uncontrolled activation of intracellular signalling cascades. These processes are orchestrated by an intricate intracellular protein machinery (McNeil and Kirchhausen 2005; Tam et al. 2010; Draeger et al. 2011); among them are the members of the annexin family of Ca^{2+}- and membrane-binding proteins protein, playing a prominent role in the repair of both mechanical and toxin-induced lesions (Draeger et al. 2011).

2 Ceramide Formation in the Wake of Plasmalemmal Damage

Ceramide can be generated at several subcellular locations. In addition to the endoplasmic reticulum where its de novo synthesis takes place, ceramide is locally produced from plasmalemmal sphingomyelin either by neutral or acid sphingomyelinases (Hannun and Obeid 2008).

In damaged cells, the ceramide production is triggered by an injury-induced sustained rise in $[Ca^{2+}]_i$ (Babiychuk et al. 2008). It is conceivable that the hydrolysis of sphingomyelin, which is localised predominantly within the outer leaflet of the plasma membrane, is triggered by the Ca^{2+}-dependent fusion of acid sphingomyelinase-containing recycling endosomes or lysosomes with the plasma membrane. This fusion leads to the emptying of the lysosomal contents and the exposure of the outer leaflet to the sphingomyelinase (Tepper et al. 2000; Grassme et al. 2001; Perrotta et al. 2010; Tam et al. 2010).

Additionally, an outside–inside transbilayer movement of sphingomyelin takes place, due to the activation of a Ca^{2+}-induced, bidirectional lipid-scrambling activity (Tepper et al. 2000) or a spontaneous transbilayer equilibration (Lopez-Montero et al. 2005); both processes are expected to be significantly accelerated in the vicinity of the plasmalemmal lesion. Within the inner leaflet of the plasmalemmal lipid bilayer, sphingomyelin can be hydrolysed by neutral sphingomyelinases, which reside in the immediate vicinity of the plasma membrane (Walev et al. 2000). Since the inner leaflet of the plasma membrane is generally sphingomyelin-poor, it is possible that the production of ceramide by neutral sphingomyelinases is spatially restricted and confined to the relatively sparse sphingomyelin in the vicinity of a membrane lesion, compared with a more widespread action of acid sphingomyelinase, acting on the sphingomyelin-rich outer leaflet of the plasma membrane.

3 Ceramide Platform in Injured Cells

In vitro studies demonstrate that the enzymatic conversion of sphingomyelin to ceramide can lead to a dramatic alteration of membrane microdomain organisation. Promoting the segregation of membrane lipids, ceramide molecules self-associate forming highly ordered domains, which exclude glycerophospholipids and coalesce into large, ceramide-rich membrane platforms (Kolesnick et al. 2000).The mechanism for the formation of ceramide platforms was demonstrated using phosphatidylcholine/sphingomyelin unilamellar vesicles. In this system the sphingomyelinase-induced generation of ceramide resulted in the formation of ceramide patches that merged rapidly into ceramide-rich platforms (Kolesnick et al. 2000; van Blitterswijk et al. 2003).

The formation of ceramide-rich platforms was demonstrated in the plasma membrane of mammalian cells using a monoclonal anticeramide antibody. These platforms were rapidly formed after CD95 or CD40 stimulation, ionising radiation, UV light, heat, bacterial or viral infection or after treatment of cells with bacterial sphingomyelinase or natural C_{16}-ceramide (Grassme et al. 2001, 2002; Abdel Shakor et al. 2004; Bezombes et al. 2004).

Using fluorescently tagged annexin A1 as a reporter for ceramide platforms, their intracellular dynamics have been visualised in living cells (Babiychuk et al. 2008). The process commences with Ca^{2+}-dependent hydrolysis of sphingomyelin and a lateral segregation of ceramide into individual platforms within the plasma membrane. Upon formation the ceramide platforms are internalised in a unique, ATP- independent manner (Babiychuk et al. 2008, 2011). This tenet accords with the suggestion that ceramide by itself induces local changes in the structure of the membrane lipid bilayer and promotes bending and budding of membranes (Gulbins and Kolesnick 2003; van Blitterswijk et al. 2003; Grassme et al. 2007). The membrane-bending properties of ceramide have been directly demonstrated in artificial liposomes (Holopainen et al. 2000). They were shown to be instrumental in physiological events such as intraluminal budding in multivesicular endosomes (Trajkovic et al. 2008). This large-scale lipid reorganisation of the plasma membrane can lead to proliferation, differentiation and—most often—to apoptosis (Morgan et al. 1986). However, in vivo studies have demonstrated that ceramide platforms can also serve as cellular signalling hubs by reversibly bundling receptor molecules, and so permit the cell to regulate and to fine-tune its intra- and intercellular communication (Gulbins and Li 2006).

4 Ceramide in Membrane Repair: Exocytosis

A plasmalemmal injury made by mechanical forces, in general, leads to a gaping membrane wound which is kept open by the tension of the submembraneous cytoskeleton. The unrestrained Ca^{2+} influx causes an immediate contraction of the subcortical actomyosin complex, which prevents the spontaneous resealing of the exposed, "sticky" membrane lipid edges. Therefore, excess membrane material needs to be delivered to the site of injury. The potential reservoir of lipids for repair is thought to originate largely from lysosomes, which are fused into giant patches before they are integrated into the plasma membrane (Ninomiya et al. 1996; Caler et al. 2001; McNeil and Steinhardt 2003). Their activation and prompt delivery is triggered by the Ca^{2+} influx through the site of injury. However, if in immediate need of material for membrane repair, the cell can also recruit other organelles, such as secretory granules (McNeil and Steinhardt 2003). Moreover, implicated in membrane resealing is also the "enlargeosome", an exocytotic organelle containing the giant protein AHNAK/desmoyokin (Borgonovo et al. 2002; Lorusso et al. 2006). Ultimately, a lipid patch closes the lesion, enables the cell to pump out excess Ca^{2+} and to re-establish cellular homeostasis (McNeil and Steinhardt 2003).

In any case, a fusion of intracellular organelles with the damaged plasma membrane is the prerequisite for the closure of a mechanically induced lesion. Lipid metabolising enzymes such as phospholipases or sphingomyelinases support membrane fusion by generating lipid fragments of suitable structures, which are compatible with the remodelling of the membrane and the different steps of the fusion process (Janmey and Kinnunen 2006; Chernomordik and Kozlov 2008).

In addition, proteins which possess membrane-binding and/or fusogenic activity (annexins, SNAREs, synaptotagmins, ferlins) are thought to take part in plasmalemmal resealing (Draeger et al. 2011). Ceramide platforms and cholesterol/sphingomyelin enriched membrane "rafts" on the cell membrane cluster receptor molecules and recruit intracellular signalling molecules to the aggregated receptors (Gulbins and Li 2006). SNARE complex proteins are clustered in lipid rafts (Chamberlain et al. 2001; Salaun et al. 2005; Gil et al. 2006), and it is possible that changes in their conformation as a result of alteration in ceramide content could affect exocytosis (Tang et al. 2007), thus influencing the repair of mechanically induced lesions.

5 Ceramide in Membrane Repair: Microparticle Shedding

In contrast to the irregularly shaped mechanical lesions, protein-lined membrane pores generated by bacterial toxins, perforins or the membrane attack complex of blood complement are characterised by their uniform size, contour and physical stability. For example, cholesterol-dependent cytolysins, such as streptolysin O or pneumolysin secreted by *Streptococcus pyogenes* or *Streptococcus pneumoniae*, assemble into complexes of 35–50 subunits within the membrane lipid bilayer, forming pores of ~30 nm in diameter (Freche et al. 2007).

Whereas mechanical lesions can be "patched" with excess membrane, the toxin pores have to be physically removed from the plasmalemma. Outward vesiculation and microvesicle shedding seem to be the strategy of choice to eliminate perforated membrane regions of immune cells perforated by the complement membrane attack complex (Morgan et al. 1987). Streptolysin O is likewise removed from the plasma membrane of human embryonic kidney cells (Babiychuk et al. 2009; Potez et al. 2011; Keyel et al. 2011) by microvesicle shedding.

Ceramide, formed in the inner lipid bilayer due to the activity of neutral sphingomyelinase, seems to play an important role in the process of microvesicle shedding (Walev et al. 2000). It is conceivable that an accelerated transfer of sphingomyelin from the outer to the inner leaflet in the vicinity of the plasmalemmal lesion results in its hydrolysis to ceramide by the membrane resident neutral sphingomyelinase. The outward membrane curvature induced by ceramide within the inner leaflet of the membrane lipid bilayer leads to the formation of the outward membrane folding, thus initiating plasmalemmal repair. In this process, ceramide is supported by the intracellular proteins: in response to a local elevation in $[Ca^{2+}]_i$, members of the annexin family of proteins translocate to the base of the fold (McNeil et al. 2006; Babiychuk et al. 2009; Potez et al. 2011). The membrane aggregation properties and fusogenic activities of the annexins—and in particular that of annexin A1—are instrumental in isolating the lesion, elevating the membrane fold, which is subsequently shed in form of a microvesicle (Babiychuk et al. 2009; Potez et al. 2011). Since the Ca^{2+} sensitivity of annexin A1-membrane binding is increased in the presence of ceramide (Babiychuk et al. 2008), the formation of ceramide within the damaged plasma membrane increases also the selectivity of annexin A1 binding to the injured plasmalemmal regions facilitating their repair (Babiychuk et al. 2009, 2010).

Local ceramide/annexin A1 accumulations situated on the inner leaflet of the plasmalemmal lipid bilayer might thus be considered as sites of ongoing repair, which predominantly occur in cells with good chances of expelling the toxin pores, hence survive the attack.

6 Ceramide in Membrane Repair: Endocytosis

In addition to extracellular shedding, Ca^{2+}-dependent exocytosis of lysosomes followed by rapid endocytosis, which have been portrayed as cooperative events, come into effect for the elimination of toxin-induced pores (Idone et al. 2008; Tam et al. 2010; Fernandes et al. 2011).

Ceramide seems to be directly involved in these processes. Accordingly an enzyme, which is capable of hydrolyzing sphingomyelin to ceramide, lysosomal acid sphingomyelinase, is released extracellularly when cells are wounded in the presence of Ca^{2+} (Tam et al. 2010). Moreover, cells, deficient in acid sphingomyelinase, fail to reseal after wounding, whereas exogenously added recombinant human acid sphingomyelinase restores endocytosis and resealing in acid sphingomyelinase-depleted cells (Tam et al. 2010). Thus, conversion of sphingomyelin to ceramide within the outer leaflet of the plasmalemmal lipid bilayer is responsible for lesion internalisation and resealing of the plasma membrane.

Correspondingly, the formation of plasmalemmal ceramide/annexin A1 platforms, followed by their internalisation, has been observed in Streptolysin O-injured cells (Babiychuk et al. 2008). However, this study also showed that the ceramide platforms were internalised in an ATP-independent manner. This implies that the translocation of the platforms from the plasma membrane to the interior of the cell occurred independently of endocytosis and concurs with the notion that the platform internalisation is driven by ceramide itself, which is capable of inducing local changes in the structure of the plasmalemmal lipid bilayer promoting its budding and inward vesiculation (Kolesnick et al. 2000; van Blitterswijk et al. 2003; Li et al. 1999; Trajkovic et al. 2008).

Likewise large-scale, calcium-activated internalisation of the plasma membrane without involvement of classical endocytic proteins has recently been described and is supposed to be closely associated with the formation of ceramide during cellular stress (Lariccia et al. 2011; Hilgemann and Fine 2011; Fine et al. 2011). This type of inward vesiculation or "massive endocytosis", which might affect up to >50 % of a cell's surface (Lariccia et al. 2011), reflects the formation of ceramide domains that develop high inward curvature and undergo spontaneous budding (Goni and Alonso 2009).

In contrast to microvesicle shedding, which is believed to be responsible for the survival of the injured cells, ceramide-induced intracellular uptake of toxin-damaged membranes seems to be restricted to the damage management in heavily perforated cells, which are pre-destined for subsequent elimination (Babiychuk et al. 2008, 2009, 2011). One obvious reason is that once inside the cell, the toxin-bearing vesicles might fuse with and thus permeabilize intracellular organelles, such as the endoplasmic reticulum and mitochondria, which would inevitably lead to apoptosis or, in the case of lysosomal fusion, even to lytic degradation.

Such forced elimination of heavily damaged cells might be beneficial for the whole organism.

Heavily permeabilized cells, which are still capable of partially neutralising plasmalemmal lesions but unable to complete the repair process, enter a prolonged phase of $[Ca^{2+}]_i$ oscillations (Babiychuk et al. 2009). It has been shown that $[Ca^{2+}]_i$ oscillations might induce a sustained cell activation followed by proliferation (Walev et al. 2001). In contrast to a physiological stimulation, the damage-induced cell proliferation is beyond the organism's control and is potentially harmful. Thus, to prevent uncontrolled proliferation, the heavily permeabilized cells are forced into apoptosis by means of a process which involves massive internalisation of ceramide platforms (Babiychuk et al. 2008, 2011). The potential functional consequences, which might arise from such a profound reorganisation of the plasmalemma elicited by $[Ca^{2+}]_i$ overload, are presumably linked with the establishment of connections between the plasmalemma and intracellular organelles, such as mitochondria, which otherwise do not come into direct contact (Babiychuk et al. 2011).

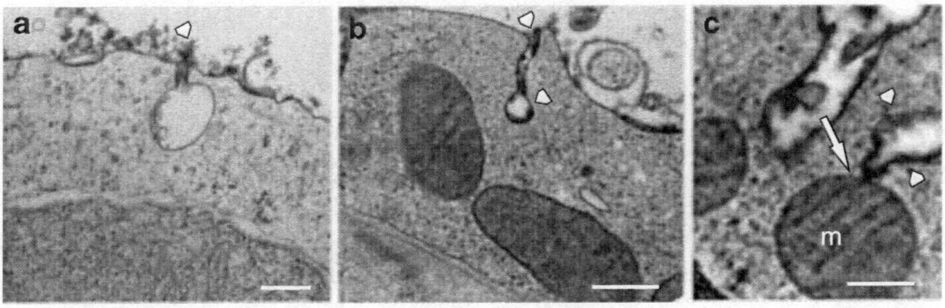

Fig. 1 Invagination of plasma membrane in toxin-damaged cells. Electron micrographs of unfixed, high-pressure frozen Jurkat cells, whose outer leaflet of the plasma membrane was pre-labelled with horseradish peroxidase-conjugated cholera toxin B (*arrowheads*) before subjecting the cells to toxin-induced damage. (**a**, **b**) Membrane invaginations extend deep into the cytoplasm. (**c**) The contact site between a plasmalemmal fold and a mitochondrion is visible (*arrow*). *m* mitochondrion. *Bars* = 0.5 μm

7 Plasmalemma-to-Mitochondria Transfer of Ceramide in Injured Cells

Ca^{2+} overload is a potent pro-apoptotic stimulus (Pedrosa Ribeiro et al. 2000; Szabadkai and Rizzuto 2004), which leads to the hydrolysis of sphingomyelin and the formation of ceramide molecules and their self-association at the plasma membrane. Intriguingly, ceramide that is generated by the activity of acid sphingomyelinase within the plasma membrane is considered to be vital for the initiation of apoptosis that is triggered by the activation of death receptors, cytotoxic drugs and environmental stress stimuli (Gulbins and Kolesnick 2003; Andrieu-Abadie and Levade 2002; Gulbins and Li 2006). The central role of acid sphingomyelinase in the induction of apoptosis has been demonstrated in acid sphingomyelinase-deficient cells and in acid sphingomyelinase-deficient mice (Kirschnek et al. 2000; Grassme et al. 2001; Paris et al. 2001).

In most cells, apoptosis is initiated by the release of mitochondrial proteins into the cytoplasm, and there is abundant evidence to suggest that an increase in permeability is caused by the appearance of pores in the outer mitochondrial membrane (Bleicken et al. 2010). Recent data suggest that generation of intracellular ceramide increases mitochondrial permeability (Siskind et al. 2002).

We have shown that invaginated ceramide platforms, initially formed within the plasma membrane of toxin-damaged cells, come into close contact with the mitochondrial outer membrane (Fig. 1). It is conceivable that ceramide is exchanged at these contact points and that this triggers the release of pro-apoptotic molecules. Inward vesiculation and vesicular transport of ceramide platforms to the mitochondria and the nuclear envelope can be the means of delivering ceramide to these intracellular membranes, which are poor in sphingomyelin and ceramide, but are implicated in ceramide-induced apoptosis (Babiychuk et al. 2011).

The modulation of mitochondrial-mediated cell death is a well-documented strategy of numerous bacterial pathogens, which balance colonisation and dissemination

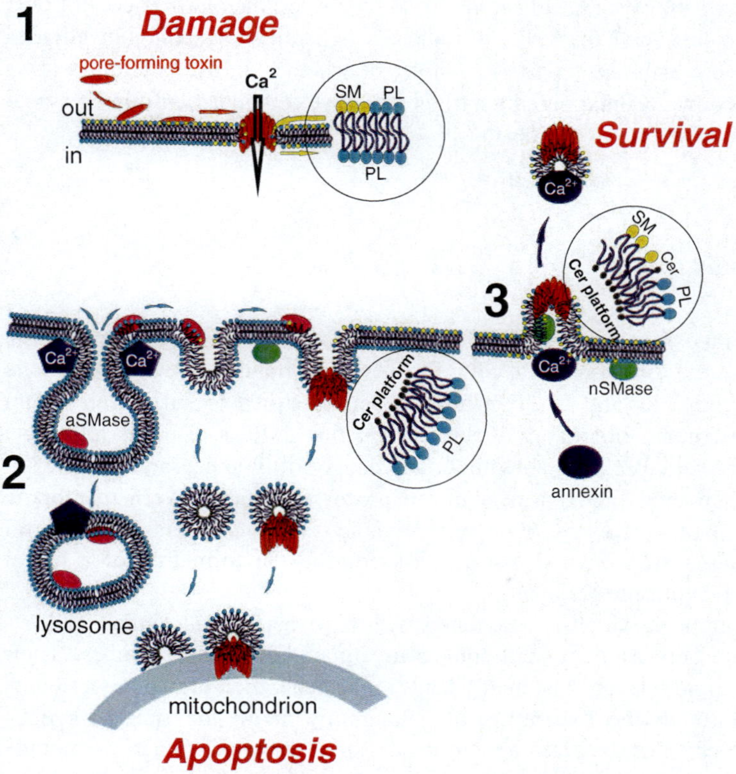

Fig. 2 (1) Perforation of the plasma membrane by bacterial pore-forming toxins. Pathogens gain access to host cells by secreting pore-forming toxins (*red ovals*, *red arrows*), which, after binding to the plasma membrane, oligomerize to form a transmembrane pore. A pore-induced elevation in $[Ca^{2+}]_i$ is sensed as an "immediate danger" signal by an injured cell. In the vicinity of the plasmalemmal lesion, an outside–inside transbilayer movement of sphingomyelin (SM, *yellow arrows*), which is localised predominantly within the outer leaflet of the plasma membrane, takes place. The inner leaflet is composed mostly of phospholipids (PL). (2) Toxin internalisation (*light blue arrows*). In heavily damaged cells, a massive $[Ca^{2+}]_i$ elevation triggers the Ca^{2+}-dependent fusion of lysosomes with the plasma membrane. This fusion leads to the emptying of the lysosomal contents and exposure of the outer leaflet to acid sphingomyelinase (aSMase, *red ovals*) (Tepper et al. 2000; Tam et al. 2010). Enzymatic conversion of sphingomyelin to ceramide leads to a dramatic alteration of membrane structure: promoting the segregation of membrane lipids, ceramide molecules coalesce into large, ceramide-rich membrane platforms (Cer platform). Negative curvature, induced by the formation of ceramide platforms in the outer leaflet of the lipid bilayer, promotes inward vesiculation of the plasma membrane resulting in the budding and internalisation of ceramide-rich vesicles. Toxin pores are removed from the plasma membrane together with internalised vesicles. Fusion of ceramide-rich vesicles with mitochondria enforces apoptosis in heavily damaged cells. (3) Microparticle shedding (*dark blue arrows*). After damage-induced transfer to the inner leaflet, sphingomyelin is hydrolysed by plasmalemma-associated neutral sphingomyelinase (nSMase, *green ovals*). The outward membrane curvature induced by ceramide within the inner leaflet of the membrane lipid bilayer leads to the formation of the outward membrane folding, thus initiating plasmalemmal repair. In this process, ceramide is supported by the intracellular proteins: in response to a local elevation in $[Ca^{2+}]_i$, proteins of the annexin family (*blue ovals*) plug membrane folds, which are subsequently shed in form of microvesicles

by manipulating pathways (Ashida et al. 2011). However, the ultimate control over these processes lies with the cell: its judicious activation of sphingomyelinases decides whether membrane repair is feasible, or whether an overwhelming toxin attack will shut down cellular defences, trigger massive ceramide platform invagination, and force the apoptotic cell death.

8 Conclusion

Plasma membrane repair is vital for the *restitutio ad integrum* of the plasma membrane and the survival of a cell. Equally important, however, is plasma membrane integrity for the fatally wounded cell, destined to suffer cell death, since only membrane integrity will ensure that the cellular disposal follows a coordinated apoptotic programme without any danger of lytic degradation.

Ceramides are vital lipid partners in the processes which govern membrane injury and cellular recovery. It is conceivable that each acid and neutral sphingomyelinases are assigned distinct tasks during the formation of a lesion and its subsequent closure (Fig. 2).

Membrane damage which is associated with an overwhelming influx of Ca^{2+} will lead to the activation of acid sphingomyelinase-containing vesicles. Their fusion with the lipid bilayer presumably leads to the generation of copious amounts of ceramide through the hydrolysis of sphingomyelin on the sphingomyelin-enriched outer leaflet of the plasma membrane. Rapid reorganisation of membrane lipids and invagination of the ceramide-containing membrane folds are followed by mitochondrial-membrane contacts and permeabilisation of the mitochondrial outer layer. It is highly likely that cells which scoop up to 50 % of their membrane surface will be doomed, however strong the efforts of valiant proteins are in their attempt to reseal the lesion.

On the other hand, circumscribed local lesions which cause only a spatially limited rise in $[Ca^{2+}]_i$ might lead to local scrambling of sphingomyelin and the generation of ceramide by membrane resident neutral sphingomyelinases. Recruitment of Ca^{2+}-dependent repair proteins and in particular of ceramide-sensitive annexin A1 might lead to the aggregation and fusion of the injured plasmalemmal sites and the containment of the lesion, which can subsequently be shed.

Should these mechanisms apply, exocytotic microvesicle shedding could be regarded as a repair process possibly leading to cellular recovery, whereas endocytotic invagination, driven by the altered steric properties of the ceramide-enriched membrane fold, would lead to apoptosis and cell death. The activation of acid and the neutral sphingomyelinases appears to draw the lottery of cellular doom.

References

Abdel Shakor AB, Kwiatkowska K, Sobota A (2004) Cell surface ceramide generation precedes and controls FcgammaRII clustering and phosphorylation in rafts. J Biol Chem 279:36778–36787

Andrieu-Abadie N, Levade T (2002) Sphingomyelin hydrolysis during apoptosis. Biochim Biophys Acta 1585:126–134

Ashida H, Mimuro H, Ogawa M, Kobayashi T, Sanada T, Kim M, Sasakawa C (2011) Host-pathogen interactions: cell death and infection: a double-edged sword for host and pathogen survival. J Cell Biol. doi:10.1083/jcb.201108081:

Babiychuk EB, Monastyrskaya K, Draeger A (2008) Fluorescent annexin A1 reveals dynamics of ceramide platforms in living cells. Traffic 9:1757–1775

Babiychuk EB, Monastyrskaya K, Potez S, Draeger A (2009) Intracellular Ca(2+) operates a switch between repair and lysis of streptolysin O-perforated cells. Cell Death Differ 16:1126–1134

Babiychuk EB, Monastyrskaya K, Potez S, Draeger A (2010) Blebbing confers resistance against cell lysis. Cell Death Differ 18:80–89

Babiychuk EB, Atanassoff AP, Monastyrskaya K, Brandenberger C, Studer D, Allemann C, Draeger A (2011) The targeting of plasmalemmal ceramide to mitochondria during apoptosis. PLoS One 6:e23706

Bezombes C, Grazide S, Garret C, Fabre C, Quillet-Mary A, Muller S, Jaffrezou JP, Laurent G (2004) Rituximab antiproliferative effect in B-lymphoma cells is associated with acid-sphingomyelinase activation in raft microdomains. Blood 104:1166–1173

Bleicken S, Classen M, Padmavathi PV, Ishikawa T, Zeth K, Steinhoff HJ, Bordignon E (2010) Molecular details of Bax activation, oligomerization, and membrane insertion. J Biol Chem 285:6636–6647

Borgonovo B, Cocucci E, Racchetti G, Podini P, Bachi A, Meldolesi J (2002) Regulated exocytosis: a novel, widely expressed system. Nat Cell Biol 4:955–962

Caler EV, Chakrabarti S, Fowler KT, Rao S, Andrews NW (2001) The Exocytosis-regulatory protein synaptotagmin VII mediates cell invasion by Trypanosoma cruzi. J Exp Med 193:1097–1104

Chamberlain LH, Burgoyne RD, Gould GW (2001) SNARE proteins are highly enriched in lipid rafts in PC12 cells: implications for the spatial control of exocytosis. Proc Natl Acad Sci USA 98:5619–5624

Chambers R (1917) Microdissection studies. I. The visible structure of cell protoplasm and death changes. Am J Physiol 43:1–12

Chernomordik LV, Kozlov MM (2008) Mechanics of membrane fusion. Nat Struct Mol Biol 15:675–683

Draeger A, Monastyrskaya K, Babiychuk EB (2011) Plasma membrane repair and cellular damage control: the annexin survival kit. Biochem Pharmacol 81:703–712

Fernandes MC, Cortez M, Flannery AR, Tam C, Mortara RA, Andrews NW (2011) Trypanosoma cruzi subverts the sphingomyelinase-mediated plasma membrane repair pathway for cell invasion. J Exp Med 208:909–921

Fine M, Llaguno MC, Lariccia V, Lin MJ, Yaradanakul A, Hilgemann DW (2011) Massive endocytosis driven by lipidic forces originating in the outer plasmalemmal monolayer: a new approach to membrane recycling and lipid domains. J Gen Physiol 137:137–154

Freche B, Reig N, van der Goot FG (2007) The role of the inflammasome in cellular responses to toxins and bacterial effectors. Semin Immunopathol 29:249–260

Gil C, Cubi R, Blasi J, Aguilera J (2006) Synaptic proteins associate with a sub-set of lipid rafts when isolated from nerve endings at physiological temperature. Biochem Biophys Res Commun 348:1334–1342

Goni FM, Alonso A (2009) Effects of ceramide and other simple sphingolipids on membrane lateral structure. Biochim Biophys Acta 1788:169–177

Grassme H, Jekle A, Riehle A, Schwarz H, Berger J, Sandhoff K, Kolesnick R, Gulbins E (2001) CD95 signaling via ceramide-rich membrane rafts. J Biol Chem 276:20589–20596
Grassme H, Jendrossek V, Bock J, Riehle A, Gulbins E (2002) Ceramide-rich membrane rafts mediate CD40 clustering. J Immunol 168:298–307
Grassme H, Riethmuller J, Gulbins E (2007) Biological aspects of ceramide-enriched membrane domains. Prog Lipid Res 46:161–170
Gulbins E, Kolesnick R (2003) Raft ceramide in molecular medicine. Oncogene 22:7070–7077
Gulbins E, Li PL (2006) Physiological and pathophysiological aspects of ceramide. Am J Physiol Regul Integr Comp Physiol 290:R11–R26
Hannun YA, Obeid LM (2008) Principles of bioactive lipid signalling: lessons from sphingolipids. Nat Rev Mol Cell Biol 9:139–150
Hilgemann DW, Fine M (2011) Mechanistic analysis of massive endocytosis in relation to functionally defined surface membrane domains. J Gen Physiol 137:155–172
Holopainen JM, Angelova MI, Kinnunen PK (2000) Vectorial budding of vesicles by asymmetrical enzymatic formation of ceramide in giant liposomes. Biophys J 78:830–838
Idone V, Tam C, Goss JW, Toomre D, Pypaert M, Andrews NW (2008) Repair of injured plasma membrane by rapid Ca^{2+}-dependent endocytosis. J Cell Biol 180:905–914
Janmey PA, Kinnunen PK (2006) Biophysical properties of lipids and dynamic membranes. Trends Cell Biol 16:538–546
Keyel PA, Heid ME, Salter RD (2011) Macrophage responses to bacterial toxins: a balance between activation and suppression. Immunol Res 50:118–123
Kirschnek S, Paris F, Weller M, Grassme H, Ferlinz K, Riehle A, Fuks Z, Kolesnick R, Gulbins E (2000) CD95-mediated apoptosis in vivo involves acid sphingomyelinase. J Biol Chem 275:27316–27323
Kolesnick RN, Goni FM, Alonso A (2000) Compartmentalization of ceramide signaling: physical foundations and biological effects. J Cell Physiol 184:285–300
Lariccia V, Fine M, Magi S, Lin MJ, Yaradanakul A, Llaguno MC, Hilgemann DW (2011) Massive calcium-activated endocytosis without involvement of classical endocytic proteins. J Gen Physiol 137:111–132
Lewis RS (2003) Calcium oscillations in T-cells: mechanisms and consequences for gene expression. Biochem Soc Trans 31:925–929
Li R, Blanchette-Mackie EJ, Ladisch S (1999) Induction of endocytic vesicles by exogenous C(6)-ceramide. J Biol Chem 274:21121–21127
Lopez-Montero I, Rodriguez N, Cribier S, Pohl A, Velez M, Devaux PF (2005) Rapid transbilayer movement of ceramides in phospholipid vesicles and in human erythrocytes. J Biol Chem 280:25811–25819
Lorusso A, Covino C, Priori G, Bachi A, Meldolesi J, Chieregatti E (2006) Annexin2 coating the surface of enlargeosomes is needed for their regulated exocytosis. EMBO J 25:5443–5456
McNeil PL, Kirchhausen T (2005) An emergency response team for membrane repair. Nat Rev Mol Cell Biol 6:499–505
McNeil PL, Steinhardt RA (2003) Plasma membrane disruption: repair, prevention, adaptation. Annu Rev Cell Dev Biol 19:697–731
McNeil AK, Rescher U, Gerke V, McNeil PL (2006) Requirement for annexin A1 in plasma membrane repair. J Biol Chem 281:35202–35207
Morgan BP (1989) Complement membrane attack on nucleated cells: resistance, recovery and non-lethal effects. Biochem J 264:1–14
Morgan BP, Luzio JP, Campbell AK (1986) Intracellular Ca2+ and cell injury: a paradoxical role of Ca2+ in complement membrane attack. Cell Calcium 7:399–411
Morgan BP, Dankert JR, Esser AF (1987) Recovery of human neutrophils from complement attack: removal of the membrane attack complex by endocytosis and exocytosis. J Immunol 138:246–253
Ninomiya Y, Kishimoto T, Miyashita Y, Kasai H (1996) Ca^{2+}-dependent exocytotic pathways in Chinese hamster ovary fibroblasts revealed by a caged-Ca^{2+} compound. J Biol Chem 271:17751–17754

Paris F, Grassme H, Cremesti A, Zager J, Fong Y, Haimovitz-Friedman A, Fuks Z, Gulbins E, Kolesnick R (2001) Natural ceramide reverses Fas resistance of acid sphingomyelinase(−/−) hepatocytes. J Biol Chem 276:8297–8305

Pedrosa Ribeiro CMP, McKay RR, Hosoki E, Bird GS, Putney JW (2000) Effects of elevated cytoplasmic calcium and protein kinase C on endoplasmic reticulum structure and function in HEK293 cells. Cell Calcium 27:175–185

Perrotta C, Bizzozero L, Cazzato D, Morlacchi S, Assi E, Simbari F, Zhang Y, Gulbins E, Bassi MT, Rosa P, Clementi E (2010) Syntaxin 4 is required for acid sphingomyelinase activity and apoptotic function. J Biol Chem 285:40240–40251

Pilzer D, Gasser O, Moskovich O, Schifferli JA, Fishelson Z (2005) Emission of membrane vesicles: roles in complement resistance, immunity and cancer. Springer Semin Immunopathol 27:375–387

Potez S, Luginbuhl M, Monastyrskaya K, Hostettler A, Draeger A, Babiychuk EB (2011) Tailored protection against plasmalemmal injury by annexins with different Ca2+ sensitivities. J Biol Chem 286:17982–17991

Salaun C, Gould GW, Chamberlain LH (2005) Lipid raft association of SNARE proteins regulates exocytosis in PC12 cells. J Biol Chem 280:19449–19453

Siskind LJ, Kolesnick RN, Colombini M (2002) Ceramide channels increase the permeability of the mitochondrial outer membrane to small proteins. J Biol Chem 277:26796–26803

Szabadkai G, Rizzuto R (2004) Participation of endoplasmic reticulum and mitochondrial calcium handling in apoptosis: more than just neighborhood? FEBS Lett 567:111–115

Tam C, Idone V, Devlin C, Fernandes MC, Flannery A, He X, Schuchman E, Tabas I, Andrews NW (2010) Exocytosis of acid sphingomyelinase by wounded cells promotes endocytosis and plasma membrane repair. J Cell Biol 189:1027–1038

Tang N, Ong WY, Zhang EM, Chen P, Yeo JF (2007) Differential effects of ceramide species on exocytosis in rat PC12 cells. Exp Brain Res 183:241–247

Tepper AD, Ruurs P, Wiedmer T, Sims PJ, Borst J, van Blitterswijk WJ (2000) Sphingomyelin hydrolysis to ceramide during the execution phase of apoptosis results from phospholipid scrambling and alters cell-surface morphology. J Cell Biol 150:155–164

Trajkovic K, Hsu C, Chiantia S, Rajendran L, Wenzel D, Wieland F, Schwille P, Brugger B, Simons M (2008) Ceramide triggers budding of exosome vesicles into multivesicular endosomes. Science 319:1244–1247

van Blitterswijk WJ, van der Luit AH, Veldman RJ, Verheij M, Borst J (2003) Ceramide: second messenger or modulator of membrane structure and dynamics? Biochem J 369:199–211

Walev I, Tappe D, Gulbins E, Bhakdi S (2000) Streptolysin O-permeabilized granulocytes shed L-selectin concomitantly with ceramide generation via neutral sphingomyelinase. J Leukoc Biol 68:865–872

Walev I, Bhakdi SC, Hofmann F, Djonder N, Valeva A, Aktories K, Bhakdi S (2001) Delivery of proteins into living cells by reversible membrane permeabilization with streptolysin-O. Proc Natl Acad Sci USA 98:3185–3190

Sphingolipids and Inflammatory Diseases of the Skin

Burkhard Kleuser and Lukasz Japtok

Contents

Abstract Mammalian skin protects our body against external assaults due to a well-organized skin barrier. The formation of the skin barrier is a complex process, in which basal keratinocytes lose their mitotic activity and differentiate to corneocytes. These corneocytes are embedded in intercellular lipid lamellae composed of ceramides, cholesterol, fatty acids, and cholesterol esters. Ceramides are the dominant lipid molecules and their reduction is connected with a transepidermal water loss and an epidermal barrier dysfunction resulting in inflammatory skin diseases. Moreover, bioactive sphingolipid metabolites like

B. Kleuser (✉) • L. Japtok
Faculty of Mathematics and Natural Science, Institute of Nutritional Science, Department of Toxicology, University of Potsdam, Arthur-Scheunert Allee 114-116, 14558 Nuthetal, Potsdam, Germany
e-mail: kleuser@uni-potsdam.de

E. Gulbins and I. Petrache (eds.), *Sphingolipids in Disease*,
Handbook of Experimental Pharmacology 216,
DOI 10.1007/978-3-7091-1511-4_18,

ceramide-1-phosphate, sphingosylphosphorylcholine, and sphingosine-1-phosphate are also involved in the biological modulation of keratinocytes and immune cells of the skin. Therefore, it is not astonishing that a dysregulation of sphingolipid metabolism has been identified in inflammatory skin diseases such as atopic dermatitis and psoriasis vulgaris. This chapter will describe not only the specific sphingolipid species and their skin functions but also the dysregulation of sphingolipid metabolism in inflammatory skin diseases.

Keywords Atopic dermatitis • Dendrtic cells • Epidermal barrier • Keratinocytes • Psoriasis vulgaris • Skin • Stratum corneum

1 Introduction

Sphingolipids are essential molecules of the mammalian epidermis. Epidermal cells, especially keratinocytes, generate and secrete enormous amounts of glucosylceramide and sphingomyelin precursors to the extracellular domains of the stratum corneum, where they are further metabolized to specific ceramide species. The arrangement of these sphingolipids to well-organized lipid lamellae is essential to protect the epidermis against an excessive water loss and moreover for the penetration of xenobiotics. Thus, the cycle of lipid precursor formation and the subsequent hydrolysis of the precursor molecules is a fine-tuned process for the arrangement of the epidermal barrier. Moreover, sphingolipids are not only structural components of the epidermal barrier but also essential molecules involved in the modulation of epidermal cells and immune cells of the skin. Prominent skin disorders, such as atopic dermatitis (AD) and psoriasis, are characterized by altered sphingolipid levels that result in a dysregulation of the epidermal barrier as well as in a modified modulation of skin cells. The role of sphingolipids in inflammatory skin diseases is the focus of this chapter.

2 Sphingolipids as Essential Structural Components of the Skin

2.1 Structure and Function of Mammalian Epidermis

Mammalian skin is the largest organ in surface area and weight that defends our body against external assaults such as physical and chemical stress, UV radiation, and invasion of immunological pathogens. Besides its protective role, skin regulates body temperature and fluid balance and participates in vitamin D_3 homeostasis (Bikle 2012; Meyer et al. 2007).

Human skin consists of two distinct layers, dermis and epidermis, which are divided by a basement membrane. The main cell type of the underlying dermis are fibroblasts which are responsible for the formation of extracellular matrix proteins like collagen and elastin and therefore contributes to the thickness of the skin.

The overlying epidermis is a stratified epithelium derived from the neuroectoderm acting as a primordial defense system due to the formation of the skin barrier (Jensen and Proksch 2009).

The formation of the skin barrier is a complex process, in which proliferating basal keratinocytes loose their mitotic activity, differentiate and migrate through four epidermal layers (stratum basale, stratum spinosum, stratum granulosum, stratum corneum) continuously adjusting to the requirements of the individual cell layer. In consequence, differentiated keratinocytes are characterized by polymerized keratin intermediate filaments and a water-resistant environment enclosed by the cross-linked cornified cell envelope with ceramide molecules bound on the cell surface (Masukawa et al. 2008). These corneocytes are implanted in intercellular lipid lamellae and attached to each other by corneodesmosomes. The lipid lamellae form a crystalline matrix composed of ceramides, cholesterol, fatty acids, cholesterol esters, and embedded structural proteins. Such a protein is filaggrin which is integrated into the lipid envelope, interacts with keratin intermediate filaments, and contributes to the skin barrier function. Additionally, filaggrin is an important protein as it aggregates the keratin fibers of the cellular cytoskeleton into bundles resulting in a collapse of corneocytes into flattened discs (Aho et al. 2012).

2.2 Ceramides Are Essential for Epidermal Barrier Function

Within the lipid lamellae, ceramides are the dominant lipid molecules (approximately 50% by weight), and they show a large molecular heterogeneity consisting of at least 11 species (Holleran et al. 2006; Masukawa et al. 2008). Moreover, some ceramides, which possess a terminal ω-hydroxy group, covalently bind to the proteins of the cornified envelope, such as involucrin, resulting in the formation of the corneocyte lipid-bound envelope (Behne et al. 2000; Uchida and Holleran 2008). Thus, it is not surprising that a reduction of ceramides in the lipid lamellae is connected with a transepidermal water loss and an epidermal barrier dysfunction resulting in a failure of protection against antigens, and moreover can result in inflammatory skin diseases (Macheleidt et al. 2002).

Ceramide synthesis is already upregulated in the suprabasal cell layer via the de novo pathway (Lampe et al. 1983). The so-formed ceramides are immediately converted to glucosylceramides and sphingomyelins (Hamanaka et al. 2002; Jensen et al. 2004). Both lipid classes are then packaged into epidermal lamellar bodies, exclusive secretory organelles of keratinocytes that are increased with the differentiation state of the epidermal cells. Glucosylceramides and sphingomyelins are carried and secreted to the interface of the stratum granulosum and the stratum corneum. At this point they are converted back to their ceramide species due to activation of the enzymes β-glucocerebrosidase and sphingomyelinase. Indeed, two sphingomyelinase isoenzymes have been detected in the skin, a lysosomal-type acid-pH optimum sphingomyelinase (aSMase) and a non-lysosomal, magnesium-dependent neutral-pH optimum sphingomyelinase (nSMase). It seems likely that both the aSMase and the nSMase are responsible for the formation of ceramides

(Uchida et al. 2000). Nevertheless, the lysosomal aSMase is the dominant enzyme that is crucial for ceramide generation and the epidermal permeability barrier (Menon et al. 1986).

It is of interest that skin ceramides possess a wide diversity of the chain length of the N-acylated fatty acids compared to other tissues (Masukawa et al. 2008). There is a variety in chain length between C16 and C30. The family of ceramide synthases (CerS) catalyzes the incorporation of the fatty acid into the dihydro-sphingosine backbone. To date, 6 family members CerS1–6 have been identified. CerS3 seems to be exclusively required for skin ceramide synthesis containing polyunsaturated fatty acids with C26 or greater chain length. Deficiency of CerS3 in mice results in complete loss of very long-chain ceramides (≥C26), lack of continuous extracellular lipid lamellae, and a non-functional cornified lipid envelope (Jennemann et al. 2012). In congruence, CerS3 exhibits skin- and testis-specific expression, which is further increased in differentiated keratinocytes (Mizutani et al. 2008). These data suggest that CerS3 is the crucial isoenzyme for the formation of the unique ceramides of the skin (Mizutani et al. 2006).

In addition, another unique ceramide species are ω-O-acyl-ceramides, which are characterized by very long-chain amide-linked fatty acids with terminal ω-hydroxyl groups that are further esterified with some other fatty acids like the essential fatty acid linoleic acid (Uchida and Holleran 2008). It is well recognized that ω-O-acyl-ceramides play a crucial role in epidermal barrier function. Thus, a diet only deficient in linoleate causes a dysfunction of the epidermal lipid barrier (Hansen and Jensen 1985). Moreover, in the stratum corneum of patients with atopic dermatitis (AD), ω-O-acyl-ceramides are significantly reduced in both lesional and non-lesional skin (Imokawa et al. 1991). Harlequin-type ichthyosis is the most severe form of congenital ichthyosis, characterized by a massive disturbance of the lipid barrier (Rajpopat et al. 2011). It is associated with a mutation in the gene for the ATP-binding cassette transporter *ABCA12* (Quazi and Molday 2011). This transporter is important for loading lipids and especially ω-O-acyl-ceramides into the lamellar bodies of the keratinocytes. This is visible in *ABCA12* knockout mice, which show decreased contents of ω-O-acyl-ceramides in the epidermal lipid barrier. These animals die shortly after birth due to a drastic water loss (Zuo et al. 2008).

Taken together, unique ceramides of the skin are essential for the formation of the well-organized epidermal permeability barrier. Thus, exogenous and endogenous factors are thought to be critical for maintaining ceramide levels in the skin. Endogenous factors regulating ceramide levels in the stratum corneum are enzymes of the de novo synthesis and moreover β-glucocerebrosidase and sphingomyelinase. For instance, Gaucher disease, which is caused by a defect in β-glucocerebrosidase, is connected with a severe epidermal barrier dysfunction (Holleran et al. 1994). Sphingomyelinase is a further endogenous factor that is critical for ceramide synthesis, and an altered activity has been discussed in inflammatory skin disorders (Jensen et al. 2004). A typical exogenous factor that can induce a disarrangement of ceramide levels in the skin is bacterial ceramidase. Indeed, bacteria that are able to secrete ceramidase have been detected in patients with atopic disorder (Ohnishi et al. 1999). Although ω-O-acyl-ceramides play a substantial role in the formation of the epidermal barrier, the detailed pathways of their biosynthesis are not well characterized.

3 Sphingolipids as Bioactive Molecules in the Skin

Although sphingolipids and in particular ceramides play an essential role in the formation of the epidermal lipid barrier and a barrier dysregulation may contribute to the development of skin diseases, recent studies indicate that sphingolipid metabolites like ceramide-1-phosphate (Cer1P), sphingosylphosphorylcholine (SPC), and sphingosine-1-phosphate (S1P) are also involved in the biological modulation of epidermal cells such as keratinocytes. Moreover, immune cells of the skin, mainly antigen-presenting cells like dendritic cells, are influenced by sphingolipid derivatives.

3.1 Ceramides and Keratinocytes

The involvement of ceramides in essential functions of keratinocytes has been well documented. In cultured keratinocytes, synthetic C6 and C2 ceramides inhibit cell proliferation and promote cell differentiation (Bektas et al. 1998; Jung et al. 1998). Cornified envelope formation, involucrin expression and transglutaminase activity are augmented in keratinocytes stimulated with cell permeable ceramides (Wakita et al. 1994). The molecular mechanisms by which ceramides enhance keratinocyte differentiation are not fully understood, but it seems likely that an enhanced gene transcription occurs in response to the sphingolipid. In addition, ceramides increase glucosylceramide synthase activity and promote expression of aSMase (Deigner et al. 2001; Liu et al. 2008). Thus, increasing ceramide levels in differentiating keratinocytes are necessary not only for lamellar membrane formation but also for corneocyte development and regulation of sphingolipid synthesis. Most interestingly, it has been indicated that ceramides also increase ABCA12 expression in the epidermal cells (Jiang et al. 2009). It has been postulated that the increase of ABCA12 expression by ceramides is mediated via peroxisome proliferator-activated receptor (PPAR) as ceramides enhance the expression of PPAR-gamma but not other PPARs. Thus, ceramides serve not only as a precursor of glucosylceramide formation but also as a bioactive molecule that will allow the keratinocyte to carry the glucosylceramide via the ABCA12 to the appropriate position within the lamellar body.

3.2 Ceramide-1-Phosphate and Keratinocytes

A major metabolite of ceramide is Cer1P, which is generated via direct phosphorylation of ceramide by the enzyme ceramide kinase (CerK). It has been supposed that Cer1P can regulate cell proliferation and apoptosis (Bornancin 2011). In keratinocytes, it has been shown that intracellularly formed Cer1P protects the epidermal cells from apoptosis (Tsuji et al. 2008). It is of interest that a link between PPARβ and Cer1P formation exists. Indeed, apoptosis is inhibited by an activation of PPARβ, and this effect is diminished in keratinocytes isolated from *CerK* null mice. Moreover, PPARβ binds to the *CerK* gene indicating that *CerK* gene expression is directly regulated by PPARβ (Tsuji et al. 2008).

3.3 Sphingosylphosphorylcholine and Keratinocytes

Although it has been reported that SPC is a strong mitogen in fibroblasts and other cell types it inhibits DNA synthesis in keratinocytes (Higuchi et al. 2001; Wakita et al. 1998). Thus, SPC releases calcium from inositol trisphosphate-sensitive or insensitive intracellular pools which is a crucial event for the differentiation process in keratinocytes. In congruence, activation of soluble and membrane-bound transglutaminases can be detected in response to SPC. This process is accompanied by an enhanced activity of cathepsin D, which is one of the crucial enzymes for either releasing soluble types from the membrane-bound transglutaminase or processing the cytosolic form leading to its activation. Moreover, SPC is a potent inducer of intercellular adhesion molecule-1 expression in human keratinocytes, in part via the stimulated secretion of tumour necrosis factor-α (TNF-α) and the activation of mitogen-activated protein kinase (Imokawa et al. 1999).

3.4 Sphingosine 1-Phosphate and Keratinocytes

S1P has also been identified as a pivotal molecule in modulation of keratinocytes. In analogy to SPC, S1P stimulates proliferation of fibroblasts but inhibits keratinocyte cell growth (Vogler et al. 2003). However, S1P does not drive keratinocytes into apoptosis, but rather induces their differentiation as assessed by transglutaminase activity. In congruence calcium influx is increased in response to S1P in human keratinocytes (Lichte et al. 2008). S1P also stimulates the migration of keratinocytes implying a role of S1P in the re-epithelialization of wounds (Vogler et al. 2003). All five S1P receptors are expressed by keratinocytes at the mRNA level. The S1P-mediated inhibition of keratinocyte growth is mediated via the activation of protein kinase Cδ (PKCδ) followed by a subsequent dephosphorylation of Akt (Schuppel et al. 2008). Moreover, the $S1P_2$ receptor is dominantly involved in the S1P-induced dephosphorylation of Akt and keratinocyte growth arrest. This is of great clinical interest as the novel immunomodulator FTY720, after being phosphorylated by sphingosine kinase, activates all of the five S1P receptors except $S1P_2$ and therefore fails to inhibit keratinocyte proliferation (Schuppel et al. 2008).

Further on, S1P parallels the effects of TGFβ which also inhibits keratinocyte proliferation while stimulating fibroblast matrix production and proliferation. Indeed, S1P also stimulates phosphorylation of Smad3 in keratinocytes (Sauer et al. 2004). Both S1P and TGFβ receptors are required for this response. Smad3 activation is essential for the stimulation of keratinocyte migration and, though to a lesser degree, also for the inhibition of keratinocyte growth providing further evidence for a cross-communication between S1P and TGFβ signalling (Sauer et al. 2004). The exact nature of this communication has remained elusive so far.

4 Sphingolipids and Skin Dendritic Cells

A multiplicity of specialized antigen-presenting cells (APCs) is located in the skin, which belongs to the family of classical dendritic cells (DCs). The most prominent DC subtypes in the skin are the Langerhans cells (LCs) (Christensen and Haase 2012; Kaplan 2010). They ferret out exogenous molecules that have penetrated the skin barrier and convey this information to the dermal lymph nodes. LCs possess the pronounced ability of phagocytosis and are phenotypically characterized by the occurrence of the intracellular organelles named Birbeck granules. These organelles function as an endosomal recycling space, and their antigen-processing activity within LCs is related to the expression of an endocytotic receptor on the cell surface, namely, Langerin (CD207). Besides LCs, dermal DCs (dDCs) comprise another subtype of DCs in the skin that contributes to an effective antigen capture. The dDCs are phenotypically divergent from the LCs but share the capacity to activate naive T cells, and it has been proposed that in the absence of LCs, dDCs can adopt the immunological defense against penetrated allergens. LCs and dDCs are the cutting point in innate and adaptive immunity (Steinman 2007). The main functions of these APCs are the capture, endosomal processing, and presentation of antigens, providing them an exclusive aptitude to provoke adaptive immune responses and to induce and control tolerance. In the skin, immature LCs/dDCs capture and process antigens, for example, allergens and microorganisms, and then migrate towards secondary lymphoid organs where they present antigens to naive T cells (Fig. 1). During migration, LCs/dDCs lose their ability to capture additional antigens and obtain the competence to present antigens to naive T cells in a well-synchronized sequence of procedures referred to as maturation. As LCs/dDCs play a substantial role in the initiation of inflammatory skin diseases, recent research indicated that sphingolipids are modulators of essential steps in LCs/DCs homeostasis including antigen uptake, migration, and maturation.

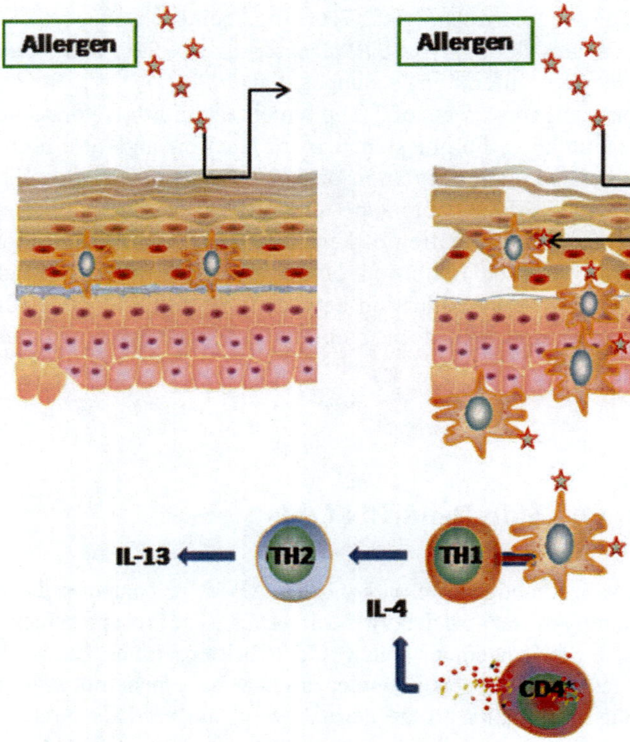

Fig. 1 A defective epidermal barrier contributes to the development of AD. A breakdown of the epidermal barrier in patients with AD permits the penetration of allergens. Skin DCs take up and present these allergens to T helper (TH) cells and recruit $CD4^+$ T cells. Activated DCs and IL-4, which are secreted by the $CD4^+$ cells, promote a shift from TH1 to TH2 cells. This is connected to the release of proinflammatory cytokines such as IL-13

4.1 Ceramides and Skin Dendritic Cells

It has been indicated that stimulation of DCs with proinflammatory molecules like LPS, TNF-α or IL-1β increased intracellular levels of ceramide (Santinha et al. 2012). This increase leads to the suggestion that a ceramide-mediated pathway is responsible for some of the functional alterations observed during the maturation process. The increased ceramide content is a result of an elevated sphingomyelinase activity leading to ceramide formation, which may regulate antigen-presenting cell function. Indeed, DCs show a downregulation of the endocytotic capacity in response to exogenously added ceramide indicating that ceramides have an important function in DCs maturation (Sallusto et al. 1996).

4.2 *S1P and Skin Dendritic Cells*

More recently, it has also been suggested that S1P has a crucial role in DC homeostasis. Thus, it has been shown that all five S1P receptors are expressed on the cell surface of DCs and the treatment of DC with S1P increased intracellular Ca^{2+} concentrations, actin remodelling, and chemotaxis (Radeke et al. 2005). Recruitment of DCs into peripheral tissues and their migration to dermal lymph nodes are under control of several chemokines such as monocyte chemotactic protein and sequential expression of their receptors. The chemotactic response towards S1P implies that this sphingolipid might play a critical function in the accumulation of DC at peripheral target sites and migration to the lymphoid organs. Indeed, in vitro cell culture assays revealed that S1P is a migratory stimulus of DCs and that the $S1P_1$ and $S1P_3$ receptor subtypes are responsible for the migratory response (Radeke et al. 2005). It seems likely that an S1P gradient plays a crucial role in guiding DCs from the peripheral skin site to the lymph nodes. It is of interest that topical administration of S1P in an animal model of contact hypersensibility inhibits the migration of skin DCs (Reines et al. 2009). The inhibition of migration after a topical treatment with S1P can be explained by either an internalization of the $S1P_1$ receptor subtype or dysregulation of the S1P gradient. Nevertheless, a recent study suggests that the $S1P_4$ receptor subtype negatively regulates migration of DCs towards S1P (Schulze et al. 2011). Besides the influence of S1P on DC migration, it has been suggested that topical treatment with S1P is also connected with an impaired antigen uptake by skin DCs. It is of interest that S1P inhibits macropinocytosis of DCs via the $S1P_2$ receptor subtype.

S1P may regulate not only the trafficking of DC but also the quality of DC-mediated T cell response. DCs treated with S1P shift the immune response of naive T cells towards a TH 2 subset. Thus, it has been indicated that stimulation of the $S1P_1$ receptor subtype inhibited LPS-induced IL-12 production thereby promoting TH-2 immune response (Schroder et al. 2011).

5 Sphingolipid Dysfunction in Atopic Dermatitis

AD is a multifactorial skin disorder which is characterized by erythematous lesions, pruritus, and xerosis. It has been well established that both environmental and genetic factors contribute to the development of AD (De Benedetto et al. 2012). Mutations in several genes encoding proteins that are involved in the formation of the epidermal barrier function have been discussed. The most prominent genetic cause leading to AD is a loss-of-function mutation in the *FLG* gene, which encodes for the structural protein filaggrin (Kubo et al. 2012). Additionally, enhanced protease activity due to mutations facilitates the cleavage of corneodesmosome junctions destroying the interactions between corneocytes. Moreover, an incomplete maturation and translocation of the lamellar granulae has been demonstrated in atopic skin. This imbalance is connected with a substantial lack of acid, enzyme, and lipid constituents of the stratum corneum resulting in an incomplete formation of the protective lamellar barrier. Consequently, the penetration of allergens through the skin is alleviated allowing an interaction of these allergens with the local skin DCs. In response to the detection of these immunological stimuli, DCs migrate to skin-associated lymph nodes and cross-communicate with T-lymphocytes. This process is connected with the terminal differentiation of DCs and the expansion and differentiation of T cells. This model is shown in Fig. 1.

It is of interest that a disturbance of sphingolipid homeostasis is involved not only in an impaired barrier formation but also in a pathophysiological alteration of DC function.

Several studies indicate that ceramide levels are decreased within the stratum corneum of involved or uninvolved skin of patients with AD compared to age-matched healthy controls (Imokawa et al. 1991). Indeed, a deficiency of this ordinary sphingolipid in the stratum corneum is a critical etiologic parameter for the dry and barrier-disrupted skin. Normally, ceramide concentrations in the stratum corneum are physiologically modulated by the equilibrium of the sphingolipid hydrolysis enzymes, β-glucocerebrosidase, sphingomyelinase, and ceramidase (Jin et al. 1994). Most interestingly, a novel sphingolipid metabolizing enzyme, termed sphingomyelin/glucosylceramide deacylase, has been identified in patients with AD. This enzyme hydrolyzes sphingomyelin or glucosylceramide at the acyl site yielding to the formation of SPC or glucosylsphingosine instead of ceramide (Higuchi et al. 2000). It has been shown that the sphingomyelin/glucosylceramide deacylase activity is elevated more than fivefold in involved stratum corneum from patients with AD compared with healthy controls (Ishibashi et al. 2003). Since the sphingomyelin/glucosylceramide deacylase has not been cloned and further identified, it is a matter of debate whether the enzyme is of bacterial origin or represents a novel enzyme in sphingolipid metabolism. Nevertheless, increases of sphingomyelin/glucosylceramide deacylase activity explain not only a diminished ceramide and sphingosine level but also an elevated SPC and glucosylsphingosine formation (Fig. 2).

It is well established that decreased levels of ceramide correlate with an impairment of stratum corneum barrier resulting in an increased transepidermal water loss. Moreover, the decreased levels of sphingosine also contribute to the development of AD. Thus, it has been discussed that vulnerability to bacterial colonization in the skin of patients with AD is connected to a reduced level of the antimicrobial molecule sphingosine (Arikawa et al. 2002).

Besides decreased levels of ceramide and sphingosine, elevated biosynthesis of SPC seems to be a critical factor in the progression of AD (Fig. 2). Additionally to the described actions, SPC significantly decreases filaggrin gene transcription in keratinocytes via a prostanoid receptor pathway implying that SPC plays a crucial role in destruction of the epidermal barrier function (Choi et al. 2010). Moreover, SPC affects functions of DCs which contribute to an inflammatory response. SPC induces chemotactic migration in DCs and improves T cell priming by an enhanced expression of costimulatory molecules on the DC cell surface such as HLA-DR,

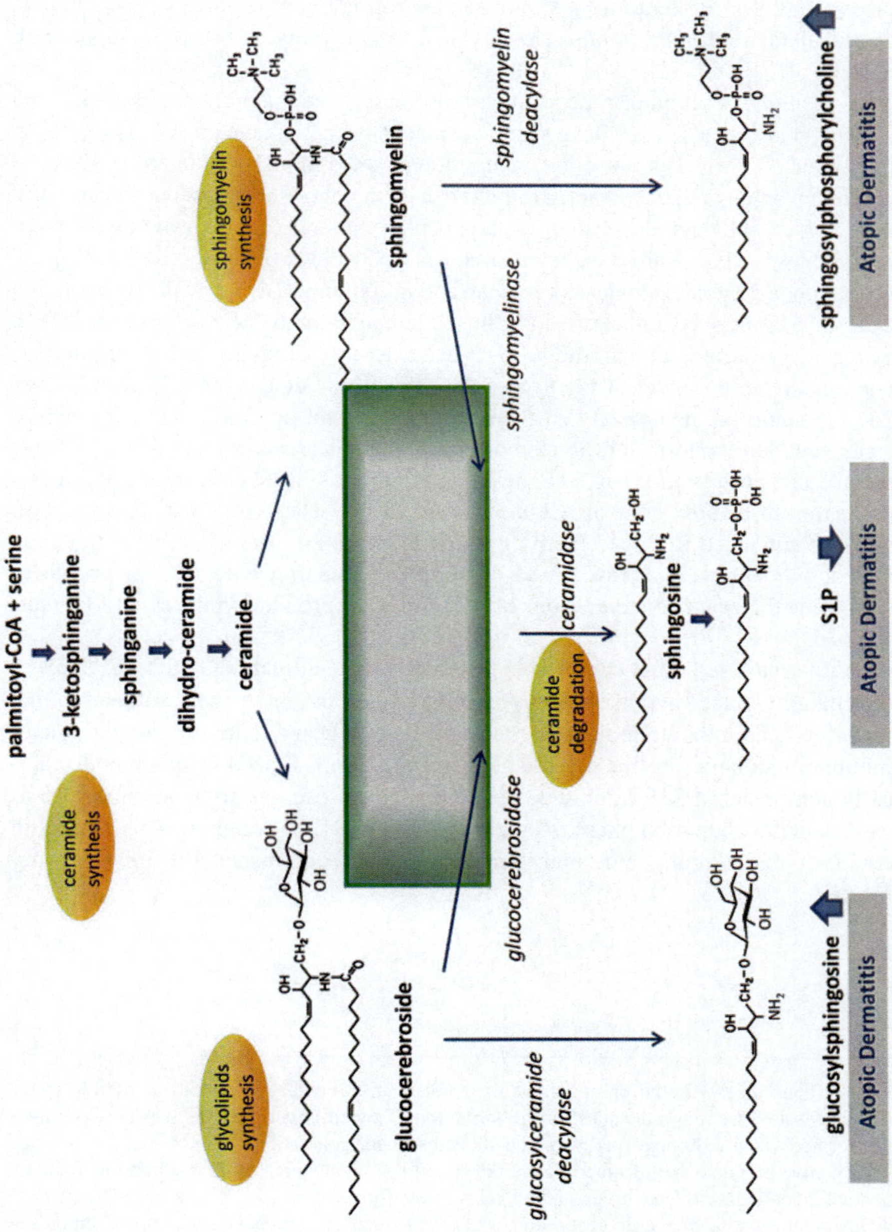

Fig. 2 (continued)

CD86, and CD83. Finally, an animal model indicated that an intradermal injection of SPC, but not sphingomyelin and sphingosine, induced scratching and an itch-associated response (Kim et al. 2008). Administration of SPC to the skin and to primary cultures of keratinocytes induced the formation of leukotriene B4 which is a crucial factor for the stimulation of neuronal primary afferents (Andoh et al. 2011).

In the light of previous reports, it has been suggested that not only ceramide and sphingosine but also S1P levels are decreased in patients with AD (Fig. 2). A decreased S1P level is not only a consequence of diminished content of the S1P precursor sphingosine; in fact, an altered S1P metabolism has been discussed in AD. S1P can be degraded either via dephosphorylation to sphingosine by different phosphatases like alkaline phosphatases, lipid phosphate phosphatases, and sphingosine phosphate phosphatases or via an activation of S1P-lyase which irreversibly cleaves S1P into hexadecenal and phosphoethanolamine (Maceyka et al. 2012). Several reports indicate an altered S1P-lyase activity in atopic lesions in humans and dogs (Baumer et al. 2011; Seo et al. 2006; Wood et al. 2009). With regard to AD, it should be mentioned that there exist quite a lot of similarities between the canine and human forms of this skin disease (Marsella and Girolomoni 2009). Thus, the mean concentration of S1P in lesional skin of dogs is significantly lower compared to healthy controls (Baumer et al. 2011). These findings are consistent with the enhanced S1P-lyase activity found in mRNA of lesional skin from human and dogs (Seo et al. 2006; Wood et al. 2009). But it would be worthwhile to determine the exact concentrations of S1P and lyase products in the skin of humans with AD.

With regard to AD, it can be speculated that topical treatment with S1P may be beneficial. Contact hypersensitivity is one of the most intensively studied animal model to examine immunological mechanisms of AD and to investigate the role of immunomodulators in this disease (Rose et al. 2012). Indeed in this model, topically administered S1P inhibited the inflammatory reaction in the sensitization as well as in the elicitation phase (Reines et al. 2009). S1P reduced the weight and cell count of the draining auricular lymph node and diminished the inflammatory reaction.

Fig. 2 (continued) Altered sphingolipid metabolism in patients with atopic dermatitis (AD). Ceramides are essential molecules for the formation of a well-organized lipid barrier. Ceramides are formed via the de novo pathway and the subsequent generation of glucosylceramides and sphingomyelin. Cleavage of glucosylceramides and sphingomyelin due to β-glucocerebrosidase and sphingomyelinase leads to ceramides responsible for the formation of the stratum corneum. AD patients show an increased generation of SPC and glucosylsphingosine, whereas ceramide and sphingosine levels are decreased. This is connected to a dysregulated lipid barrier. Moreover, S1P levels are decreased due to an enhanced degradation via S1P-lyase activity. In consequence the modulatory role of S1P is decreased

6 Sphingolipid Dysfunction in Psoriasis Vulgaris

Psoriasis is a common, chronic inflammatory skin disease which is present in approximately 2% of the population. The clinical symptoms are the results of a hyperproliferative epidermis consisting of premature keratinocytes and incomplete cornification. This leads to a thickened epidermis with typical elongated rete ridges. Moreover, the skin disease is characterized by an infiltration of T cells (Nickoloff et al. 2007). Besides the contribution of genetic alterations in the development of psoriasis, it is still a matter of debate whether the skin disorder is either an epithelial or an immunological disease. But there exist accumulating evidence that the origin of psoriasis is due to a dysregulated immune system (Lowes et al. 2007). Therefore, it is not astonishing that drugs targeting the immune system (e.g. corticosteroids and biologics) are beneficial for the treatment of this skin disease. In this context, it should be mentioned that novel approaches using monoclonal antibodies directed against IL-12, IL-17 and IL-23 (Leonardi et al. 2008, 2012) are in the focus of interest in recent psoriasis research (Di Cesare et al. 2009).

Nevertheless, the severity of barrier dysfunction in skin of psoriatic patients is comparative to the disease phenotype. Indeed, a variety of studies propose an altered lipid homeostasis in psoriasis (Motta et al. 1993, 1994a, b, 1995). It is still unresolved whether these modulations contribute to the pathogenesis of the disease or represent an outcome of the multifaceted skin disorder. A different pattern in ceramide distribution in psoriatic plaques compared with normal control skin has been detected.

It has been suggested that deficiency in the expression of prosaposin may be responsible for the altered ceramide levels in patients suffering from psoriasis (Alessandrini et al. 2001). Saposines or so-called sphingolipid activator proteins (SAPs) are nonenzymatic proteins which are required for the lysosomal breakdown of certain sphingolipids by hydrolase enzymes (Tamargo et al. 2012). Thus, the activity of β-glucocerebrosidase as well as of sphingomyelinase has been shown to be stimulated by saposines. Decreased levels of saposine precursor molecules have been detected in the skin of patients with psoriasis vulgaris. In congruence a diminished β-glucocerebrosidase mRNA has been reported in whole skin punch biopsies from lesional sites in patients with psoriasis vulgaris, when compared with skin from healthy, non-affected controls (Alessandrini et al. 2001). Immunohistochemical localization of β-glucocerebrosidase and sphingomyelinase confirmed a reduced expression of these enzymes on the protein level. In analogy, prosaposin-deficient animals retain glucosylceramide species in the stratum corneum which is connected to an epidermal barrier dysfunction (Doering et al. 1999). Thus, altered glucosylceramide and sphingomyelin levels could play an essential factor for the development of psoriasis.

Apart from the importance of ceramides as structural component of the stratum corneum, the action of sphingolipid derivatives as bioactive molecules must be taken into account. This became visible when it was indicated that calcitriol, which is used for the treatment of psoriasis vulgaris, affects sphingolipid metabolism

(Okazaki et al. 1989). Calcitriol has been identified to inhibit cell growth of keratinocytes and to induce their differentiation (Geilen et al. 1996, 1997). It has been implicated that the calcitriol-induced differentiation is mediated, at least in part, by the formation of ceramides. However, ceramides have also been identified to induce apoptosis in keratinocytes. On the contrary, physiological concentrations of calcitriol did not induce apoptosis in keratinocytes despite the formation of ceramides (Manggau et al. 2001). Moreover, calcitriol possesses a cytoprotective action and made keratinocytes resistant to apoptosis. Thus, a further metabolism to S1P occurred in response to calcitriol (Manggau et al. 2001). Most interestingly, S1P did protect not only keratinocytes from apoptosis but in analogy to calcitriol inhibited cell growth of the epidermal cells and promoted their differentiation indicating a pivotal role of S1P in skin homeostasis (Vogler et al. 2003). In fact, examination of S1P metabolism in psoriatic skin lesions indicated an increased degradation of S1P (Mechtcheriakova et al. 2007). Most interestingly, an upregulation of the sphingosine phosphate phosphates type 2 has been detected in the lesional skin of all patients compared to the corresponding non-lesional skin.

Due to its properties, it would be of interest to further elucidate whether S1P is a promising candidate in the therapy of psoriasis vulgaris.

7 Conclusion

Recent studies concerning the metabolic activities localized within the stratum corneum have indicated that the epidermis by the means is not an inert accumulation of dead corneocytes embedded in a static lipid-enriched matrix. It has been indicated that the generation of the lipid barrier is a complex process in which a variety of enzymes are involved that catalyzes the formation of specific lipids. Especially ceramides play a substantial role in the formation of an organized structure of the lipid barrier. Thus, it is not astonishing that a dysregulation in sphingolipid formation contributes to disease pathogenesis. Additionally, defining the action of lipid transport proteins, such as ABCA12, as a crucial parameter of epidermal lipid topology in both normal and diseased skin, is a challenging area of research. Moreover, sphingolipid metabolites have also been identified as bioactive molecules involved in the modulation of epidermal cells and immune cells of the skin. Studies of epidermal lipidomics in addition with transgenic/knockout experiments will further clarify the functional role of sphingolipids in skin homeostasis.

Thus, therapeutic approaches based on topical administration or systemic delivery of sphingolipid metabolites or sphingomimetic molecules abolish dysregulation of sphingolipid metabolism and restore its homeostasis. These functions can be beneficial in avoiding pathophysiological immune responses maybe beneficial in avoiding pathophysiological immune responses.

References

Aho S, Harding CR, Lee JM, Meldrum H, Bosko CA (2012) Regulatory role for the profilaggrin N-terminal domain in epidermal homeostasis. J Invest Dermatol 132(10):2376–2385

Alessandrini F, Stachowitz S, Ring J, Behrendt H (2001) The level of prosaposin is decreased in the skin of patients with psoriasis vulgaris. J Invest Dermatol 116:394–400

Andoh T, Haza S, Saito A, Kuraishi Y (2011) Involvement of leukotriene B4 in spontaneous itch-related behaviour in NC mice with atopic dermatitis-like skin lesions. Exp Dermatol 20:894–898

Arikawa J, Ishibashi M, Kawashima M, Takagi Y, Ichikawa Y, Imokawa G (2002) Decreased levels of sphingosine, a natural antimicrobial agent, may be associated with vulnerability of the stratum corneum from patients with atopic dermatitis to colonization by Staphylococcus aureus. J Invest Dermatol 119:433–439

Baumer W, Rossbach K, Mischke R, Reines I, Langbein-Detsch I, Luth A, Kleuser B (2011) Decreased concentration and enhanced metabolism of sphingosine-1-phosphate in lesional skin of dogs with atopic dermatitis: disturbed sphingosine-1-phosphate homeostasis in atopic dermatitis. J Invest Dermatol 131:266–268

Behne M, Uchida Y, Seki T, de Montellano PO, Elias PM, Holleran WM (2000) Omega-hydroxyceramides are required for corneocyte lipid envelope (CLE) formation and normal epidermal permeability barrier function. J Invest Dermatol 114:185–192

Bektas M, Dullin Y, Wieder T, Kolter T, Sandhoff K, Brossmer R, Ihrig P, Orfanos CE, Geilen CC (1998) Induction of apoptosis by synthetic ceramide analogues in the human keratinocyte cell line HaCaT. Exp Dermatol 7:342–349

Bikle DD (2012) Vitamin D and the skin: physiology and pathophysiology. Rev Endocr Metab Disord 13:3–19

Bornancin F (2011) Ceramide kinase: the first decade. Cell Signal 23:999–1008

Choi H, Kim S, Kim HJ, Kim KM, Lee CH, Shin JH, Noh M (2010) Sphingosylphosphorylcholine down-regulates filaggrin gene transcription through NOX5-based NADPH oxidase and cyclooxygenase-2 in human keratinocytes. Biochem Pharmacol 80:95–103

Christensen AD, Haase C (2012) Immunological mechanisms of contact hypersensitivity in mice. APMIS 120:1–27

De Benedetto A, Kubo A, Beck LA (2012) Skin barrier disruption: a requirement for allergen sensitization? J Invest Dermatol 132:949–963

Deigner HP, Claus R, Bonaterra GA, Gehrke C, Bibak N, Blaess M, Cantz M, Metz J, Kinscherf R (2001) Ceramide induces aSMase expression: implications for oxLDL-induced apoptosis. FASEB J 15:807–814

Di Cesare A, Di Meglio P, Nestle FO (2009) The IL-23/Th17 axis in the immunopathogenesis of psoriasis. J Invest Dermatol 129:1339–1350

Doering T, Holleran WM, Potratz A, Vielhaber G, Elias PM, Suzuki K, Sandhoff K (1999) Sphingolipid activator proteins are required for epidermal permeability barrier formation. J Biol Chem 274:11038–11045

Geilen CC, Bektas M, Wieder T, Orfanos CE (1996) The vitamin D3 analogue, calcipotriol, induces sphingomyelin hydrolysis in human keratinocytes. FEBS Lett 378:88–92

Geilen CC, Bektas M, Wieder T, Kodelja V, Goerdt S, Orfanos CE (1997) 1alpha,25-dihydroxyvitamin D3 induces sphingomyelin hydrolysis in HaCaT cells via tumor necrosis factor alpha. J Biol Chem 272:8997–9001

Hamanaka S, Hara M, Nishio H, Otsuka F, Suzuki A, Uchida Y (2002) Human epidermal glucosylceramides are major precursors of stratum corneum ceramides. J Invest Dermatol 119:416–423

Hansen HS, Jensen B (1985) Essential function of linoleic acid esterified in acylglucosylceramide and acylceramide in maintaining the epidermal water permeability barrier. Evidence from feeding studies with oleate, linoleate, arachidonate, columbinate and alpha-linolenate. Biochim Biophys Acta 834:357–363

Higuchi K, Hara J, Okamoto R, Kawashima M, Imokawa G (2000) The skin of atopic dermatitis patients contains a novel enzyme, glucosylceramide sphingomyelin deacylase, which cleaves the N-acyl linkage of sphingomyelin and glucosylceramide. Biochem J 350(Pt 3):747–756

Higuchi K, Kawashima M, Takagi Y, Kondo H, Yada Y, Ichikawa Y, Imokawa G (2001) Sphingosylphosphorylcholine is an activator of transglutaminase activity in human keratinocytes. J Lipid Res 42:1562–1570

Holleran WM, Ginns EI, Menon GK, Grundmann JU, Fartasch M, McKinney CE, Elias PM, Sidransky E (1994) Consequences of beta-glucocerebrosidase deficiency in epidermis. Ultrastructure and permeability barrier alterations in Gaucher disease. J Clin Invest 93:1756–1764

Holleran WM, Takagi Y, Uchida Y (2006) Epidermal sphingolipids: metabolism, function, and roles in skin disorders. FEBS Lett 580:5456–5466

Imokawa G, Abe A, Jin K, Higaki Y, Kawashima M, Hidano A (1991) Decreased level of ceramides in stratum corneum of atopic dermatitis: an etiologic factor in atopic dry skin? J Invest Dermatol 96:523–526

Imokawa G, Takagi Y, Higuchi K, Kondo H, Yada Y (1999) Sphingosylphosphorylcholine is a potent inducer of intercellular adhesion molecule-1 expression in human keratinocytes. J Invest Dermatol 112:91–96

Ishibashi M, Arikawa J, Okamoto R, Kawashima M, Takagi Y, Ohguchi K, Imokawa G (2003) Abnormal expression of the novel epidermal enzyme, glucosylceramide deacylase, and the accumulation of its enzymatic reaction product, glucosylsphingosine, in the skin of patients with atopic dermatitis. Lab Invest 83:397–408

Jennemann R et al (2012) Loss of ceramide synthase 3 causes lethal skin barrier disruption. Hum Mol Genet 21:586–608

Jensen JM, Proksch E (2009) The skin's barrier. G Ital Dermatol Venereol 144:689–700

Jensen JM, Folster-Holst R, Baranowsky A, Schunck M, Winoto-Morbach S, Neumann C, Schutze S, Proksch E (2004) Impaired sphingomyelinase activity and epidermal differentiation in atopic dermatitis. J Invest Dermatol 122:1423–1431

Jiang YJ, Uchida Y, Lu B, Kim P, Mao C, Akiyama M, Elias PM, Holleran WM, Grunfeld C, Feingold KR (2009) Ceramide stimulates ABCA12 expression via peroxisome proliferator-activated receptor delta in human keratinocytes. J Biol Chem 284:18942–18952

Jin K, Higaki Y, Takagi Y, Higuchi K, Yada Y, Kawashima M, Imokawa G (1994) Analysis of beta-glucocerebrosidase and ceramidase activities in atopic and aged dry skin. Acta Derm Venereol 74:337–340

Jung EM, Griner RD, Mann-Blakeney R, Bollag WB (1998) A potential role for ceramide in the regulation of mouse epidermal keratinocyte proliferation and differentiation. J Invest Dermatol 110:318–323

Kaplan DH (2010) In vivo function of Langerhans cells and dermal dendritic cells. Trends Immunol 31:446–451

Kim HJ, Kim H, Han ES, Park SM, Koh JY, Kim KM, Noh MS, Kim JJ, Lee CH (2008) Characterizations of sphingosylphosphorylcholine-induced scratching responses in ICR mice using naltrexon, capsaicin, ketotifen and Y-27632. Eur J Pharmacol 583:92–96

Kubo A, Nagao K, Amagai M (2012) Epidermal barrier dysfunction and cutaneous sensitization in atopic diseases. J Clin Invest 122:440–447

Lampe MA, Williams ML, Elias PM (1983) Human epidermal lipids: characterization and modulations during differentiation. J Lipid Res 24:131–140

Leonardi CL, Kimball AB, Papp KA, Yeilding N, Guzzo C, Wang Y, Li S, Dooley LT, Gordon KB (2008) Efficacy and safety of ustekinumab, a human interleukin-12/23 monoclonal antibody, in patients with psoriasis: 76-week results from a randomised, double-blind, placebo-controlled trial (PHOENIX 1). Lancet 371:1665–1674

Leonardi C, Matheson R, Zachariae C, Cameron G, Li L, Edson-Heredia E, Braun D, Banerjee S (2012) Anti-interleukin-17 monoclonal antibody ixekizumab in chronic plaque psoriasis. N Engl J Med 366:1190–1199

Lichte K, Rossi R, Danneberg K, ter Braak M, Kurschner U, Jakobs KH, Kleuser B, Meyer zu Heringdorf D (2008) Lysophospholipid receptor-mediated calcium signaling in human keratinocytes. J Invest Dermatol 128:1487–1498

Liu YY et al (2008) A role for ceramide in driving cancer cell resistance to doxorubicin. FASEB J 22:2541–2551

Lowes MA, Bowcock AM, Krueger JG (2007) Pathogenesis and therapy of psoriasis. Nature 445:866–873

Maceyka M, Harikumar KB, Milstien S, Spiegel S (2012) Sphingosine-1-phosphate signaling and its role in disease. Trends Cell Biol 22:50–60

Macheleidt O, Kaiser HW, Sandhoff K (2002) Deficiency of epidermal protein-bound omega-hydroxyceramides in atopic dermatitis. J Invest Dermatol 119:166–173

Manggau M, Kim DS, Ruwisch L, Vogler R, Korting HC, Schafer-Korting M, Kleuser B (2001) 1Alpha,25-dihydroxyvitamin D3 protects human keratinocytes from apoptosis by the formation of sphingosine-1-phosphate. J Invest Dermatol 117:1241–1249

Marsella R, Girolomoni G (2009) Canine models of atopic dermatitis: a useful tool with untapped potential. J Invest Dermatol 129:2351–2357

Masukawa Y et al (2008) Characterization of overall ceramide species in human stratum corneum. J Lipid Res 49:1466–1476

Mechtcheriakova D et al (2007) Sphingosine 1-phosphate phosphatase 2 is induced during inflammatory responses. Cell Signal 19:748–760

Menon GK, Grayson S, Elias PM (1986) Cytochemical and biochemical localization of lipase and sphingomyelinase activity in mammalian epidermis. J Invest Dermatol 86:591–597

Meyer T, Stockfleth E, Christophers E (2007) Immune response profiles in human skin. Br J Dermatol 157(Suppl 2):1–7

Mizutani Y, Kihara A, Igarashi Y (2006) LASS3 (longevity assurance homologue 3) is a mainly testis-specific (dihydro)ceramide synthase with relatively broad substrate specificity. Biochem J 398:531–538

Mizutani Y, Kihara A, Chiba H, Tojo H, Igarashi Y (2008) 2-Hydroxy-ceramide synthesis by ceramide synthase family: enzymatic basis for the preference of FA chain length. J Lipid Res 49:2356–2364

Motta S, Monti M, Sesana S, Caputo R, Carelli S, Ghidoni R (1993) Ceramide composition of the psoriatic scale. Biochim Biophys Acta 1182:147–151

Motta S, Sesana S, Monti M, Giuliani A, Caputo R (1994a) Interlamellar lipid differences between normal and psoriatic stratum corneum. Acta Derm Venereol Suppl (Stockh) 186:131–132

Motta S, Monti M, Sesana S, Mellesi L, Ghidoni R, Caputo R (1994b) Abnormality of water barrier function in psoriasis. Role of ceramide fractions. Arch Dermatol 130:452–456

Motta S, Sesana S, Ghidoni R, Monti M (1995) Content of the different lipid classes in psoriatic scale. Arch Dermatol Res 287:691–694

Nickoloff BJ, Qin JZ, Nestle FO (2007) Immunopathogenesis of psoriasis. Clin Rev Allergy Immunol 33:45–56

Ohnishi Y, Okino N, Ito M, Imayama S (1999) Ceramidase activity in bacterial skin flora as a possible cause of ceramide deficiency in atopic dermatitis. Clin Diagn Lab Immunol 6:101–104

Okazaki T, Bell RM, Hannun YA (1989) Sphingomyelin turnover induced by vitamin D3 in HL-60 cells. Role in cell differentiation. J Biol Chem 264:19076–19080

Quazi F, Molday RS (2011) Lipid transport by mammalian ABC proteins. Essays Biochem 50:265–290

Radeke HH, von Wenckstern H, Stoidtner K, Sauer B, Hammer S, Kleuser B (2005) Overlapping signaling pathways of sphingosine 1-phosphate and TGF-beta in the murine Langerhans cell line XS52. J Immunol 174:2778–2786

Rajpopat S et al (2011) Harlequin ichthyosis: a review of clinical and molecular findings in 45 cases. Arch Dermatol 147:681–686

Reines I, Kietzmann M, Mischke R, Tschernig T, Luth A, Kleuser B, Baumer W (2009) Topical application of sphingosine-1-phosphate and FTY720 attenuate allergic contact dermatitis reaction through inhibition of dendritic cell migration. J Invest Dermatol 129:1954–1962

Rose L, Schneider C, Stock C, Zollner TM, Docke WD (2012) Extended DNFB-induced contact hypersensitivity models display characteristics of chronic inflammatory dermatoses. Exp Dermatol 21:25–31

Sallusto F, Nicolo C, De Maria R, Corinti S, Testi R (1996) Ceramide inhibits antigen uptake and presentation by dendritic cells. J Exp Med 184:2411–2416

Santinha DR, Marques DR, Maciel EA, Simoes CS, Rosa S, Neves BM, Macedo B, Domingues P, Cruz MT, Domingues MR (2012) Profiling changes triggered during maturation of dendritic cells: a lipidomic approach. Anal Bioanal Chem 403:457–471

Sauer B, Vogler R, von Wenckstern H, Fujii M, Anzano MB, Glick AB, Schafer-Korting M, Roberts AB, Kleuser B (2004) Involvement of Smad signaling in sphingosine 1-phosphate-mediated biological responses of keratinocytes. J Biol Chem 279:38471–38479

Schroder M, Richter C, Juan MH, Maltusch K, Giegold O, Quintini G, Pfeilschifter JM, Huwiler A, Radeke HH (2011) The sphingosine kinase 1 and S1P1 axis specifically counteracts LPS-induced IL-12p70 production in immune cells of the spleen. Mol Immunol 48:1139–1148

Schulze T, Golfier S, Tabeling C, Rabel K, Graler MH, Witzenrath M, Lipp M (2011) Sphingosine-1-phospate receptor 4 (S1P(4)) deficiency profoundly affects dendritic cell function and TH17-cell differentiation in a murine model. FASEB J 25:4024–4036

Schuppel M, Kurschner U, Kleuser U, Schafer-Korting M, Kleuser B (2008) Sphingosine 1-phosphate restrains insulin-mediated keratinocyte proliferation via inhibition of Akt through the S1P2 receptor subtype. J Invest Dermatol 128:1747–1756

Seo EY, Park GT, Lee KM, Kim JA, Lee JH, Yang JM (2006) Identification of the target genes of atopic dermatitis by real-time PCR. J Invest Dermatol 126:1187–1189

Steinman RM (2007) Dendritic cells: understanding immunogenicity. Eur J Immunol 37(Suppl 1): S53–S60

Tamargo RJ, Velayati A, Goldin E, Sidransky E (2012) The role of saposin C in Gaucher disease. Mol Genet Metab 106:257–263

Tsuji K, Mitsutake S, Yokose U, Sugiura M, Kohama T, Igarashi Y (2008) Role of ceramide kinase in peroxisome proliferator-activated receptor beta-induced cell survival of mouse keratinocytes. FEBS J 275:3815–3826

Uchida Y, Holleran WM (2008) Omega-O-acylceramide, a lipid essential for mammalian survival. J Dermatol Sci 51:77–87

Uchida Y, Hara M, Nishio H, Sidransky E, Inoue S, Otsuka F, Suzuki A, Elias PM, Holleran WM, Hamanaka S (2000) Epidermal sphingomyelins are precursors for selected stratum corneum ceramides. J Lipid Res 41:2071–2082

Vogler R, Sauer B, Kim DS, Schafer-Korting M, Kleuser B (2003) Sphingosine-1-phosphate and its potentially paradoxical effects on critical parameters of cutaneous wound healing. J Invest Dermatol 120:693–700

Wakita H, Tokura Y, Yagi H, Nishimura K, Furukawa F, Takigawa M (1994) Keratinocyte differentiation is induced by cell-permeant ceramides and its proliferation is promoted by sphingosine. Arch Dermatol Res 286:350–354

Wakita H, Matsushita K, Nishimura K, Tokura Y, Furukawa F, Takigawa M (1998) Sphingosylphosphorylcholine stimulates proliferation and upregulates cell surface-associated plasminogen activator activity in cultured human keratinocytes. J Invest Dermatol 110:253–258

Wood SH, Clements DN, Ollier WE, Nuttall T, McEwan NA, Carter SD (2009) Gene expression in canine atopic dermatitis and correlation with clinical severity scores. J Dermatol Sci 55:27–33

Zuo Y, Zhuang DZ, Han R, Isaac G, Tobin JJ, McKee M, Welti R, Brissette JL, Fitzgerald ML, Freeman MW (2008) ABCA12 maintains the epidermal lipid permeability barrier by facilitating formation of ceramide linoleic esters. J Biol Chem 283:36624–36635

Sphingolipids in Obesity, Type 2 Diabetes, and Metabolic Disease

S.B. Russo, J.S. Ross, and L.A. Cowart

Contents

Abstract Metabolic disease, including obesity and type 2 diabetes, constitutes a major emerging health crisis in Western nations. Although the symptoms and clinical pathology and physiology of these conditions are well understood, the molecular mechanisms underlying the disease process have largely remained obscure. Sphingolipids, a lipid class with both signaling and structural properties, have recently emerged as key players in most major tissues affected by diabetes and are required components in the molecular etiology of this disease. Indeed, sphingolipids have been shown to mediate loss of insulin sensitivity, to promote the characteristic diabetic proinflammatory state, and to induce cell death and dysfunction in important organs such as the pancreas and heart. Furthermore, plasma sphingolipid levels are emerging as potential biomarkers for the decompensation of insulin resistance to frank type 2 diabetes. Despite these discoveries, the roles of specific sphingolipid species and sphingolipid metabolic pathways remain

S.B. Russo • J.S. Ross • L.A. Cowart (✉)
Department of Biochemistry and Molecular Biology, Ralph H. Johnson VA Medical Center, The Medical University of South Carolina, 173 Ashley Ave, MSC509 Charleston, SC, USA
e-mail: cowartl@musc.edu

E. Gulbins and I. Petrache (eds.), *Sphingolipids in Disease*, Handbook of Experimental Pharmacology 216,
DOI 10.1007/978-3-7091-1511-4_19, © Springer-Verlag Wien 2013

obscure, and newly developed experimental approaches must be employed to elucidate the detailed molecular mechanisms necessary for rational drug development and other clinical applications.

Keywords lipotoxicity • cardiomyopathy • insulin resistance

1 Introduction

Type 2 diabetes is an increasingly important healthcare concern in Western nations, with incidence on the rise in both Europe and the United States (Passa 2002; Wei et al. 2011). This disease, which is diagnosed based on blood plasma hyperglycemia, has been linked to lipid overload and abdominal obesity and may synergize with these conditions to promote negative clinical outcomes (Maheux et al. 1994; Nussey and Whitehead 2001; Pontiroli and Camisasca 2002; Grundy et al. 2004; Banegas et al. 2007; Qiao and Nyamdorj 2010). This connection is significant, as the incidence of obesity is also increasing, with greater than 65 % of individuals classified as overweight or obese in both the United States and the European Union (Martinez et al. 1999; Baskin et al. 2005; Fry and Finley 2005; Samad et al. 2006; Mueller-Riemenschneider et al. 2008). This chapter summarizes current knowledge of the sphingolipid-dependent manifestations of obesity and diabetes and points to areas for future study and potential clinical utility.

2 Pathogenesis of Type 2 Diabetes Mellitus

Diabetes develops not as a single-organ disease but as a syndrome progressively affecting multiple organ systems. Because diabetes is silent early on and often is only discovered after onset of complications, it has been difficult to tease out a complete natural history of this disease. However, careful epidemiological and laboratory studies have revealed a general outline of disease progression.

It is believed that insulin resistance is the primary defect in type 2 diabetes mellitus, preceding frank hyperglycemia (Cavaghan et al. 2000). In many diabetics, the initial cause of insulin resistance may be hyperlipidemia consequent to abdominal obesity (Pouliot et al. 1990). Excess free fatty acids promote insulin resistance in peripheral tissues, including the skeletal muscle and liver; at first, pancreatic β-cells are able to compensate for this by increasing insulin secretion (Bevilacqua et al. 1987; Hirose et al. 1996; Roden et al. 1996). This compensated state, in which hyperinsulinemia effectively controls plasma glucose levels, may persist for years and, in fact, may never progress to overt diabetes mellitus (Leahy 2008). Over time, the ability of the pancreas to produce high levels of insulin may be exhausted due both to aberrant cell signaling and to β-cell loss to apoptosis (Hirose et al. 1996; Pick et al. 1998). This decompensation results in the hyperglycemia observed in type 2 diabetes, which occurs as a result of both continually elevated hepatic glucose output and reduced peripheral glucose uptake and utilization. At this point, the hyperglycemia itself damages pancreatic β-cells, accelerating the course of β-cell insufficiency and disease progression (Maedler et al. 2003).

Type 2 diabetes is not merely a disease of the pancreas. Rather, insulin resistance and dysregulation of plasma glucose and free fatty acid levels can produce severe dysfunction in multiple organs, including the heart, kidneys, liver, and eyes (as discussed in the following sections), as well as the brain and peripheral nervous system. Insidiously, tissue damage can begin before noticeable decompensation of insulin resistance, causing patients to suffer silent tissue damage long prior to diagnosis (Nerpin et al. 2008; Dinh et al. 2010, 2011; Manchanayake et al. 2011). Through this multisystemic onslaught, diabetes slowly saps patients of both longevity and quality of life. Thus, it is critical to understand the mechanisms underlying this disease and to develop rational treatment regimens to halt this cascade of dysfunction.

3 Sphingolipid Metabolism and Regulation

Sphingolipids are a class of structural and signaling molecules incorporating a unique dihydrosphingosine backbone, which may also be present in its desaturated form as sphingosine (Gault et al. 2010). These lipids play important roles in determining membrane biophysical properties, topology, and integrity and in numerous cellular signaling processes, particularly apoptosis and proliferation (Zheng et al. 2006; Hannun and Obeid 2008). Sphingolipid-dependent mechanisms of pathogenesis have been demonstrated in many disease contexts, and sphingolipids have emerged as important players in obesity and diabetes (Shimabukuro et al. 1998; Summers and Nelson 2005; Zheng et al. 2006; Hannun and Obeid 2008). Importantly, sphingolipid metabolism is highly interconnected by a well-characterized network of enzymes, making it possible to study mechanisms of sphingolipid production and signaling with great precision (Gault et al. 2010). This metabolic system, which we recently reviewed (Brice and Cowart 2011), is described in detail in several other chapters.

Given the great complexity of sphingolipid metabolism, it follows that its regulation may also be complex or multifactorial. Indeed, it has been demonstrated that sphingolipid synthesis can be regulated at both the substrate level and through signaling in response to cellular stimuli. In perhaps the most intuitive manner of regulation, which is extremely relevant to the study of type 2 diabetes, fatty acids impact sphingolipid levels through both substrate supply and regulation of the enzymes of sphingolipid metabolism (Shimabukuro et al. 1998; Hu et al. 2009, 2011). Initially, oversupply of saturated fatty acids directly stimulates sphingolipid synthesis via substrate supply (Merrill Jr et al. 1985; Shimabukuro et al. 1998). This excess of fatty acids provides additional acyl-CoA moieties for de novo production of sphingoid bases by serine palmitoyltransferase (SPT) and *N*-acylation of sphingoid bases by CerS. In a complementary manner, fatty acid supply also regulates expression of the enzymes of sphingolipid metabolism, including SPT, dihydroceramide desaturase 1 (Des1), and sphingosine kinase 1 (SphK1), thus upregulating sphingolipid production in conditions of steatosis (lipid overload) (Hu et al. 2009, 2011). Strikingly, regulation of these enzymes by fatty acids appears to occur in a chain length- and saturation-dependent manner, with unsaturated fatty acids protecting against upregulation of SphK1, SphK2, and Des1 (Hu et al. 2009, 2011).

Sphingolipid metabolism is also regulated in a substrate-independent manner by other components of the diabetic physiologic phenotype. Indeed, sphingolipid metabolism can be activated by several aspects of diabetic physiology, including the presence of inflammatory cytokines, growth factors, and oxidative stress (Goldkorn et al. 1998; Memon et al. 1998; Levy et al. 2006, 2009; Samad et al. 2006; Sultan et al. 2006; Aerts et al. 2007). In vivo, these conditions were all observed in the adipose tissue of genetically obese, diabetic *ob/ob* mice (Samad et al. 2006). Compared to their lean counterparts, genetically obese mice demonstrated significantly higher expression of enzymes promoting ceramide and sphingosine/sphingosine-1-phosphate accumulation, including acid and neutral sphingomyelinase, alkaline and acid ceramidase, and SPT. In contrast, levels of enzymes that would route ceramide to more complex metabolites, including glucosylceramide synthase and *N*-acetylneuraminylgalactosylceramide (GM3) synthase, were downregulated in the diabetic mice. This indicated that in the diabetic state, expression of sphingolipid-metabolizing enzymes is recalibrated toward generation of proinflammatory and proapoptotic species. Furthermore, direct exposure of HepG2 hepatocellular carcinoma cells to the inflammatory cytokines TNF-α and IL-1 directly stimulated SPT expression and activity in vitro, providing a direct mechanistic link between these inflammation and sphingolipid synthesis (Memon et al. 1998). Other work has suggested a similar stimulatory effect of inflammatory cytokines on the sphingolipid recycling pathway, in which ceramides are deacylated to form sphingosine and then reacylated (Sultan et al. 2006). Likewise, it has been shown that oxidative stress promotes ceramide accumulation in other disease contexts, particularly through stimulation of neutral sphingomyelinase 2 (NSMase2) and the recycling pathway (Goldkorn et al. 1998; Levy et al. 2006, 2009; Sultan et al. 2006). Thus, the conditions present in diabetes stimulate sphingolipid production and increased expression of the enzymes of sphingolipid metabolism.

4 Adipose Tissue in Obesity and Diabetes

Adipose tissue primarily functions to store free fatty acids (FFA) after food intake and to release FFA in the fasting state (Hajer et al. 2008). However, with the rise in obesity, adipose tissue has emerged as an important tissue in the metabolism of lipids and glucose and possibly the induction of insulin resistance and type 2 diabetes (Samad et al. 2006; Hajer et al. 2008; Bonzon-Kulichenko et al. 2009). Adipose has been demonstrated to act in an endocrine fashion by producing a multitude of factors that include a variety of proinflammatory cytokines and chemokines collectively known as adipokines (Mohamed-Ali et al. 1997; Christiansen et al. 2005; Hotamisligil 2006; Samad et al. 2006, 2011), and the production of these inflammatory mediators is believed to mediate the initiation of insulin resistance in distal tissues such as skeletal muscle and liver.

A proposed mechanism for obesity-induced insulin resistance in distal tissues is by increased intracellular lipid accumulation. Induction of the expression and production of adipokines results in increased lipolysis in adipocytes and elevated release of FFA. Adipocytes at this point lose their ability to store FFA and as a result promote the accumulation of lipids such as liver, skeletal muscle, and pancreas. The inappropriate storage of lipids in these tissues results in the activation of signaling pathways that interfere with insulin signaling and ultimately contribute to insulin resistance in adipose, muscle, and liver. Collectively, these actions have been termed "lipotoxicity." A key outcome of lipotoxicity is insulin resistance, which precedes type 2 diabetes (Turpin et al. 2009).

Indeed, significant changes in the sphingolipids produced by the adipose tissue have been reported in genetically obese (ob/ob) mice (Samad et al. 2006). Compared to their lean counterparts, ob/ob mice demonstrated significantly higher levels of expression of acid sphingomyelinases (ASMase), NSMase, and SPT. When the levels of sphingolipids within the adipose tissue were measured, it was found that there were decreased levels of total sphingomyelin and ceramide and increased levels of sphingosine. These data suggest that sphingomyelin may be hydrolyzed to ceramide and further converted to sphingosine. Consistent with this, the expression of both alkaline and acid ceramidases was increased in the adipose tissue of ob/ob mice over their lean counterparts. Analysis of plasma sphingolipids from lean and ob/ob mice found that there were increases in sphingomyelin, ceramide, sphingosine, and S1P (Samad et al. 2006). There was no detectable S1P in the adipose tissue of these mice. These data support studies that suggest increased sphingolipid levels in the plasma of obese mice are associated with increased cardiovascular risk in obesity (Auge et al. 2004; Hojjati et al. 2005). Not only have studies evaluated sphingolipid metabolism in genetically obese mice, they have evaluated sphingolipid metabolism in diet-induced obesity models as well. In this study, C57BL/6J mice were maintained on a high-fat (60 % calories from fat) or a low-fat (10 % calories from fat) diet for 16 weeks. In the mice fed with the high-fat diet (HFD), ceramide levels in the adipose and the plasma were increased as evidenced by increases of the enzymes SPT, ASMase, and NSMase (Shah et al. 2008). Studies evaluating changes to sphingolipid metabolism in human adipose tissue remain scarce and have demonstrated that ceramide and triacylglycerides are elevated in the adipose tissue of obese individuals (Kolak et al. 2007). Other studies have demonstrated that inflammation in human adipose tissue affects sphingolipid metabolism in other tissues such as skeletal muscle (Lam et al. 2011).

The data demonstrate that sphingolipid synthesis is disrupted in adipose tissue in response to increased fatty acid influx. As mentioned above, the source of excess FFA delivery from adipose to other tissues such as skeletal muscle leads to loss of insulin sensitivity. In fact, additional roles for sphingolipids in these disease processes are also emerging.

5 Sphingolipids in Skeletal Muscle

The sensitivity of skeletal muscle to insulin is critical for the removal of glucose from the bloodstream as it accounts for approximately 40 % of total body mass and 80 % of whole body insulin-stimulated glucose disposal (Gorski et al. 2002; Bruni and Donati 2008). It is for these reasons that skeletal muscle is believed to contribute most significantly to the glucose intolerance associated with nutrient oversupply and obesity.

In insulin-sensitive tissues, insulin promotes the uptake of glucose (Summers 2006). Insulin signaling is initiated through the activation of the insulin receptor. Insulin receptor substrates (IRS proteins) recruit and activate the phosphatidylinositol 3-kinase (PI3K) (Keller and Lienhard 1994; Summers 2006), which in turn produces phospholipid phosphatidylinositol (3,4,5)-triphosphate (PIP3), a bioactive lipid that mediates recruitment of the cytosolic enzyme Akt/protein kinase B (PKB) to the plasma membrane (Keller and Lienhard 1994). Phosphorylation of the Thr308 in the activation loop and Ser473 in the hydrophobic C-terminal domain by PIP3 is required for full activation of Akt/PKB (Vanhaesebroeck and Alessi 2000; Summers 2006).

Akt/PKB promotes the uptake of glucose in skeletal muscle by stimulating translocation of Glucose Transporter 4 (GluT4) resulting in an increase in the rate of glucose flux into the tissue, creating a state of decreased circulating glucose and stimulating its conversion to energy storage molecules such as glycogen (Hajduch et al. 1998, 2001; Whiteman et al. 2002). Akt/PKB response to insulin can be perturbed by stimuli such as TNF-α, glucocorticoids, and prolonged exposure to long-chain fatty acids such as palmitate, all of which have been implicated in insulin resistance (Chavez et al. 2003; Holland et al. 2007). Intriguingly, a common feature of these stimuli is their ability also to promote the accumulation of sphingolipids (i.e., ceramide) (Schmitz-Peiffer et al. 1999; Teruel et al. 2001; Holland et al. 2007).

Exposure of skeletal muscle to FFA results in the development and accumulation of intramuscular triacylglycerol (IMTG) and other fatty-acid-derived molecules (i.e., diacylglycerol and ceramide) (Itani et al. 2002; Moro et al. 2008; Lipina and Hundal 2011). The accumulation of IMTG in skeletal muscle has been associated with endurance-trained athletes with significant sensitivity to insulin, indicating that the accumulation of IMTG is not the sole mediator of insulin resistance in skeletal muscle (Goodpaster et al. 2001; Bruce et al. 2004); however, it has been demonstrated that exposure of skeletal muscle to exogenous palmitate, the most abundant fatty acid in plasma, impaired insulin signaling by preventing insulin-induced phosphorylation of Akt/PKB (Hajduch et al. 2001; Moro et al. 2008). Several studies have demonstrated that the insulin-desensitizing effects of palmitate can be mimicked through the acute application of ceramide and ceramide analogues

(Schmitz-Peiffer et al. 1999; Teruel et al. 2001; Chavez et al. 2003). And, moreover, inhibitors of sphingolipid biosynthesis protect cells from the insulin-desensitizing effects of palmitate (Chavez et al. 2003).

The accumulation of IMTG does not appear to be pathogenic by itself but reflective of lipid oversupply, inducing an increase in free palmitate; however, whether IMTG serve as a source of substrate for sphingolipid synthesis is unknown, but in general, sequestration of cell lipids into TAG is protective (Dimitrios et al. 2010; Amati et al. 2011).

6 Sphingolipids in the Diabetic Heart

Heart disease is a major clinical concern in type 2 diabetes: patients with diabetes are several times more likely to die of heart disease than nondiabetics (Ho et al. 1993; Kucharska-Newton et al. 2010). Significantly, diabetics develop a distinct cardiomyopathy that is independent of traditional risk factors (e.g., hypertension and ischemia/reperfusion) (Guha et al. 2008). This condition, termed *diabetic cardiomyopathy*, is characterized by systolic and diastolic dysfunction, reduced contractility, fibrosis, hypertrophy, and a profound shift away from utilization of glucose as a substrate (Boudina and Abel 2010). Strikingly, severity of insulin resistance was associated with both the level of myocardial steatosis and the degree of cardiac remodeling (Utz et al. 2011), while the amount of myocardial steatosis was itself an independent predictor of left-ventricular systolic dysfunction in adult humans (Rijzewijk et al. 2008). In fact, induction of myocardial steatosis is used to replicate the phenotype of diabetic cardiomyopathy in mouse models (discussed below). Together, these observations suggested the potential contribution of lipid overload to the development of diabetic cardiomyopathy.

Similarly to what was seen in pancreatic β-cells and other tissues, excess saturated fatty acids were shown to induce both metabolic alterations and apoptosis in cardiomyocytes (de Vries et al. 1997; Hickson-Bick et al. 2000; Sparagna et al. 2000). In isolated neonatal rat cardiac myocytes, it was shown that palmitate induced a dramatic reduction in fatty acid oxidation, as compared to cells treated with oleate (Hickson-Bick et al. 2000). This downregulation of fatty acid oxidation was thought to occur due to reduced AMPK activity and subsequent accumulation of malonyl coenzyme A (CoA), which in turn inhibits fatty acid oxidation through CPT-1. The reduction in fatty acid oxidation shunted fatty acyl-CoAs into non-oxidative pathways: triglyceride and ceramide levels both doubled in response to palmitate, but not oleate, treatment (Hickson-Bick et al. 2000; Sparagna et al. 2000). This increase in ceramide was accompanied by a significant increase in DNA laddering as well as more than twofold increase in caspase 3 activity in palmitate-treated cells (Hickson-Bick et al. 2000). These results were suggestive of a toxic role for ceramides, as had been demonstrated a short time before in pancreatic β-cell lipotoxicity (Shimabukuro et al. 1998) and in TNF-α-mediated cardiomyocyte toxicity (Krown et al. 1996).

Ceramide was first directly implicated in cardiomyocyte apoptosis through the use of a cell-permeable artificial ceramide; this finding was confirmed and extended using a combination of inhibitor studies and ceramide treatments. Initially, it was shown that treatment with C_2-ceramide, a cell-permeable synthetic ceramide, induced neonatal cardiomyocyte apoptosis in a caspase 3- and 8-dependent manner (Wang et al. 2000). A subsequent, conceptually important study confirmed that de novo sphingolipid synthesis was required for induction of lipotoxicity in isolated adult rat ventricular cardiomyocytes. In this system, it was shown that inhibition of ceramide synthesis attenuated lipotoxic and glucolipotoxic TUNEL staining and completely ameliorated lipotoxic myofibrillar degeneration (Dyntar et al. 2001). This approach was important for two reasons: first, primary adult cardiomyocytes are postmitotic and, therefore, a better model of the adult heart than neonatal cells, and, second, the use of an inhibitor provided redundancy to the short-chain ceramide analogue-based approach, which has caveats that will be discussed later in this chapter. Therefore, through the use of complementary approaches, these experiments demonstrated that an increase in ceramide levels is both necessary and sufficient to induce lipotoxic cell death and myofibrillar degeneration in cardiomyocytes.

Further studies extended the known role for ceramide in development of the diabetic cardiomyopathy phenotype, including induction of heart failure markers, downregulation of glucose oxidation, and impairment of contractility. Numerous aspects of the response of cardiomyocytes to ceramide were worked out in AC-16 cells, which are an immortalized human ventricular cardiomyocyte-derived cell line (Park et al. 2008). For example, it was shown that C_6-ceramide treatment of AC-16 cells induced upregulation of the heart failure markers ANF and BNP (Park et al. 2008). Suggesting a role for ceramide in lipotoxic downregulation of glucose oxidation, C_6-ceramide treatment also suppressed expression of the GLUT-4 glucose transporter and potentiated expression of PDK4, which inhibits the pyruvate dehydrogenase complex. Ceramide has also been implicated in the lipotoxic impairment of cardiac contractility. Specifically, it was shown that treatment with either palmitate or C_6-ceramide induced β-adrenergic dysfunction in AC-16 cells (Drosatos et al. 2011). Strikingly, the induction of β-adrenergic dysfunction by palmitate was completely ameliorated by co-treatment with the de novo sphingolipid synthesis inhibitor myriocin. This sphingolipid-dependent impairment of β-adrenergic function was thought to occur as a result of activation of PKCα and PKCδ. However, it should be noted that treatment with diacylglycerol was shown to produce the same effects, and it is likely that sphingolipids and diacylglycerol work in concert in this process. Together, these studies showed that, in an immortalized cardiomyocyte cell system, increased cellular ceramide was sufficient to induce expression of heart failure markers, changes in glucose uptake and metabolism, and β-adrenergic dysfunction, and de novo sphingolipid synthesis was required for induction of β-adrenergic dysfunction by palmitate. Critically, results obtained in isolated cell systems have been recapitulated and extended in vivo. One model that has provided a great deal of information is the LpL^{GPI} mouse, which expresses a heart-specific membrane-bound form of lipoprotein lipase and thus develops profound cardiac steatosis (Yagyu et al. 2003). Cardiomyocytes bearing this transgene accumulated ceramides, sphingomyelin, and diacylglycerol (Park et al. 2008). Importantly, while myriocin treatment prevented the increase in cardiac ceramides and sphingomyelin, diacylglycerol levels remained elevated; this allowed the authors to tease out ceramide-specific effects from those of diacylglycerol. The results of this study were striking. It was shown that many of the pathological outcomes in the LpL^{GPI} mouse, including gross cardiac hypertrophy, increased left-ventricular systolic diameter, reduced cardiac efficiency, and increased mortality, were completely prevented by myriocin treatment, which inhibits the first step of de novo sphingolipid synthesis. Glucose oxidation and palmitate oxidation were also restored to normal levels by myriocin treatment, as was expression of molecular markers of heart failure (ANF and BNP), hypertrophy (GSK3β phosphorylation), and insulin signaling (AKT phosphorylation). These sphingolipid-dependent outcome effects were also attenuated upon crossing with a mouse heterozygous null for the serine palmitoyltransferase subunit SPTLC1, which also displays impaired de novo sphingolipid biosynthesis. Thus, the role of sphingolipids in the development of lipotoxic cardiac hypertrophy and heart failure was demonstrated in vivo.

Another transgenic model system provided additional support for the role of sphingolipids in diabetic cardiomyopathy. The *ob*/*ob* mouse is a very popular leptin-deficient model of obesity and type 2 diabetes; these mice develop marked cardiac hypertrophy, dysfunction, and steatosis (Barouch et al. 2006; Dobrzyn et al. 2010). Paralleling results in the LpL^{GPI} mouse, these animals dramatically accumulate ceramide and overexpress SPT, the initial enzyme of de novo sphingolipid synthesis (Dobrzyn et al. 2010). However, it was shown that loss of stearoyl-CoA desaturase 1 (SCD1), which catalyzes synthesis of monounsaturated fatty acids, not only protected *ob*/*ob* mice from obesity but also reduced SPT expression and ceramide to normal levels and prevented the hypertrophic cardiac phenotype. This study thereby demonstrated that a mutation that indirectly normalized ceramide levels prevented the development of diabetic cardiomyopathy.

While most studies have focused on the role of ceramide in the heart, other work has revealed a protective role for sphingosine-1-phosphate, similar to that demonstrated in other lipotoxic cell models. In isolated primary cardiomyocytes, treatment with sphingosine-1-phosphate protected against palmitate-induced cell death as effectively as myriocin treatment (Holland et al. 2011). Additionally, overexpression of sphingosine kinase 1 protected diabetic mice against the development of cardiomyopathy (Ma et al. 2007). Intriguingly, though, treatment with sphingosine-1-phosphate promoted hypertrophy of neonatal cardiomyocytes, possibly through signaling of the EDG1/S1PR1 receptor (Robert et al. 2001). It remains unknown whether this effect is recapitulated in adult cardiomyocytes. Therefore, although sphingosine-1-phosphate signaling appears to play a protective role in the lipotoxic heart, its function has yet to be completely understood.

The role for sphingolipids other than ceramide and sphingosine-1-phosphate in the diabetic heart remains almost totally uncharacterized. Two studies have indicated that administration of a ganglioside preparation protected atrial function in genetically obese, diabetic mice and prevented reduction in cardiac norepinephrine concentration in diabetes (Prosdocimi et al. 1987; Tessari et al. 1988); potential actions of other sphingolipids in the diabetic heart remain undescribed. Significantly, many studies addressing the roles of sphingolipids in the heart do so by inhibiting de novo sphingolipid synthesis. In addition to reducing ceramide, though, these strategies can reduce levels of downstream metabolites of ceramide (Miyake et al. 1995). Thus, because of the known roles for sphingolipid species other than ceramide and sphingosine-1-phosphate in other diabetic organs, this represents a major gap in knowledge of sphingolipid functions in the diabetic heart and presents a caveat to attribution of cellular outcomes to ceramide per se.

Although numerous studies have addressed the roles of ceramide in diabetic cardiomyopathy, few have addressed the mechanisms by which sphingolipid levels are regulated in diabetes. While general mechanisms parallel those described elsewhere in this chapter, a few cardiac-specific observations are worthy of mention. As discussed above, the prevailing theory of lipotoxicity proposes that accumulation of toxic lipid species, including ceramide, results from the inability of cells to sequester free fatty acids adequately as neutral lipid (Cnop et al. 2001). To test this hypothesis, diacylglycerol acyltransferase 1 (DGAT1), which catalyzes the formation of triacylglycerol from diacylglycerol, was overexpressed in a cardiac-specific manner in mice (Liu et al. 2009). This resulted in accumulation of triacylglycerols and a concomitant reduction in ceramide and diacylglycerol levels. Furthermore, when these animals were crossed with a transgenic strain that developed lipotoxic cardiomyopathy, the resulting mice displayed improved heart function and reduced mortality as well as more normal expression of markers of heart failure, oxidative stress, and apoptosis. Strikingly, statistical analyses revealed a significant correlation between cardiac ceramide levels and the level of cardiac function in a mixed cohort of transgenic and wild-type mice. This study thus provided support for the hypothesis that lipid overload induces steatosis by overwhelming cells' capacity to sequester fatty acids as neutral lipid.

Other work has addressed a potential role for the transcription factor PPARα in regulation of ceramide levels (Baranowski et al. 2007). In this study, treatment of wild-type rats with a PPARα agonist (WY-14643) synergized with high-fat diet to increased SPT activity in a more dramatic manner than observed with diet alone. Likewise, ceramide levels also increased markedly upon administration of WY-14643 in high-fat diet, and the fatty acyl chain distributions of both ceramide and sphingomyelin were also altered. This latter finding suggests that PPARα could potentially act on other components of the sphingolipid synthetic pathway besides SPT. Strikingly, other studies have implicated PPARα in heart failure [reviewed in Finck (2004) and Madrazo and Kelly (2008)], thus raising the question of whether these effects are mediated by regulation of sphingolipid levels by PPARα. Together, these findings suggest that PPARα-mediated signaling may augment the observed substrate-based effects of lipid oversupply to promote de novo sphingolipid synthesis.

In summary, these findings demonstrate roles for sphingolipids in the pathogenesis of cardiac lipotoxicity and diabetic cardiomyopathy. Ceramide was implicated not only in the induction of cardiomyocyte apoptosis but also in insulin resistance, altered cardiac substrate utilization, hypertrophy, β-adrenergic dysfunction, and impaired cardiac function. In contrast, there are hints that sphingosine-1-phosphate may protect cardiomyocytes from apoptosis while, perhaps, potentiating hypertrophy. As of this writing, no published studies have addressed potential roles for other sphingolipids, including glycosphingolipids, sulfatide, and ceramide-1-phosphate, in the diabetic heart. Furthermore, the contributions of particular enzymes of sphingolipid metabolism and individual *N*-acyl chain lengths have been utterly neglected. Additionally, it should be noted that, while much of the literature on sphingolipids in the diabetic heart focuses on their roles in diabetic cardiomyopathy, they also play important roles in injury following ischemia and reperfusion (I/R) (Maulik et al. 1993; Bielawska et al. 1997; Theilmeier et al. 2006) and in defects in electrical conductance (Constable et al. 2003) and autonomic nervous function (Tessari et al. 1988; Davis et al. 2006), all of which are major concerns in the diabetic heart. Finally, the role of sphingolipids in diabetic glucotoxicity in the heart has been abjectly neglected, despite its potential importance. Thus, while sphingolipids have been shown to play multiple roles in the pathogenesis of cardiac lipotoxicity, much remains to be learned about their regulation, specific mechanisms of action, and broader significance in the diabetic heart.

7 Sphingolipids in the Diabetic Pancreas

Pancreatic dysfunction is the hallmark disease outcome in diabetes. In insulin resistance, excess plasma free fatty acids initially stimulate insulin secretion by the pancreas, both directly and in response to insulin resistance in peripheral tissues, resulting in a compensatory hyperinsulinemic state (Milburn Jr et al. 1995). However, a gradual decline in insulin release occurs, in part, due to decreased insulin sensitivity in the pancreatic β-cells themselves; this effect has been shown to be mediated by excess uptake of free fatty acids from the plasma (Lee et al. 1994). The resulting loss of compensation results in frank hyperglycemia, which synergizes with the hyperlipidemic state to promote molecular pathology and cell death, as described later in this section.

The process of fatty-acid-mediated β-cell dysfunction and, ultimately, death occurs through *lipotoxicity*. The ability of lipid oversupply to increase ceramide levels in the diabetic pancreas has long been appreciated (Shimabukuro et al. 1998). Initially, it was observed that both DNA laddering (an indicator of apoptosis) and ceramides increased in the pancreatic islets from obese, diabetic rats, compared to lean controls. Furthermore, treatment of isolated islets with free fatty acids recapitulated this DNA laddering and ceramide production in vitro. Strikingly, it was found that only free fatty acids that promoted ceramide synthesis were able to cause lipotoxicity and insulin resistance in vitro, supporting a link between ceramides and islet lipotoxicity (Maedler et al. 2001, 2003). Substantiating this relationship, treatment with a membrane-permeable synthetic ceramide, C_2-ceramide, directly potentiated DNA laddering (Shimabukuro et al. 1998). Intriguingly, this study also noted that islet cells from obese, diabetic rats were more vulnerable to free fatty-acid-mediated lipotoxicity, suggesting that changes in metabolism or protein expression in the diabetic pancreas augmented susceptibility of these cells to lipotoxicity. Together, these findings suggested both that ceramide plays a key role in lipotoxic β-cell apoptosis and that the diabetic pancreas displays an increased vulnerability to toxicity from excess plasma free fatty acids, possibly due to the already-heavy fatty acid loading of these cells.

Glucotoxicity has also been shown to play an important role in pancreatic damage after decompensation of insulin resistance; this process, too, seems to be governed by sphingolipids. Most directly, glucose-stimulated cytotoxicity was shown to be ameliorated by inhibition of ceramide synthase (Maedler et al. 2003). Unfortunately, the effect of this treatment on downstream, complex sphingolipid species was not determined. This point may be significant because, as described later in this section, ceramides may be adorned with carbohydrate headgroups to form complex sphingolipids, which have distinct signaling and membrane roles from those of simple ceramide. Regardless of this point, though, it appears that oversupplies of glucose and free fatty acids work in concert to promote ceramide synthesis in the diabetic pancreas. In fact, neither excess glucose nor a low dose of palmitate alone was able to induce ceramide production profoundly in INS832/13 or INS-1 cells, but the two conditions in combination strongly potentiated this effect (El-Assaad et al. 2010; Veret et al. 2011). In contrast, excess glucose and palmitate each provoked ceramide production in isolated rat islet cells, while the combination of both stimulated an even stronger increase in ceramide levels (Kelpe et al. 2003). This suggests a synergy between excess glucose and excess fatty acids in the induction of sphingolipid-dependent diabetic β-cell toxicity. These findings bring to light several unanswered questions, such as whether glucose and fatty acid oversupply act on distinct metabolic routes of sphingolipid production and clearance, whether these two routes of toxicity promote production of different sphingolipid metabolites, and whether they act on the same or complementary lipotoxic pathways.

Further studies more fully characterized the sphingolipid-dependent molecular features of pancreatic functional deterioration, including apoptosis, endoplasmic reticulum (ER) stress, and insulin resistance. For example, increased intracellular ceramides stimulated several components of the apoptotic cascade, such as cytochrome C release into the cytosol, a reduction in Bcl-2 expression, and loss of mitochondrial membrane potential (Maedler et al. 2003; Veluthakal et al. 2005). Broadly, it was found that ceramide acted to promote apoptosis, in part, by inducing caspase activation (Lupi et al. 2002); studies in the MIN6 insulinoma cell line and in glucolipotoxicity in INS-1 cells specifically implicated caspases 3/7 in sphingolipid-mediated cell death (Boslem et al. 2011; Veret et al. 2011). Ceramide accumulation also induced lipotoxic ER stress, as indicated by increased expression of CHOP, in MIN6 cells (Boslem et al. 2011). Finally, ceramide was directly implicated in insulin resistance at a molecular level, in addition to its nonspecific actions of reducing β-cell proliferation and promoting apoptosis (Maedler et al. 2001). In particular, ceramide accumulation reduced proinsulin mRNA levels via reduced transcription in INS-1 and isolated rat pancreatic islet cells (Kelpe et al. 2003; Guo et al. 2010). This may have been mediated by attenuation of ERK phosphorylation (Guo et al. 2010). Thus, elevated sphingolipid loads in the diabetic pancreas promote insulin resistance and pancreatic failure through a suite of specific molecular outcomes. These findings lead to a larger question of how sphingolipid synthesis is upregulated in the diabetic pancreas, thereby allowing all of these downstream effects to ensue.

As discussed above, lipid oversupply perturbs sphingolipid homeostasis by diverse mechanisms in many organs and cell types, and this is also true in diabetic pancreas. For example, it has been shown that expression of SPT is upregulated in the diabetic pancreas (Shimabukuro et al. 1998). This upregulation may be recapitulated in vitro by oversupply of the fatty acid palmitate as well as by other factors, including exposure to elevated levels of hepatocyte growth factor (HGF), which is overproduced in diabetes (Shimabukuro et al. 1998; Gonzalez-Pertusa et al. 2010). Other routes and enzymes of sphingolipid synthesis have been shown to be regulated by saturated fatty acids in skeletal muscle, as discussed above, and may also play important roles in governance of sphingolipid levels in the diabetic pancreas (Hu et al. 2009). Thus, the diabetic pancreas harbors elevated sphingolipid levels due to both excess substrate supply and upregulation of the enzymes of sphingolipid synthesis. However, while most studies have focused on bulk sphingolipid levels and general routes of metabolism, the roles of individual metabolites and enzymes in this system have remained undetermined. Largely because the distinct functions of individual *N*-acyl chain lengths of ceramide and specific (dihydro)ceramide synthase (CerS) enzymes have only recently been appreciated, only one study thus far has attempted to implicate an individual CerS isoform in β-cell lipotoxicity: Véret et al. proposed that CerS4 is responsible for glucolipotoxicity in INS-1 cells (Veret et al. 2011). Indeed, overexpression of CerS4 induced caspase 3/7 activity, while siRNA-mediated knockdown of CerS4 partially attenuated caspase activity in response to glucose and fatty acid overload. This work clearly demonstrated a role for CerS4 in glucolipotoxicity; however, additional studies will be required to define or to rule out contributions from other CerS isoforms. There are a few reasons for this. First, although the ceramide chain length profiles in the manuscript are suggestive, they are not sufficiently clear-cut to implicate a specific isoform. Second, the effect of overexpressing or knocking down other CerS isoforms was not tested. Third, and finally, cells were presented with only palmitate in the media. Beta cells encounter a mixed profile of fatty acids in vivo, which are utilized differentially for *N*-acylation by individual CerS isoforms, and exposure to this characteristic mixture of substrates may reveal roles for other CerS isoforms in a physiologic context. Thus, although CerS4 has been implicated in β-cell lipotoxicity, roles for other CerS isoforms have not been ruled out. Furthermore and importantly, no studies have identified a role for specific ceramide species in β-cell lipoapoptosis. Thus, significant work will be required to elucidate the mechanisms by which diabetes and lipotoxicity upregulate sphingolipid levels in the pancreas, as well as the significance of specific metabolic pathways.

Although most studies discussing sphingolipids in pancreatic lipotoxicity focus on ceramide, ceramide itself is not the only sphingolipid that accumulates during fatty acid overload. Indeed, many downstream metabolites of ceramide are modulated by manipulation of ceramide synthesis, and one must be careful before concluding that it is ceramide, per se, that is responsible for particular molecular outcomes. For example, glucosylceramide was shown to increase in MIN6 insulinoma cells treated with palmitate, and this accumulation was, in fact, much more robust than that of simple ceramide (Boslem et al. 2011). Furthermore, this may have important functional implications, as inhibition of glucosylceramide synthase, which may presumably increase its metabolic precursor, ceramide, partially preserved pancreatic insulin secretion in Zucker diabetic fatty rats (Aerts et al. 2007). This suggests a potential role for glucosylceramide, in particular, and sphingolipids other than ceramide, in general, in pancreatic insulin resistance. An additional example of this principle involves sulfatide (3′-sulfogalactosylceramide), a glycosphingolipid present in the islets of Langerhans in the pancreas (Buschard et al. 2002). Like its metabolic precursor ceramide, this lipid has also been implicated in diabetes and insulin resistance. Mechanistically, sulfatide is involved in proper folding of proinsulin, preservation of insulin crystals, and monomerization of stored insulin hexamers (Osterbye et al. 2001). Furthermore, C16:0 sulfatide, specifically, was shown to be required for stabilization of insulin granules (Blomqvist et al. 2003). C16:0 sulfatide also appears to be required for appropriate management of insulin secretion in normal rat pancreatic islets; it was shown to attenuate insulin secretion by reducing sensitivity of the ATP-dependent potassium channel to inhibition by ATP (Buschard et al. 2002, 2006). Significantly, this particular sulfatide species was deficient in the pancreas of several rodent models of type 2 diabetes (Blomqvist et al. 2003), and administration of sulfatide appeared to ameliorate some aspects of diabetic pathology in vivo. For example, administration of C16:0 sulfatide, specifically, to Zucker fatty rats, which express a defective leptin receptor, reduced fasting insulin and improved the first-phase insulin response and glucose-stimulated insulin response (Blomqvist et al. 2005). More generally, sulfatide was shown to protect isolated islet cells from apoptosis, iNOS expression, and nitric oxide secretion induced by cytokines; this protective effect was not specific to C16:0 sulfatide (Roeske-Nielsen et al. 2010). These findings indicate that sulfatide exerts a protective effect in pancreatic islets via normalization of insulin secretion and protection from inflammation. It is still unknown whether these protective effects are also relevant to the context of fatty acid oversupply, and the mechanism by which C16:0 sulfatide is reduced in the diabetic pancreas remains unknown.

Finally, sphingosine-1-phosphate, the product of sphingosine kinase, has been suggested as an antiapoptotic lipid in inflammation-mediated β-cell death. In particular, supplementation with sphingosine-1-phosphate protected isolated pancreatic islets from IL-1β-induced apoptosis (Rutti et al. 2009), and exposure to both IL-1β and TNF-α upregulated sphingosine kinase activity in INS-1 cells and isolated rat pancreatic islets (Mastrandrea et al. 2005). It was suggested that this increase in activity was primarily mediated by sphingosine kinase 2 (SPHK2), based on in vitro activity assays. These results indicate that, in contrast to ceramide, sphingosine-1-phosphate acts to protect β-cells from the pathological inflammatory state present in diabetes and the metabolic syndrome; indeed, the balance of ceramide and sphingosine-1-phosphate levels has been suggested as a clinically important determinant of β-cell fate (Jessup et al. 2011).

In summary, these findings illustrate an important role for sphingolipids in the pathogenesis of the diabetic pancreas. Fatty acid and glucose oversupply both stimulate ceramide production, which promotes apoptosis and insulin resistance, as well as synthesis of downstream metabolites of ceramide. Like ceramide, glucosylceramide may act in a pro-pathogenic manner by potentiating insulin resistance. In contrast, sulfatide normalizes insulin secretion, and both sulfatide and sphingosine-1-phosphate protect cells from the harmful effects of inflammation. These foundational findings lead to numerous further questions about the detailed mechanisms of sphingolipid action in the diabetic pancreas. For example, it remains unknown which ceramide species play active roles in pathology, which pathways (i.e., salvage or de novo synthesis) generate the bulk of these pathogenic lipids in vivo, and which specific enzyme isoforms are involved. Functions of sphingosine-1-phosphate and other metabolites also remain incompletely characterized. Finally, the roles of sphingolipids in diabetic glucotoxicity are severely understudied and may be distinct but synergistic with those observed in lipotoxicity. Thus, several sphingolipid species play key, but contrasting, roles in β-cells during diabetes, and further studies are needed to clarify them and to understand the underlying molecular mechanisms.

8 Emerging Concepts in the Sphingolipid-Diabetes Axis: Liver, Kidney, and Retina

While many of the changes in sphingolipid synthesis and signaling have been somewhat well defined in tissues such as skeletal muscle, heart, and pancreas, there are several tissues that have yet to be thoroughly examined. One such tissue is the liver. Located between the intestinal tract and the systemic circulation, the liver plays an important role in the metabolism and storage of dietary fats as well as playing a central role in glucose and lipid metabolism (Coleman and Lee 2004; Yosuke et al. 2011). During times of feeding, hepatocytes increase their uptake of glucose and synthesis of glycogen (Yosuke et al. 2011). The liver also acts to synthesize triacylglycerides (TAG) through β-oxidation of fatty acids obtained from overnutrition or adipocyte cell death and/or inflammation (Coleman and Lee 2004; Bijl et al. 2009). In obesity, the combination of increased fatty acid flux from dysfunctional adipose tissue, coupled with induced fatty acid synthesis in the liver, results in the accumulation of TAG-rich lipid droplets in the cytosol (Preiss and

Sattar 2008; Bijl et al. 2009; Deevska et al. 2009) and ultimately the development of fatty liver and lipotoxicity (Deevska et al. 2009). The development of these pathologies can lead to loss of insulin sensitivity, which in the liver prevents the suppression of glucose generation, signaling the pancreas to increase insulin production resulting in hyperinsulinemia (Brown and Goldstein 2008; Bijl et al. 2009).

As a result of TAG accumulation, lipid synthesis and metabolism are disrupted and perturbations of these pathways have been demonstrated to contribute to loss of insulin sensitivity and the development of hyperglycemia (Deevska et al. 2009; Yosuke et al. 2011). For example, treatment of hepatocytes with palmitate promoted an increase in TAG. TAG accumulation was attenuated by the inhibition of the activity of acid sphingomyelinase (aSMase), which generates ceramide by hydrolyzing sphingomyelin derived from the recycling/endocytic pathway (Deevska et al. 2009). This was confirmed in mice lacking functional aSMase as well as functional LDL receptors. These animals demonstrated that when maintained on a high-fat diet, there were no increases in TAG content in the cytosol of hepatocytes. They did, however, demonstrate increases in de novo sphingolipid synthesis metabolites sphinganine, dihydroceramide, and ceramide, as well as the activity SPT (Deevska et al. 2009). These animals were also protected against diet-induced hyperglycemia and insulin resistance.

Sphingolipid synthesis in the kidney is also adversely affected by the development of insulin resistance in other distal tissues. Hyperglycemia and insulin resistance induced by fatty acid oversupply to the liver result in dramatic alterations in kidney function, sphingolipid profiles, and the development of diabetic nephropathy. Indeed, complex glycosphingolipids have been indicated in the development of diabetic nephropathy, particularly where changes in glomerular sialic acids and/or sialidase activity correlate with the onset of proteinuria (Cohen-Forterre et al. 1984; Baricos et al. 1986; Cardenas et al. 1991; Mather and Siskind 2011). Ceramide has also been implicated in renal injury as its accumulation is a result of nephropathy induced by the consumption of a high-fat diet (Boini et al. 2010). While this study pointed to ceramide as the causative agent in high-fat diet-induced renal injury and nephropathy, glycosphingolipid levels were not assessed. As ceramide is a necessary for glycosphingolipid synthesis, it is possible that the synthesis of glycosphingolipids is also increased and may play a role (Boini et al. 2010). Indeed, the link between glucose metabolism, insulin resistance, and glycosphingolipid synthesis in the kidney has been established (Zhao et al. 2007), indicating glycosphingolipids in renal injury and the development of nephropathy.

The retina plays a significant role in vision and is adversely affected in type 2 diabetes; however, the effects of changes in sphingolipid metabolism have not been thoroughly evaluated. Studies have indicated that the accumulation of ceramide in response to fatty acid oversupply results in increased apoptosis in retinal pericytes and induction of diabetic retinopathy (Cacicedo et al. 2005; Fox et al. 2006). While these studies provide significant evidence, more has to be done to ascertain the mechanism by which sphingolipid synthesis is perturbed in the retina.

9 Sphingolipids as Biomarkers of Diabetes

In addition to their direct roles in the pathogenesis of diabetes, sphingolipids have also emerged as potential biomarkers of diabetes and the metabolic syndrome. It has long been appreciated that sphingolipid levels are perturbed in the blood plasma of diabetic patients. The first such finding indicated that glycosphingolipids were elevated in the plasma of some groups of diabetic human patients (Kremer et al. 1975). This was later confirmed by a study demonstrating that glucosylceramide was elevated in the plasma of diabetics (Serlie et al. 2007). Similarly, plasma ceramides have emerged as potential biomarkers in diabetes. For example, plasma ceramides were elevated in humans with type 2 diabetes and correlated negatively with the rate of insulin-stimulated glucose disposal, indicating a relationship between plasma ceramide levels and insulin resistance (Haus et al. 2009). Furthermore, the concentration of ceramide in the blood plasma positively correlated with levels of the proinflammatory cytokine IL-6, which has been shown to promote insulin resistance (de Mello et al. 2009). Additionally, it was shown in human patients that sphingomyelin levels in the plasma membranes of erythrocytes positively correlated with circulating insulin levels (Zeghari et al. 2000). These results were echoed in an animal model of diabetes: sphingomyelin, ceramide, sphingosine, and sphingosine-1-phosphate were all elevated in the plasma of *ob*/*ob* mice, which are leptin-deficient, diabetic, and obese (Samad et al. 2006). Thus, multiple studies have suggested that sphingolipid levels in the plasma correlate with specific diagnostic criteria for diabetes and the metabolic syndrome in both mice and humans.

Despite these highly suggestive observations, relatively few studies have directly analyzed the predictive value of sphingolipid levels for diabetes. In one such study, sphingomyelin content of erythrocyte plasma membranes was shown to be an independent predictor of fasting insulin levels, degree of insulin resistance, and level of glucose tolerance (Candiloros et al. 1996). Indeed, higher levels of membrane sphingomyelin correlated with increased fasting insulin, insulin resistance, and glucose tolerance in both normal and diabetic human patients.

Another recent study analyzed the potential of uncommon sphingosine bases as biomarkers of type 2 diabetes and the metabolic syndrome (Othman et al. 2011). In this study, plasma sphingolipids were deacylated, and the relative contributions of different sphingoid bases to the total pool were quantified in a small, homogenous population. The authors found that the concentrations of the uncommon sphingoid bases deoxysphinganine and deoxysphingosine, which incorporate the amino acid alanine rather than serine, were increased in patients with metabolic syndrome, with or without type 2 diabetes. Additionally, it was found that the levels of these compounds had a significant positive predictive value for the metabolic syndrome. Furthermore, the concentration of C_{16}-sphingosine, which is derived from myristoyl-CoA rather than palmitoyl-CoA, was decreased in patients with both diabetes and the metabolic syndrome, but not in patients with metabolic syndrome alone. It was also determined that levels of C_{16}-sphingosine had significant predictive value in differentiating patients with type 2 diabetes from prediabetic and control patients. Thus, it was suggested that levels of alanine-derived sphingoid bases are a potential biomarker for metabolic syndrome, with or without type 2 diabetes, while levels of C_{16}-sphingosine could be used to detect the transition from compensated insulin resistance to type 2 diabetes.

Together, these studies demonstrate the potential value of sphingolipids, and particularly uncommon sphingolipids, as biomarkers of diabetes and the metabolic syndrome. However, although strong correlations exist between sphingolipid levels and parameters of these diseases, few studies have performed the statistical analyses necessary to determine their predictive value, including ROC curves, which determine specificity and sensitivity of a potential biomarker. Additionally, further studies will be required to validate the findings of Othman et al. (2011) and Candiloros et al. (1996). These studies would need to incorporate larger and more diverse populations and, ideally, follow a prospective design. Furthermore, studies of ceramides and complex sphingolipids could benefit from analysis of individual *N*-acyl chain lengths. Thus, while plasma sphingolipids display much potential as biomarkers of diabetes, substantial work remains to be done before these tests reach the clinic.

10 Summary and Future Directions

Diabetes presents as a multifactorial disorder affecting multiple organ systems. Although broader disease outcomes (e.g., liver disease, heart failure, kidney failure) are complex and diverse, the underlying cellular and molecular etiologies flow through some common channels. One major theme seems to be sphingolipid-mediated insulin resistance and cell death, consequent to lipotoxicity. Indeed, a great deal of work has been done to link de novo sphingolipid synthesis and individual sphingolipid metabolites to particular molecular events. Broadly, ceramide has been implicated as a toxic lipid promoting insulin resistance, metabolic derangement, and cell death in skeletal muscle, heart, and pancreas (Shimabukuro et al. 1998; Schmitz-Peiffer et al. 1999; Park et al. 2008). In contrast, sphingosine-1-phosphate has been tentatively shown to exert protective effects in these same organ systems, despite its promotion of inflammation (Rutti et al. 2009; Holland et al. 2011). However, these general findings merely reveal an entire network of questions about the specific molecular mechanisms of disease, especially regarding which pathways, enzyme isoforms, and sphingolipid metabolites are pathogenic or protective. Fortunately, new tools have emerged that will allow researchers to uncover precise aspects of these mechanisms.

Previously, the nature of the available experimental tools limited the resolution to which sphingolipid pathways could be teased apart. For many years, ceramide and its metabolites could only be measured by techniques such as TLC, which revealed changes only in bulk levels of these lipids. However, recent developments now allow researchers to examine detailed changes in sphingolipid *N*-acyl chain length profiles using quantitative mass spectrometry-based approaches (Bielawski et al. 2010). Results in other disease contexts, particularly cancer, have demonstrated that alterations in levels and proportions of ceramide *N*-acyl chain lengths are functionally important (Senkal et al. 2010, 2011), and it may now be determined whether the lipotoxic effects of ceramide occur due to increases in bulk ceramide levels or to shifts toward particular chain length profiles. Furthermore, mass spectrometry-based labeling strategies, such as the use of ^{13}C-palmitate, allowed investigators to trace the specific metabolic routes that are invoked in their experimental systems, distinguishing between de novo synthesis and salvage pathways, for instance (Hu et al. 2009). These findings would lead to a better understanding of the underlying molecular mechanisms of ceramide-mediated lipotoxicity.

Critically, determining the nature of any changes in sphingolipid *N*-acyl profiles would help to reveal which CerS isoforms mediate the lipotoxic effects of ceramide and its downstream metabolites, as each CerS isoform produces a unique *N*-acyl chain length distribution (Pewzner-Jung et al. 2006; Laviad et al. 2008). Once again, newly developed experimental approaches enable investigators to resolve the cellular functions of particular CerS isoforms. While myriocin, which inhibits SPT, or fumonisin B1, which inhibits all isoforms of CerS, can function as powerful tools to screen for sphingolipid dependence, they reveal very little about involvement of specific sphingolipid metabolic pathways (Wang et al. 1991; Miyake et al. 1995). In contrast, new approaches based on DNA transfection, RNAi, and knockout animals allow precise insight into the roles of specific CerS isoforms in cellular functions and disease, with the caveat that dysregulation of individual isoforms can perturb global CerS expression patterns (Mullen et al. 2011a, b). These approaches may also be applied to other enzymes of SL synthesis that are present in more than one isoform (e.g., ceramidase, sphingosine kinase, sphingomyelinase). Thoughtful employment of these strategies would do much to advance the field beyond a binary sphingolipid-dependent/sphingolipid-independent understanding of diabetic molecular and cellular phenotypes. Successful identification of specific enzymes for therapeutic targeting would also provide rationale for developing inhibitors selective for specific CerS isoforms, a strategy already well underway for sphingosine kinases (ref some Sk1 vs. SK2 inhibitor papers).

Notwithstanding the need for specific inhibitors, due to the highly interconnected nature of sphingolipid metabolism, any approach based on inhibiting or perturbing these pathways should be interpreted with care. In particular, many studies consider sensitivity to myriocin or fumonisin B1 to indicate a dependence on ceramide itself, excluding potential roles for subsequent metabolites. In fact, although such results are suggestive, they can be misleading; inhibition of de novo sphingolipid synthesis can reduce not only levels of ceramide but of its downstream metabolites (Saito et al. 2005; Mullen et al. 2011a, b). Thus, conclusions based on inhibitor studies actually indicate a dependence on either ceramide or its downstream metabolites. Similar caution should be used when employing knockdown or overexpression strategies to manipulate specific enzyme isoforms. Lipid profiles produced by individual isoforms may partition distinctly either at a metabolic level, due to differential acyl chain length preferences of downstream enzymes, or at the organelle level, due to different localization of particular enzymes of sphingolipid metabolism [reviewed in Brice and Cowart (2011)]. Furthermore, knockdown of individual CerS isoforms has been shown to perturb regulation of nontargeted isoforms (Mullen et al. 2011a, b). Thus, results of inhibitor, knockdown, or overexpression experiments should be interpreted cognizant of potential off-target effects on sphingolipid synthesis and metabolism.

Similar caution should be exerted toward experiments using C_2- and C_6-ceramide, which have served as important tools in determining the roles of sphingolipids in diabetes. This is because these artificial, short-chain ceramides have different biophysical properties than endogenous ceramides and may induce distinct cellular outcomes (Gidwani et al. 2003; Nybond et al. 2005), including lipotoxic outcomes through mechanisms that are not physiologically relevant. For example, treatment with C_2-ceramide can provoke mitochondrial depolarization and cytochrome C release in cardiomyocytes (Sparagna et al. 2000; Di Paola et al. 2004; Parra et al. 2008); these results would obviously suggest a role for ceramide. However, these cellular events actually precede ceramide accumulation in the context of palmitate-induced neonatal rat cardiomyocyte lipotoxicity, suggesting that this portion of the lipotoxic cascade is not ceramide-dependent in this system (Sparagna et al. 2000). Similarly, while pyridinium ceramides can provide important clues to intracellular functions of ceramide, it should be kept in mind that these compounds preferentially target to the mitochondria, which could potentially cause their functions to diverge from those of natural ceramides (Dindo et al. 2006). Furthermore, the activity of downstream enzymes, particularly the ceramidases, could be quite different toward endogenous ceramides than toward either short-chain or pyridinium-containing ceramides, which have modified *N*-acyl chains. Therefore, although treatment with cell-permeable ceramide analogues can activate multiple pathologic pathways and can provide important insights, other modes of experimentation are required to confirm that these mechanisms of pathology are consistent with those induced by endogenous sphingolipids.

In conclusion, it has been shown that sphingolipids play a multiplicity of both harmful and protective roles in the pathogenesis of the metabolic syndrome and diabetes, but these roles are, as yet, only crudely understood. A number of important questions remain unanswered regarding the sphingolipid-diabetes axis, and tools are now available to investigate mechanisms that would have been inscrutable only a decade ago. In particular, it remains to be shown which enzyme isoforms of sphingolipid metabolism are responsible for potentiating specific aspects of the cellular diabetic phenotype. Furthermore, the individual sphingolipid species involved in these processes are almost totally unknown. It is unknown which specific sphingolipid *N*-acyl chain lengths or classes of chain lengths are important in cellular pathology; just as critically, the roles of complex sphingolipids and, in many tissues, the potently bioactive lipid sphingosine-1-phosphate remain neglected. Furthermore, the vast majority of studies have focused on the roles of sphingolipids in lipotoxicity; this may owe to the intuitive substrate-level connection. However, glucotoxicity is also an important player in the pathogenesis of diabetes, and the literature has provided a few intriguing hints of a synergistic relationship between sphingolipid-dependent glucotoxicity and lipotoxicity that beg for further investigation. Finally, the possibility of using circulating sphingolipids as clinical biomarkers is only now emerging. To develop this area, larger-scale studies with appropriate statistical approaches will need to be conducted. Furthermore, the physiological basis linking levels of particular circulating sphingolipids to diabetes remains to be determined. Thus, while much fundamental research has already implicated sphingolipids in diabetes in a very general way, this field of study blossoms both with questions regarding specific mechanisms of action and with new technologies and strategies to address them.

Acknowledgments This work was supported by Grant Number F30DK092125 from the NIDDK (to S.B.R.), GAANN fellowship in Lipidomics and Systems Biology (to J.S.R), a Merit Award from the Department of Veterans Affairs (to L.A.C.), and the NIH COBRE in Lipidomics and Pathobiology at MUSC (to L.A.C.).

References

Aerts JM, Ottenhoff R, Powlson AS, Grefhorst A, van Eijk M, Dubbelhuis PF, Aten J, Kuipers F, Serlie MJ, Wennekes T, Sethi JK, O'Rahilly S, Overkleeft HS (2007) Pharmacological inhibition of glucosylceramide synthase enhances insulin sensitivity. Diabetes 56(5):1341

Amati F, Dube J, Alvarez-Carnero E, Edreira M, Chomentowski P, Coen P, Switzer G, Bickel P, Stefanovic-Racic M, Toledo F, Goodpaster B (2011) Skeletal muscle triglycerides, diacylglycerides, and ceramids in insulin resistance: another paradox in endurance-trained athletes? Diabetes 60(10):2588–2597

Auge N, Maupas-Schwalm F, Elbaz M, Thiers J, Waysbort A, Itohara S, Krell H, Salvayre R, Negre-Salvayre A (2004) Role for matrix metalloproteinase-2 in oxidized low-density lipoprotein-induced activation of sphingomyelin/ceramide pathway and smooth muscle cell proliferation. Circulation 110:571–578

Banegas JR, Lopez-Garcia E, Graciani A, Guallar-Castillon P, Gutierrez-Fisac JL, Alonso J, Rodriguez-Artalejo F (2007) Relationship between obesity, hypertension and diabetes, and health-related quality of life among the elderly. Eur J Cardiovasc Prev Rehabil 14(3):456–462

Baranowski M, Blachnio A, Zabielski P, Gorski J (2007) PPARalpha agonist induces the accumulation of ceramide in the heart of rats fed high-fat diet. J Physiol Pharmacol 58(1):57

Baricos W, Cortez-Schwartz S, Shah S (1986) Renal neuraminidase. Characterization in normal rat kidney and measurement in experimentally induced nephrotic syndrome. Biochem J 239(3):705–710

Barouch LA, Gao D, Chen L, Miller KL, Xu W, Phan AC, Kittleson MM, Minhas KM, Berkowitz DE, Wei C, Hare JM (2006) Cardiac myocyte apoptosis is associated with increased DNA damage and decreased survival in murine models of obesity. Circ Res 98(1):119

Baskin M, Ard J, Franklin F, Allison D (2005) Prevalence of obesity in the United States. Obes Rev 6:5–7

Bevilacqua S, Bonadonna R, Buzzigoli G, Boni C, Ciociaro D, Maccari F, Giorico MA, Ferrannini E (1987) Acute elevation of free fatty acid levels leads to hepatic insulin resistance in obese subjects. Metabolism 36(5):502

Bielawska AE, Shapiro JP, Jiang L, Melkonyan HS, Piot C, Wolfe CL, Tomei LD, Hannun YA, Umansky SR (1997) Ceramide is involved in triggering of cardiomyocyte apoptosis induced by ischemia and reperfusion. Am J Pathol 151(5):1257

Bielawski J, Pierce JS, Snider J, Rembiesa B, Szulc ZM, Bielawska A (2010) Sphingolipid analysis by high performance liquid chromatography-tandem mass spectrometry (HPLC-MS/MS). Adv Exp Med Biol 688:46–59

Bijl N, Sokolovic M, Vrins C, Langeveld M, Moerland P, Ottenhoff R, Roomen Cv, Claessen N, Boot R, Aten J, Groen A, Aerts J, Eijk Mv (2009) Modulation of glycosphingolipid metabolism significantly improves hepatic insulin sensitivity and reverses hepatic steatosis in mice. Hepatology 50(5):1431–1441

Blomqvist M, Osterbye T, Mansson JE, Horn T, Buschard K, Fredman P (2003) Selective lack of the C16:0 fatty acid isoform of sulfatide in pancreas of type II diabetic animal models. Apmis 111(9):867

Blomqvist M, Carrier M, Andrews T, Pettersson K, Mansson JE, Rynmark BM, Fredman P, Buschard K (2005) In vivo administration of the C16:0 fatty acid isoform of sulfatide increases pancreatic sulfatide and enhances glucose-stimulated insulin secretion in Zucker fatty (fa/fa) rats. Diabetes Metab Res Rev 21(2):158

Boini K, Zhang C, Xia M, Poklis J, Li P (2010) Role of sphingolipid mediator ceramide in obesity and renal injury in mice fed a high-fat diet. J Pharmacol Exp Ther 334(3):839–846

Bonzon-Kulichenko E, Schwudke D, Gallardo N, Molto E, Fernandez-Agullo T, Shevchenko A, Andres A (2009) Central leptin regulates total ceramide content and sterol regulatory element binding protein-1C proteolytic maturation in rat white adipose tissue. Endocrinology 150(1):169–178

Boslem E, MacIntosh G, Preston AM, Bartley C, Busch AK, Fuller M, Laybutt DR, Meikle PJ, Biden TJ (2011) A lipidomic screen of palmitate-treated MIN6 beta-cells links sphingolipid metabolites with endoplasmic reticulum (ER) stress and impaired protein trafficking. Biochem J 435(1):267

Boudina S, Abel ED (2010) Diabetic cardiomyopathy, causes and effects. Rev Endocr Metab Disord 11(1):31

Brice SE, Cowart LA (2011) Sphingolipid metabolism and analysis in metabolic disease. Adv Exp Med Biol 721:1–17

Brown M, Goldstein J (2008) Selective versus total insulin resistance: a pathogenic paradox. Cell Metab 7:95–96

Bruce C, Kriketos A, Cooney G, Hawley J (2004) Disassociation of muscle triglyceride content and insulin sensitivity after exercise training in patients with type 2 diabetes. Diabetologia 47:23–30

Bruni P, Donati C (2008) Pleiotropic effects of sphingolipids in skeletal muscle. Cell Mol Life Sci 65:3725–3736

Buschard K, Hoy M, Bokvist K, Olsen HL, Madsbad S, Fredman P, Gromada J (2002) Sulfatide controls insulin secretion by modulation of ATP-sensitive K(+)-channel activity and Ca(2+)-dependent exocytosis in rat pancreatic beta-cells. Diabetes 51(8):2514

Buschard K, Blomqvist M, Mansson JE, Fredman P, Juhl K, Gromada J (2006) C16:0 sulfatide inhibits insulin secretion in rat beta-cells by reducing the sensitivity of KATP channels to ATP inhibition. Diabetes 55(10):2826

Cacicedo J, Benjachareowong S, Chou E, Ruderman N, Ido Y (2005) Palmitate-induced apoptosis in cultured bovine retinal pericytes: role for NAD(P)H oxidase, oxidant stress, and ceramide. Diabetes 54(6):1838–1845

Candiloros H, Zeghari N, Ziegler O, Donner M, Drouin P (1996) Hyperinsulinemia is related to erythrocyte phospholipid composition and membrane fluidity changes in obese nondiabetic women. J Clin Endocrinol Metab 81(8):2912

Cardenas A, Schadeck C, Bernard A, Lauwerys R (1991) Depletion of sialic acid without changes in sialidase activity in glmeruli of uninephrectomized diabetic rats. Biochem Med Metab Biol 46(3):416–421

Cavaghan MK, Ehrmann DA, Polonsky KS (2000) Interactions between insulin resistance and insulin secretion in the development of glucose intolerance. J Clin Invest 106(3):329

Chavez JA, Knotts TA, Wang LP, Li G, Dobrowsky RT, Florant GL, Summers SA (2003) A role for ceramide, but not diacylglycerol, in the antagonism of insulin signal transduction by saturated fatty acids. J Biol Chem 278:10297–10303

Christiansen T, Richelsen B, Bruun J (2005) Monocyte chemoattractant protein-1 is produced in isolated adipocytes, associated with adiposity and reduced after weight loss in morbid obese subjects. Int J Obes 29(1):146–150

Cnop M, Hannaert JC, Hoorens A, Eizirik DL, Pipeleers DG (2001) Inverse relationship between cytotoxicity of free fatty acids in pancreatic islet cells and cellular triglyceride accumulation. Diabetes 50(8):1771

Cohen-Forterre L, Mozere G, Andre J, Sternberg M (1984) Studies on kidney sialidase in normal and diabetic rats. Biochim Biophys Acta 800(1):138–145

Coleman R, Lee D (2004) Enzymes of triacylglycerol synthesis and their regulation. Prog Lipid Res 43:134–176

Constable PD, Smith GW, Rottinghaus GE, Tumbleson ME, Haschek WM (2003) Fumonisin-induced blockade of ceramide synthase in sphingolipid biosynthetic pathway alters aortic input impedance spectrum of pigs. Am J Physiol Heart Circ Physiol 284(6):H2034

Davis S, Deo SH, Barlow M, Yoshishige D, Farias M, Caffrey JL (2006) The monosialosyl ganglioside GM-1 reduces the vagolytic efficacy of delta2-opioid receptor stimulation. Am J Physiol Heart Circ Physiol 291(5):H2318

de Mello VD, Lankinen M, Schwab U, Kolehmainen M, Lehto S, Seppanen-Laakso T, Oresic M, Pulkkinen L, Uusitupa M, Erkkila AT (2009) Link between plasma ceramides, inflammation and insulin resistance: association with serum IL-6 concentration in patients with coronary heart disease. Diabetologia 52(12):2612

de Vries JE, Vork MM, Roemen TH, de Jong YF, Cleutjens JP, van der Vusse GJ, van Bilsen M (1997) Saturated but not mono-unsaturated fatty acids induce apoptotic cell death in neonatal rat ventricular myocytes. J Lipid Res 38(7):1384

Deevska G, Rozenova K, Giltiay N, Chambers M, White J, Boyanovsky B, Wei J, Daugherty A, Smart E, Reid M, Merrill JAH, Nikolova-Karakashian M (2009) Acid sphingomyelinase deficiency prevents diet-induced hepatic triacylglycerol accumulation and hyperglycemia. J Biol Chem 284:8359–8368

Di Paola M, Zaccagnino P, Montedoro G, Cocco T, Lorusso M (2004) Ceramide induces release of pro-apoptotic proteins from mitochondria by either a Ca2+ –dependent or a Ca2+ –independent mechanism. J Bioenerg Biomembr 36(2):165

Dimitrios I, Drosatos K, Hiyama Y, Goldberg I, Zannis V (2010) MicroRNA-370 controls the expression of MicroRNA-122 and CPT1a and affects lipid metabolism. J Lipid Res 51(6):1513–1523

Dindo D, Dahm F, Szulc Z, Bielawska A, Obeid LM, Hannun YA, Graf R, Clavien PA (2006) Cationic long-chain ceramide LCL-30 induces cell death by mitochondrial targeting in SW403 cells. Mol Cancer Ther 5(6):1520–1529

Dinh W, Lankisch M, Nickl W, Scheyer D, Scheffold T, Kramer F, Krahn T, Klein RM, Barroso MC, Futh R (2010) Insulin resistance and glycemic abnormalities are associated with deterioration of left ventricular diastolic function: a cross-sectional study. Cardiovasc Diabetol 9:63

Dinh W, Lankisch M, Nickl W, Gies M, Scheyer D, Kramer F, Scheffold T, Krahns T, Sause A, Futh R (2011) Metabolic syndrome with or without diabetes contributes to left ventricular diastolic dysfunction. Acta Cardiol 66(2):167

Dobrzyn P, Dobrzyn A, Miyazaki M, Ntambi JM (2010) Loss of stearoyl-CoA desaturase 1 rescues cardiac function in obese leptin-deficient mice. J Lipid Res 51(8):2202

Drosatos K, Bharadwaj KG, Lymperopoulos A, Ikeda S, Khan R, Hu Y, Agarwal R, Yu S, Jiang H, Steinberg SF, Blaner WS, Koch WJ, Goldberg IJ (2011) Cardiomyocyte lipids impair beta-adrenergic receptor function via PKC activation. Am J Physiol Endocrinol Metab 300(3):E489

Dyntar D, Eppenberger-Eberhardt M, Maedler K, Pruschy M, Eppenberger HM, Spinas GA, Donath MY (2001) Glucose and palmitic acid induce degeneration of myofibrils and modulate apoptosis in rat adult cardiomyocytes. Diabetes 50(9):2105

El-Assaad W, Joly E, Barbeau A, Sladek R, Buteau J, Maestre I, Pepin E, Zhao S, Iglesias J, Roche E, Prentki M (2010) Glucolipotoxicity alters lipid partitioning and causes mitochondrial dysfunction, cholesterol, and ceramide deposition and reactive oxygen species production in INS832/13 ss-cells. Endocrinology 151(7):3061

Finck BN (2004) The role of the peroxisome proliferator-activated receptor alpha pathway in pathological remodeling of the diabetic heart. Curr Opin Clin Nutr Metab Care 7(4):391

Fox T, Han X, Kelly S, Merrill JAH, Martin R, Anderson R, Gardner T, Kester M (2006) Diabetes alters sphingolipid metabolism in the retina: a potential mechanism of cell death in diabetic retinopathy. Diabetes 55(12):3573–3580

Fry J, Finley W (2005) The prevalence and costs of obesity in the EU. Proc Nutr Soc 64(3):359–362

Gault CR, Obeid LM, Hannun YA (2010) An overview of sphingolipid metabolism: from synthesis to breakdown. Adv Exp Med Biol 688:1–23

Gidwani A, Brown HA, Holowka D, Baird B (2003) Disruption of lipid order by short-chain ceramides correlates with inhibition of phospholipase D and downstream signaling by FcepsilonRI. J Cell Sci 116(Pt 15):3177

Goldkorn T, Balaban N, Shannon M, Chea V, Matsukuma K, Gilchrist D, Wang H, Chan C (1998) H_2O_2 acts on cellular membranes to generate ceramide signaling and initiate apoptosis in tracheobronchial epithelial cells. J Cell Sci 111(Pt 21):3209–3220

Gonzalez-Pertusa JA, Dube J, Valle SR, Rosa TC, Takane KK, Mellado-Gil JM, Perdomo G, Vasavada RC, Garcia-Ocana A (2010) Novel proapoptotic effect of hepatocyte growth factor: synergy with palmitate to cause pancreatic {beta}-cell apoptosis. Endocrinology 151(4):1487

Goodpaster B, He J, Watkins S, Kelley D (2001) Skeletal muscle lipid content and insulin resistance: evidence for a paradox in endurance-trained athletes. J Clin Endocrinol Metab 86:5755–5761

Gorski J, Dobrzyn A, Zendzian-Piotrowska M (2002) The sphingomyelin-signaling pathway in skeletal muscles and its role in regulation of glucose uptake. Ann NY Acad Sci 967:236–248

Grundy SM, Bryan Brewer JH, Cleeman JI, Sidney J, Smith C, Lenfant C (2004) Definition of metabolic syndrome: report of the National Heart, Lung, and Blood Institute/American Heart Association conference on scientific issues related to definition. Circulation 109:433–438

Guha A, Harmancey R, Taegtmeyer H (2008) Nonischemic heart failure in diabetes mellitus. Curr Opin Cardiol 23(3):241

Guo J, Qian Y, Xi X, Hu X, Zhu J, Han X (2010) Blockage of ceramide metabolism exacerbates palmitate inhibition of pro-insulin gene expression in pancreatic beta-cells. Mol Cell Biochem 338(1–2):283

Hajduch E, Alessi D, Hemmings B, Hundal H (1998) Constitutive activation of protein kinase B alpha by membrane targeting promotes glucose and system A amino acid transport, protein synthesis, and inactivation of glycogen synthase kinase 3 in L6 muscle cells. Diabetes 47:1006–1013

Hajduch E, LItherland G, Hundal H (2001) Protein kinase B (PKB/Akt)-a key regulator of glucose transport? FEBS Lett 492:199–203

Hajer GR, Haeften TWv, Visseren FL (2008) Adipose tissue dysfunction in obesity, diabetes, and vascular disease. Eur Heart J 29:2959–2971

Hannun YA, Obeid LM (2008) Principles of bioactive lipid signalling: lessons from sphingolipids. Nat Rev Mol Cell Biol 9(2):139–150

Haus JM, Kashyap SR, Kasumov T, Zhang R, Kelly KR, Defronzo RA, Kirwan JP (2009) Plasma ceramides are elevated in obese subjects with type 2 diabetes and correlate with the severity of insulin resistance. Diabetes 58(2):337

Hickson-Bick DL, Buja LM, McMillin JB (2000) Palmitate-mediated alterations in the fatty acid metabolism of rat neonatal cardiac myocytes. J Mol Cell Cardiol 32(3):511

Hirose H, Lee YH, Inman LR, Nagasawa Y, Johnson JH, Unger RH (1996) Defective fatty acid-mediated beta-cell compensation in Zucker diabetic fatty rats. Pathogenic implications for obesity-dependent diabetes. J Biol Chem 271(10):5633

Ho KK, Pinsky JL, Kannel WB, Levy D (1993) The epidemiology of heart failure: the Framingham study. J Am Coll Cardiol 22(4 Suppl A):6A

Hojjati M, Li Z, Zhou H, Tang S, Huan C, Ooi E, Lu S, Jiang X (2005) Effect of myriocin on plasma sphingolipid metabolism and atherosclerosis in apoE-deficient mice. J Biol Chem 280:10284–10289

Holland W, Brozinick J, Wang L, Hawkins E, Sargent S, Liu Y, Narra K, Hoehn K, Knotts T, Siesky A, Nelson D, Karathanasis S, Foentenot G, Birnbaum M, Summers S (2007) Inhibition of ceramide synthesis ameliorates glucocorticoid-, saturated-fat-, and obesity-induced insulin resistance. Cell Metab 5:167–179

Holland WL, Miller RA, Wang ZV, Sun K, Barth BM, Bui HH, Davis KE, Bikman BT, Halberg N, Rutkowski JM, Wade MR, Tenorio VM, Kuo MS, Brozinick JT, Zhang BB, Birnbaum MJ, Summers SA, Scherer PE (2011) Receptor-mediated activation of ceramidase activity initiates the pleiotropic actions of adiponectin. Nat Med 17(1):55

Hotamisligil G (2006) Inflammation and metabolic disorders. Nature 444(7121):860–867

Hu W, Bielawski J, Samad F, Merrill AH Jr, Cowart LA (2009) Palmitate increases sphingosine-1-phosphate in C2C12 myotubes via upregulation of sphingosine kinase message and activity. J Lipid Res 50(9):1852

Hu W, Ross J, Geng T, Brice SE, Cowart LA (2011) Differential regulation of dihydroceramide desaturase by palmitate versus monounsaturated fatty acids: implications for insulin resistance. J Biol Chem 286(19):16596

Itani S, Ruderman N, Schmieder F, Boden G (2002) Lipid-induced insulin resistance in human muscle is associated with changes in diacylglycerol, protien kinase C, and IkB-alpha. Diabetes 51:2005–2011

Jessup CF, Bonder CS, Pitson SM, Coates PT (2011) The sphingolipid rheostat: a potential target for improving pancreatic islet survival and function. Endocr Metab Immune Disord Drug Targets 11(4):262–272

Keller S, Lienhard G (1994) Insulin signalling the role of insulin receptor substrate 1. Trends Cell Biol 4:115–119

Kelpe CL, Moore PC, Parazzoli SD, Wicksteed B, Rhodes CJ, Poitout V (2003) Palmitate inhibition of insulin gene expression is mediated at the transcriptional level via ceramide synthesis. J Biol Chem 278(32):30015

Kolak M, Westerbacka J, Velagapudi V, Wagsater D, Yetukuri L, Makkonen J, Rissanen A, Hakkinen A, Lindell M, Bergholm R, Hamsten A, Eriksson P, Fisher R, Oresic M, Yki-Jarvinen H (2007) Adipose tissue inflammation and increased ceramide content characterize subjects with high liver fat content independent of obesity.". Diabetes 56(8):1960–1968

Kremer GJ, Atzpodien W, Schnellbacher E (1975) Plasma glycosphingolipids in diabetics and normals. Klin Wochenschr 53(13):637

Krown KA, Page MT, Nguyen C, Zechner D, Gutierrez V, Comstock KL, Glembotski CC, Quintana PJ, Sabbadini RA (1996) Tumor necrosis factor alpha-induced apoptosis in cardiac myocytes. Involvement of the sphingolipid signaling cascade in cardiac cell death. J Clin Invest 98(12):2854

Kucharska-Newton AM, Couper DJ, Pankow JS, Prineas RJ, Rea TD, Sotoodehnia N, Chakravarti A, Folsom AR, Siscovick DS, Rosamond WD (2010) Diabetes and the risk of sudden cardiac death, the Atherosclerosis Risk in Communities study. Acta Diabetol 47(Suppl 1):161

Lam Y, Hatzinikolas G, Weir J, Janovska A, McAinch A, Game P, Meikle P, Wittert G (2011) Insulin-stimulated glucose uptake and pathways regulating energy metabolism in skeletal muscle cells: the effects of subcutaneous and visceral fat, and long-chain saturated, n-3 and n-6 polyunsaturated fatty acids. Biochim Biophys Acta 1811(7–8):468–475

Laviad EL, Albee L, Pankova-Kholmyansky I, Epstein S, Park H, Merrill AH Jr, Futerman AH (2008) Characterization of ceramide synthase 2: tissue distribution, substrate specificity, and inhibition by sphingosine 1-phosphate. J Biol Chem 283(9):5677–5684

Leahy JL (2008) Pathogenesis of type 2 diabetes mellitus. In: Feinglos MN, Bethel MA (eds) Type 2 diabetes mellitus. Humana, Amstgerdam, p 17

Lee Y, Hirose H, Ohneda M, Johnson JH, McGarry JD, Unger RH (1994) Beta-cell lipotoxicity in the pathogenesis of non-insulin-dependent diabetes mellitus of obese rats: impairment in adipocyte-beta-cell relationships. Proc Natl Acad Sci USA 91(23):10878

Levy M, Castillo SS, Goldkorn T (2006) nSMase2 activation and trafficking are modulated by oxidative stress to induce apoptosis. Biochem Biophys Res Commun 344(3):900–905

Levy M, Khan E, Careaga M, Goldkorn T (2009) Neutral sphingomyelinase 2 is activated by cigarette smoke to augment ceramide-induced apoptosis in lung cell death. Am J Physiol Lung Cell Mol Physiol 297(1):L125–L133

Lipina C, Hundal H (2011) Sphingolipids: agents provoccateurs in the pathogenesis of insulin resistance. Diabetologia 54:1596–1607

Liu L, Shi X, Bharadwaj KG, Ikeda S, Yamashita H, Yagyu H, Schaffer JE, Yu YH, Goldberg IJ (2009) DGAT1 expression increases heart triglyceride content but ameliorates lipotoxicity. J Biol Chem 284(52):36312

Lupi R, Dotta F, Marselli L, Del Guerra S, Masini M, Santangelo C, Patane G, Boggi U, Piro S, Anello M, Bergamini E, Mosca F, Di Mario U, Del Prato S, Marchetti P (2002) Prolonged exposure to free fatty acids has cytostatic and pro-apoptotic effects on human pancreatic islets: evidence that beta-cell death is caspase mediated, partially dependent on ceramide pathway, and Bcl-2 regulated. Diabetes 51(5):1437

Ma MM, Chen JL, Wang GG, Wang H, Lu Y, Li JF, Yi J, Yuan YJ, Zhang QW, Mi J, Wang LS, Duan HF, Wu CT (2007) Sphingosine kinase 1 participates in insulin signalling and regulates glucose metabolism and homeostasis in KK/Ay diabetic mice. Diabetologia 50(4):891

Madrazo JA, Kelly DP (2008) The PPAR trio: regulators of myocardial energy metabolism in health and disease. J Mol Cell Cardiol 44(6):968

Maedler K, Spinas GA, Dyntar D, Moritz W, Kaiser N, Donath MY (2001) Distinct effects of saturated and monounsaturated fatty acids on beta-cell turnover and function. Diabetes 50(1):69

Maedler K, Oberholzer J, Bucher P, Spinas GA, Donath MY (2003) Monounsaturated fatty acids prevent the deleterious effects of palmitate and high glucose on human pancreatic beta-cell turnover and function. Diabetes 52(3):726

Maheux P, Jeppesen J, Sheu WH, Hollenbeck CB, Clinkingbeard C, Greenfield MS, Chen YD, Reaven GM (1994) Additive effects of obesity, hypertension, and type 2 diabetes on insulin resistance. Hypertension 24(6):695–698

Manchanayake J, Chitturi S, Nolan C, Farrell GC (2011) Postprandial hyperinsulinemia is universal in non-diabetic patients with nonalcoholic fatty liver disease. J Gastroenterol Hepatol 26(3):510

Martinez J, Kearney J, Kafatos A, Paquet S, Martinez-Gonzalez M (1999) Variables independently associated with self-reported obesity in the European Union. Public Health Nutr 2(1A):125–133

Mastrandrea LD, Sessanna SM, Laychock SG (2005) Sphingosine kinase activity and sphingosine-1 phosphate production in rat pancreatic islets and INS-1 cells: response to cytokines. Diabetes 54(5):1429

Mather A, Siskind LJ (2011) Glycosphingolipids and kidney disease. Adv Exp Med Biol 721:121–138

Maulik N, Das DK, Gogineni M, Cordis GA, Avrova N, Denisova N (1993) Reduction of myocardial ischemic reperfusion injury by sialylated glycosphingolipids, gangliosides. J Cardiovasc Pharmacol 22(1):74

Memon RA, Holleran WM, Moser AH, Seki T, Uchida Y, Fuller J, Shigenaga JK, Grunfeld C, Feingold KR (1998) Endotoxin and cytokines increase hepatic sphingolipid biosynthesis and produce lipoproteins enriched in ceramides and sphingomyelin. Arterioscler Thromb Vasc Biol 18(8):1257–1265

Merrill AH Jr, Nixon DW, Williams RD (1985) Activities of serine palmitoyltransferase (3-ketosphinganine synthase) in microsomes from different rat tissues. J Lipid Res 26(5):617

Milburn JL Jr, Hirose H, Lee YH, Nagasawa Y, Ogawa A, Ohneda M, BeltrandelRio H, Newgard CB, Johnson JH, Unger RH (1995) Pancreatic beta-cells in obesity. Evidence for induction of functional, morphologic, and metabolic abnormalities by increased long chain fatty acids. J Biol Chem 270(3):1295

Miyake Y, Kozutsumi Y, Nakamura S, Fujita T, Kawasaki T (1995) Serine palmitoyltransferase is the primary target of a sphingosine-like immunosuppressant, ISP-1/myriocin. Biochem Biophys Res Commun 211(2):396

Mohamed-Ali V, Goodrick S, Rawesh A, Katz D, Miles J, Yudkin J, Klein S, Coppack S (1997) Subcutaneous adipose tissue releases interleukin-6, but not tumor necrosis factor-alpha, in vivo. J Clin Endocrinol Metab 82:4196–4200

Moro C, Bajpeyi S, Smith S (2008) Determinants of intramyocellular triglyceride turnover: implications for insulin sensitivity. Am J Physiol Endocrinol Metab 294:E203–E213

Mueller-Riemenschneider F, Reinhold T, Berghofer A, Willich S (2008) Health-economic burden of obesity in Europe. Eur J Epidemiol 23(8):499–509

Mullen TD, Jenkins RW, Clarke CJ, Bielawski J, Hannun YA, Obeid LM (2011a) Ceramide synthase-dependent ceramide generation and programmed cell death: involvement of salvage pathway in regulating postmitochondrial events. J Biol Chem 286(18):15929–15942

Mullen TD, Spassieva S, Jenkins RW, Kitatani K, Bielawski J, Hannun YA, Obeid LM (2011b) Selective knockdown of ceramide synthases reveals complex interregulation of sphingolipid metabolism. J Lipid Res 52(1):68–77

Nerpin E, Riserus U, Ingelsson E, Sundstrom J, Jobs M, Larsson A, Basu S, Arnlov J (2008) Insulin sensitivity measured with euglycemic clamp is independently associated with glomerular filtration rate in a community-based cohort. Diabetes Care 31(8):1550

Nussey S, Whitehead SA (2001) Endocrinology: an integrated approach. Bios, Oxford, UK

Nybond S, Bjorkqvist YJ, Ramstedt B, Slotte JP (2005) Acyl chain length affects ceramide action on sterol/sphingomyelin-rich domains. Biochim Biophys Acta 1718(1–2):61

Osterbye T, Jorgensen KH, Fredman P, Tranum-Jensen J, Kaas A, Brange J, Whittingham JL, Buschard K (2001) Sulfatide promotes the folding of proinsulin, preserves insulin crystals, and mediates its monomerization. Glycobiology 11(6):473

Othman A, Rutti MF, Ernst D, Saely CH, Rein P, Drexel H, Porretta-Serapiglia C, Lauria G, Bianchi R, von Eckardstein A, Hornemann T (2011) Plasma deoxysphingolipids: a novel class of biomarkers for the metabolic syndrome? Diabetologia 55(2):421–31

Park TS, Hu Y, Noh HL, Drosatos K, Okajima K, Buchanan J, Tuinei J, Homma S, Jiang XC, Abel ED, Goldberg IJ (2008) Ceramide is a cardiotoxin in lipotoxic cardiomyopathy. J Lipid Res 49(10):2101

Parra V, Eisner V, Chiong M, Criollo A, Moraga F, Garcia A, Hartel S, Jaimovich E, Zorzano A, Hidalgo C, Lavandero S (2008) Changes in mitochondrial dynamics during ceramide-induced cardiomyocyte early apoptosis. Cardiovasc Res 77(2):387

Passa P (2002) Diabetes trends in Europe. Diabetes Metab Res Rev 18(Suppl 3):S3–S8

Pewzner-Jung Y, Ben-Dor S, Futerman AH (2006) When do Lasses (longevity assurance genes) become CerS (ceramide synthases)?: insights into the regulation of ceramide synthesis. J Biol Chem 281(35):25001–25005

Pick A, Clark J, Kubstrup C, Levisetti M, Pugh W, Bonner-Weir S, Polonsky KS (1998) Role of apoptosis in failure of beta-cell mass compensation for insulin resistance and beta-cell defects in the male Zucker diabetic fatty rat. Diabetes 47(3):358

Pontiroli AE, Camisasca R (2002) Additive effect of overweight and type 2 diabetes in the appearance of coronary heart disease but not of stroke: a cross-sectional study. Acta Diabetol 39(2):83–90

Pouliot MC, Despres JP, Nadeau A, Tremblay A, Moorjani S, Lupien PJ, Theriault G, Bouchard C (1990) Associations between regional body fat distribution, fasting plasma free fatty acid levels and glucose tolerance in premenopausal women. Int J Obes 14(4):293

Preiss B, Sattar N (2008) Non-alcoholic fatty liver disease: an overview of prevalence, diagnosis, pathogenesis, and treatment considerations. Clin Sci 115:141–150

Prosdocimi M, Paro M, Travagli RA, Tessari F (1987) Alterations of the vegetative nervous system, experimental diabetes and pharmacological use of gangliosides. Funct Neurol 2(4):559

Qiao Q, Nyamdorj R (2010) The optimal cutoff values and their performance of waist circumference and waist-to-hip ratio for diagnosing type II diabetes. Eur J Clin Nutr 64(1):23–29

Rijzewijk LJ, van der Meer RW, Smit JW, Diamant M, Bax JJ, Hammer S, Romijn JA, de Roos A, Lamb HJ (2008) Myocardial steatosis is an independent predictor of diastolic dysfunction in type 2 diabetes mellitus. J Am Coll Cardiol 52(22):1793

Robert P, Tsui P, Laville MP, Livi GP, Sarau HM, Bril A, Berrebi-Bertrand I (2001) EDG1 receptor stimulation leads to cardiac hypertrophy in rat neonatal myocytes. J Mol Cell Cardiol 33(9):1589

Roden M, Price TB, Perseghin G, Petersen KF, Rothman DL, Cline GW, Shulman GI (1996) Mechanism of free fatty acid-induced insulin resistance in humans. J Clin Invest 97(12):2859

Roeske-Nielsen A, Dalgaard LT, Mansson JE, Buschard K (2010) The glycolipid sulfatide protects insulin-producing cells against cytokine-induced apoptosis, a possible role in diabetes. Diabetes Metab Res Rev 26(8):631

Rutti S, Ehses JA, Sibler RA, Prazak R, Rohrer L, Georgopoulos S, Meier DT, Niclauss N, Berney T, Donath MY, von Eckardstein A (2009) Low- and high-density lipoproteins modulate function, apoptosis, and proliferation of primary human and murine pancreatic beta-cells. Endocrinology 150(10):4521

Saito M, Cooper TB, Vadasz C (2005) Ethanol-induced changes in the content of triglycerides, ceramides, and glucosylceramides in cultured neurons. Alcohol Clin Exp Res 29(8):1374–1383

Samad F, Hester KD, Yang G, Hannun YA, Bielawski J (2006) Altered adipose and plasma sphingolipid metabolism in obesity: a potential mechanism for cardiovascular and metabolic risk. Diabetes 55:2579–2587

Samad F, Badeanlou L, Shah C, Yang G (2011) Adipose tissue and ceramide biosynthesis in the pathogenesis of obesity. Adv Exp Med Biol 721:67–86

Schmitz-Peiffer C, Craig D, Biden T (1999) Ceramide generation is sufficient to account for the inhibition of the insulin-stimulated PKB pathway in C2C12 skeletal muscle cells pretreated with palmitate. J Biol Chem 274:24202–24210

Senkal CE, Ponnusamy S, Bielawski J, Hannun YA, Ogretmen B (2010) Antiapoptotic roles of ceramide-synthase-6-generated C16-ceramide via selective regulation of the ATF6/CHOP arm of ER-stress-response pathways. FASEB J 24(1):296

Senkal CE, Ponnusamy S, Manevich Y, Meyers-Needham M, Saddoughi SA, Mukhopadyay A, Dent P, Bielawski J, Ogretmen B (2011) Alteration of ceramide synthase 6/C16-ceramide induces activating transcription factor 6-mediated endoplasmic reticulum (ER) stress and apoptosis via perturbation of cellular Ca2+ and ER/Golgi membrane network. J Biol Chem 286(49):42446

Serlie MJ, Meijer AJ, Groener JE, Duran M, Endert E, Fliers E, Aerts JM, Sauerwein HP (2007) Short-term manipulation of plasma free fatty acids does not change skeletal muscle concentrations of ceramide and glucosylceramide in lean and overweight subjects. J Clin Endocrinol Metab 92(4):1524

Shah C, Yang G, Lee I, Bielawski J, Hannun YA, Samad F (2008) Protection from high fat diet-induced increase in ceramide in mice lacking plaminogen activator inhibitor 1. J Biol Chem 283:13538–13548

Shimabukuro M, Zhou Y-T, Levi M, Unger R (1998) Fatty acid-induced beta cell apoptosis: a link between obesity and diabetes. Proc Natl Acad Sci USA 95:2498–2502

Sparagna GC, Hickson-Bick DL, Buja LM, McMillin JB (2000) A metabolic role for mitochondria in palmitate-induced cardiac myocyte apoptosis. Am J Physiol Heart Circ Physiol 279(5): H2124

Sultan I, Senkal CE, Ponnusamy S, Bielawski J, Szulc Z, Bielawska A, Hannun YA, Ogretmen B (2006) Regulation of the sphingosine-recycling pathway for ceramide generation by oxidative stress, and its role in controlling c-Myc/Max function. Biochem J 393(Pt 2):513–521

Summers SA (2006) Ceramides in insulin resistance and lipotoxicity. Prog Lipid Res 45:42–72

Summers S, Nelson D (2005) A role for sphingolipids in producting the common features of type 2 diabetes, metabolic syndrome X, and Cushing's syndrome. Diabetes 54:591–602

Teruel T, Hernandez R, Lorenzo M (2001) Ceramide mediates insulin resistance by tumor necrosis factor-alpha in brown adipocytes by maintaining Akt in an inactive dephosphorylated state. Diabetes 50:2563–2571

Tessari F, Travagli RA, Zanoni R, Prosdocimi M (1988) Effects of long-term diabetes and treatment with gangliosides on cardiac sympathetic innervation: a biochemical and functional study in mice. J Diabet Complications 2(1):34

Theilmeier G, Schmidt C, Herrmann J, Keul P, Schafers M, Herrgott I, Mersmann J, Larmann J, Hermann S, Stypmann J, Schober O, Hildebrand R, Schulz R, Heusch G, Haude M, von Wnuck Lipinski K, Herzog C, Schmitz M, Erbel R, Chun J, Levkau B (2006) High-density lipoproteins and their constituent, sphingosine-1-phosphate, directly protect the heart against ischemia/reperfusion injury in vivo via the S1P3 lysophospholipid receptor. Circulation 114(13):1403

Turpin SM, Ryall JG, Southgate R, Darby I, Hevener AL, Febbraio MA, Kemp BE, Lynch GS, Watt MJ (2009) Examination of 'lipotoxicity' in skeletal muscle of high-fat fed and ob/ob mice. J Physiol 587(7):1598–1605

Utz W, Engeli S, Haufe S, Kast P, Hermsdorf M, Wiesner S, Pofahl M, Traber J, Luft FC, Boschmann M, Schulz-Menger J, Jordan J (2011) Myocardial steatosis, cardiac remodelling and fitness in insulin-sensitive and insulin-resistant obese women. Heart 97(19):1585

Vanhaesebroeck B, Alessi D (2000) The PI3K-PDK1 connection: more than just a road to PKB. Biochem J 346(3):561–576

Veluthakal R, Palanivel R, Zhao Y, McDonald P, Gruber S, Kowluru A (2005) Ceramide induces mitochondrial abnormalities in insulin-secreting INS-1 cells: potential mechanisms underlying ceramide-mediated metabolic dysfunction of the beta cell. Apoptosis 10(4):841

Veret J, Coant N, Berdyshev EV, Skobeleva A, Therville N, Bailbe D, Gorshkova I, Natarajan V, Portha B, Le Stunff H (2011) Ceramide synthase 4 and de novo production of ceramides with specific N-acyl chain lengths are involved in glucolipotoxicity-induced apoptosis of INS-1 beta-cells. Biochem J 438(1):177

Wang E, Norred WP, Bacon CW, Riley RT, Merrill AH Jr (1991) Inhibition of sphingolipid biosynthesis by fumonisins. Implications for diseases associated with Fusarium moniliforme. J Biol Chem 266(22):14486–14490

Wang J, Zhen L, Klug MG, Wood D, Wu X, Mizrahi J (2000) Involvement of caspase 3- and 8-like proteases in ceramide-induced apoptosis of cardiomyocytes. J Card Fail 6(3):243

Wei GS, Coady SA, Goff DC Jr, Brancati FL, Levy D, Selvin E, Vasan RS, Fox CS (2011) Blood pressure and the risk of developing diabetes in african americans and whites: ARIC, CARDIA, and the framingham heart study. Diabetes Care 34(4):873–879

Whiteman E, Cho H, Birnbaum M (2002) Role of Akt/protein kinase B in metabolism. Trends Endocrinol Metab 13:444–451

Yagyu H, Chen G, Yokoyama M, Hirata K, Augustus A, Kako Y, Seo T, Hu Y, Lutz EP, Merkel M, Bensadoun A, Homma S, Goldberg IJ (2003) Lipoprotein lipase (LpL) on the surface of cardiomyocytes increases lipid uptake and produces a cardiomyopathy. J Clin Invest 111 (3):419

Yosuke O, Seki E, Kodama Y, Suetsugu A, Miura K, Adachi M, Ito H, Shiratori Y, Banno Y, Olefsky J, Nagaki M, Moriwaki H, Brenner D, Seishima M (2011) Acid sphingomyelinase regulates glucose and lipid metabolism in hepatocytes through Akt activation and AMP-activated protein kinase suppression. FASEB J 25(4):1133–1144

Zeghari N, Younsi M, Meyer L, Donner M, Drouin P, Ziegler O (2000) Adipocyte and erythrocyte plasma membrane phospholipid composition and hyperinsulinemia: a study in nondiabetic and diabetic obese women. Int J Obes Relat Metab Disord 24(12):1600

Zhao H, Przybylska M, Wu I, Siegel C, Komarnitsky S, Yew N, Cheng S (2007) Inhibiting glycosphingolipid synthesis improves glycemic control and insulin sensitivity in animal models of type 2 diabetes. Diabetes 56(5):1210–1218

Zheng W, Kollmeyer J, Symolon H, Momin A, Munter E, Wang E, Kelly S, Allegood JC, Liu Y, Peng Q, Ramaraju H, Sullards MC, Cabot M, Merrill AH Jr (2006) Ceramides and other bioactive sphingolipid backbones in health and disease: lipidomic analysis, metabolism and roles in membrane structure, dynamics, signaling and autophagy. Biochim Biophys Acta 1758 (12):1864–1884

Part IV
Sphingolipids in Neuro-psychiatry and Muscle Diseases

Neuronal Forms of Gaucher Disease

Einat B. Vitner and Anthony H. Futerman

Contents

Abstract Gaucher disease is an inherited metabolic disease caused by the defective activity of the lysosomal enzyme, glucosylceramidase (GlcCerase), which is responsible for the last step in the degradation of complex glycosphingolipids. As a result, glucosylceramide (GlcCer) accumulates intracellularly. Little is known about the mechanisms by which GlcCer accumulation leads to Gaucher disease, particularly for the types of the disease in which severe neuropathology occurs. We now summarize recent advances in this area and in particular focus in the biochemical and cellular pathways that may cause neuronal defects. Most recent work has taken advantage of newly available mouse models, which mimic to a large extent human disease progression. Finally, we discuss observations of a genetic link between Gaucher disease and Parkinson's disease and discuss how this link has stimulated research into the basic biology of the previously underappreciated glycosphingolipid, GlcCer.

E.B. Vitner • A.H. Futerman (✉)
Department of Biological Chemistry, Weizmann Institute of Science, Rehovot 76100, Israel
e-mail: tony.futerman@weizmann.ac.il

E. Gulbins and I. Petrache (eds.), *Sphingolipids in Disease*,
Handbook of Experimental Pharmacology 216,
DOI 10.1007/978-3-7091-1511-4_20, © Springer-Verlag Wien 2013

Keywords Gaucher disease • Glucosylceramidase • Glucosylceramide • Neurodegeneration

1 Glucosylceramide in Gaucher Disease

Glucosylceramide (GlcCer) and galactosylceramide (GalCer) are the simplest glycosphingolipids (GSLs) in mammalian cells. Both are synthesized by the glycosylation of ceramide, with GlcCer synthesized on the cytosol surface of the Golgi apparatus (Futerman and Pagano 1991) and GalCer synthesized mainly on the lumenal surface of the endoplasmic reticulum (ER) (Gault et al. 2010). GlcCer and GalCer are the precursors for more complex GSLs, which are synthesized by sequential steps of glycosylation. Once synthesized, GSLs are targeted to the outer leaflet of the plasma membrane, where they perform a number of critical functions (Merrill 2011). Subsequent to their internalization via endocytosis, GSLs are degraded in lysosomes by glycosidases. GalCer is cleaved to ceramide and galactose via the action of galactosylceramidase (galactocerebroside, β-galactosidase), and GlcCer is cleaved to ceramide and glucose via the action of glucosylceramidase (glucocerebrosidase, GlcCerase). Genetic deficiencies in either of these enzymes lead to an inherited disease: Krabbe disease in the case of defective GalCer degradation (Suzuki 2003) and Gaucher disease (GD) in the case of defective GlcCer degradation. This chapter focuses on GD.

GD is the most common lysosomal storage disease (LSD), and type 1 GD (the non-neuronopathic form, see below) occurs with an incidence of about 1 in 40,000–60,000 in the general population and 1 in 500–1,000 among Ashkenazi Jews (Horowitz and Zimran 1994). GlcCerase is highly susceptible to changes in its protein sequence since >200 mutations have been described. The X-ray structure of GlcCerase has been resolved (Dvir et al. 2003), which has lead to mechanistic understanding of the mode of GlcCer hydrolysis (Brumshtein et al. 2007). The GlcCerase structure consists of a characteristic $(\alpha/\beta)_8$ (TIM) barrel containing the catalytic residues, designated as domain III, and two smaller domains, I and II, which are composed mainly of β-sheets. Although their biological role is unknown, mutations in both domains I and II can cause GD. Three loops, which control access to the active site, have been shown to display alternative conformations (Premkumar et al. 2005). No 3D structures of any of the mutations causing GD, including the most common mutation, N370S, were available until recently, when two independent studies appeared simultaneously. One used X-ray crystallography to solve the structure experimentally (Wei et al. 2011), whereas the other used molecular dynamics (Offman et al. 2010). Both studies analyzed the structure at physiological and lysosomal pH, but the molecular dynamics study also examined the structure of the apoenzyme and that of a complex with *N*-butyldeoxynojirimycin, a putative chemical chaperone (Sawkar et al. 2006). Based on these studies, suggestions were made as to how the N370S mutation affects GlcCer hydrolysis.

Below, we discuss how the accumulation of this simple GSL, GlcCer, leads to the diverse and complex pathologies that have been described since GD was first discovered more than a century ago by the French physician, Philippe Gaucher (Gaucher 1882). In particular, we focus on the forms of the disease which display overt neurological symptoms. After introducing disease classification, we discuss the limited availability of mouse models (which has been a bottleneck for basic research aimed at delineating disease pathology), recent progress in determining the cellular and biochemical pathways that are responsible for disease pathology, particularly in the brain, and finally discuss very recent clinical findings that imply a genetic connection between Parkinson's disease and GD. The latter issue has pushed GD research to the forefront of studies on neurodegenerative disorders and has served as a reminder that the study of even the most simple and esoteric GSLs can lead to significant advances in research fields that were previously thought to be beyond the reach of sphingolipids.

2 Clinical Classification of GD and Treatment Options

GD is nowadays commonly divided into three subtypes based on age of onset and on signs of nervous system involvement. Type 1, classically defined as the chronic non-neuronopathic form, is the most common (>90 % of patients). The major symptoms of type 1 GD patients are enlargement of the spleen and liver (hepatosplenomegaly), anemia, thrombocytopenia, bone pain, and skeletal lesions (Cox and Schofield 1997). GlcCer accumulation is mainly restricted to macrophages, known as "Gaucher cells," which are the classical hallmark of the disease. However, although type 1 is defined by the absence of nervous system involvement, an increasing number of reports have emerged documenting neurological manifestations in patients with type 1 GD (Pastores et al. 2003).

Types 2 and 3, the neuronopathic forms of GD (nGD), are classified according to the time of onset and rate of progression of neurological symptoms. Type 2, the acute neuronopathic form, is very rare (1 % of patients) and usually refers to children who display neurological abnormalities before 6 months of age and die by age of 2–4 years (Sidransky et al. 1996). The subacute, chronic neuronopathic form (type 3) is also relatively rare (5 % of patients), and patients present with similar symptoms to those observed in type 2 but with a later onset and decreased severity. More than 200 different mutations have been identified in the GlcCerase gene (Beutler et al. 2005; Dvir et al. 2003) with no major correlation detected between genotype and phenotype. Prediction of the clinical course of the disease cannot usually be made on the basis of mutational analysis, with some patients having the same mutation severely affected, whereas others are asymptomatic (Azuri et al. 1998); attempting to understand the different susceptibilities of different individuals to the same GlcCerase mutation is one of the most exciting and unexplored areas in GD research.

Currently available treatments for GD include enzyme replacement therapy and substrate reduction therapy (Futerman et al. 2004; Platt and Jeyakumar 2008). Although enzyme replacement therapy has a marked effect on non-neurological manifestations, it has no demonstrable effect on neurological abnormalities since it does not cross the blood–brain barrier (Vellodi et al. 2009). Other potential treatments include cell and gene therapy as well as pharmacological chaperones (Platt and Lachmann 2009).

Although many years have passed since GD was first described, little is known about the molecular mechanisms leading from GlcCer accumulation to neurodegeneration and/or neuronal cell death. One major reason for this was the absence, until recently, of genuine nGD mouse models that faithfully mimic nGD symptoms.

3 nGD Mouse Models

The first attempt to generate an animal model for use in GD research involved administration of an active site-directed inhibitor of GlcCerase (Legler and Bieberich 1988; Premkumar et al. 2005), conduritol-β-epoxide (CBE) (Kanfer et al. 1975). Daily intraperitoneal injections of mice using relatively large doses of CBE for 3 weeks resulted in >90 % inhibition of GlcCerase activity and GlcCer accumulation in the spleen, liver, and brain, which could be reversed upon termination of CBE treatment (Stephens et al. 1978). Surprisingly, despite the fact that this chemically induced model is efficient and relatively inexpensive, it has not been widely used.

Generation of the first genetic mouse model of GD was based on production of a null GlcCerase allele (the $gba^{-/-}$ mouse). However, this mouse died soon after birth due to a skin permeability disorder (Tybulewicz et al. 1992) and was therefore not useful to study the long-term effects of GlcCer accumulation. Another GD mouse model, the L444P mouse, which carries a mutation that most commonly leads to nGD in humans, did not accumulate significant GlcCer levels and did not display central nervous system (CNS) pathology (Mizukami et al. 2002). More recently, two new nGD mouse models were generated. One of these displayed a marked reduction in GlcCerase activity in all tissues except the skin (the K14-*loxp–Neo–loxp* (lnl)/lnl mouse) (Enquist et al. 2007), and the other was generated by restricting the GlcCerase deficiency to the CNS (Enquist et al. 2007), in particular in neural and glial cells but not microglia. In this model, GlcCer levels were elevated in the brain but not in the spleen or liver (Enquist et al. 2007). These mice exhibited rapid motor dysfunction associated with severe neurodegeneration and neuronal loss and developed paralysis by 21 days of age (Enquist et al. 2007). Some of these features are reminiscent of neuropathological findings in nGD patients (Enquist et al. 2007; Wong et al. 2004). These mice were the first genetically induced mouse models that faithfully recapitulated neuropathological and biochemical aspects of nGD in human patients, paving the way for investigation of basic pathogenic mechanisms and of potential treatments.

4 Cellular and Biochemical Pathways in GD

Using the mouse models described above, along with various in vitro models, such as hippocampal neurons incubated with CBE (Korkotian et al. 1999; Pelled et al. 2000), a number of clues about possible pathological mechanisms in nGD have been obtained (Vitner et al. 2010a and b). However, some pathological mechanisms have been proposed based on studies in human skin fibroblasts obtained from GD patients, which may not be a suitable model for nGD research, since they may not share pathways affected in neuronal tissues, and, moreover, there is no evidence that fibroblasts from GD patients actually accumulate significant GlcCer levels. By way of example, ER stress together with activation of the unfolded protein response (UPR) was suggested to play a key role in cell death in nGD, and the UPR was proposed to be a common mediator for apoptosis in LSDs (Wei et al. 2008); this data was based on analysis of the UPR in a variety of human skin fibroblasts from LSD patients, including from nGD patients. In contrast, when the UPR was studied in either cultured neurons or astrocytes or in brain regions from nGD mouse models, even at late symptomatic stages, no evidence of UPR was observed (Farfel-Becker et al. 2009). This example serves as a note of caution that conclusions concerning pathological mechanisms in neuronal forms of LSDs, such as nGD, need to be derived from studies performed with appropriate tissues, cell types, or animal models.

4.1 Calcium Homeostasis in nGD

A number of studies have shown a mechanistic link between GlcCer accumulation in nGD and calcium homeostasis. Calcium plays an important role in regulating a variety of neuronal processes. The ER is the major intracellular Ca^{2+} store in neurons (Henkart 1980). Ca^{2+} is released from the ER to the cytosol via two types of Ca^{2+}-gated Ca^{2+}-release channel known as the ryanodine receptor (RyR) (Fill and Copello 2002) and the inositol 1,4,5-trisphosphate-gated Ca^{2+}-release channel known as the IP_3 receptor (IP_3R) (Mikoshiba 1997).

In 1999, the first correlation was demonstrated between altered Ca^{2+} homeostasis and neuronal cell death in nGD (Korkotian et al. 1999). Upon incubation of neurons with CBE, enhanced Ca^{2+} release was observed from intracellular stores in response to caffeine, an agonist of the RyR, which resulted in increased sensitivity to neurotoxic agents, especially glutamate. Inhibition of SL synthesis or addition of exogenously added GlcCerase abolished the deleterious effect of GlcCer, as did preincubation with antagonistic concentrations of ryanodine (Pelled et al. 2000). Moreover, GlcCer was shown to directly interact with and modulate the activity of the RyR. When added exogenously to rat brain microsomes, GlcCer itself had no effect on Ca^{2+} release but significantly and specifically augmented Ca^{2+} release induced by RyR agonists (Lloyd-Evans et al. 2003). In addition, elevated agonist-induced Ca^{2+} release was detected in human temporal lobe brain microsomes that were obtained post-mortem from GD patients (Pelled et al. 2005). Together, these findings suggested that defective Ca^{2+} homeostasis may be a mechanism responsible for at least some of the neuropathophysiology in nGD.

4.2 Gaucher Disease and Inflammation

Most studies on inflammation in GD have been performed on type 1 GD. GlcCer storage in macrophages leads to macrophage activation and release of both cytokines and chitinase chitotriosidase, the latter serving as a useful clinical biomarker for the disease (Jmoudiak and Futerman 2005). Levels of interleukin-1β (IL-1β), interleukin-1 receptor antagonist, IL-6, tumor necrosis factor α (TNFα), and soluble IL-2 receptor (sIL-2R) were elevated in the serum of GD patients (Barak et al. 1999), as were CD14 and M-CSF (Hollak et al. 1997).

Limited studies have been performed on neuroinflammation in nGD. Hong et al. found elevation of the proinflammatory cytokines, IL1-α IL1-β, IL-6, and TNFα in fetal brains of *gba* null mice as well as elevation in ROS (Hong et al. 2006). However, since *gba* null mice die before significant CNS damage (Willemsen et al. 1995), this model is limited in its application. A recent study demonstrated elevation of TNFα and IL-6 mRNA in the midbrain of a GD-like mouse model (V394L/V394L + saposin C−/−) (Sun et al. 2011). However, since this mouse model is not a bona fide nGD model, as GlcCer levels are only moderately elevated (Farfel-Becker et al. 2011a; Sun et al. 2010), the relevance of these findings to nGD is uncertain.

We recently demonstrated cathepsin D elevation in microglia using the nestin-flox/flox nGD mouse (Vitner et al. 2010b). Cathepsin mRNA expression was elevated by up to ~10-fold, with the time course of the increase correlating with the progression of disease severity. Significant changes in cathepsin D distribution in the brain were detected, with cathepsin D elevated in areas where neuronal loss, astrogliosis, and microgliosis were observed, such as in layer V of the cerebral cortex, the lateral globus pallidus, and in various nuclei in the thalamus, brain regions known to be affected in nGD. Cathepsin D elevation was greatest in microglia and also noticeable in astrocytes. The distribution of cathepsin D was altered in neurons in a manner consistent with its release from the lysosome to the cytosol. These data suggested the involvement of cathepsins in the neuropathology of nGD, which is consistent with a crucial role for reactive microglia in neuronal degeneration.

In addition, significant changes in levels of inflammatory mediators were demonstrated in the brain of these mice (Vitner et al. 2012). Levels of mRNA expression of IL-1β, TNF-α, TNF-α receptor, macrophage colony-stimulating factor, and transforming growth factor-β were elevated by up to ~30-fold. The most

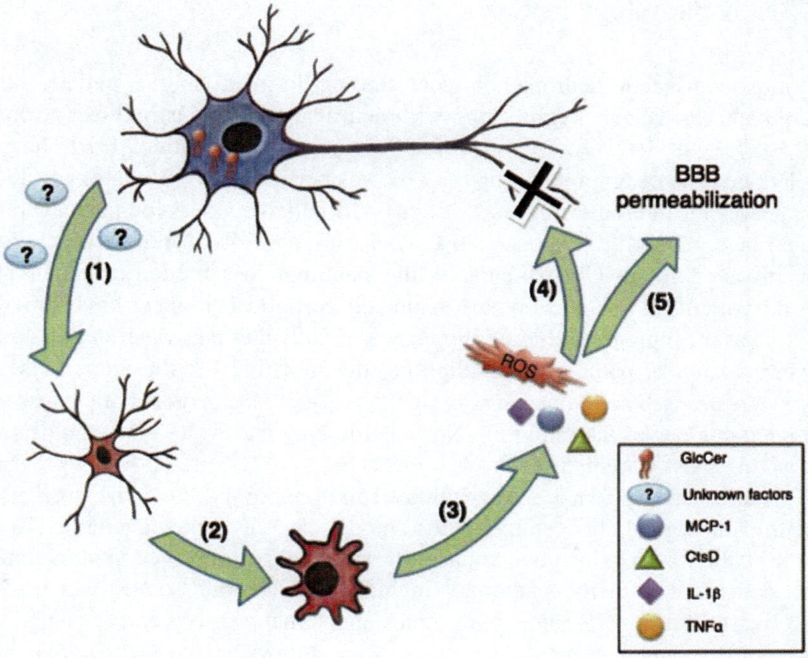

Fig. 1 Possible pathological cascades in nGD. Upon neuronal GlcCer accumulation, neurons signal to the surrounding microglia by an unknown mechanism, (1) and as a result, resting microglia becomes activated (2). Activated microglia initiate a neuroinflammatory cascade involving elevation of cytokines and neurotoxic agents such as IL-1β, TNFα, and reactive oxygen/nitrogen species (3), which results in neuronal cell death (4) and BBB permeabilization (5). The persistence of GlcCer accumulation in neurons, neuronal cell death, and continuous glial activation results in chronic inflammation, which further enhances neuronal cell loss (6)

significant elevation was detected for the chemokines CCL2, CCL3, and CCL5. Blood–brain barrier disruption was also evident in nGD mice. Extensive elevation of nitrotyrosine, a hallmark of peroxynitrite formation, was also observed, consistent with oxidative damage caused by macrophage/microglia activation. Based on these results, we suggested that once a critical threshold of GlcCer storage is reached in neurons, a signaling cascade is triggered that activates microglia, which in turn release inflammatory cytokines that amplify the inflammatory response, contributing to neuronal death (Fig. 1). The involvement of neuroinflammation in neuronal cell death in nGD was further supported by findings showing that microglial activation and astrocytosis are spatially and temporally correlated with selective neuronal loss (Farfel-Becker et al. 2011b). Interestingly, both neuroinflammation and neuronal loss in defined areas were detected before the onset of noticeable symptoms implicating these mechanisms in early disease pathogenesis.

5 Brain Pathology in GD

The most consistent neuropathological finding in nGD is the periadventitial accumulation of Gaucher cells along with neuronal loss and astrogliosis (Conradi et al. 1984, 1988, 1991; Kaye et al. 1986; Schiffmann and Vellodi 2007). The most detailed analysis of brain pathology in GD was performed by Wong et al. (2004). After analyzing 14 brains of types 1, 2, and 3 GD patients, it was found that cerebral cortical layers 3 and 5, hippocampal CA2-4, and layer 4b of the calcarine cortex were involved in all GD patients. While neuronal loss predominated in nGD patients, patients classified as type 1 displayed astrogliosis in these areas. Interestingly, adjacent regions, including hippocampal CA1 and calcarine lamina 4a and 4c, were spared of pathology, highlighting the specificity of the vulnerability of selective neurons. The brainstem was significantly gliotic in most patients, as well as the red nucleus of the midbrain. Nonspecific gray and white matter gliosis was observed in most patients.

A detailed and systematic description of the temporal and spatial progression of neuropathological changes in a mouse model of nGD is now available (Farfel-Becker et al. 2011b), allowing comparisons with changes observed in human nGD. Thus, consistent with the pathology in human nGD brain, cortical layer V is severely affected in nGD mice, with motor and somatosensory cortex being more affected than caudal cortical regions, such as the visual cortex. Human nGD patients show selective loss of pyramidal neurons in CA2–CA4, but not in the CA1 regions of the hippocampus (Wong et al. 2004), and a similar pattern was observed in nGD mice with respect to microglial activation, although this occurred late in disease progression and was not as pronounced as in other brain areas (Farfel-Becker et al. 2011b). Substantia nigra and red nucleus pathology were observed in both the human and in mouse nGD brain (Farfel-Becker et al. 2011b; Wong et al. 2004).

Neuropathological changes were also indentified in nGD mice in the substantia nigra reticulata (SNR) and in the reticulotegmental nucleus of the pons (RtTg), which may be related to the abnormal eye movements observed in humans, and in the cochlear nucleus and inferior colliculus, which may be related to the abnormal auditory brainstem responses in human patients; both phenomena are early neurological signs in nGD patients (Farfel-Becker et al. 2011b). Many components of the somatosensory system, including the sensory trigeminal nucleus SP5 and the thalamic nucleus VPM/VPL, showed significant early pathology in mice (Farfel-Becker et al. 2011b). Another anatomical characteristic of the structures that display microglial activation in nGD mice is that many of them send mossy fibers to the cerebellum. These structures include the pontine nuclei (which receives its major input from cortical layer V), the RtTg, the vestibular nuclei, the lateral reticular nucleus (LRt), the dorsal column nuclei, the trigeminal nuclei, and the spinal cord, which all displayed early microglial activation in mice (Farfel-Becker et al. 2011b).

6 GD and Parkinson's Disease

Of enormous importance to research on GD, and specifically to nGD, is the recent suggestion of a genetic link between Parkinson's disease (PD) and GD. Clinical observations, genetic and family studies, and neuropathological examinations suggested that alterations in GlcCerase may increase the susceptibility for PD. The first suggestion of a possible relationship between GD and PD was based on observations of PD in patients and of relatives of GD patients (Goker-Alpan et al. 2004; Halperin et al. 2006; Tayebi et al. 2003). This was further supported by several studies documenting a genetic association between PD and GlcCerase mutations in several population groups (Eblan et al. 2006; Gan-Or et al. 2008; Kalinderi et al. 2009; Mao et al. 2010; Rogaeva and Hardy 2008; Sidransky et al. 2009; Ziegler et al. 2007). Recently, the frequency of GlcCerase mutations in an ethnically diverse group of PD patients was determined. GlcCerase mutations in PD patients were screened in 16 centers around the world and demonstrated a strong association between GlcCerase mutations and PD (Sidransky et al. 2009). Either one of the two GlcCerase mutations, L444P and N370S, was found in 15 % of PD patients and 3 % of controls among Ashkenazi Jewish subjects. Among non-Ashkenazi Jewish subjects, either mutation was found in 3 % of PD patients, while in controls it was <1 %. The odds ratio for any GlcCerase mutation in PD patients versus controls was 5.43. Moreover, patients who had a GlcCerase mutation presented earlier PD symptoms than patients not carrying the mutation (Goker-Alpan et al. 2008) and were more likely to have affected relatives. Most studies did not detect any other significant difference in clinical manifestations and disease progression between GlcCerase carriers and controls (Velayati et al. 2010), but two studies noted atypical clinical PD manifestations and an earlier onset in patients with GlcCerase mutations (Goker-Alpan et al. 2010; Sidransky et al. 2009).

Neuropathological and histopathological examinations also supported this link. Brains from patients with early-onset Parkinsonism and type 1 GD showed α-synuclein inclusions in neurons from regions CA2-4 in the hippocampus. Those inclusions were similar to Lewy bodies found in the brainstem of PD patients. In one patient, loss of substantia nigra neurons was detected along with brainstem-type Lewy bodies (Wong et al. 2004). Even though type 1 GD is classically characterized by lack of CNS involvement, neurological involvement, including PD, has been found in type 1 patients (Biegstraaten et al. 2008; Capablo et al. 2008); indeed, a recent study documented 60 patients with GD type 1 that exhibit adult onset Parkinsonism (Alonso-Canovas et al. 2010).

Immunofluorescence and immunohistochemical analyses have also been performed on brain tissues from GD patients and carriers with Parkinsonism. All of the samples with GlcCerase mutations showed Lewy body pathology, and GlcCerase and α-synuclein were detected in Lewy body inclusions. Among patients with GlcCerase mutations, the mean number of Lewy bodies that stained positive for GlcCerase was 75 % but only 5 % in controls (Goker-Alpan et al. 2010). Imaging with positron emission tomography (Poupetova et al. 2010) showed presynaptic dopaminergic deficits, while transcranial ultrasound showed nigral hyperechogenicity (Saunders-Pullman et al. 2010). Substantia nigra hyperechogenicity is related to a high iron levels in the tissue and increasing ROS production by cells in the substantia nigra (Berg et al. 2002). These two features are consistent with Parkinsonism (Saunders-Pullman et al. 2010). Hypometabolism in the supplemental motor area has also been suggested as a clinical characteristic of GD carriers experiencing Parkinsonism (Kono et al. 2010).

Some biochemical studies have been performed to attempt to examine the link between PD and GD. Reduced GlcCerase activity was observed in the cerebrospinal fluid of PD patients (Balducci et al. 2007). A single injection of CBE to a mouse resulted in elevated nigral α-synuclein levels after 48 h (Manning-Bog et al. 2009). α-Synuclein aggregation was observed in brains from some nGD mice but not others (Xu et al. 2011), and there was a difference in α-synuclein concentrations in mice carrying one versus two D409V alleles (Cullen et al. 2011).

An extensive literature review (Shachar et al. 2011) identified case reports of appearance of α-synuclein deposits in the brain and of substantia nigra pathology that suggested that additional LSDs might be associated with PD. These findings indicate that the search for biochemical and cellular pathways that link PD with LSDs should not be limited exclusively to changes that occur in GD, such as changes in GlcCerase activity or in GlcCer levels, but rather include changes that might be common to a wide variety of LSDs. Indeed, a mechanistic link between GD and PD was suggested in a study showing that lysosomal GlcCer accumulation stabilizes α-synuclein oligomers and that α-synuclein inhibits lysosomal trafficking of GlcCerase in synucleinopathies (Mazzulli et al. 2011). Likewise, a direct physical interaction between specific residues in α-synuclein and GlcCerase in the lysosome was suggested (Yap et al. 2011), and CNS expression of GlcCerase corrected α-synuclein pathology and memory in a mouse model of a GD-related synucleinopathy (Kinghorn 2011). Thus, although there is little doubt that a connection (both genetic and biochemical) exists between GD and PD, which might be explained mechanistically by some of the examples given above, the possibility of a wider association between other LSDs and PD should drive the field to examine other possible biochemical mechanisms that may involve common pathological pathways in various LSDs.

7 Conclusions

Although GlcCer is the simplest GSL, its accumulation in GD, and in particular in the brain of nGD patients, results in unexpectedly complex pathologies. One of the main issues remaining to be resolved is the identification of the initial event(s) that triggers these pathological cascades. Our discovery of a role for dysfunctional ER calcium stores may be a right step in this direction, but more proximal events still await identification. Does lysosomal GlcCer accumulation affect general lysosomal function? If so, what specific mechanisms might be involved? Does the lysosome send a distress signal once a threshold level of GlcCer accumulates, and if so, what are the signaling mechanisms? How are other intracellular organelles affected? Are the inflammatory pathways common to other LSDs or specific to GD? Why are certain brain areas susceptible to inflammation, and does this correlate with susceptibility of certain neuronal populations upon GlcCer accumulation? These, and other questions, need to be addressed in order to obtain a more extensive picture of the role of GlcCer in GD pathology and to interrupt disease onset and/or progression.

Acknowledgments Work in the Futerman laboratory on nGD has been generously supported for the past decade by the Children's Gaucher Research Fund.

References

Alonso-Canovas A, Katschnig P, Tucci A, Carecchio M, Wood NW, Edwards M et al (2010) Atypical parkinsonism with apraxia and supranuclear gaze abnormalities in type 1 Gaucher disease. Expanding the spectrum: case report and literature review. Mov Disord 25(10):1506–1509

Azuri J, Elstein D, Lahad A, Abrahamov A, Hadas-Halpern I, Zimran A (1998) Asymptomatic Gaucher disease implications for large-scale screening. Genet Test 2(4):297–299

Balducci C, Pierguidi L, Persichetti E, Parnetti L, Sbaragli M, Tassi C et al (2007) Lysosomal hydrolases in cerebrospinal fluid from subjects with Parkinson's disease. Mov Disord 22(10):1481–1484

Barak V, Acker M, Nisman B, Kalickman I, Abrahamov A, Zimran A et al (1999) Cytokines in Gaucher's disease. Eur Cytokine Netw 10(2):205–210

Berg D, Roggendorf W, Schroder U, Klein R, Tatschner T, Benz P et al (2002) Echogenicity of the substantia nigra: association with increased iron content and marker for susceptibility to nigrostriatal injury. Arch Neurol 59(6):999–1005

Beutler E, Gelbart T, Scott CR (2005) Hematologically important mutations: Gaucher disease. Blood Cells Mol Dis 35(3):355–364

Biegstraaten M, van Schaik IN, Aerts JM, Hollak CE (2008) 'Non-neuronopathic' Gaucher disease reconsidered. Prevalence of neurological manifestations in a Dutch cohort of type I Gaucher disease patients and a systematic review of the literature. J Inherit Metab Dis 31(3):337–349

Brumshtein B, Greenblatt HM, Butters TD, Shaaltiel Y, Aviezer D, Silman I et al (2007) Crystal structures of complexes of N-butyl- and N-nonyl-deoxynojirimycin bound to acid beta-glucosidase: insights into the mechanism of chemical chaperone action in Gaucher disease. J Biol Chem 282(39):29052–29058

Capablo JL, Saenz de Cabezon A, Fraile J, Alfonso P, Pocovi M, Giraldo P (2008) Neurological evaluation of patients with Gaucher disease diagnosed as type 1. J Neurol Neurosurg Psychiatry 79(2):219–222

Conradi NG, Sourander P, Nilsson O, Svennerholm L, Erikson A (1984) Neuropathology of the Norrbottnian type of Gaucher disease. Morphological and biochemical studies. Acta Neuropathol 65(2):99–109

Conradi NG, Kalimo H, Sourander P (1988) Reactions of vessel walls and brain parenchyma to the accumulation of Gaucher cells in the Norrbottnian type (type III) of Gaucher disease. Acta Neuropathol 75(4):385–390

Conradi N, Kyllerman M, Mansson JE, Percy AK, Svennerholm L (1991) Late-infantile Gaucher disease in a child with myoclonus and bulbar signs: neuropathological and neurochemical findings. Acta Neuropathol 82(2):152–157

Cox TM, Schofield JP (1997) Gaucher's disease: clinical features and natural history. Baillieres Clin Haematol 10(4):657–689

Cullen V, Sardi SP, Ng J, Xu YH, Sun Y, Tomlinson JJ et al (2011) Acid beta-glucosidase mutants linked to Gaucher disease, Parkinson disease, and Lewy body dementia alter alpha-synuclein processing. Ann Neurol 69(6):940–953

Dvir H, Harel M, McCarthy AA, Toker L, Silman I, Futerman AH et al (2003) X-ray structure of human acid-beta-glucosidase, the defective enzyme in Gaucher disease. EMBO Rep 4(7):704–709

Eblan MJ, Nguyen J, Ziegler SG, Lwin A, Hanson M, Gallardo M et al (2006) Glucocerebrosidase mutations are also found in subjects with early-onset parkinsonism from Venezuela. Mov Disord 21(2):282–283

Enquist IB, Lo Bianco C, Ooka A, Nilsson E, Mansson JE, Ehinger M et al (2007) Murine models of acute neuronopathic Gaucher disease. Proc Natl Acad Sci USA 104(44):17483–17488

Farfel-Becker T, Vitner E, Dekel H, Leshem N, Enquist IB, Karlsson S et al (2009) No evidence for activation of the unfolded protein response in neuronopathic models of Gaucher disease. Hum Mol Genet 18(8):1482–1488

Farfel-Becker T, Vitner EB, Futerman AH (2011a) Animal models for Gaucher disease research. Dis Model Mech 4(6):746–752

Farfel-Becker T, Vitner EB, Pressey SN, Eilam R, Cooper JD, Futerman AH (2011b) Spatial and temporal correlation between neuron loss and neuroinflammation in a mouse model of neuronopathic Gaucher disease. Hum Mol Genet 20(7):1375–1386

Fill M, Copello JA (2002) Ryanodine receptor calcium release channels. Physiol Rev 82(4):893–922

Futerman AH, Pagano RE (1991) Determination of the intracellular sites and topology of glucosylceramide synthesis in rat liver. Biochem J 280(Pt 2):295–302

Futerman AH, Sussman JL, Horowitz M, Silman I, Zimran A (2004) New directions in the treatment of Gaucher disease. Trends Pharmacol Sci 25(3):147–151

Gan-Or Z, Giladi N, Rozovski U, Shifrin C, Rosner S, Gurevich T et al (2008) Genotype-phenotype correlations between GBA mutations and Parkinson disease risk and onset. Neurology 70(24):2277–2283

Gaucher PCE (1882) De l'Epithélioma Primitif de la Rate: Hypertrophie Idiopathique de la Rate sans Leuce´mie, Ph.D. thesis. University of Paris, Paris

Gault CR, Obeid LM, Hannun YA (2010) An overview of sphingolipid metabolism: from synthesis to breakdown. Adv Exp Med Biol 688:1–23

Goker-Alpan O, Schiffmann R, LaMarca ME, Nussbaum RL, McInerney-Leo A, Sidransky E (2004) Parkinsonism among Gaucher disease carriers. J Med Genet 41(12):937–940

Goker-Alpan O, Lopez G, Vithayathil J, Davis J, Hallett M, Sidransky E (2008) The spectrum of parkinsonian manifestations associated with glucocerebrosidase mutations. Arch Neurol 65(10):1353–1357

Goker-Alpan O, Stubblefield BK, Giasson BI, Sidransky E (2010) Glucocerebrosidase is present in alpha-synuclein inclusions in Lewy body disorders. Acta Neuropathol 120(5):641–649

Halperin A, Elstein D, Zimran A (2006) Increased incidence of Parkinson disease among relatives of patients with Gaucher disease. Blood Cells Mol Dis 36(3):426–428

Henkart M (1980) Identification and function of intracellular calcium stores in neurons. Fed Proc 39(10):2776–2777

Hollak CE, Evers L, Aerts JM, van Oers MH (1997) Elevated levels of M-CSF, sCD14 and IL8 in type 1 Gaucher disease. Blood Cells Mol Dis 23(2):201–212

Hong YB, Kim EY, Jung SC (2006) Upregulation of proinflammatory cytokines in the fetal brain of the Gaucher mouse. J Korean Med Sci 21(4):733–738

Horowitz M, Zimran A (1994) Mutations causing Gaucher disease. Hum Mutat 3(1):1–11

Jmoudiak M, Futerman AH (2005) Gaucher disease: pathological mechanisms and modern management. Br J Haematol 129(2):178–188

Kalinderi K, Bostantjopoulou S, Paisan-Ruiz C, Katsarou Z, Hardy J, Fidani L (2009) Complete screening for glucocerebrosidase mutations in Parkinson disease patients from Greece. Neurosci Lett 452(2):87–89

Kanfer JN, Legler G, Sullivan J, Raghavan SS, Mumford RA (1975) The Gaucher mouse. Biochem Biophys Res Commun 67(1):85–90

Kaye EM, Ullman MD, Wilson ER, Barranger JA (1986) Type 2 and type 3 Gaucher disease: a morphological and biochemical study. Ann Neurol 20(2):223–230

Kinghorn KJ (2011) Pathological looping in the synucleinopathies: investigating the link between Parkinson's disease and Gaucher disease. Dis Model Mech 4(6):713–715

Kono S, Ouchi Y, Terada T, Ida H, Suzuki M, Miyajima H (2010) Functional brain imaging in glucocerebrosidase mutation carriers with and without parkinsonism. Mov Disord 25(12):1823–1829

Korkotian E, Schwarz A, Pelled D, Schwarzmann G, Segal M, Futerman AH (1999) Elevation of intracellular glucosylceramide levels results in an increase in endoplasmic reticulum density and in functional calcium stores in cultured neurons. J Biol Chem 274(31):21673–21678

Legler G, Bieberich E (1988) Active site directed inhibition of a cytosolic beta-glucosidase from calf liver by bromoconduritol B epoxide and bromoconduritol F. Arch Biochem Biophys 260(1):437–442

Lloyd-Evans E, Pelled D, Riebeling C, Bodennec J, de-Morgan A, Waller H et al (2003) Glucosylceramide and glucosylsphingosine modulate calcium mobilization from brain microsomes via different mechanisms. J Biol Chem 278(26):23594–23599

Manning-Bog AB, Schule B, Langston JW (2009) Alpha-synuclein-glucocerebrosidase interactions in pharmacological Gaucher models: a biological link between Gaucher disease and parkinsonism. Neurotoxicology 30(6):1127–1132

Mao XY, Burgunder JM, Zhang ZJ, An XK, Zhang JH, Yang Y et al (2010) Association between GBA L444P mutation and sporadic Parkinson's disease from Mainland China. Neurosci Lett 469(2):256–259

Mazzulli JR, Xu YH, Sun Y, Knight AL, McLean PJ, Caldwell GA et al (2011) Gaucher disease glucocerebrosidase and alpha-synuclein form a bidirectional pathogenic loop in synucleinopathies. Cell 146(1):37–52

Merrill AH Jr (2011) Sphingolipid and glycosphingolipid metabolic pathways in the era of sphingolipidomics. Chem Rev 111(10):6387–6422

Mikoshiba K (1997) The InsP3 receptor and intracellular Ca2+ signaling. Curr Opin Neurobiol 7(3):339–345

Mizukami H, Mi Y, Wada R, Kono M, Yamashita T, Liu Y et al (2002) Systemic inflammation in glucocerebrosidase-deficient mice with minimal glucosylceramide storage. J Clin Invest 109 (9):1215–1221

Offman MN, Krol M, Silman I, Sussman JL, Futerman AH (2010) Molecular basis of reduced glucosylceramidase activity in the most common Gaucher disease mutant, N370S. J Biol Chem 285(53):42105–42114

Pastores GM, Barnett NL, Bathan P, Kolodny EH (2003) A neurological symptom survey of patients with type I Gaucher disease. J Inherit Metab Dis 26(7):641–645

Pelled D, Shogomori H, Futerman AH (2000) The increased sensitivity of neurons with elevated glucocerebroside to neurotoxic agents can be reversed by imiglucerase. J Inherit Metab Dis 23 (2):175–184

Pelled D, Trajkovic-Bodennec S, Lloyd-Evans E, Sidransky E, Schiffmann R, Futerman AH (2005) Enhanced calcium release in the acute neuronopathic form of Gaucher disease. Neurobiol Dis 18(1):83–88

Platt FM, Jeyakumar M (2008) Substrate reduction therapy. Acta Paediatr Suppl 97(457):88–93

Platt FM, Lachmann RH (2009) Treating lysosomal storage disorders: current practice and future prospects. Biochim Biophys Acta 1793(4):737–745

Poupetova H, Ledvinova J, Berna L, Dvorakova L, Kozich V, Elleder M (2010) The birth prevalence of lysosomal storage disorders in the Czech Republic: comparison with data in different populations. J Inherit Metab Dis 33(4):387–396

Premkumar L, Sawkar AR, Boldin-Adamsky S, Toker L, Silman I, Kelly JW et al (2005) X-ray structure of human acid-beta-glucosidase covalently bound to conduritol-B-epoxide. Implications for Gaucher disease. J Biol Chem 280(25):23815–23819

Rogaeva E, Hardy J (2008) Gaucher and Parkinson diseases: unexpectedly related. Neurology 70(24):2272–2273

Saunders-Pullman R, Hagenah J, Dhawan V, Stanley K, Pastores G, Sathe S et al (2010) Gaucher disease ascertained through a Parkinson's center: imaging and clinical characterization. Mov Disord 25(10):1364–1372

Sawkar AR, D'Haeze W, Kelly JW (2006) Therapeutic strategies to ameliorate lysosomal storage disorders–a focus on Gaucher disease. Cell Mol Life Sci 63(10):1179–1192

Schiffmann R, Vellodi A (2007) Neuoronopathic Gaucher disease. In: Futerman AH, Zimran A (eds) Gaucher disease. CRC, Boca Raton, pp 175–196

Shachar T, Lo Bianco C, Recchia A, Wiessner C, Raas-Rothschild A, Futerman AH (2011) Lysosomal storage disorders and Parkinson's disease: Gaucher disease and beyond. Mov Disord 26(9):1593–1604

Sidransky E, Fartasch M, Lee RE, Metlay LA, Abella S, Zimran A et al (1996) Epidermal abnormalities may distinguish type 2 from type 1 and type 3 of Gaucher disease. Pediatr Res 39(1):134–141

Sidransky E, Nalls MA, Aasly JO, Aharon-Peretz J, Annesi G, Barbosa ER et al (2009) Multicenter analysis of glucocerebrosidase mutations in Parkinson's disease. N Engl J Med 361(17):1651–1661

Stephens MC, Bernatsky A, Burachinsky V, Legler G, Kanfer JN (1978) The Gaucher mouse: differential action of conduritol B epoxide and reversibility of its effects. J Neurochem 30 (5):1023–1027

Sun Y, Liou B, Ran H, Skelton MR, Williams MT, Vorhees CV et al (2010) Neuronopathic Gaucher disease in the mouse: viable combined selective saposin C deficiency and mutant glucocerebrosidase (V394L) mice with glucosylsphingosine and glucosylceramide accumulation and progressive neurological deficits. Hum Mol Genet 19(6):1088–1097

Sun Y, Ran H, Liou B, Quinn B, Zamzow M, Zhang W et al (2011) Isofagomine in vivo effects in a neuronopathic Gaucher disease mouse. PLoS One 6(4):e19037

Suzuki K (2003) Globoid cell leukodystrophy (Krabbe's disease): update. J Child Neurol 18(9):595–603

Tayebi N, Walker J, Stubblefield B, Orvisky E, LaMarca ME, Wong K et al (2003) Gaucher disease with parkinsonian manifestations: does glucocerebrosidase deficiency contribute to a vulnerability to parkinsonism? Mol Genet Metab 79(2):104–109

Tybulewicz VL, Tremblay ML, LaMarca ME, Willemsen R, Stubblefield BK, Winfield S et al (1992) Animal model of Gaucher's disease from targeted disruption of the mouse glucocerebrosidase gene. Nature 357(6377):407–410

Velayati A, Yu WH, Sidransky E (2010) The role of glucocerebrosidase mutations in Parkinson disease and Lewy body disorders. Curr Neurol Neurosci Rep 10(3):190–198

Vellodi A, Tylki-Szymanska A, Davies EH, Kolodny E, Bembi B, Collin-Histed T et al (2009) Management of neuronopathic Gaucher disease: revised recommendations. J Inherit Metab Dis 32(5):660–664

Vitner EB, Dekel H, Zigdon H, Shachar T, Farfel-Becker T, Eilam R et al (2010a) Altered expression and distribution of cathepsins in neuronopathic forms of Gaucher disease and in other sphingolipidoses. Hum Mol Genet 19(18):3583–3590

Vitner EB, Platt FM, Futerman AH (2010b) Common and uncommon pathogenic cascades in lysosomal storage diseases. J Biol Chem 285(27):20423–20427

Vitner EB, Farfel-Becker T, Eilam R, Biton I, Futerman AH (2012) Contribution of brain inflammation to neuronal cell death in neuronopathic forms of Gaucher disease. Brain 135:1724–1735

Wei H, Kim SJ, Zhang Z, Tsai PC, Wisniewski KE, Mukherjee AB (2008) ER and oxidative stresses are common mediators of apoptosis in both neurodegenerative and non-neurodegenerative lysosomal storage disorders and are alleviated by chemical chaperones. Hum Mol Genet 17(4):469–477

Wei RR, Hughes H, Boucher S, Bird JJ, Guziewicz N, Van Patten SM et al (2011) X-ray and biochemical analysis of N370S mutant human acid beta-glucosidase. J Biol Chem 286 (1):299–308

Willemsen R, Tybulewicz V, Sidransky E, Eliason WK, Martin BM, LaMarca ME et al (1995) A biochemical and ultrastructural evaluation of the type 2 Gaucher mouse. Mol Chem Neuropathol 24(2–3):179–192

Wong K, Sidransky E, Verma A, Mixon T, Sandberg GD, Wakefield LK et al (2004) Neuropathology provides clues to the pathophysiology of Gaucher disease. Mol Genet Metab 82(3):192–207

Xu YH, Sun Y, Ran H, Quinn B, Witte D, Grabowski GA (2011) Accumulation and distribution of alpha-synuclein and ubiquitin in the CNS of Gaucher disease mouse models. Mol Genet Metab 102(4):436–447

Yap TL, Gruschus JM, Velayati A, Westbroek W, Goldin E, Moaven N et al (2011) Alpha-synuclein interacts with Glucocerebrosidase providing a molecular link between Parkinson and Gaucher diseases. J Biol Chem 286(32):28080–28088

Ziegler SG, Eblan MJ, Gutti U, Hruska KS, Stubblefield BK, Goker-Alpan O et al (2007) Glucocerebrosidase mutations in Chinese subjects from Taiwan with sporadic Parkinson disease. Mol Genet Metab 91(2):195–200

Sphingolipids in Neuroinflammation

Laura Davies, Klaus Fassbender, and Silke Walter

Contents

Abstract Sphingolipids, the main component of cellular membranes, are cellular 'jack-of-all-trades', influencing a variety of functions including signal transduction, cell activation, membrane fluidity and cell–cell interactions.

In the last few years, sphingolipids have begun to be investigated in the pathophysiology of major diseases of the brain, e.g. multiple sclerosis and dementia. Modulation of neuroinflammatory responses, such as lymphocyte behaviour, is a chance to intervene in the pathways that cause disease. There is much research still to be done in this field, but the prospect of treating previously untreatable medical conditions compels us onwards. Here, we review the current knowledge of the link between sphingolipids and neuroinflammation.

Keywords sphingolipids • neuroinflammation • neuroinflammatory diseases • Multiple Sclerosis

L. Davies • K. Fassbender • S. Walter (✉)
Department of Neurology, Saarland University Hospital, Kirrbergerstr, 66424 Homburg, Germany
e-mail: silke.walter@uks.eu

E. Gulbins and I. Petrache (eds.), *Sphingolipids in Disease*,
Handbook of Experimental Pharmacology 216,
DOI 10.1007/978-3-7091-1511-4_21,

1 Introduction

Sphingolipids are increasingly implicated as important players in the pathophysiology of a variety of neuroinflammatory diseases, particularly multiple sclerosis, but also others such as stroke, dementia and inflammatory neuropathies. Recently, a series of discoveries have confirmed the hypothesis linking sphingolipid function and neuroinflammation. The precise role of sphingolipids in neuroinflammation, however, is complex acting via diverse mechanisms such as apoptosis, astrogliosis, leukocyte activation, leukocyte trafficking and associated receptor clustering (Jana and Pahan 2010). Much of what we know in this field has been discovered within the last 15 years and will remain a veritable treasure chest for future studies and developments. However, although there is still much to learn, progress in understanding and transfer of sphingolipid targeting treatment in numerous neuroinflammatory conditions has already begun.

2 Neuroinflammation Mechanisms

Neuroinflammation, much like inflammation in other organ systems, is the response of tissue to injurious stimuli, e.g. infections, trauma, toxins and autoimmune reactions (Crutcher et al. 2006). These injurious stimuli either are passed on via the peripheral circulation or develop directly in the nervous system (Streit et al. 2004; Rivest 2009). Neuroinflammation includes both nervous system damage and subsequent repair and regeneration (Frank-Cannon et al. 2009). Neuroinflammatory diseases are classified either according to timing (i.e. acute vs. chronic inflammatory processes) or according to localisation (i.e. central vs. peripheral nervous system inflammation).

A common pathophysiological hallmark is the release of proinflammatory mediators, such as TNFα, interleukin-(IL)6, IL-1β, nitric oxide and ROS production; this is the result of a variety of mechanisms and inflammatory cells and is involved in mediating nervous tissue damage (Rao et al. 2012). For example, acute traumatic nerve injury induces TNFα production by activated Schwann cells, which subsequently alters neuronal function resulting in both acute pain and the development of chronic neuropathic pain (Myers and Shubayev 2011). Neuroinflammation with NF-κB activation and subsequent production of inflammatory mediators such as IL-6, TNFα, COX-2 or iNOS is an integral component of the commonly occurring diabetes-induced chronic neuropathy which results in sensory deficits (Cameron and Cotter 2008; Edwards et al. 2008; Kumar et al. 2012).

The acute autoimmune neuropathy, Guillain–Barré syndrome (demyelinating and/or axonal-affecting subtypes), shows an immense inflammatory cell infiltration into the peripheral nerves consisting mainly of Th1 and Th17 cells, as well as macrophages.

The inflammatory attack leads to a block or impairment of impulse conduction with a clinical presentation of paresis and, in severe cases, respiratory failure (Kieseier et al. 2004; Hughes and Cornblath 2005; Langrish et al. 2005; Yang et al. 2008).

However, aside from the aforementioned neuroinflammatory mechanisms in peripheral nervous system diseases, most of the research focuses on neuroinflammation in the central nervous system (CNS).

Inflammation and immunity play key roles in the pathophysiology of most of the CNS diseases. Examples of neuroinflammatory diseases include acute diseases, such as bacterial, viral and parasitic infections; traumatic CNS injury; stroke and chronic diseases such as multiple sclerosis (MS); Alzheimer's disease; Parkinson's disease; Huntington's disease; and amyotrophic lateral sclerosis (McGeer and McGeer 2001; McColl et al. 2007; Wang and Doré 2007; Frank-Cannon et al. 2009). In the majority of cases, the magnitude of CNS inflammation is strongly dependent on the integrity of the blood–brain barrier.

Since the late nineteenth century, the CNS has been considered to be an immunoprivileged organ, insulated from the rest of the body by the so-called blood–brain barrier, protecting against invading pathogens (Ribatti et al. 2006). When there is blood–brain barrier compromise, such as in stroke, MS or acute CNS infection, leukocytes infiltrate from the peripheral circulation and act in concert with the resident innate immune cells of the CNS. This sustained local inflammatory reaction is orchestrated by the local production of cytokines, chemokines, antibodies that aggravate blood–brain barrier disruption and increased leukocyte recruitment from the periphery mediated by adhesion molecules on both endothelium and lymphocytes (Rivest 2009; Jana and Pahan 2010; Barth et al. 2012). Analysis of the cerebrospinal fluid of patients with inflammatory neurological diseases (including MS) revealed an enrichment of Th1-polarised memory T cells that were capable of differentiating into effector cells after antigen exposure. T cells are locally recruited by inducible chemokines (Giunti et al. 2003), which link the immune and nervous system, and are thought to confer a number of leukocyte functions during inflammation (Ransohoff 2012).

In general, a consequence of brain inflammation is reactive gliosis typified by glial cell hypertrophy and proliferation, activation of astrocytes and microglia and sustained release of inflammatory mediators as well as increased oxidative and nitrosative stress (Barth et al. 2012). In particular, microglial activation is considered to be crucial in the development of several chronic neuroinflammatory CNS diseases that lack the extensive infiltration of blood-derived leukocytes (Streit et al. 2004). For example, activated microglia detected in the substantia nigra in Parkinson's disease are thought to be the cause of nerve cell death (McGeer et al. 1988; Witton 2007). Inflammation caused by microglial activation has also been proposed as a neurotoxic mechanism for other neurodegenerative diseases, most notably Alzheimer's disease. Several clinical studies show a reduced risk of developing Alzheimer's disease following the use of nonsteroidal anti-inflammatory drugs (NSAIDs; In t'Veld et al. 2001; Etminan et al. 2003).

3 Sphingolipids: Major Players in Neuroinflammation

Sphingolipids are one of the main components of cell membranes, and their metabolism is a critical step involved in a number of cellular processes (Gulbins and Lee 2006). Nervous system cell subtypes, in particular, have a vast repertoire of sphingolipids; the highest concentrations of gangliosides can be found in the grey matter of the CNS, and four of five known sphingosine 1-phosphate (S1P) receptors are expressed in one or more CNS cell types (Choi and Chun 2013; Yu et al. 2012).

It has been demonstrated that the deletion of sphingosine kinase or sphingosine-1-phosphate (S1P) receptor reduces astroglial proliferation and gliosis (Wu et al. 2008). A knockdown of neutral sphingomyelinase (nSMase) decreases glial activation in mouse cortex in an in vivo model, strongly suggesting that sphingolipids are involved in glial activation and play an important role in modulation of neuroinflammation. NSMase knockdown was also able to prevent the induction of proinflammatory molecules (TNFα, iNOS, IL-1β and IL-6) and the activation of NF-κB (Jana and Pahan 2010).

FTY720, the S1P agonist, when applied to TNFα-treated human astrocytes, reduced the secretion of monocyte chemotactic protein-1, indicating attenuating effects of the proinflammatory response of astrocytes (Van Doorn et al. 2010). Conversely, the addition of S1P to activated microglia resulted in increased mRNA expression levels of TNFα, IL-1β and iNOS and production of TNFα and nitric oxide (Nayak et al. 2010). Furthermore, it has been shown that LPS treatment increases sphingosine kinase 1 (SphK1) mRNA and protein expression in microglia and that the suppression of SphK1 results in decreased mRNA expression of TNFα, IL-1β and iNOS and release of TNFα and nitric oxide (Nayak et al. 2010).

Additionally, stimulation of T cells or B cells (via the co-stimulatory receptors, CD28 and CD40, respectively) results in the activation of acid sphingomyelinase (aSMase) and release of ceramide (Stoffel et al. 1998). LFA-1, a key adhesion molecule at the blood–brain barrier, also triggers release of ceramide, possibly through aSMase (Rosenman et al. 1993). Ceramide released by nSMase is also involved in the stimulation of lymphocytes via the adhesion molecule, L-selectin (Phong et al. 2003).

4 Sphingolipids in Acute Neuroinflammatory Diseases

The prototype of acute neuroinflammatory CNS diseases is meningoencephalitis caused by bacteria, virus, fungi or parasites. Recently, an involvement of sphingolipids in disease pathogenesis has been reported; specifically, measles virus has been shown to cause ceramide accumulation in human T cells through the activation of nSMase and aSMase (Gassert et al. 2009), which is important for measles virus uptake by dendritic cells (Avota et al. 2011). In patients with HIV-associated neurological disorders, a dysregulation of sphingomyelin and ceramide has been suggested (Ben-David and Futerman 2010; Huang et al. 2011).

Based on the observation that plasma S1P is decreased in children with cerebral malaria, FTY720 (a S1P agonist) treatment in a murine model has been demonstrated to improve survival, decrease plasma IFNγ levels, reduce soluble ICAM-1, increase angiopoietin-1 and decrease blood–brain barrier leakage (Finney et al. 2011). An in vitro study using human microvascular endothelial cells showed that the endocytosis of Cryptococcus neoformans, the cause of a devastating meningoencephalitis, is mediated by lipid rafts involving CD44 (Huang et al. 2011).

In traumatic spinal cord injury, two clinical trials showed that treating patients with GM-1 gangliosides resulted in a faster neurological recovery (Geisler et al. 1991, 2001); this, however, has not yet been fully transferred to daily clinical practice. Investigations in murine models of spinal cord injury also hint towards an involvement of ceramide in disease pathophysiology (Pannu et al. 2004; Cuzzocrea et al. 2009). Intrathecal injections of ceramide biosynthesis inhibitors attenuated neuropathic pain and suppressed microglial activation in the spinal cord (Kobayashi et al. 2012).

Stroke, which affects 16 million people worldwide every year, is one of the most important acute neurological diseases (Strong et al. 2007). There is increasing evidence of an involvement of ceramide in ischemia-induced mitochondrial injury (Novgorodov and Gudz 2011). Furthermore, FTY720 treatment has been shown to display neuroprotective effects with reduced ischaemic lesion size in mouse and rat models of stroke (Czech et al. 2009; Hasegawa et al. 2010). Similar observations have been made upon treatment with sphingomyelin analogues which inhibit nSMase activity (Soeda et al. 2004).

Guillain–Barré syndrome is currently the best described inflammatory disease of the peripheral nervous system. For many years, this disease has been known to be associated with sphingolipids; specifically, patients develop autoantibodies to different gangliosides (GM1, GD1a, GT1a and GQ1b) depending on the subtype of the disease (Yuki and Hartung 2012). There are several clues that hint at a beneficial role of gangliosides in Guillain–Barré syndrome (Köhne et al. 2012). For example, internalisation of anti-ganglioside antibodies into rat neuronal cells attenuated complement-mediated neuronal toxicity (Fewou et al. 2012). Furthermore, challenge of rat Schwann cells with phosphorylated FTY720, which induced production of reactive oxygen species associated with cell apoptosis and GalNAcT/ST-II double knockout mice, lacking most of the gangliosides, develops significant and progressive neuropathies (Tajima et al. 2009).

5 Sphingolipids in Chronic Neuroinflammatory Diseases

Huntington's disease is one of the most common genetic neurological diseases with accumulation of the huntingtin protein in neurons, resulting in inflammation (Frank-Cannon et al. 2009). Lipid alterations in the brains of patients with Huntington's disease were already described in 1967 (Borri et al. 1967), but it was only 45 years later that a sphingolipid-modulating therapy has been shown to have beneficial effects in a murine model of the disease. Intraventricular treatment of mutant huntingtin mice, with ganglioside GM1, induced phosphorylation of the huntingtin protein and attenuated its toxicity, restoring normal motor function in already symptomatic mice (Di Pardo et al. 2012).

Another devastating neurodegenerative disease is amyotrophic lateral sclerosis characterised by progressive degeneration of motor neurons and neuroinflammation driven by resident microglia and intracerebrally invading macrophages (Frank-Cannon et al. 2009). Abnormalities in sphingolipid metabolism have been reported in the transgenic mouse model of the disease (SOD mutant mice), as well as in amyotrophic lateral sclerosis patients. Oxidative stress increased the accumulation of sphingomyelin and ceramide, while pharmacologic inhibition of sphingomyelin synthesis not only prevented this accumulation but also protected motor neurons against oxidative stress (Cutler et al. 2002). Several case reports point towards seroconversion with development of anti-GM-1 antibodies during the disease course of at least some amyotrophic lateral sclerosis patients (Haggiag et al. 2004). However, no therapeutic approach addressing sphingolipids in amyotrophic lateral sclerosis is available for patients.

In contrast, therapeutic approaches focusing on sphingolipids have been elaborated in preclinical studies of Parkinson's disease (Orr et al. 2002). As in most of the neurological diseases, the majority of sphingolipid research in this field focuses on gangliosides, the most abundantly expressed sphingolipid in the nervous system (Yu et al. 2009). For example, Parkinson's disease patients with tremor dominance showed increased circulating anti-GM1 antibodies (Zappia et al. 2002). Mutant mice with a disrupted ganglioside gene expressed a greatly elevated amount of aggregated α-synuclein in their substantia nigra, resulting in intensive disease symptoms, while administration of a membrane-permeable GM1 derivative attenuated the symptoms (Wu et al. 2011; Yu et al. 2012). A 5-year treatment of Parkinson's disease patients with GM1 resulted in reduced motor symptoms (Schneider et al. 2010).

Extensive information is available for the role of sphingolipids in the closely related neurodegenerative disease, Alzheimer's disease, which is addressed in a separate chapter.

For the most part, sphingolipid research in neuroinflammatory diseases is still in its infancy. However, research in Multiple Sclerosis (MS), the prototype disease of chronic neuroinflammation, is much more advanced and has been translated to clinical practice. Specifically, the non-selective S1P receptor modulator FTY720 (Gilenya®) has already been approved as the first oral MS medication for 2 years (Brinkmann et al. 2010). In two large multicentre trials (FREEDOMS; Kappos et al. 2010 and TRANSFORMS; Cohen et al. 2010), FTY720 was shown to reduce the rate of relapse by 60 %. FTY720 was initially believed to target S1P receptors on T cells, resulting in receptor internalisation and a redistribution of T cells to secondary lymphoid organs with subsequent reduction of circulating auto-aggressive lymphocytes (Cohen and Chun 2011). Different subtypes of S1P receptors have been described to be upregulated in both MS and the animal model of MS, experimental autoimmune encephalomyelitis (Foster et al. 2009; Van Doorn et al. 2010). Increasingly, information regarding direct CNS effects is surfacing.

Experimental autoimmune encephalomyelitis was attenuated in CNS mutant mice lacking S1P1 on astrocytes (Choi et al. 2011). Furthermore, FTY720 treatment prior to TNFα stimulation of human MS lesion-derived astrocytes reduced ceramide production, expression of aSMase mRNA and subsequent monocytic transendothelial migration (Van Doorn et al. 2012). Oligodendrocyte injury is a major hallmark of MS. There are also hints towards a detrimental role of astrocytic ceramide in oligodendroglial death (Kim et al. 2012).

6 Conclusion and Perspective

In summary, this chapter illustrates the involvement of sphingolipids in neuroinflammatory disorders. While most experimental data are still far from being translated to clinical practice, modulation of the S1P receptor in MS has already become a major clinical application of sphingolipid research in the last few years. However, it can be expected that future exploration of the role of sphingolipids in other neuroinflammatory diseases will become of similar clinical relevance.

References

Avota E, Gulbins E, Schneider-Schaulies S (2011) DC-SIGN mediated sphingomyelinase-activation and ceramide generation is essential for enhancement of viral uptake in dendritic cells. PLoS Pathog 7(2):e1001290

Barth BM, Gustafson SJ, Kuhn TB (2012) Neutral sphingomyelinase activation precedes NADPH-oxidase-dependent damage in neurons exposed to the proinflammatory cytokine tumor necrosis factor-α. Neurosci Res 90:229–242

Ben-David O, Futerman HA (2010) The role of the ceramide acyl chain length in neurodegeneration: involvement of ceramide synthases. Neuromol Med 12:341–350

Borri PF, Op den Velde WM, Hooghwinkel GJ, Bruyn GW (1967) Biochemical studies in Huntington's chorea. VI. Composition of striatal neutral lipids, phospholipids, glycolipids, fatty acids, and amino acids. J Neurol 17(2):172–178

Brinkmann V, Billich A, Baumruker T, Heining P, Schmouder R, Francis G, Aradhye S, Burtin P (2010) Fingolimod (FTY729): discovery and development of an oral drug to treat multiple sclerosis. Nat Rev Drug Discov 9:833–897

Cameron NE, Cotter MA (2008) Pro-inflammatory mechanisms in diabetic neuropathy: focus on the nuclear factor kappa B pathway. Curr Drug Targets Jan 9(1):60–67

Choi JW, Chun J (2013) Lysophospholipids and their receptors in the central nervous system. Biochim Biophys Acta 1831(1):20–32

Choi JW, Gardell SE, Herr DR, Rivera R, Lee CW, Noguchi K, Teo ST, Yung YC, Lu M, Kennedy G, Chun J (2011) FTY720 (fingolimod) efficacy in an animal model of multiple sclerosis requires astrocyte sphingosine 1-phosphate receptor 1 (S1P1) modulation. Proc Natl Acad Sci USA 108(2):751–756

Cohen JA, Chun J (2011) Mechanisms of fingolimod's efficacy and adverse effects in multiple sclerosis. Ann Neurol 69:759–777

Cohen JA, Barkhof F, Comi G, Hartung HP, Khatri BO, Montalban X, Pelletier J, Capra R, Gallo P, Izquierdo G, Tiel-Wilck K, de Vera A, Jin J, Stites T, Wu S, Aradhye S, Kappos L, TRANSFORMS Study Group (2010) Oral fingolimod or intramuscular interferon for relapsing multiple sclerosis. N Engl J Med 362(5):402–415

Crutcher KA, Gendelman HE, Kipnis J, Perez-Polo JR, Perry VH, Popovich PG, Weaver LC (2006) Debate: "is increasing neuroinflammation beneficial for neural repair?". J Neuroimmune Pharmacol 1(3):195–211

Cutler RG, Pedersen WA, Camandola S, Rothstein JD, Mattson MP (2002) Evidence that accumulation of ceramides and cholesterol esters mediates oxidative stress-induced death of motor neurons in amyotrophic lateral sclerosis. Ann Neurol 52(4):448–457

Cuzzocrea S, Deigner HP, Genovese T, Mazzon E, Esposito E, Crisafulli C, Di Paola R, Bramanti P, Matuschak G, Salvemini D (2009) Inhibition of ceramide biosynthesis ameliorates pathological consequences of spinal cord injury. Shock 31(6):634–644

Czech B, Pfeilschifter W, Mazaheri-Omrani N, Strobel MA, Kahles T, Neumann-Haefelin T, Rami A, Huwiler A, Pfeilschifter J (2009) The immunomodulatory sphingosine 1-phosphate analog FTY720 reduces lesion size and improves neurological outcome in a mouse model of cerebral ischemia. Biochem Biophys Res Commun 389(2):251–256

Di Pardo A, Maglione V, Alpaugh M, Horkey M, Atwal RS, Sassone J, Ciammola A, Steffan JS, Fouad K, Truant R, Sipione S (2012) Ganglioside GM1 induces phosphorylation of mutant huntingtin and restores normal motor behavior in Huntington disease mice. Proc Natl Acad Sci USA 109(9):3528–3533

Edwards JL, Vincent AM, Cheng HT, Feldman EL (2008) Diabetic neuropathy: mechanisms to management. Pharmacol Ther 120(1):1–34

Etminan M, Gill S, Samii A (2003) Effect of non-steroidal anti-inflammatory drugs on risk of Alzheimer's disease: systematic review and meta-analysis of observational studies. BMJ 327(7407):128

Fewou SN, Rupp A, Nickolay LE, Carrick K, Greenshields KN, Pediani J, Plomp JJ, Willison HJ (2012) Anti-ganglioside antibody internalization attenuates motor nerve terminal injury in a mouse model of acute motor axonal neuropathy. J Clin Invest 122(3):1037–1051

Finney CAM, Hawkes CA, Kain DC, Dhabangi A, Musoke C, Cserti-Gazdewich C, Oravecz T, Liles WC, Kain KC (2011) S1P is associated with protection in human and experimental cerebral malaria. Mol Med 17(7–8):717–725

Foster CA, Mechtcheriakova D, Storch MK, Balatoni B, Howard LM, Bornancin F, Wlachos A, Sobanov J, Kinnunen A, Baumruker T (2009) FTY720 rescue therapy in the dark agouti rat model of experimental autoimmune encephalomyelitis: expression of central nervous system genes and reversal of blood–brain damage. Brain Pathol 19:254–266

Frank-Cannon TC, Alto LT, McAlpine FE, Tansey MG (2009) Does neuroinflammation fan the flame in neurodegenerative diseases? Mol Neurodegener 4:47

Gassert E, Avota E, Harms H, Krohne G, Gulbins E, Schneider-Schaulies S (2009) Induction of membrane ceramides: a novel strategy to interfere with T lymphocyte cytoskeletal reorganisation in viral immunosuppression. PLoS Pathog 5(10):e1000623

Geisler FH, Dorsey FC, Coleman WP (1991) Recovery of motor function after spinal-cord injury–a randomized, placebo-controlled trial with GM-1 ganglioside. N Engl J Med 324(26):1829–1838

Geisler FH, Coleman WP, Grieco G, Poonian D, Sygen Study Group (2001) The Sygen multicenter acute spinal cord injury study. Spine 26(24 Suppl):S87–S98

Giunti D, Borsellino G, Benelli R, Marchese M, Capello E, Valle MT, Pedemonte E, Noonan D, Albini A, Bernardi G, Mancardi GL, Battistini L, Uccelli A (2003) Phenotypic and functional analysis of T cells homing into the CSF of subjects with inflammatory diseases of the CNS. J Leukoc Biol 73(5):584–590

Gulbins E, Lee PL (2006) Physiological and pathophysiological aspects of ceramide. Am J Physiol Regul Integr Comp Physiol 290:11–26

Haggiag S, Steiner-Birmanns B, Wirguin I, Sicsic C, Brenner T, Steiner I (2004) Seroconversion of anti-GM1 antibodies in patients with amyotrophic lateral sclerosis. J Neurol 63(4):755–756

Hasegawa Y, Suzuki H, Sozen T, Rolland W, Zhang JH (2010) Activation of sphingosine 1-phosphate receptor-1 by FTY720 is neuroprotective after ischemic stroke in rats. Stroke 41(2):368–374

Huang SH, Long M, Wu CH, Kwon-Chung KJ, Chang YC, Chi F, Lee S, Jong A (2011) Invasion of Cryptococcus neoformans into human brain microvascular endothelial cells is mediated through the lipid rafts-endocytic pathway via the dual specificity tyrosine phosphorylation-regulated kinase 3 (DYRK3). J Biol Chem 286(40):34761–34769

Hughes RA, Cornblath DR (2005) Guillain-Barré syndrome. Lancet 366(9497):1653–1666

In t'Veld BA, Ruitenberg A, Hofman A (2001) Nonsteroidal anti-inflammatory drugs and the risk of Alzheimer's disease. N Engl J Med 345:1515–1521

Jana A, Pahan K (2010) Sphingolipids in multiple sclerosis. Neuromolecular Med 12(4):351–361

Kappos L, Radue EW, O'Connor P, Polman C, Hohlfeld R, Calabresi P, Selmaj K, Agoropoulou C, Leyk M, Zhang-Auberson L, Burtin P, FREEDOMS Study Group (2010) A placebo-controlled trial of oral fingolimod in relapsing multiple sclerosis. N Engl J Med 362(5):387–401

Kieseier BC, Kiefer R, Gold R, Hemmer B, Willison HJ, Hartung HP (2004) Advances in understanding and treatment of immune-mediated disorders of the peripheral nervous system. Muscle Nerve 30(2):131–156

Kim S, Steelman AJ, Zhang Y, Kinney HC, Li J (2012) Aberrant upregulation of astroglial ceramide potentiates oligodendrocyte injury. Brain Pathol 22(1):41–57

Kobayashi Y, Kiguchi N, Maeda T, Ozaki M, Kishioka S (2012) The critical role of spinal ceramide in the development of partial sciatic nerve ligation-induced neuropathic pain in mice. Biochem Biophys Res Commun 421(2):318–322

Köhne AS, Stettner M, Jangouk P, Dehmel T, Hartung HP, Lehmann HC, Kieseier BC (2012) Fingolimod impedes Schwann cell-mediated myelination: implications for the treatment of immune neuropathies? Effect of Fingolimod on Schwann cells. Arch Neurol 2:1–10

Kumar A, Negi G, Sharma SS (2012) Suppression of NF-κB and NF-κB regulated oxidative stress and neuroinflammation by BAY 11–7082 (IκB phosphorylation inhibitor) in experimental diabetic neuropathy. Biochimie 94(5):1158–1165

Langrish CL, Chen Y, Blumenschein WM, Mattson J, Basham B, Sedgwick JD, McClanahan T, Kastelein RA, Cua DJ (2005) IL-23 drives a pathogenic T cell population that induces autoimmune inflammation. Exp Med 201(2):233–240

McColl BW, Allan SW, Rothwell NJ (2007) Systemic inflammation and stroke: aetiology, pathology and targets for therapy. Biochem Soc Trans 35:1163–1165

McGeer PL, McGeer EG (2001) Inflammation, autotoxicity and Alzheimer disease. Neurobiol Aging 22:799–809

McGeer PL, Itagaki S, Akiyama H, McGeer EG (1988) Rate of cell death in parkinsonism indicates active neuropathological process. Ann Neurol 24(4):574–576

Myers RR, Shubayev VI (2011) The ology of neuropathy: an integrative review of the role of neuroinflammation and TNF-α axonal transport in neuropathic pain. J Peripher Nerv Syst 16(4):277–286

Nayak D, Huo Y, Kwang WXT, Pushparaj PN, Kumar SD, Ling EA, Dheen ST (2010) Sphingosine kinase 1 regulates the expression of proinflammatory cytokines and nitric oxide in activated microglia. Neuroscience 166:132–144

Novgorodov SA, Gudz TI (2011) Ceramide and mitochondria in ischemic brain injury. Int J Biochem Mol Biol 2(4):347–361

Orr CF, Rowe DB, Halliday GM (2002) An inflammatory review of Parkinson's disease. Prog Neurobiol 68(5):325–340

Pannu R, Won JS, Khan M, Singh AK, Singh I (2004) A novel role of lactosylceramide in the regulation of lipopolysaccharide/interferon-gamma-mediated inducible nitric oxide synthase gene expression: implications for neuroinflammatory diseases. J Neurosci 24(26):5942–5954

Phong MC, Gutwein P, Kadel S, Hexel K, Altevogt P, Linderkamp O, Brenner B (2003) Molecular mechanisms of L-selectin-induced co-localization in rafts and shedding. Biochem Biophys Res Commun 300:563–569

Ransohoff RM (2012) Chemokines and chemokine receptors: standing at the crossroads of immunobiology and neurobiology. Immunity 31:711–721

Rao JS, Kellom M, Kim HW, Rapoport SI, Reese EA (2012) Neuroinflammation and synaptic loss. Neurochem Res 37(5):903–910

Ribatti D, Nico B, Crivellato E, Artico M (2006) Development of the blood–brain barrier: a historical point of view. Anat Rec 289B:3–8

Rivest S (2009) Regulation of innate immune responses in the brain. Nat Rev Immunol 9(6):429–430

Rosenman SJ, Ganji AA, Tedder TF, Gallatin WM (1993) Syn-capping of human T lymphocyte adhesion/activation molecules and their redistribution during interaction with endothelial cells. J Leukoc Biol 53:1–10

Schneider JS, Sendek S, Daskalakis C, Cambi F (2010) GM1 ganglioside in Parkinson's disease: results of a five year open study. J Neurol Sci 292(1–2):45–51

Soeda S, Tsuji Y, Ochiai T, Mishima K, Iwasaki K, Fujiwara M, Yokomatsu T, Murano T, Shibuya S, Shimeno H (2004) Inhibition of sphingomyelinase activity helps to prevent neuron death caused by ischemic stress. Neurochem Int 45(5):619–626

Stoffel B, Bauer P, Nix M, Geres K, Stoffel W (1998) Ceramide-independent CD28 and TCR signaling but reduced IL-2 secretion in T cells of acid sphingomyelinase-deficient mice. Eur J Immunol 28:874–880

Streit WJ, Mrak RE, Griffin WST (2004) Microglia and neuroinflammation: a pathological perspective. J Neuroinflammation 1(1):14

Strong K, Mathers C, Bonita R (2007) Preventing stroke: saving lives around the world. Lancet Neurol 6(2):182–187

Tajima O, Egashira N, Ohmi Y (2009) Reduced motor and sensory functions and emotional response in GM3-only mice: emergence from early stage of life and exacerbation with aging. Behav Brain Res 198:74–82

Van Doorn R, Van Horssen J, Verzijl D, Witte M, Ronken E, Van Het Hof B, Lakeman K, Dijkstra CD, Van Der Valk P, Reijerkerk A, Alewijnse AE, Peters SLM, De Vries HE (2010) Sphingosine 1-Phosphate receptor 1 and 3 are upregulated in multiple sclerosis lesions. Glia 58:1465–1476

Van Doorn R, Nijland PG, Dekker N, Witte ME, Lopes-Pinheiro MA, van Het Hof B, Kooij G, Reijerkerk A, Dijkstra C, van van der Valk P, van Horssen J, de Vries HE (2012) Fingolimod attenuates ceramide-induced blood–brain barrier dysfunction in multiple sclerosis by targeting reactive astrocytes. Acta Neuropathol 124(3):397–410

Wang J, Doré S (2007) Inflammation after intracerebral hemorrhage. J Cereb Blood Flow Metab 27:894–908

Witton PS (2007) Inflammation as a causative factor in the aetiology of Parkinson's disease. Br J Pharmacol 150(8):963–976

Wu YP, Mizugishi K, Bektas M, Sandhoff R, Proia RL (2008) Sphingosine kinase 1/S1P receptor signaling axis controls glial proliferation in mice with Sandhoff disease. Hum Mol Genet 17:2257–2264

Wu G, Lu ZH, Kulkarni N, Amin R, Ledeen RW (2011) Mice lacking major brain gangliosides develop parkinsonism. Neurochem Res 36(9):1706–1714

Yang Y, Xu J, Niu Y, Bromberg JS, Ding Y (2008) T-bet and eomesodermin play critical roles in directing T cell differentiation to Th1 versus Th17. J Immunol 181(12):8700–8710

Yu RK, Nakatani Y, Yanagisawa M (2009) The role of glycosphingolipid metabolism in the developing brain. J Lipid Res 50(Suppl):S440–S445

Yu RK, Tsai YT, Ariga T (2012) Functional roles of gangliosides in neurodevelopment: an overview of recent advances. Neurochem Res 37(6):1230–1244

Yuki N, Hartung HP (2012) Guillain-Barré syndrome. N Engl J Med 366(24):2294–2304

Zappia M, Crescibene L, Bosco D, Arabia G, Nicoletti G, Bagalà A, Bastone L, Napoli ID, Caracciolo M, Bonavita S, Di Costanzo A, Gambardella A, Quattrone A (2002) Anti-GM1 ganglioside antibodies in Parkinson's disease. Acta Neurol Scand 106(1):54–57

Sphingolipids in Psychiatric Disorders and Pain Syndromes

C. Mühle, M. Reichel, E. Gulbins, and J. Kornhuber

Contents

Abstract Despite the high prevalence and devastating impact of psychiatric disorders, little is known about their etiopathology. In this review, we provide an overview on the participation of sphingolipids and enzymes responsible for their metabolism in mechanisms underlying psychiatric disorders. We focus on the pathway from sphingomyelin to proapoptotic ceramide and the subsequent metabolism of ceramide to sphingosine, which is in turn phosphorylated to yield anti-apoptotic sphingosine-1-phosphate (S1P).

The sphingomyelinase/ceramide system has been linked to effects of reactive oxygen species and proinflammatory cytokines in the central nervous system as well as to synaptic transmission. Compared to ubiquitously expressed acid

C. Mühle • M. Reichel • J. Kornhuber (✉)
Department of Psychiatry and Psychotherapy, Friedrich-Alexander-University of Erlangen, Schwabachanlage 6, 91054 Erlangen, Germany
e-mail: johannes.kornhuber@uk-erlangen.de

E. Gulbins
Department of Molecular Biology, University of Duisburg-Essen, Hufelandstrasse 55, 45122 Essen, Germany

E. Gulbins and I. Petrache (eds.), *Sphingolipids in Disease*,
Handbook of Experimental Pharmacology 216,
DOI 10.1007/978-3-7091-1511-4_22,

sphingomyelinase, acid and neutral ceramidase and neutral sphingomyelinase are highly active in brain regions. Depressed patients show elevated plasma ceramide levels and increased activities of acid sphingomyelinase which is functionally inhibited by many anti-depressive drugs. Exposure to alcohol is associated with an activation of acid and neutral sphingomyelinase observed in cell culture, mouse models and in alcohol-dependent patients and with increased concentrations of ceramide in various organs.

Levels of sphingomyelin and ceramide are altered in erythrocytes and post-mortem brain tissues of schizophrenic patients in addition to changes in expression patterns for serine palmitoyltransferase and acid ceramidase leading to impaired myelination. After induction of anxiety-like behavior in animal models, higher serum levels of S1P were reported to lead to neurodegeneration. Correspondingly, S1P infusion appeared to increase anxiety-like behavior. Significantly upregulated levels of the endogenous ceramide catabolite *N,N*-dimethylsphingosine were observed in rat models of allodynia. Conversely, rats injected intrathecally with *N,N*-dimethylsphingosine developed mechanical allodynia. Moreover, S1P has been implicated in spinal nociceptive processing.

The increasing interest in lipidomics and improved analytical methods led to growing insight into the connection between psychiatric and neurological disorders and sphingolipid metabolism and may once provide new targets and strategies for therapeutic intervention.

Keywords Ceramide • Pain syndromes • Psychiatric disorders • Sphingolipids • Sphingomyelinase

1 Introduction

Psychiatric disorders are among the most prevalent human disorders and contribute substantially to the burden of disease, particularly in high-income countries (Prince et al. 2007). The leading contributors to this burden are unipolar and bipolar affective disorder, substance-use and alcohol-use disorders, schizophrenia, and dementia. Despite the devastating impact of psychiatric disorders, little is known about their etiology and pathophysiology or their underlying molecular mechanisms. Generally, psychiatric disorders are believed to arise by a complex interplay among biological, psychological, and social factors. Research during the past 100 years has revealed that genetic factors and pathological changes in neuro transmitter metabolism play important roles, and these findings support the concept that psychiatric disorders are at least in part of physical origin (Mann 1999; Tandon and McGuffin 2002). Recent studies have identified other biological aspects of psychiatric disorders, such as the activation of the hypothalamic–pituitary–adrenal axis and the immune system and the deleterious effects of oxidative stress, cytokine signaling, neuronal cell loss, and hippocampal atrophy (Müller and Ackenheil 1998; Ng et al. 2008; Sapolsky 2000).

In this chapter, we explore the possibility that sphingolipids (SLs) play a central role in the emergence of psychiatric disorders. The generation of excess bioactive SLs in response to external stimuli may be a missing link between known etiologically factors and the pathophysiology that underlies some of the disorders. We first describe the distribution of SL enzymes in the brain and give examples of how SL metabolism can affect neuronal functioning. We then summarize the empirical data associated with SL metabolism in psychiatric disorders unrelated to neurodegeneration. The closing section presents an integrative view on the mechanisms that may link SL abnormalities to psychiatric disorders.

2 SL Enzymes: Distribution and Involvement in Neuronal Functions

Lipids, the group of molecular components with the second highest concentration in the human brain (surpassed only by that of adipose tissue), provide the structures and substrates necessary for neuronal excitability and for complex signal transduction processes at the plasma membranes. A network of specialized and compartmentalized enzymes regulates the levels of biologically active lipids and thereby couples various extra- and intracellular signals to SL-mediated responses.

The enzymes of SL metabolism play a crucial role in maintaining the balance between proapoptotic ceramide and anti-apoptotic sphingosine-1-phosphate (S1P), known as the "rheostat" (Hait et al. 2006; Spiegel et al. 1998). Moreover, the conversion by these enzymes of membrane microdomains that have high concentrations of SLs and cholesterol immersed in a phospholipid-rich environment to ceramide-enriched regions leads to the release of cholesterol and the consequent alteration of plasma membrane fluidity. These changes allow the selective recruitment of receptors or ion channels via lipid–protein or lipid–lipid interactions or the exclusion of such proteins from these microdomains and are therefore important for coordinating signal transduction, e.g., in stress responses (Posse de Chaves and Sipione 2010; Zhang et al. 2009). The recently discovered direct and highly specific interaction of exclusively one sphingomyelin species (SM18) with the transmembrane domain of a protein expands the spectrum of SLs to the function as potential cofactors to regulate transmembrane proteins (Contreras et al. 2012). Ceramide can be formed de novo by successive reactions starting from serine and palmitoyl-CoA involving serine palmitoyltransferase as the first enzyme or from the hydrolysis of cerebrosides such as galactosylceramide. However, we concentrate in this chapter mainly on the pathway from sphingomyelin (SM) to ceramide and the subsequent metabolism of ceramide to sphingosine, which is in turn phosphorylated to yield S1P (Ogretmen and Hannun 2004).

Several isoenzymatic forms with different pH optima are able to catalyze the hydrolysis of SM to ceramide and phosphorylcholine, a reaction that is coupled to the conversion of diacylglycerols to phosphatidylcholines. Acid sphingomyelinase (human: ASM, mouse: Asm; for easier reading both are abbreviated with ASM), a chiefly lysosomal constituent, is widely distributed in mammalian tissues (Weinreb et al. 1968) and can also be found as a Zn^{2+}-dependent secretory isoform (Schissel et al. 1998). Despite different cellular trafficking, both enzymes arise from the same gene *SMDP1*. In contrast, Mg^{2+}-stimulated neutral sphingomyelinase (NSM, later specified as nSMase2) is localized more predominantly in cells of neural origin and enriched in gray matter (Rao and Spence 1976). Early examination of the somatotopographical distribution of ASM and NSM in human brains revealed a 30-fold gradient of NSM activity from the globus pallidus over substantia nigra and putamen to the thalamus and cortical structures and finally to white matter (Spence et al. 1979). In contrast, ASM was found to be distributed ubiquitously in all rat tissues analyzed (with highest activity in the liver) and expressed evenly in human brain regions, differing at most by a factor of two (Spence et al. 1979). The increasing activity of NSM during development in correlation with neuronal maturation (Spence and Burgess 1978) further suggests a specific role of this enzyme in the physiological functions of the brain.

A considerable heterogeneity in the molecular weight of membrane-bound active NSM in the brains of healthy humans was suggested by the identification of a 740-kDa variant in one individual compared with the mass of 160 kDa that was previously determined using the radiation inactivation method (Levade et al. 1986). By applying various combinations of extracting agents during the purification process from bovine brain, multiple forms of NSM enzymes differing in their chromatographic profiles and biochemical properties (e.g., pH dependence, Mg^{2+} requirements, and effects of detergents) have been characterized (Jung et al. 2000). Following the identification of a ubiquitously expressed mammalian NSM (nSMase1) with a remote similarity to the secreted bacterial enzymes (Tomiuk et al. 1998), Hofmann et al. (2000) succeeded in cloning and analyzing a mammalian brain-specific NSM (named nSMase2) with a different domain structure and only marginal sequence similarity to other sphingomyelinases. In immunofluorescence, this isoform presented with a neuron-specific punctate perinuclear localization. Based on sequence homology to a Mg^{2+}-independent NSM purified from bovine brain, a third mammalian NSM gene (designated nSMase3) with low homology to nSMase 1 and 2 has been cloned and characterized (Krut et al. 2006). Its ubiquitously expressed mRNA is translated to an active tumor necrosis factor (TNF)-responsive enzyme that modulates SL levels and is linked to the cellular stress response (Corcoran et al. 2008; Krut et al. 2006).

As expected from its pH optimum, alkaline sphingomyelinase activity has not been reported for brain tissues but has rather been found predominantly in the digestive system of humans and animals, with considerable tissue variability (Duan et al. 1996).

Analogously to sphingomyelinases, ceramidases differ in their pH optima for the breakdown of ceramide to sphingosine and fatty acid, and multiple isoforms encoded by corresponding genes have been identified. A lysosomal acid ceramidase (AC) was first isolated from rat brain (Gatt 1963) and was later purified and cloned from various species and tissues. AC also catalyzes the reverse reaction, i.e., free fatty acid-dependent ceramide synthesis (Okino et al. 2003). Ceramidase activity at pH 9 was detected in human cerebellum, indicating the existence of an alkaline isoform (Sugita et al. 1975).

The hydrolytic activity of rat tissue homogenates on *N*-oleoyl- and *N*-palmitoyl-sphingosine at pH optima of 5 and 8 was described in detail by Spence et al. (1986). Compared with NSM, this ceramidase activity was less tissue specific and higher at pH 8, but activities in the brain stem, cerebrum, and cerebellum again exceeded those in other organs such as liver, spleen, lung, and muscle (with the exception of kidney). A distinct membrane-bound non-lysosomal ceramidase with a broad pH optimum in the neutral to alkaline range, independent of cations, was later purified from rat brain and shown to display specific kinetic characteristics (El Bawab et al. 1999). In addition to this neutral ceramidase (NC) that had a molecular weight of 90 kDa, Thayyullathil et al. (2011) purified and characterized a novel neutral ceramidase from rat brain, termed RBCDase II, which had a larger apparent weight of 110 kDa, unrelated to differences in the glycosylation pattern. This enzyme was inhibited by nucleotides such as TMP and UMP and differed in its metal and phospholipid dependence from the previously known isoform.

In agreement with these findings, we have observed comparable activities of ASM in brain regions and peripheral organs in adult rats. In sharp contrast, the activities of NSM in the hippocampus, striatum, and prefrontal cortex were more than 20-fold higher than in the liver, spleen, and lung. Similarly, AC and NC activities were approximately 15-fold and sixfold higher, respectively, in these brain tissues than in the other organs (C. Mühle et al. unpublished).

The high concentration of lipids as components of lipid membranes emphasizes their pivotal roles in neuronal function and signal transduction, particularly in the case of ceramide and SM. These SLs and the enzymes that control their levels may be involved in the etiopathogenesis of psychiatric disorders and represent targets for therapeutic approaches. NSM, AC, and NC have appeared more attractive to researchers as relevant enzymes initially because of their distinctive high activities in brain regions. However, the more ubiquitously expressed ASM enzyme may prove to participate significantly in physiological and pathophysiological mechanisms in neuronal processes, as exemplified by the finding that microparticle shedding from glial cells is associated with rapid activation of ASM (Bianco et al. 2009).

Several reports indicate that the effects of proinflammatory cytokines in the central nervous system involve SLs, particularly the activation of the SMase/ceramide pathway. Interleukin (IL)-1β inhibits the firing rate of anterior hypothalamic neurons, and this inhibition involves the activation of NSM and concomitant ceramide production (Davis et al. 2006). The treatment of primary hippocampal neurons with TNF-α leads to increased surface localization of the *N*-methyl-D-aspartate (NMDA) receptor subunit NR1 and to a specific clustering of phosphorylated NR1 in membrane domains. Genetic and pharmacological inhibition studies demonstrate that the TNF-α-induced changes involve the activation of nSMase2 and the rapid generation of ceramide (Wheeler et al. 2009). The pivotal role of the SMase/ceramide system in synaptic transmission was confirmed by the direct application of SMase or cell-permeable ceramide to hippocampal slices (Norman et al. 2010; Yang 2000). The effects of ceramide on neuron excitability can be blocked by inhibitors of sphingosine kinase and are mimicked by direct application of S1P, indicating that the further metabolization of sphingomyelinase-generated ceramide plays a critical role (Norman et al. 2010). This is also implied by a screen for synaptic function in *Drosophila*, which identified a mutant called *slug-a-bed* (*slab*). The *slab* gene encodes ceramidase, and null *slab* mutants are characterized by presynaptic impairment, impaired synaptic vesicle fusion, loss of readily releasable vesicles, and reduced FM1-43 dye uptake in presynaptic terminals (Rohrbough et al. 2004). Sánchez et al. (2001) demonstrated that neurotransmission itself can result in increased ceramide production. In primary astrocytes, the activation of CB1 receptors by the application of Δ^9-tetrahydrocannabinol (THC) resulted in increased SM hydrolysis, most likely through the action of NSM.

SLs are an integral component of plasma membranes, and their local composition can help to organize transmembrane proteins such as receptors and transporters (Zhang et al. 2009). Several reports indicate that signal transduction via monoamine transmitters is also affected by this mechanism. The maximal serotonin (5-hydroxytryptamine, 5-HT) binding to HT_7 receptors was decreased by the application of fumonisin B1 (an inhibitor of de novo synthesis of ceramide) or of PDMP (an inhibitor of glucosylceramide synthesis) without affecting the total level of the receptor (Sjögren and Svenningsson 2007). The ligand-binding activity of 5-HT_{1A} receptor was altered by sphingomyelinase treatment (Jafurulla et al. 2008). The genetic or pharmacological inhibition of NSM in PC12 cells resulted in decreased dopamine uptake, whereas exposure to cell-permeable C6-ceramide increased dopamine uptake in a concentration-dependent manner (Kim et al. 2010). Ceramide was also shown to reversibly alter the function of the dopamine transporter in rat striatal synaptosomes, resulting in decreased transport of dopamine and increased transport of serotonin (Riddle et al. 2003).

Early signs for an association of SLs and psychiatric disorders arose from the observation that patients with genetic disorders involving enzymes of sphingolipid metabolism presented with psychiatric manifestations. In Tay-Sachs disease, a neuronal storage disorder due to mutations in the gene for β-hexosaminidase A, psychosis, mood disorders, and cognitive impairment were reported for more than 20 % of adult-onset patients (MacQueen et al. 1998). Depression is also common in Anderson-Fabry disease, an X-linked lysosomal storage disorder with accumulating deposition of glycosphingolipids caused by a deficiency of the enzyme alpha-galactosidase A. Among 184 responders to a questionnaire, severe depression was prevalent in 36 % of male and 22 % of female patients in the United Kingdom with a high proportion of previously undiagnosed cases (Cole et al. 2007).

Due to the association of SL with various complex psychiatric and neurological diseases, their circulating levels could present important intermediate phenotypes that point to variants within genes encoding enzymes involved in SL metabolism. First results of a genome-wide association study in European populations identified genetic determinants for the concentration of SM, ceramide, and other SLs explaining up to 10 % of the genetic variation in each trait. Strongest associations were found in or near 7 genes functionally involved in ceramide biosynthesis or trafficking including those for serine palmitoyl transferase and ceramide synthase (Hicks et al. 2009). Polymorphisms in the gene encoding sphingomyelin synthase 1 have been associated with traits such as the heritability of coffee consumption (Sulem et al. 2011) and the personality dimension openness (Terracciano et al. 2010). Further candidate loci including several mapping into SL pathways were identified in a meta-analysis of five European family-based genome-wide association studies on plasma levels of various SL species and their proportions and underline the strong genetic control of key components of SL metabolism (Demirkan et al. 2012). Testing for such polymorphisms associated with additional risk or protective effects might help to illuminate the genetic basis for the development of diseases.

Thus, although research in lipidomics has been lagging behind research in the genomics and proteomics areas, increasing interest in and the development of improved methods for lipid analysis have led to growing insight into the connection between psychiatric disorders and mechanisms of lipid metabolism. The experimental manipulation of these mechanisms may provide new targets and strategies for therapeutic intervention in psychiatric disorders.

3 The Roles of SLs in Specific Psychiatric Disorders and Pain Syndromes

3.1 *Depression*

Major depression is a severe, chronic, and often life-threatening illness with a lifetime prevalence of more than 10 % (Belmaker and Agam 2008). Key symptoms of depression include depressed mood and a loss of interest and pleasure; in addition, concentration deficits and feelings of worthlessness, weight loss, and insomnia may develop. Major depression may be triggered by psychological stress and is associated with dysfunction of the hypothalamic–pituitary–adrenal axis (Belmaker and Agam 2008). The disorder is also an important risk factor for the development of cardiovascular disease (Musselman et al. 1998). Furthermore, patients with major depression exhibit elevations in the serum levels of proinflammatory cytokines (Howren et al. 2009).

Interestingly, antidepressive drugs such as desipramine and imipramine are prototypical functional inhibitors of ASM. The functional inhibition of ASM by these drugs may be related to their antidepressive effect. Only a few functional inhibitors of ASM have been known so far. We screened a large number of small molecules licensed for medical use for their effect on ASM and thereby discovered many novel functional inhibitors of ASM (Kornhuber et al. 2008, 2011). Several of these novel inhibitors are antidepressant drugs. The molecular structures of these compounds are very diverse and include mono-, di-, tri- and tetracyclic compounds. A virtual screening approach revealed that functional inhibitors of ASM are not evenly distributed among the groups of drugs licensed for medical use; rather, they are enriched in only a few groups of drugs, including the antidepressant drugs. There are several antidepressant drugs that do not functionally inhibit ASM (Kornhuber et al. 2011). However, every tricyclic antidepressant drug investigated so far does inhibit ASM activity. We constructed inhibition curves for some of these drugs and found that the functional inhibition of ASM occurs within a concentration range that is usually achieved in patients during antidepressive drug therapy (Kornhuber et al. unpublished).

In a pilot study, we detected enhanced ASM activity in peripheral blood mononuclear cells of depressed patients compared with controls and a correlation between ASM activity and the degree of depression (Kornhuber et al. 2005). An important question is whether or not this marker is also altered in the brain tissue of depressed patients. Measurements of ASM activity in the central nervous system and lungs following subchronic application of amitriptyline or following activation of the ASM in the spleen and central nervous system 30 min after whole body irradiation (12 Gy) revealed a correlation of ASM activity in the peripheral and central nervous systems (Gulbins et al. unpublished). These findings support the concept that ASM activity in the periphery can be used to determine the enzyme activity in the brain. A clinical study in older people revealed that several plasma ceramide species were increased in patients with a recent depression compared to a group with no or past depression (Gracia-Garcia et al. 2011). Significant associations of plasma phospho- and SL species with depression and anxiety symptoms were found in a lipidomic study of 742 Dutch people. Absolute levels of phosphatidylcholine O-36:4 and the ratio of SM species 23:1 and 16:0 were related to depressive symptom scores (Demirkan et al. 2012). These reports provide further evidence of an activation of the ASM/ceramide system in major depressive disorders.

3.2 Schizophrenia

The first evidence for an altered SL composition in schizophrenia arose from the postmortem biochemical analysis of brain lobes from patients with neurological diseases, including one subject with schizophrenia (Cherayil 1969). This early observation of reduced levels of cerebrosides and sulfatides was confirmed much later in thalami of 18 patients with chronic schizophrenia (Schmitt et al. 2004). Increased levels of phosphatidylserine in the left thalamic gray matter were assumed to indicate cell death (Schmitt et al. 2004) because phosphatidylserine acts as a potent activator of neutral sphingomyelinase (Sawai and Hannun 1999) and exposure of phosphatidylserine at the cell's surface has been reported to stimulate phagocytosis of apoptotic cells (Chang et al. 2000). A reduction in the concentration of phosphatidylcholine (Schmitt et al. 2004) has previously been observed in the caudate nucleus of schizophrenics (Yao et al. 2000) and detected by in vivo phosphorus 31 magnetic resonance spectroscopy in other brain regions (Fukuzako et al. 1999). Together with reduced levels of SM and galactocerebrosides 1 and 2 (Schmitt et al. 2004), these data suggest disturbed lipid metabolism and myelination in schizophrenia and support the membrane phospholipid concept of schizophrenia (Horrobin 1998).

Several studies indicate altered patterns of gangliosides, SLs, and phospholipids in erythrocyte membranes of schizophrenic patients. In acute schizophrenics without neuroleptic treatment, the fractions of GM3 and GD3 gangliosides (the primary glycolipids of neuronal membranes) were increased in comparison with normal subjects (Haselhorst et al. 1988). The erythrocyte membranes of unmedicated patients with schizophrenia or schizoaffective disorder contained significantly higher levels of SM but lower levels of phosphatidylethanolamine (PE) and phosphatidylinositol bisphosphate compared to those of normal subjects (Keshavan et al. 1993). The reduced levels of phosphatidylcholine in erythrocyte ghost membranes of lithium-responsive as well as nonresponsive schizophrenic-like patients (Hitzemann et al. 1984) were shown to be associated with reduced phospholipid methylation activity for the conversion of PE to phosphatidylcholine (Hitzemann et al. 1985). Increased SM and decreased PE levels were also found in 32 schizophrenic patients treated with haloperidol, particularly in those with predominantly negative symptoms (Ponizovsky et al. 2001). Interestingly, the SM/PE ratio was correlated inversely with positive symptom scale scores and was correlated directly and more strongly with negative symptom scale scores.

Demisch et al. (1992) analyzed SM and various phospholipids in platelets from medication-free patients with schizoaffective disorder and detected twice as much phosphatidylinositol and increased levels of phosphatidylcholine compared with controls. The largest reduction in esterification rates of arachidonic acid into different phospholipids was found for patients with schizoaffective disorders, followed by those with schizophreniform and major depressive disorders. In contrast, the rates were unchanged for patients with chronic schizophrenia (Demisch et al. 1992).

Methodological advances in high-throughput profiling techniques based on ultra-performance liquid chromatography-mass spectrometry allowed the comparison of postmortem lipid levels in the prefrontal cortex of subjects (15 each) with schizophrenia, bipolar disorder, and no disorder (control) (Schwarz et al. 2008). Significantly elevated ceramide levels were detected in white matter from the patients with both of the neuropsychiatric disorders regardless of antipsychotic treatment. In contrast to the similar pattern of alteration of free fatty acids observed in both gray and white matter, phosphatidylcholines were significantly increased in the gray matter but somewhat decreased in the white matter of schizophrenic patients. Statistically significant alterations of free fatty acid and ceramide concentrations were found in red blood cell samples from both drug-treated and drug-naïve, first-onset patients (Schwarz et al. 2008). Further differences in the periphery were observed for the skin lipid composition in first-episode antipsychotic-naïve schizophrenia patients. Compared to matched controls, the total ceramide fraction of the stratum corneum lipids was significantly reduced with inverse alterations of several single ceramide classes (Smesny et al. 2012).

In addition to these biochemical findings, gene expression studies in mice and humans provide evidence for the disruption of SL metabolism in schizophrenia. Narayan et al. have analyzed prefrontal cortex tissues from schizophrenic patients with short, intermediate, and long duration of disorder and matched controls on a custom microarray for glycobiology-related genes (Narayan et al. 2009). Widespread changes in expression patterns, particularly for genes related to glycosphingolipid biosynthesis and SL metabolism, were identified primarily for subjects within 5 years of the initial diagnosis. The validation of seven associated genes (*UGT8*, *SGPP1*, *GALC*, *B4GALT6*, *SPTLC2*, *ASAH1*, and *GAL3ST1*) in patients with early-stage illness by quantitative PCR demonstrated a reduction of expression levels that was significant for all of the genes except *ASAH1* (which encodes AC). The decreased expression of *UGT8* (which encodes an enzyme responsible for converting ceramide to galactosylceramide) and of *GAL3ST1* (which encodes an enzyme for further metabolization to sulfatides) may result in an overall decrease in the levels of galactocerebroside and sulfatide, major lipid components of the myelin sheath. The impaired myelination would in turn lead to oligodendrocyte dysfunction and present as white matter deficits that are characteristic of schizophrenic patients (Kubicki et al. 2005). Consistent with this concept, Narayan et al. have also detected decreased expression of myelin/oligodendrocyte-related genes by in situ hybridization in several white matter regions of mouse brains following chronic treatment with haloperidol (Narayan et al. 2007). In chronic schizophrenia cases, only *SPTLC2* was differentially expressed, indicating the occurrence of possible compensatory mechanisms in disease progression (Narayan et al. 2009).

The AC encoding gene *ASAH1* located in one of the most important regions with strong evidence for linkage to schizophrenia (Lewis et al. 2003) was again identified as a potential candidate in the Han Chinese population: expression in peripheral leukocytes was down-regulated in microarray and quantitative PCR analysis in chronic schizophrenia subjects with a positive family history. Moreover, two single nucleotide polymorphisms within *ASAH1* were significantly associated with schizophrenia (Zhang et al. 2012).

Altogether, perturbations of SL levels including SM, PE, phosphatidylcholine, and galactocerebrosides have been identified in erythrocyte membranes and postmortem brains of schizophrenics. Increased brain levels of proapoptotic ceramide may contribute to the typical observation of reduced white matter. Recent gene expression studies have found contrary patterns for genes involved in SL metabolism depending on the duration of illness which point towards decreased myelination for patients at early stages of the disease.

3.3 Alcohol Dependence

The acute or chronic consumption of alcohol is associated with multiple and often severe changes in lipid metabolism (Baraona and Lieber 1979; Reitz 1979). Several research groups have shown that acute or chronic ethanol exposure increases the ceramide levels in cultured cells of neuronal origin and in the livers of ethanol-fed mice (Deaciuc et al. 2000; Liangpunsakul et al. 2010; Pascual et al. 2003; Saito et al. 2005; Zhao et al. 2011). In most of these studies, a concomitant increase was observed in the activity of ASM, NSM, or both enzymes (Deaciuc et al. 2000; Liu et al. 2000; Pascual et al. 2003). Ethanol-induced cell death, activation of stress-related kinases, or increased caspase-3 activity was mimicked by the administration of C2-ceramide and prevented by treatment either with the functional ASM inhibitor desipramine or with myriocin, an inhibitor of de novo ceramide synthesis (Liu et al. 2000; Pascual et al. 2003; Saito et al. 2005).

The molecular consequences of ethanol-induced ceramide production and its relevance for the occurrence of alcohol-related complications were further analyzed in cultured cells and in experimental mouse models. The treatment of rat hepatoma H4IIEC3 cells with ethanol increased the activity of ceramide-dependent protein phosphatase 2A (PP2A) and inhibited the phosphorylation of AMP-activated protein kinase (AMPK) (Liangpunsakul et al. 2010). The pretreatment of cells with the ASM inhibitor imipramine or fumonisin B, but not with myriocin or GW4869, prevented the activation of PP2A and the inhibitory effect of ethanol on AMPK phosphorylation (Liangpunsakul et al. 2010). In mice, imipramine treatment partially prevented the ethanol-induced increase of ceramide and the activation of PP2A during a 4-week period. Ethanol-fed mice developed sign of steatosis and impaired glucose tolerance, but these symptoms were improved by co-treatment with imipramine (Liangpunsakul et al. 2012). A critical role for ASM in the development of hepatic failure was also demonstrated using ASM-deficient mice, and the occurrence of alcohol-induced liver disease involves increased susceptibility of hepatocytes to TNF-α and reactive oxygen species that is supposed to be at least in part mediated by ASM-induced ceramide production (Fernández et al. 2008; Marí et al. 2004). A similar sensitization mechanism is implicated in the neurotoxic effect of ethanol (DeVito et al. 2000).

In alcohol-dependent patients, ASM activity is increased compared to healthy controls. High levels of cellular ASM activity in peripheral blood cells was associated with acute intoxication, whereas the activity of secreted ASM (S-ASM) in blood plasma was more related to the chronic consumption of alcohol (Reichel et al. 2010, 2011). S-ASM activity declined gradually in patients that withdrew from alcohol, and it correlated with the levels of carbohydrate-deficient transferrin (CDT). CDT is an established biomarker of chronic alcohol consumption and in particular indicates high-risk drinking (Conigrave et al. 2002). In a pilot study, the sensitivity of S-ASM as a marker of problematic drinking was similar to CDT (Reichel et al. 2011). Further analyses revealed that the levels of S1P are also increased in alcohol-dependent patients (Reichel et al. unpublished). The analysis of brain tissue from alcohol-dependent patients revealed a significant decrease of long-chain fatty acids in the sphingomyelins in the gray matter of the cortex and a significant or slight decrease of 18:0 SM in the white matter of the cortex, the cerebellum, and the medulla oblongata (Lesch et al. 1973). A similar reduction was observed for long-chain fatty acids of cerebrosides and phosphatidylethanolamines. However, because most of the alcohol-dependent patients died as a result of alcoholic cirrhosis, it is not clear whether the SL changes were directly related to alcohol consumption or were secondary to the occurrence of liver failure. Another comparative analysis of the lipid class composition in alcohol-dependent patients vs. control subjects did not reveal significant differences (Olsson et al. 1996).

Taken together, these findings suggest that acute or chronic alcohol consumption is a cellular stress factor that results in increased levels of ceramide in various organs. Increased ceramide levels may in turn cause cell death in liver or brain but may also facilitate more subtle cellular changes in neurons and astrocytes, leading to symptoms of neuronal dysfunction. For example, a disturbance in SL metabolism may contribute to the emergence of alcohol-related somatic and neuropsychiatric disorders. Another important question is whether SL metabolism is involved in the development of alcohol addiction per se. It was shown recently that the repeated administration of morphine increased the activity of serine palmitoyltransferase, ceramide synthase, and ASM, resulting in increased levels of ceramide. The pharmacological inhibition of ceramide synthesis not only attenuated the occurrence of nitroxidative stress and neuroimmune activation but also inhibited the development of morphine antinociceptive tolerance (Ndengele et al. 2009). A similar mechanism may be involved in drug habituation and contribute to the development of substance- and alcohol-use disorders.

3.4 Anxiety

Only a few studies have addressed the role of SLs in anxiety disorders, but none of these investigations specifically addressed phobias. In a large lipidomic study in Dutch families, relative SM 23:1 and phosphocholine O-36:4 plasma levels and ratios of different phospho- and SL species were associated with anxiety symptoms measured by the Hospital Anxiety Scale (Demirkan et al. 2012). Some animal studies provided evidence for the lack of an association with SLs. For example, the administration of ganglioside GM1 to male rats with electrolytic lesions of the left entorhinal cortex resulted in a moderate improvement in the performance of working memory in a Morris water maze task. However, no difference in anxiety levels in a plus maze was observed between animals treated with saline vs. ganglioside GM1 (Glasier et al. 1995).

In a mouse model, the precocious weaning of pups, which is known to increase anxiety-related behavior, led to an altered accumulation of galactosylceramide as part of myelinated fibers in the amygdala of 5-week-old male animals (Ono et al. 2008). In contrast, galactosylceramide was not increased in the prefrontal cortex or hippocampus of early-weaned males nor in any of the brain regions of early-weaned females.

The induction of anxiety-like behavior in a chronic stress model involving the immobilization of rats for 6 h daily for 3 weeks lead to increased serum levels of S1P (which is thought to act as a stress signal molecule) and concomitant signs of neuronal cell damage in the cortex and hippocampus (Jang et al. 2008). The repeated restraint of rats for 7 days combined with the application of electric shock resulted in similarly elevated levels of S1P, while base levels of sphingosine remained unaltered (Jang et al. 2011). The microinfusion of S1P into the lateral cerebroventricle for 7 days caused neurodegenerative changes with higher expression of NR1, as well as iNOS and glial fibrillary acidic protein (GFAP) (Jang et al. 2008), and a significant decrease in the expression of tyrosine hydroxylase (the rate-limiting enzyme in catecholamine synthesis) in the amygdala but not in the cortex (Jang et al. 2011). The S1P-infused animals spent less time in the open arms of an elevated plus maze, indicating increased anxiety (Jang et al. 2011).

A single intraperitoneal injection of ganglioside GM1 had a dose-related protective effect against ethanol-induced sensorimotor deficit in mice. The regular administration of GM1 in combination with an ethanol-containing liquid diet alleviated anxiety-like behavior in an elevated plus maze, which is associated with ethanol withdrawal (Wallis et al. 1995). Gilmore et al. suggested that ethanol intoxication is reduced by nonspecific stimulant properties of gangliosides or by their sialic acid constituent (Gilmore et al. 1991). Pretreatment with either substance reduced ethanol-induced deficits in locomotion, nose-poke exploration, and rearing and crossings in dark–light preference tests. Twenty-two percent of the pretreated animals displayed increased staircase climbing, indicating lower anxiety levels (Gilmore et al. 1991).

In summary, induction of anxiety-like behavior in animal models has been associated with increased S1P levels leading to neurodegeneration. Correspondingly, S1P infusion appears to increase anxiety-like behavior, whereas application of GM1 decreases anxiety levels.

3.5 Pain Syndromes

Indications of a possible association of pain with SL metabolism originated from studies of the most common hereditary disorder of peripheral sensory neurons, hereditary sensory neuropathy type I (HSN1). The gene *SPTLC1*, which encodes serine palmitoyltransferase, has been mapped to the HSN1 locus on chromosome 9. Missense mutations that have been identified in all affected members of 11 HSN1 families are predictive of increased enzymatic activity leading to increased de novo glucosylceramide synthesis and eventually massive cell death during neural tube closure (Dawkins et al. 2001). A possible alternative explanation is that the mutant serine palmitoyltransferase shifts from its canonical substrate L-serine to L-alanine resulting in the increased formation of neurotoxic deoxysphingolipids (dSL).

Consistent with a modulatory effect of amino acids on the generation of dSL in cell culture, transgenic mice expressing the mutant enzyme and fed a diet enriched with L-serine improved in measures of motor and sensory performance, whereas animals fed a diet enriched with L-alanine displayed severe peripheral neuropathy. Similarly, oral L-serine supplementation reduced the production of dSL in patients with HSN1 (Garofalo et al. 2011)

In an untargeted metabolomics study of the chemical basis of neuropathic pain, tissues and plasma from rats subjected to tibial nerve transection (TNT), a well-established model for inducing allodynia, were profiled by mass spectrometry (Patti et al. 2012). The dysregulation of SM-ceramide metabolism was prominent in the ipsilateral dorsal horn, including a more than threefold upregulation of ceramide and phosphatidylcholine levels. A fourfold increase in the concentration of platelet-activating factor, a metabolite reported to prevent SMase-evoked alterations in SM and ceramide levels possibly via accelerating the SM cycle (Latorre et al. 1999), was detected in the dorsal horn of treated rats. The observed upregulation of AC expression in dorsal root ganglia following TNT supports the concept of cellular membrane degradation and reorganization of central termination areas of primary sensory neurons following peripheral nerve injury. Levels of sphingosine and *N*,*N*-dimethylsphingosine (DMS), an endogenous catabolite of ceramide, were found to be significantly upregulated more than twofold (Patti et al. 2012).

Endogenous DMS is known to increase intracellular Ca^{2+} concentrations in astrocytes as well as to inhibit glutamate uptake (Lee et al. 2007). The latter, in turn, results in excessive activation of NMDA receptors in spinal sensory synapses and has been associated with neuropathic pain (Nie and Weng 2010). In agreement with this proposed mechanism, rats injected intrathecally with DMS at physiological concentrations developed mechanical allodynia in the hind paw that persisted for at least 3 days (Patti et al. 2012). Stronger GFAP staining in the spinal cord of these DMS-treated animals compared to controls indicated an increase in astrocyte activation which has been shown to modulate hypersensitivity via the release of proinflammatory cytokines such as IL-1β (Watkins et al. 2001). Treatment of cultured astrocytes with DMS also triggered the release of IL-1β (Patti et al. 2012) that has been associated with chronic pain (Alexander et al. 2005; Wolf et al. 2006) and of monocyte chemoattractant protein-1 (MCP-1) (Patti et al. 2012), a chemokine that was reported to contribute to hyperalgesia and central sensitization (Gao et al. 2009). These data imply a role for the endogenous ceramide catabolite DMS in nociception via the production of inflammatory mediators.

Although DMS is also an inhibitor of sphingosine kinase and blocks S1P production in cells (Edsall et al. 1998), the physiological concentrations of DMS are substantially lower than the inhibition constant, and S1P levels did not differ in TNT-treated vs. sham-operated rats (Patti et al. 2012). Thus, the alternative mechanism of DMS-mediated neuron sensitization remains to be elucidated.

However, involvement of S1P was demonstrated by dose-dependently reduced nociceptive behavior of rats in a formalin assay following intrathecal application of the S1P receptor agonist FTY720 (Coste et al. 2008b). S1P has been implicated in spinal nociceptive processing through the inhibition of the synthesis of a key messenger, neuronal cyclic AMP (Coste et al. 2008a). In acute and inflammatory pain models, S1P concentrations were selectively and significantly reduced in the cerebrospinal fluid of adult rats following injection of formalin or zymosan, respectively. The pharmacological inhibition of sphingosine kinases that convert sphingosine to S1P was found to decrease basal pain thresholds in a formalin assay in mice. In contrast, the intrathecal application of S1P or dihydro-S1P reduced pain-related nociceptive behavior by abolishing cAMP-dependent NMDA receptor phosphorylation in the outer laminae of the dorsal horn (Coste et al. 2008a).

Acute and chronic severe pain can be effectively treated by opiates such as morphine; however, the use of opiates is limited by the development of analgesic tolerance. Salvemini's group recently identified ceramide as a key signaling molecule in the development of morphine antinociceptive tolerance in a mouse model. Following chronic morphine administration, the enzymatic activities of serine palmitoyltransferase, ceramide synthase, and ASM in the spinal cord were significantly increased and led to ceramide accumulation, nitroxidative stress, and neuroimmune activation (Ndengele et al. 2009). The upregulation of ceramide in the dorsal horn in response to chronic morphine uptake in mice was associated with significant neuronal apoptosis, whereas the inhibition of ceramide biosynthesis by fumonisin B1 treatment reduced both apoptosis and the development of antinociceptive tolerance as determined by the tail flick test (Bryant et al. 2009). In a subsequent study using a rat model, repeated morphine administration caused upregulation of ceramide in both spinal astrocytes and microglia (but not neurons) and increased S1P levels. The coadministration of ceramide and S1P inhibitors blocked the development of hyperalgesia and tolerance (Muscoli et al. 2010). On the other hand, the intraplantar injection of ceramide in rats resulted in the development of time-dependent thermal hyperalgesia related to an increase of TNF-α, which is known to stimulate S1P production via activation of sphingosine kinase. Further experiments using enzyme inhibitors or antibodies provide evidence for the involvement of S1P and the S1P receptor 1 subtype as the downstream signaling pathway in ceramide-induced hyperalgesia (Doyle et al. 2011).

Efforts to explain the pathomechanism underlying the aberrant pain and sensory processing in fibromyalgia included a search for autoantibodies and led to the identification of anti-serotonin as well as anti-ganglioside and anti-phospholipid antibodies (reviewed in Dadabhoy et al. 2008). Klein et al. detected antibodies against gangliosides in the sera of 74 % of patients with clinically well-defined fibromyalgia syndrome (Klein et al. 1992). The same group confirmed the high incidence of a defined autoantibody pattern in fibromyalgia in two cross-sectional studies (Klein et al. 1992; Klein and Berg 1995). The observation of such antibodies in patients' family members suggests a genetic predisposition to the disorder (Klein et al. 1992). However, the applicability of these findings is not yet clear. Another group was unable to reproduce the finding of increased levels of autoantibodies against ganglioside GM1, although levels of anti-serotonin and anti-thromboplastin antibodies were higher in fibromyalgia patients than in control subjects (Werle et al. 2001).

Although a number of studies have shed light upon various components of the SL network and its enzymes with a focus on ceramide, S1P and DMS, and their various ways of action with respect to nociception, most reports are based on animal models, and a whole clear picture required for the development of therapeutic agents is still lacking.

4 Sphingomyelinase-Induced Ceramide Generation: A Core Feature of Psychiatric Disorders?

Sphingomyelinase-induced ceramide generation may contribute to the emergence of complex and multifactorial psychiatric disorders (Kornhuber et al. 2009; Gulbins et al. unpublished). In this model, several stress-related factors including psychoactive substances, inflammatory cytokines, reactive oxygen species, and infections may induce activation of sphingomyelinases (Grassmé et al. 2003, 2005; Haimovitz-Friedman et al. 1997; Hofmeister et al. 1997; Nalivaeva et al. 2000; Reichel et al. 2010, 2011; Scheel-Toellner et al. 2004; Wong et al. 2000; Zhang et al. 2008) and result in increased levels of ceramide in various brain areas. This model may account (to varying degrees) for several psychiatric disorders besides depression (Table 1). Alcohol dependence, with its high burden of inflammatory cytokines and high levels of oxidative stress, may be a paradigm of an SMase-/ceramide-related disorders with somatic and psychiatric manifestations (Crews et al. 2006; Wu and Cederbaum 2009). Anxiety and schizophrenia have been similarly linked to inflammatory cytokines and to oxidative stress that may result in increased ceramide levels (Bitanihirwe and Woo 2011; de Miranda et al. 2011; Potvin et al. 2008). The molecular mechanisms whereby increased levels of ceramide lead to cellular and neuronal malfunctions are not yet understood. These mechanisms most likely involve profound changes in the biochemistry and biophysics of the plasma membrane, which could explain the spreading of a single stress signal to a variety of downstream targets. Ceramide molecules undergo spontaneous assembly into ceramide-enriched membrane domains that serve to trap and cluster or to exclude receptors, associated signaling molecules and ion channels (Zhang et al. 2009). Increased ceramide levels also affect the curvature of the plasma membrane, thereby facilitating the budding and fusion of vesicles, e.g., in endocytotic and exocytotic processes or in microparticle shedding (Bianco et al. 2009; Holopainen et al. 2000; Perrotta et al. 2010; Rogasevskaia and Coorssen 2006; Zha et al. 1998).

The exact cellular and molecular consequences of the activated SMase/ceramide pathway in psychiatric disorders still need to be defined. At the molecular level, certain critical targets of ceramide-dependent signaling, such as PP2A and the Akt serine/threonine kinase, have been identified (Mora et al. 2002; Qin et al. 2012; Wolff et al. 1994). Pathological changes in Akt signal transduction were reported for patients with familial schizophrenia (Emamian et al. 2004). Lithium, a pharmacological agent used for the treatment of several psychiatric disorders, regulates Akt/glycogen synthase kinase 3 (GSK3) signaling (Beaulieu et al. 2008; Mora et al. 2002). Another interesting target of ceramide in psychiatric disorders is

Table 1 Evidence from clinical studies, animal models, and cell culture experiments supporting the role of activated SMase/ceramide pathways in psychiatric disorders

Psychiatric disorder	Findings	References
Depression	ASM activity ↑ in peripheral blood cells of patients	Kornhuber et al. (2005)
	Cer ↑ in plasma levels of patients	Gracia-Garcia et al. (2011)
	altered plasma levels of SM and PC species in patients	Demirkan et al. (2012)
Schizophrenia	SM ↓; Cer ↑ in white matter of patients	Schmitt et al. (2004), Schwarz et al. (2008)
	SM ↑, Cer ↓ in erythrocytes	Hitzemann et al. (1984), Schwarz et al. (2008)
	Cer ↓ in skin of first-episode patients	Smesny et al. (2012)
	SPTLC2 expression ↑ in chronic illness	Narayan et al. (2009)
	SPTLC2 expression ↓, *ASAH1* expression ↓ in early illness	Narayan et al. (2009)
	ASAH1 expression ↓ in chronic illness	Zhang et al. (2012)
Alcohol dependence	Cer ↑ in cell culture experiment	Pascual et al. (2003), Saito et al. (2005), Liangpunsakul et al. (2010)
	Cer ↑ in mouse model	Deaciuc et al. (2000), Liangpunsakul et al. (2010, 2012)
	ASM activity ↑ in cell culture experiment	Pascual et al. (2003), Liu et al. (2000)
	ASM activity ↑ in mouse model	Deaciuc et al. (2000)
	ASM activity ↑ in patients	Reichel et al. (2010, 2011)
	NSM activity ↑ in cell culture experiment	Pascual et al. (2003), Liu et al. (2000)
Anxiety	S1P ↑ in serum of rat model	Jang et al. (2008)
	S1P induces anxiety in rat model	Jang et al. (2011)
	altered plasma levels of SM and PC species in patients	Demirkan et al. (2012)
Pain syndromes	Cer ↑, Sph ↑, and AC activity ↑ in rat model of neuropathic pain	Patti et al. (2012)
	S1P ↓ in cerebrospinal fluid in rat models of pain	Coste et al. (2008a)
	S1P reduces pain-related behavior in mouse model	Coste et al. (2008a)
	S1P agonist reduces nociceptive behavior in rat model of pain	Coste et al. (2008b)
	Inhibition of SK lowers basal pain threshold in mouse model	Coste et al. (2008a)
	Cer increases thermal hyperalgesia in rat model	Doyle et al. (2011)
	Cer and S1P inhibitors reduce development of hyperalgesia and tolerance	Muscoli et al. (2010)
	Cer ↑, S1P ↑ after repeated morphine application in mouse/rat model	Bryant et al. (2009), Muscoli et al. (2010)
	SPT activity ↑, CS activity ↑, ASM activity ↑ after chronic morphine application in mouse model	Ndengele et al. (2009)

AC acid ceramidase, *ASM* acid sphingomyelinase, *ASAH1* acid ceramidase gene, *Cer* ceramide, *CS* ceramide synthase, *NSM* neutral sphingomyelinase, PC phosphatidylcholine, *S1P* sphingosine-1-phosphate, *SK* sphingosine kinase, *SM* sphingomyelin, *Sph* sphingosine, *SPT* serine palmitoyltransferase, *SPTLC2* serine palmitoyltransferase gene

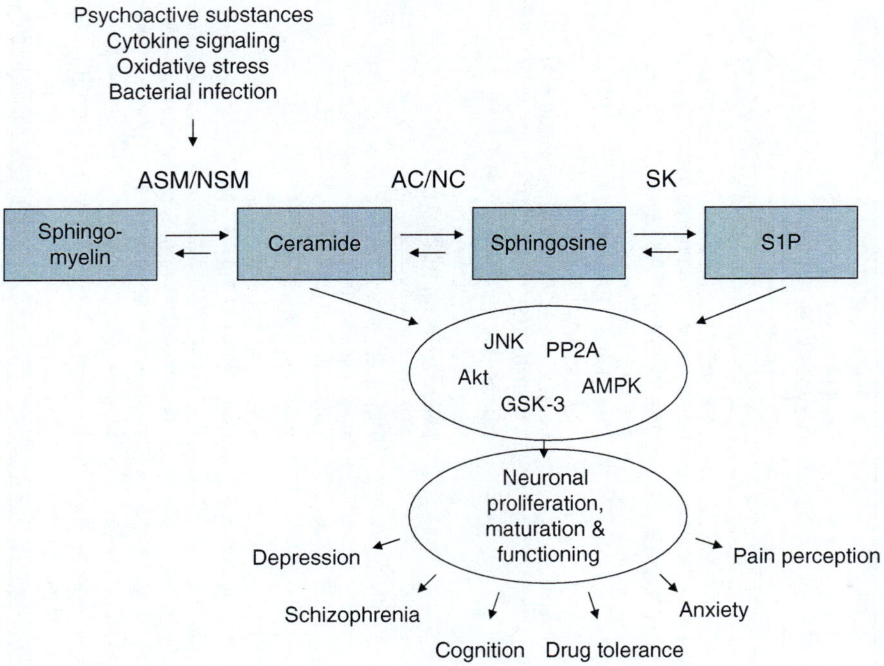

Fig. 1 A model of the activation of SMase/ceramide pathways as a molecular link in psychiatric disorders. Stress factors (e.g., alcohol consumption or oxidative stress) known to be involved in the emergence of psychiatric disorders activate the hydrolysis of SM and result in increased levels of ceramide and its metabolites Sph and S1P. The increased ceramide levels in turn interfere with the regulation of key molecules and lead to impaired neuronal plasticity and finally to neuronal dysfunction. *AC* acid ceramidase, AMPK AMP-activated protein kinase, *ASM* acid sphingomyelinase, GSK-3 glycogen synthase kinase 3, JNK c-Jun N-terminal kinase, PP2A protein phosphatase 2A, *NC* neutral ceramidase, *NSM* neutral sphingomyelinase, *S1P* sphingosine-1-phosphate, *SK* sphingosine kinase

postsynaptic density protein (PSD)-95. Decreased PSD-95 levels were reported for patients with schizophrenia and bipolar disorder (Toro et al. 2005). The PSD-95 levels can be in turn upregulated by inhibition of nSMase2, most likely as a result of decreased ceramide levels (Tabatadze et al. 2010). In addition, a similar approach to inhibit ceramide increased AMPA receptor numbers, which may play a role in depression (Alt et al. 2006).

The studies reviewed here indicate that the activation of SMase/ceramide pathways is a common motif in psychiatric disorders and a missing link in our understanding of how cellular stress is translated into behavioral changes. The model presented in Fig. 1 provides a framework for the potential development of new therapeutic agents.

References

Alt A, Nisenbaum ES, Bleakman D, Witkin JM (2006) A role for AMPA receptors in mood disorders. Biochem Pharmacol 71:1273–88

Alexander GM, van Rijn MA, van Hilten JJ, Perreault MJ, Schwartzman RJ (2005) Changes in cerebrospinal fluid levels of pro-inflammatory cytokines in CRPS. Pain 116:213–219

Baraona E, Lieber CS (1979) Effects of ethanol on lipid metabolism. J Lipid Res 20:289–315

Beaulieu JM, Marion S, Rodriguiz RM, Medvedev IO, Sotnikova TD, Ghisi V, Wetsel WC, Lefkowitz RJ, Gainetdinov RR, Caron MG (2008) A β-arrestin 2 signaling complex mediates lithium action on behavior. Cell 132:125–136

Belmaker RH, Agam G (2008) Major depressive disorder. N Engl J Med 358:55–68

Bianco F, Perrotta C, Novellino L, Francolini M, Riganti L, Menna E, Saglietti L, Schuchman EH, Furlan R, Clementi E, Matteoli M, Verderio C (2009) Acid sphingomyelinase activity triggers microparticle release from glial cells. EMBO J 28:1043–1054

Bitanihirwe BKY, Woo TUW (2011) Oxidative stress in schizophrenia: an integrated approach. Neurosci Biobehav Rev 35:878–893

Bryant L, Doyle T, Chen Z, Cuzzocrea S, Masini E, Vinci MC, Esposito E, Mazzon E, Petrusca DN, Petrache I, Salvemini D (2009) Spinal ceramide and neuronal apoptosis in morphine antinociceptive tolerance. Neurosci Lett 463:49–53

Chang GHF, Barbaro NM, Pieper RO (2000) Phosphatidylserine-dependent phagocytosis of apoptotic glioma cells by normal human microglia, astrocytes, and glioma cells. Neuro Oncol 2:174–183

Cherayil GD (1969) Estimation of glycolipids in four selected lobes of human brain in neurological diseases. J Neurochem 16:913–920

Cole AL, Lee PJ, Hughes DA, Deegan PB, Waldek S, Lachmann RH (2007) Depression in adults with Fabry disease: a common and under-diagnosed problem. J Inherit Metab Dis. 30:943–951

Conigrave KM, Degenhardt LJ, Whitfield JB, Saunders JB, Helander A, Tabakoff B (2002) CDT, GGT, and AST as markers of alcohol use: the WHO/ISBRA collaborative project. Alcohol Clin Exp Res 26:332–339

Contreras FX, Ernst AM, Haberkant P, Björkholm P, Lindahl E, Gönen B, Tischer C, Elofsson A, von Heijne G, Thiele C, Pepperkok R, Wieland F, Brügger B (2012) Molecular recognition of a single sphingolipid species by a protein's transmembrane domain. Nature 481:525–529

Corcoran CA, He Q, Ponnusamy S, Ogretmen B, Huang Y, Sheikh MS (2008) Neutral sphingomyelinase-3 is a DNA damage and nongenotoxic stress-regulated gene that is deregulated in human malignancies. Mol Cancer Res 6:795–807

Coste O, Brenneis C, Linke B, Pierre S, Maeurer C, Becker W, Schmidt H, Gao W, Geisslinger G, Scholich K (2008a) Sphingosine 1-phosphate modulates spinal nociceptive processing. J Biol Chem 283:32442–32451

Coste O, Pierre S, Marian C, Brenneis C, Angioni C, Schmidt H, Popp L, Geisslinger G, Scholich K (2008b) Antinociceptive activity of the S1P-receptor agonist FTY720. J Cell Mol Med 12:995–1004

Crews FT, Bechara R, Brown LA, Guidot DM, Mandrekar P, Oak S, Qin L, Szabo G, Wheeler M, Zou J (2006) Cytokines and alcohol. Alcohol Clin Exp Res 30:720–730

Dadabhoy D, Crofford LJ, Spaeth M, Russell IJ, Clauw DJ (2008) Biology and therapy of fibromyalgia. Evidence-based biomarkers for fibromyalgia syndrome. Arthritis Res Ther 10:211

Davis CN, Tabarean I, Gaidarova S, Behrens MM, Bartfai T (2006) IL-1β induces a MyD88-dependent and ceramide-mediated activation of Src in anterior hypothalamic neurons. J Neurochem 98:1379–1389

Dawkins JL, Hulme DJ, Brahmbhatt SB, Auer-Grumbach M, Nicholson GA (2001) Mutations in SPTLC1, encoding serine palmitoyltransferase, long chain base subunit-1, cause hereditary sensory neuropathy type I. Nat Genet 27:309–312 de Miranda AS, Lacerda-Queiroz N, de Carvalho VM, Rodrigues DH, Rachid MA, Quevedo J, Teixeira AL (2011) Anxiety-like behavior and proinflammatory cytokine levels in the brain of C57BL/6 mice infected with Plasmodium berghei (strain ANKA). Neurosci Lett 491:202–206

Deaciuc IV, Nikolova-Karakashian M, Fortunato F, Lee EY, Hill DB, McClain CJ (2000) Apoptosis and dysregulated ceramide metabolism in a murine model of alcohol-enhanced lipopolysaccharide hepatotoxicity. Alcohol Clin Exp Res 24:1557–1565

Demirkan A, van Duijn CM, Ugocsai P, Isaacs A, Pramstaller PP, Liebisch G, Wilson JF, Johansson A, Rudan I, Aulchenko YS, Kirichenko AV, Janssens ACJW, Jansen RC, Gnewuch C, Domingues FS, Pattaro C, Wild SH, Jonasson I, Polasek O, Zorkoltseva IV, Hofman A, Karssen LC, Struchalin M, Floyd J, Igl W, Biloglav Z, Broer L, Pfeufer A, Pichler I, Campbell S, Zaboli G, Kolcic I, Rivadeneira F, Huffman J, Hastie ND, Uitterlinden A, Franke L, Franklin CS, Vitart V, Nelson CP, Preuss M, Bis JC, O'Donnell CJ, Franceschini N, Witteman JCM, Axenovich T, Oostra BA, Meitinger T, Hicks AA, Hayward C, Wright AF, Gyllensten U, Campbell H, Schmitz G (2012) Genome-wide association study identifies novel loci associated with circulating phospho- and sphingolipid concentrations. PLoS Genet 8:e1002490

Demirkan A, Isaacs A, Ugocsai P, Liebisch G, Struchalin M, Rudan I, Wilson JF, Pramstaller PP, Gyllensten U, Campbell H, Schmitz G, Oostra BA, Van Duijn CM (2012) Plasma phosphatidylcholine and sphingomyelin concentrations are associated with depression and anxiety symptoms in a Dutch family-based lipidomics study. J Psychiatr Res pii:S0022-3956(12) 00341-X

Demisch L, Heinz K, Gerbaldo H, Kirsten R (1992) Increased concentrations of phosphatidylinositol (PI) and decreased esterification of arachidonic acid into phospholipids in platelets from patients with schizoaffective disorders or atypic phasic psychoses. Prostaglandins Leukot Essent Fatty Acids 46:47–52

DeVito WJ, Stone S, Shamgochian M (2000) Ethanol increases the neurotoxic effect of tumor necrosis factor-alpha in cultured rat astrocytes. Alcohol Clin Exp Res 24:82–92

Doyle T, Chen Z, Obeid LM, Salvemini D (2011) Sphingosine-1-phosphate acting via the S1P receptor is a downstream signaling pathway in ceramide-induced hyperalgesia. Neurosci Lett 499:4–8

Duan RD, Hertervig E, Nyberg L, Hauge T, Sternby B, Lillienau J, Farooqi A, Nilsson Å (1996) Distribution of alkaline sphingomyelinase activity in human beings and animals. Tissue and species differences. Dig Dis Sci 41:1801–1806

Edsall LC, Van Brocklyn JR, Cuvillier O, Kleuser B, Spiegel S (1998) N, N-Dimethylsphingosine is a potent competitive inhibitor of sphingosine kinase but not of protein kinase C: modulation of cellular levels of sphingosine 1-phosphate and ceramide. Biochemistry 37:12892–12898

El Bawab S, Bielawska A, Hannun YA (1999) Purification and characterization of a membrane-bound nonlysosomal ceramidase from rat brain. J Biol Chem 274:27948–27955

Emamian ES, Hall D, Birnbaum MJ, Karayiorgou M, Gogos JA (2004) Convergent evidence for impaired AKT1-GSK3β signaling in schizophrenia. Nat Genet 36:131–137

Fernández A, Colell A, Garcia-Ruiz C, Fernandez-Checa JC (2008) Cholesterol and sphingolipids in alcohol-induced liver injury. J Gastroenterol Hepatol 23(Suppl 1):S9–S15

Fukuzako H, Fukuzako T, Hashiguchi T, Kodama S, Takigawa M, Fujimoto T (1999) Changes in levels of phosphorus metabolites in temporal lobes of drug-naive schizophrenic patients. Am J Psychiatry 156:1205–1208

Gao YJ, Zhang L, Samad OA, Suter MR, Yasuhiko K, Xu ZZ, Park JY, Lind AL, Ma Q, Ji RR (2009) JNK-induced MCP-1 production in spinal cord astrocytes contributes to central sensitization and neuropathic pain. J Neurosci 29:4096–4108

Garofalo K, Penno A, Schmidt BP, Lee HJ, Frosch MP, von Eckardstein A, Brown RH, Hornemann T, Eichler FS (2011) Oral L-serine supplementation reduces production of neurotoxic deoxysphingolipids in mice and humans with hereditary sensory autonomic neuropathy type 1. J Clin Invest 121:4735–4745

Gatt S (1963) Enzymic hydrolysis and synthesis of ceramides. J Biol Chem 238:3131–3133

Gilmore N, Cherian L, Klemm WR (1991) Ganglioside or sialic acid attenuates ethanol-induced decrements in locomotion, nose-poke exploration, and anxiety, but not body temperature. Prog Neuropsychopharmacol Biol Psychiatry 15:91–104

Glasier MM, Sutton RL, Stein DG (1995) Effects of unilateral entorhinal cortex lesion and ganglioside GM1 treatment on performance in a novel water maze task. Neurobiol Learn Mem 64:203–214

Gracia-Garcia P, Rao V, Haughey NJ, Banduru VVR, Smith G, Rosenberg PB, Lobo A, Lyketsos CG, Mielke MM (2011) Elevated plasma ceramides in depression. J Neuropsychiatry Clin Neurosci 23:215–218

Grassmé H, Jendrossek V, Riehle A, von KG, Berger J, Schwarz H, Weller M, Kolesnick R, Gulbins E (2003) Host defense against Pseudomonas aeruginosa requires ceramide-rich membrane rafts. Nat Med 9:322–330

Grassmé H, Riehle A, Wilker B, Gulbins E (2005) Rhinoviruses infect human epithelial cells via ceramide-enriched membrane platforms. J Biol Chem 280:26256–26262

Haimovitz-Friedman A, Cordon-Cardo C, Bayoumy S, Garzotto M, McLoughlin M, Gallily R, Edwards CK III, Schuchman EH, Fuks Z, Kolesnick R (1997) Lipopolysaccharide induces disseminated endothelial apoptosis requiring ceramide generation. J Exp Med 186:1831–1841

Hait NC, Oskeritzian CA, Paugh SW, Milstien S, Spiegel S (2006) Sphingosine kinases, sphingosine 1-phosphate, apoptosis and diseases. Biochem Biophys Acta 1758:2016–2026

Haselhorst U, Schenk H, Beyer I, Uebelhack R, Franke E, Kielstein V (1988) Abnormality of gangliosides in erythrocyte membranes of schizophrenic patients. Clin Physiol Biochem 6:281–284

Hicks AA, Pramstaller PP, Johansson A, Vitart V, Rudan I, Ugocsai P, Aulchenko Y, Franklin CS, Liebisch G, Erdmann J, Jonasson I, Zorkoltseva IV, Pattaro C, Hayward C, Isaacs A, Hengstenberg C, Campbell S, Gnewuch C, Janssens ACJW, Kirichenko AV, König IR, Marroni F, Polasek O, Demirkan A, Kolcic I, Schwienbacher C, Igl W, Biloglav Z, Witteman JCM, Pichler I, Zaboli G, Axenovich TI, Peters A, Schreiber S, Wichmann HE, Schunkert H, Hastie N, Oostra BA, Wild SH, Meitinger T, Gyllensten U, van Duijn CM, Wilson JF, Wright A, Schmitz G, Campbell H (2009) Genetic determinants of circulating sphingolipid concentrations in European populations. PLoS Genet 5:e1000672

Hitzemann R, Hirschowitz J, Panini A, Mark C, Garver D (1984) Membranes, methylation and lithium responsive psychoses. Nutr Health 3:153–162

Hitzemann R, Mark C, Hirschowitz J, Garver D (1985) Characteristics of phospholipid methylation in human erythrocyte ghosts: relationship(s) to the psychoses and affective disorders. Biol Psychiatry 20:397–407

Hofmann K, Tomiuk S, Wolff G, Stoffel W (2000) Cloning and characterization of the mammalian brain-specific, Mg^{2+}-dependent neutral sphingomyelinase. Proc Natl Acad Sci USA 97:5895–5900

Hofmeister R, Wiegmann K, Korherr C, Bernardo K, Krönke M, Falk W (1997) Activation of acid sphingomyelinase by interleukin-1 (IL-1) requires the IL-1 receptor accessory protein. J Biol Chem 272:27730–27736

Holopainen JM, Angelova MI, Kinnunen PK (2000) Vectorial budding of vesicles by asymmetrical enzymatic formation of ceramide in giant liposomes. Biophys J 78:830–838

Horrobin DF (1998) The membrane phospholipid hypothesis as a biochemical basis for the neurodevelopmental concept of schizophrenia. Schizophr Res 30:193–208

Howren MB, Lamkin DM, Suls J (2009) Associations of depression with C-reactive protein, IL-1, and IL-6: a meta-analysis. Psychosom Med 71:171–186

Jafurulla M, Pucadyil TJ, Chattopadhyay A (2008) Effect of sphingomyelinase treatment on ligand binding activity of human $serotonin_{1A}$ receptors. Biochim Biophys Acta 1778:2022–2025

Jang S, Suh SH, Yoo HS, Lee YM, Oh S (2008) Changes in iNOS, GFAP and NR1 expression in various brain regions and elevation of sphingosine-1-phosphate in serum after immobilized stress. Neurochem Res 33:842–851

Jang S, Kim D, Lee Y, Moon S, Oh S (2011) Modulation of sphingosine 1-phosphate and tyrosine hydroxylase in the stress-induced anxiety. Neurochem Res 36:258–267

Jung SY, Suh JH, Park HJ, Jung KM, Kim MY, Na DS, Kim DK (2000) Identification of multiple forms of membrane-associated neutral sphingomyelinase in bovine brain. J Neurochem 75:1004–1014

Keshavan MS, Mallinger AG, Pettegrew JW, Dippold C (1993) Erythrocyte membrane phospholipids in psychotic patients. Psychiatry Res 49:89–95

Kim SK, Ahn KH, Ji JE, Choi JM, Jeon HJ, Jung SY, Jung KM, Kim DK (2010) Neutral sphingomyelinase 2 induces dopamine uptake through regulation of intracellular calcium. Cell Signal 22:865–870

Klein R, Berg PA (1995) High incidence of antibodies to 5-hydroxytryptamine, gangliosides and phospholipids in patients with chronic fatigue and fibromyalgia syndrome and their relatives: evidence for a clinical entity of both disorders. Eur J Med Res 1:21–26

Klein R, Bänsch M, Berg PA (1992) Clinical relevance of antibodies against serotonin and gangliosides in patients with primary fibromyalgia syndrome. Psychoneuroendocrinology 17:593–598

Kornhuber J, Medlin A, Bleich S, Jendrossek V, Henkel AW, Wiltfang J, Gulbins E (2005) High activity of acid sphingomyelinase in major depression. J Neural Transm 112:1583–1590

Kornhuber J, Tripal P, Reichel M, Terfloth L, Bleich S, Wiltfang J, Gulbins E (2008) Identification of new functional inhibitors of acid sphingomyelinase using a structure–property-activity relation model. J Med Chem 51:219–237

Kornhuber J, Reichel M, Tripal P, Groemer TW, Henkel AW, Mühle C, Gulbins E (2009) The role of ceramide in major depressive disorder. Eur Arch Psychiatry Clin Neurosci 259(Suppl 2): S199–S204

Kornhuber J, Muehlbacher M, Trapp S, Pechmann S, Friedl A, Reichel M, Mühle C, Terfloth L, Groemer TW, Spitzer GM, Liedl KR, Gulbins E, Tripal P (2011) Identification of novel functional inhibitors of acid sphingomyelinase. PLoS One 6:e23852

Krut O, Wiegmann K, Kashkar H, Yazdanpanah B, Krönke M (2006) Novel tumor necrosis factor-responsive mammalian neutral sphingomyelinase-3 is a C-tail-anchored protein. J Biol Chem 281:13784–13793

Kubicki M, McCarley RW, Shenton ME (2005) Evidence for white matter abnormalities in schizophrenia. Curr Opin Psychiatry 18:121–134

Latorre E, Aragonés MD, Fernández I, Catalán RE (1999) Platelet-activating factor modulates brain sphingomyelin metabolism. Eur J Biochem 262:308–314

Lee YK, Kim HL, Kim YL, Im DS (2007) Multiple actions of dimethylsphingosine in 1321N1 astrocytes. Mol Cells 23:11–16

Lesch P, Schmidt E, Schmidt FW (1973) Effects of chronic alcohol abuse on the fatty acid composition of major lipids in the human brain. Hepatocerebral degeneration II. Z Klin Chem Klin Biochem 11:159–166

Levade T, Salvayre R, Potier M, Douste-Blazy L (1986) Interindividual heterogeneity of molecular weight of human brain neutral sphingomyelinase determined by radiation inactivation method. Neurochem Res 11:1131–1138

Liangpunsakul S, Sozio MS, Shin E, Zhao Z, Xu Y, Ross RA, Zeng Y, Crabb DW (2010) Inhibitory effect of ethanol on AMPK phosphorylation is mediated in part through elevated ceramide levels. Am J Physiol Gastrointest Liver Physiol 298:G1004–G1012

Liangpunsakul S, Rahmini Y, Ross RA, Zhao Z, Xu Y, Crabb DW (2012) Imipramine blocks ethanol-induced ASMase activation, ceramide generation, and PP2A activation, and ameliorates hepatic steatosis in ethanol-fed mice. Am J Physiol Gastrointest Liver Physiol 302:G515–G523

Liu JJ, Wang JY, Hertervig E, Cheng Y, Nilsson A, Duan RD (2000) Activation of neutral sphingomyelinase participates in ethanol-induced apoptosis in Hep G2 cells. Alcohol Alcohol 35:569–573

MacQueen GM, Rosebush PI, Mazurek MF (1998) Neuropsychiatric aspects of the adult variant of Tay-Sachs disease. J Neuropsychiatry Clin Neurosci 10:10–19
Mann JJ (1999) Role of the serotonergic system in the pathogenesis of major depression and suicidal behavior. Neuropsychopharmacology 21:99S–105S
Marí M, Colell A, Morales A, Pañeda C, Varela-Nieto I, Garcia-Ruíz C, Fernández-Checa JC (2004) Acidic sphingomyelinase downregulates the liver-specific methionine adenosyltransferase 1A, contributing to tumor necrosis factor-induced lethal hepatitis. J Clin Invest 113:895–904
Mora A, Sabio G, Risco AM, Cuenda A, Alonso JC, Soler G, Centeno F (2002) Lithium blocks the PKB and GSK3 dephosphorylation induced by ceramide through protein phosphatase-2A. Cell Signal 14:557–562
Müller N, Ackenheil M (1998) Psychoneuroimmunology and the cytokine action in the CNS: implications for psychiatric disorders. Prog Neuropsychopharmacol Biol Psychiatry 22:1–33
Muscoli C, Doyle T, Dagostino C, Bryant L, Chen Z, Watkins LR, Ryerse J, Bieberich E, Neumman W, Salvemini D (2010) Counter-regulation of opioid analgesia by glial-derived bioactive sphingolipids. J Neurosci 30:15400–15408
Musselman DL, Evans DL, Nemeroff CB (1998) The relationship of depression to cardiovascular disease: epidemiology, biology, and treatment. Arch Gen Psychiatry 55:580–592
Nalivaeva NN, Rybakina EG, Pivanovich IY, Kozinets IA, Shanin SN, Bartfai T (2000) Activation of neutral sphingomyelinase by IL-1beta requires the type 1 interleukin 1 receptor. Cytokine 12:229–232
Narayan S, Kass KE, Thomas EA (2007) Chronic haloperidol treatment results in a decrease in the expression of myelin/oligodendrocyte-related genes in the mouse brain. J Neurosci Res 85:757–765
Narayan S, Head SR, Gilmartin TJ, Dean B, Thomas EA (2009) Evidence for disruption of sphingolipid metabolism in schizophrenia. J Neurosci Res 87:278–288
Ndengele MM, Cuzzocrea S, Masini E, Vinci MC, Esposito E, Muscoli C, Petrusca DN, Mollace V, Mazzon E, Li D, Petrache I, Matuschak GM, Salvemini D (2009) Spinal ceramide modulates the development of morphine antinociceptive tolerance via peroxynitrite-mediated nitroxidative stress and neuroimmune activation. J Pharmacol Exp Ther 329:64–75
Ng F, Berk M, Dean O, Bush AI (2008) Oxidative stress in psychiatric disorders: evidence base and therapeutic implications. Int J Neuropsychopharmacol 11:851–876
Nie H, Weng HR (2010) Impaired glial glutamate uptake induces extrasynaptic glutamate spillover in the spinal sensory synapses of neuropathic rats. J Neurophysiol 103:2570–2580
Norman E, Cutler RG, Flannery R, Wang Y, Mattson MP (2010) Plasma membrane sphingomyelin hydrolysis increases hippocampal neuron excitability by sphingosine-1-phosphate mediated mechanisms. J Neurochem 114:430–439
Ogretmen B, Hannun YA (2004) Biologically active sphingolipids in cancer pathogenesis and treatment. Nat Rev Cancer 4:604–616
Okino N, He X, Gatt S, Sandhoff K, Ito M, Schuchman EH (2003) The reverse activity of human acid ceramidase. J Biol Chem 278:29948–29953
Olsson NU, Harding AJ, Harper C, Salem N Jr (1996) High-performance liquid chromatography method with light-scattering detection for measurements of lipid class composition: analysis of brains from alcoholics. J Chromatogr B Biomed Appl 681:213–218
Ono M, Kikusui T, Sasaki N, Ichikawa M, Mori Y, Murakami-Murofushi K (2008) Early weaning induces anxiety and precocious myelination in the anterior part of the basolateral amygdala of male Balb/c mice. Neuroscience 156:1103–1110
Pascual M, Valles SL, Renau-Piqueras J, Guerri C (2003) Ceramide pathways modulate ethanol-induced cell death in astrocytes. J Neurochem 87:1535–1545
Patti GJ, Yanes O, Shriver LP, Courade JP, Tautenhahn R, Manchester M, Siuzdak G (2012) Metabolomics implicates altered sphingolipids in chronic pain of neuropathic origin. Nat Chem Biol 8:232–234

Perrotta C, Bizzozero L, Cazzato D, Morlacchi S, Assi E, Simbari F, Zhang Y, Gulbins E, Bassi MT, Rosa P, Clementi E (2010) Syntaxin 4 is required for acid sphingomyelinase activity and apoptotic function. J Biol Chem 285:40240–40251

Ponizovsky AM, Modai I, Nechamkin Y, Barshtein G, Ritsner MS, Yedgar S, Lecht S, Bergelson LD (2001) Phospholipid patterns of erythrocytes in schizophrenia: relationships to symptomatology. Schizophr Res 52:121–126

Posse de Chaves E, Sipione S (2010) Sphingolipids and gangliosides of the nervous system in membrane function and dysfunction. FEBS Lett 584:1748–1759

Potvin S, Stip E, Sepehry AA, Gendron A, Bah R, Kouassi E (2008) Inflammatory cytokine alterations in schizophrenia: a systematic quantitative review. Biol Psychiatry 63:801–808

Prince M, Patel V, Saxena S, Maj M, Maselko J, Phillips MR, Rahman A (2007) No health without mental health. Lancet 370:859–877

Qin J, Berdyshev E, Poirier C, Schwartz NB, Dawson G (2012) Neutral sphingomyelinase 2 deficiency increases hyaluronan synthesis by up-regulation of hyaluronan synthase 2 through decreased ceramide production and activation of Akt. J Biol Chem 287(17):13620–13632

Rao BG, Spence MW (1976) Sphingomyelinase activity at pH 7.4 in human brain and a comparison to activity at pH 5.0. J Lipid Res 17:506–515

Reichel M, Greiner E, Richter-Schmidinger T, Yedibela Ö, Tripal P, Jacobi A, Bleich S, Gulbins E, Kornhuber J (2010) Increased acid sphingomyelinase activity in peripheral blood cells of acutely intoxicated patients with alcohol dependence. Alcohol Clin Exp Res 34:46–50

Reichel M, Beck J, Mühle C, Rotter A, Bleich S, Gulbins E, Kornhuber J (2011) Activity of secretory sphingomyelinase is increased in plasma of alcohol-dependent patients. Alcohol Clin Exp Res 35:1852–1859

Reitz RC (1979) The effects of ethanol ingestion on lipid metabolism. Prog Lipid Res 18:87–115

Riddle EL, Rau KS, Topham MK, Hanson GR, Fleckenstein AE (2003) Ceramide-induced alterations in dopamine transporter function. Eur J Pharmacol 458:31–36

Rogasevskaia T, Coorssen JR (2006) Sphingomyelin-enriched microdomains define the efficiency of native Ca^{2+}-triggered membrane fusion. J Cell Sci 119:2688–2694

Rohrbough J, Rushton E, Palanker L, Woodruff E, Matthies HJ, Acharya U, Acharya JK, Broadie K (2004) Ceramidase regulates synaptic vesicle exocytosis and trafficking. J Neurosci 24:7789–7803

Saito M, Saito M, Cooper TB, Vadasz C (2005) Ethanol-induced changes in the content of triglycerides, ceramides, and glucosylceramides in cultured neurons. Alcohol Clin Exp Res 29:1374–1383

Sánchez C, Rueda D, Ségui B, Galve-Roperh I, Levade T, Guzmán M (2001) The CB_1 cannabinoid receptor of astrocytes is coupled to sphingomyelin hydrolysis through the adaptor protein fan. Mol Pharmacol 59:955–959

Sapolsky RM (2000) Glucocorticoids and hippocampal atrophy in neuropsychiatric disorders. Arch Gen Psychiatry 57:925–935

Sawai H, Hannun YA (1999) Ceramide and sphingomyelinases in the regulation of stress responses. Chem Phys Lipids 102:141–147

Scheel-Toellner D, Wang K, Craddock R, Webb PR, McGettrick HM, Assi LK, Parkes N, Clough LE, Gulbins E, Salmon M, Lord JM (2004) Reactive oxygen species limit neutrophil life span by activating death receptor signaling. Blood 104:2557–2564

Schissel SL, Keesler GA, Schuchman EH, Williams KJ, Tabas I (1998) The cellular trafficking and zinc dependence of secretory and lysosomal sphingomyelinase, two products of the acid sphingomyelinase gene. J Biol Chem 273:18250–18259

Schmitt A, Wilczek K, Blennow K, Maras A, Jatzko A, Petroianu G, Braus DF, Gattaz WF (2004) Altered thalamic membrane phospholipids in schizophrenia: a postmortem study. Biol Psychiatry 56:41–45

Schwarz E, Prabakaran S, Whitfield P, Major H, Leweke FM, Koethe D, McKenna P, Bahn S (2008) High throughput lipidomic profiling of schizophrenia and bipolar disorder brain tissue reveals alterations of free fatty acids, phosphatidylcholines, and ceramides. J Proteome Res 7:4266–4277

Sjögren B, Svenningsson P (2007) Depletion of the lipid raft constituents, sphingomyelin and ganglioside, decreases serotonin binding at human 5-$HT_{7(a)}$ receptors in HeLa cells. Acta Physiol 190:47–53

Smesny S, Schmelzer CE, Hinder A, Kohler A, Schneider C, Rudzok M, Schmidt U, Milleit B, Milleit C, Nenadic I, Sauer H, Neubert RH, Fluhr JW (2012) Skin ceramide alterations In first-episode schizophrenia indicate abnormal sphingolipid metabolism. Schizophr Bull doi:10.1093/schbul/sbs058

Spence MW, Burgess JK (1978) Acid and neutral sphingomyelinases of rat brain. Activity in developing brain and regional distribution in adult brain. J Neurochem 30:917–919

Spence MW, Burgess JK, Sperker ER (1979) Neutral and acid sphingomyelinases: somatotopographical distribution in human brain and distribution in rat organs. A possible relationship with the dopamine system. Brain Res 168:543–551

Spence MW, Beed S, Cook HW (1986) Acid and alkaline ceramidases of rat tissues. Biochem Cell Biol 64:400–404

Spiegel S, Cuvillier O, Edsall LC, Kohama T, Menzeleev R, Olah Z, Olivera A, Pirianov G, Thomas DM, Tu Z, Van Brocklyn JR, Wang F (1998) Sphingosine-1-phosphate in cell growth and cell death. Ann NY Acad Sci 845:11–18

Sugita M, Willians M, Dulaney JT, Moser HW (1975) Ceramidase and ceramide synthesis in human kidney and cerebellum. Description of a new alkaline ceramidase. Biochim Biophys Acta 398:125–131

Sulem P, Gudbjartsson DF, Geller F, Prokopenko I, Feenstra B, Aben KK, Franke B, den Heijer M, Kovacs P, Stumvoll M, Mägi R, Yanek LR, Becker LC, Boyd HA, Stacey SN, Walters GB, Jonasdottir A, Thorleifsson G, Holm H, Gudjonsson SA, Rafnar T, Björnsdottir G, Becker DM, Melbye M, Kong A, Tönjes A, Thorgeirsson T, Thorsteinsdottir U, Kiemeney LA, Stefansson K (2011) Sequence variants at CYP1A1-CYP1A2 and AHR associate with coffee consumption. Hum Mol Genet 20:2071–2077

Tabatadze N, Savonenko A, Song H, Bandaru VVR, Chu M, Haughey NJ (2010) Inhibition of neutral sphingomyelinase-2 perturbs brain sphingolipid balance and spatial memory in mice. J Neurosci Res 88:2940–2951

Tandon K, McGuffin P (2002) The genetic basis for psychiatric illness in man. Eur J Neurosci 16:403–407

Terracciano A, Sanna S, Uda M, Deiana B, Usala G, Busonero F, Maschio A, Scally M, Patriciu N, Chen WM, Distel MA, Slagboom EP, Boomsma DI, Villafuerte S, Sliwerska E, Burmeister M, Amin N, Janssens AC, van Duijn CM, Schlessinger D, Abecasis GR, Costa PT Jr (2010) Genome-wide association scan for five major dimensions of personality. Mol Psychiatry 15:647–656

Thayyullathil F, Chathoth S, Hago A, Patel M, Szulc ZM, Hannun Y, Galadari S (2011) Purification and characterization of a second type of neutral ceramidase from rat brain: a second more hydrophobic form of rat brain ceramidase. Biochim Biophys Acta 1811:242–252

Tomiuk S, Hofmann K, Nix M, Zumbansen M, Stoffel W (1998) Cloned mammalian neutral sphingomyelinase: functions in sphingolipid signaling? Proc Natl Acad Sci USA 95:3638–3643

Toro C, Deakin JF (2005) NMDA receptor subunit NRI and postsynaptic protein PSD-95 in hippocampus and orbitofrontal cortex in schizophrenia and mood disorder. Schizophr Res 80:323–30

Wallis CJ, Rezazadeh SM, Lal H (1995) GM1 ganglioside reduces ethanol intoxication and the development of ethanol dependence. Alcohol 12:573–580

Watkins LR, Milligan ED, Maier SF (2001) Glial activation: a driving force for pathological pain. Trends Neurosci 24:450–455

Weinreb NJ, Brady RO, Tappel AL (1968) The lysosomal localization of sphingolipid hydrolases. Biochim Biophys Acta 159:141–146

Werle E, Fischer HP, Müller A, Fiehn W, Eich W (2001) Antibodies against serotonin have no diagnostic relevance in patients with fibromyalgia syndrome. J Rheumatol 28:595–600

Wheeler D, Knapp E, Bandaru VVR, Wang Y, Knorr D, Poirier C, Mattson MP, Geiger JD, Haughey NJ (2009) Tumor necrosis factor-α-induced neutral sphingomyelinase-2 modulates synaptic plasticity by controlling the membrane insertion of NMDA receptors. J Neurochem 109:1237–1249

Wolf G, Gabay E, Tal M, Yirmiya R, Shavit Y (2006) Genetic impairment of interleukin-1 signaling attenuates neuropathic pain, autotomy, and spontaneous ectopic neuronal activity, following nerve injury in mice. Pain 120:315–324

Wolff RA, Dobrowsky RT, Bielawska A, Obeid LM, Hannun YA (1994) Role of ceramide-activated protein phosphatase in ceramide-mediated signal transduction. J Biol Chem 269:19605–19609

Wong ML, Xie B, Beatini N, Phu P, Marathe S, Johns A, Gold PW, Hirsch E, Williams KJ, Licinio J, Tabas I (2000) Acute systemic inflammation up-regulates secretory sphingomyelinase in vivo: a possible link between inflammatory cytokines and atherogenesis. Proc Natl Acad Sci USA 97:8681–8686

Wu D, Cederbaum AI (2009) Oxidative stress and alcoholic liver disease. Semin Liver Dis 29:141–154

Yang SN (2000) Ceramide-induced sustained depression of synaptic currents mediated by ionotropic glutamate receptors in the hippocampus: an essential role of postsynaptic protein phosphatases. Neuroscience 96:253–258

Yao JK, Leonard S, Reddy RD (2000) Membrane phospholipid abnormalities in postmortem brains from schizophrenic patients. Schizophr Res 42:7–17

Zha X, Pierini LM, Leopold PL, Skiba PJ, Tabas I, Maxfield FR (1998) Sphingomyelinase treatment induces ATP-independent endocytosis. J Cell Biol 140:39–47

Zhang Y, Li X, Carpinteiro A, Gulbins E (2008) Acid sphingomyelinase amplifies redox signaling in Pseudomonas aeruginosa-induced macrophage apoptosis. J Immunol 181:4247–4254

Zhang Y, Li X, Becker KA, Gulbins E (2009) Ceramide-enriched membrane domains–structure and function. Biochim Biophys Acta 1788:178–183

Zhang H, Li D, Su Y, Jiang S, Xu Y, Jiang K, Cui D (2012) Identification of the *N*-acylsphingosine amidohydrolase 1 gene (ASAH1) for susceptibility to schizophrenia in a Han Chinese population. World J Biol Psychiatry 13:106–113

Zhao Z, Yu M, Crabb D, Xu Y, Liangpunsakul S (2011) Ethanol-induced alterations in fatty acid-related lipids in serum and tissues in mice. Alcohol Clin Exp Res 35:229–234

Role of Sphingosine 1-Phosphate in Skeletal Muscle Cell Biology

Paola Bruni and Chiara Donati

Contents

Abstract Studies performed in the last fifteen years have clearly established that the bioactive sphingolipid sphingosine 1-phosphate (S1P) affects various different biological properties of myogenic precursor cells as well as physiological features of adult skeletal muscle. Noticeably, in myogenic precursor cells multiple growth factors and cytokines cross-communicate with S1P axis and the engagement of distinct S1P receptor subtypes appears to be crucially implicated in transmitting specific biological effects. This paper summarizes current research findings and discloses the potential for new therapeutics designed to alter S1P signaling with the aim of improving skeletal muscle repair.

Keywords myoblasts • skeletal muscle • sphingosine 1-phosphate • sphingosine kinase • S1P receptor

P. Bruni (✉) • C. Donati
Dipartimento di Scienze Biochimiche, Sperimentali e Cliniche, Università di Firenze, Viale GB Morgagni 50, 50134 Firenze, Italy
e-mail: paola.bruni@unifi.it

E. Gulbins and I. Petrache (eds.), *Sphingolipids in Disease*, Handbook of Experimental Pharmacology 216,
DOI 10.1007/978-3-7091-1511-4_23,

1 Skeletal Muscle Structure and Cell Biology

A little less than one-half of the body's mass is composed of skeletal muscle, with most muscles linked to bones by tendons through which the forces and movements developed during contractions are transmitted to the skeleton. The smallest contractile unit of skeletal muscle is the muscle fibre or myofibre, which is a long cylindrical cell that contains many nuclei, mitochondria and sarcomeres. Contraction, defined as the activation of muscle fibres with a tendency of the fibres to shorten, occurs when an increase in the cytosolic calcium concentration triggers a series of molecular events that includes the binding of calcium to the muscle-regulatory proteins, the interaction of myosin cross-bridges with actin filaments and the production of the cross-bridge working stroke. Skeletal muscles vary considerably in size, shape and arrangement of myofibres; moreover, they often contain a mixture of three distinct types of myofibres, which differ by the myosin isoforms expressed and the consequential contractile characteristics (Geeves and Holmes 1999). Type 1 myofibres are slow twitch and fatigue resistant, type 2A myofibres are fast twitch and moderately fatigue resistant and type 2B myofibres are fast twitch and not fatigue resistant. The percentage of individual types of myofibres within skeletal muscles is not fixed and can change throughout the life.

Skeletal muscles exert also additional physiological roles relevant for the homeostasis of human beings. Indeed, since muscle contractions require energy and whenever energy is used in the body, it is converted in part to heat, working skeletal muscles maintain body temperature by releasing heat. Moreover, skeletal muscle plays a key role in the overall regulation of whole body metabolism since the fed state is one of the main target tissues of insulin action, largely accounting for the hormone-dependent removal of excess glucose from the bloodstream after a meal, whereas in the prolonged fasting state, tissue protein breakdown supports the energy requirements of the body. Skeletal muscle is characterised by high degree of plasticity. When a muscle remains in disuse for a long period, the rate of degradation of contractile proteins becomes greater than the rate of replacement, resulting in muscle atrophy. This defect may occur as a result of lack of nutrition, loss of nerve supply, microgravity, ageing, systemic disease, prolonged immobilisation or disuse. Conversely, exercise and nutrition favour hypertrophy, in which the rate of synthesis is much higher than the rate of degradation of contractile proteins, leading to an increase in the size of muscle due to enlargement of existing cell fibres. This unique property permits skeletal muscle adaptation to distinct metabolic and physiological needs.

Injury to skeletal muscles may occur as a result of disease, such as dystrophy; exposure to myotoxic agents, such as bupivacaine or lidocaine; sharp or blunt trauma, such as punctures or contusions; ischemia, such as that which occurs with transplantation; exposure to excessively hot or cold temperatures and most commonly the muscles' own contractions. Regardless of the factors responsible, the manner in which the injuries are manifested appears to be the same, varying only in severity. In addition, the processes of fibre repair and regeneration appear to follow a common pathway regardless of the nature of the injurious event. Indeed, the process of skeletal muscle repair relies almost exclusively on tissue-resident stem cells, called satellite cells, which are normally quiescent but become activated upon skeletal muscle injury, giving rise to proliferating myogenic precursor cells that eventually differentiate and fuse to form multinucleated myofibres (Chargé and Rudnicki 2003; Tedesco et al. 2010). Skeletal muscle repair involves a myriad of signalling cues not fully defined at present, whose action must be precisely orchestrated to ensure its successful accomplishment. Many pathological conditions are consequent to alterations in skeletal muscle regeneration, including a number of different degenerative diseases such as muscular dystrophies as well as sarcopenia, characteristic of ageing, and cachexia; distinctive of sepsis and cancer (Karpati and Molnar 2008). Detailed comprehension of the basic mechanisms of skeletal muscle repair is necessary to pinpoint new molecular targets for interventions aimed at improving the process beneficial for a number of different diseases.

2 S1P in the Regulation of Growth, Differentiation and Motility of Skeletal Muscle Cells

As outlined above, regeneration of skeletal muscle is carried out by satellite cells. Due to their limited availability and the restricted number of experimental approaches to investigate their biological features in vivo, myoblastic cell lines, such as C2C12 which is derived from murine muscle satellite cells, are widely utilised to study in vitro skeletal muscle growth and differentiation. Upon mitogen depletion, a large subpopulation of C2C12 cells undergoes terminal differentiation forming myotubes; another small subpopulation called reserve cells remains undifferentiated and shares many characteristics with muscle-resident stem cells, while a third subset of cells undergoes apoptosis (Yoshida et al. 1998).

Since the role played by several species of sphingolipids in skeletal muscle cell biology and physiology has been subject of a recent review (Bruni and Donati 2008), we will be mainly focus our attention onto S1P. Recent data support a key role of sphingomyelin metabolism and S1P in the activation of satellite cells. By employing lysenin, a protein that specifically binds to surface sphingomyelin, Nagata et al. (2006a) showed that in C2C12 reserve cells, which model myogenic cell quiescence (Yoshida et al. 1998) as well as in satellite cells, sphingomyelin levels in the plasma membrane of noncycling myogenic cells are high but then fall as they activate. A subsequent study from the same authors (Nagata et al. 2006b) has demonstrated that a small pool of sphingomyelin at the inner leaflet of the plasma membrane is implicated in the signalling of satellite cells, since sphingomyelin levels are maintained when satellite cell activation occurs in the presence of GW4869, inhibitor of neutral SMase. Since, in the same report, S1P was found to mediate the entry of satellite cells into the cell cycle, the authors concluded that the breakdown of the sphingomyelin pool at the inner leaflet of plasma membrane occurring upon their activation is functional to S1P formation.

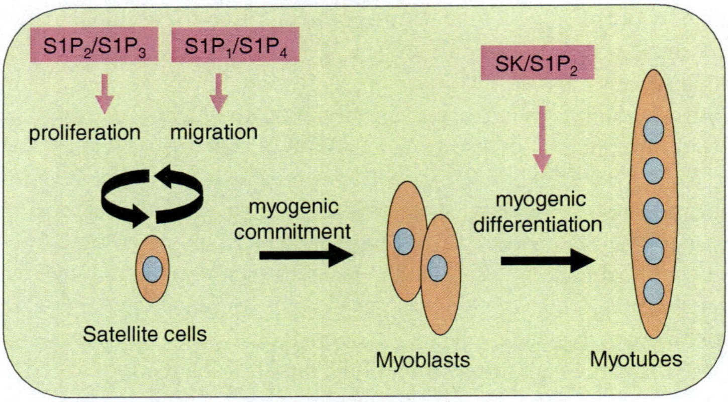

Fig. 1 Role of SK/S1P axis and S1P receptors in the activation of satellite cells and myogenic differentiation

The potential role of S1P was further demonstrated in the same study since the endogenous metabolite dimethylsphingosine, utilised as SK inhibitor, severely impaired the regeneration of cardiotoxin-treated skeletal muscle. The involvement of SK/S1P axis in this crucial biological process requires, however, further study given that dimethylsphingosine is rather unspecific, i.e., it has been recently shown to be responsible for suppression of osteoclastogenesis by a mechanism independent from SK inhibition (Kim et al. 2007).

Pilot studies on the responsiveness of C2C12 myoblasts to challenge with S1P showed that this bioactive lipid triggers numerous signalling pathways by activating phospholipase D (Meacci et al. 1999) and monomeric GTPase RhoA (Meacci et al. 2000) and increasing cytosolic calcium (Meacci et al. 2002). Since S1P action primarily depends on receptor ligation, the expression pattern of S1P receptors was also examined: $S1P_1$, $S1P_2$ and $S1P_3$ were found expressed in C2C12 myoblasts (Meacci et al. 1999). Intriguingly, the expression of $S1P_2$, which is upstream of phospholipase D activation, was found to be highly diminished throughout myoblast differentiation and substantially abolished in myotubes, suggesting a role for S1P signalling via $S1P_2$ in the molecular events required to accomplish cell differentiation (Meacci et al. 2003). Actually, subsequent studies identified S1P as negative regulator of serum-induced cell proliferation and powerful activator of myogenic differentiation of C2C12 cells (Donati et al. 2005). Notably, only $S1P_2$ was identified as mediator of S1P biological action (Fig. 1). The key role of $S1P_2$ in myogenic differentiation was further strengthened by the observation that ectopic expression of this protein accelerated the onset of differentiated phenotype (Donati et al. 2005) and by a subsequent report showing that a synthetic derivative of ceramide named K6PC-5, reported to activate sphingosine kinase (SK)-1 in keratinocytes (Kwon et al. 2007), promoted myogenic differentiation of C2C12 myoblasts in a $S1P_2$-dependent manner (Bernacchioni et al. 2011). Interestingly, the S1P-induced regulation of the gap junctional protein connexin-43 has been identified as downstream event implicated in the promyogenic effect of the sphingolipid (Squecco et al. 2006).

Actin remodelling and stress fibre formation in response to S1P generate a mechanical tension to the plasma membrane of C2C12 cells, activate stretch-activated channels, and trigger calcium-dependent signals, thus influencing the phenotypic maturation of myoblasts (Formigli et al. 2007). Likewise, in the same cell line, it has been reported that transient receptor potential canonical 1 channel (TRPC1), downstream to S1P challenge, upregulates the Ca^{2+}-sensitive protease and *m*-calpain, then affecting the regulation of classical PKCs and myoblast fusion (Meacci et al. 2010).

Experimental evidence has also been recently provided for a physiological role of SK1, one of the two isoforms of S1P-synthesising enzymes, in the regulation of myoblast proliferation and differentiation. SK1 protein content and S1P formation were found to be enhanced in myoblasts that became confluent as well as in differentiating cells (Meacci et al. 2008). Moreover, enforced expression of SK1 reduced the myoblast proliferation rate, enhanced the expression of myogenic differentiation markers and anticipated the onset of differentiated muscle phenotype. Conversely, downregulation of SK1 by specific silencing or overexpression of the catalytically inactive SK1 significantly increased cell growth and delayed the beginning of myogenesis. The biological role exerted by SK1 in myoblasts appears therefore distinct from that reported in the majority of other cells, where it plays a pivotal role of in the promotion of cell proliferation (Taha et al. 2006). Furthermore, stimulation of myogenesis in SphK1-overexpressing myoblasts was abrogated when $S1P_2$ was silenced (Meacci et al. 2008), reinforcing the notion that this receptor type is critical for eliciting S1P-mediated effects in these cells.

Upon serious skeletal muscle injury, satellite cells from neighbouring intact muscles must be highly motile in order to be recruited to the site of muscle lesion; however, subsequently their cell motility must be negatively regulated in order to establish the proper stable cell–cell contacts and fuse with pre-existing fibres. Interestingly, a recent study reported that along with a promyogenic effect, S1P exerts also an anti-migratory action on C2C12 myoblasts (Becciolini et al. 2006). The lipid mediator reduces the directional cell motility and fully abrogates the chemotactic response to insulin-like growth factor-1. Given the major role of inducer of satellite cell motility exerted by insulin-like growth factor-1 (Suzuki et al. 2000), the finding that S1P is capable of fully counteracting its action may have important implications in the in vivo regulation of exact satellite cell positioning to cause efficient cell fusion. The anti-migratory response to S1P requires ligation to $S1P_2$, and the subsequent activation of RhoA (Becciolini et al. 2006), similarly to what is observed in other cell systems, such as glioblastoma tumour cells (Lepley et al. 2005). Intriguingly, it has been recently observed that prostacyclin shares with S1P the unique property of inhibiting the migration of myoblasts (Bondesen et al. 2007).

In murine primary satellite cells isolated from tibialis anterior single muscle fibres, S1P has been very recently demonstrated to positively stimulate proliferation in a phosphatidylinositol 3-kinase-dependent manner as well as migration. By employing selective S1P receptor agonists and antagonists and silencing individual S1P receptors, the mitogenic action of S1P in satellite cells was shown to depend on $S1P_2$ and $S1P_3$, while $S1P_1$ and, for the first time, $S1P_4$, detected in these cells, were demonstrated to have a role in mediating the S1P migratory effect (Fig. 1). On the contrary, $S1P_2$ was found to negatively regulate cell migration (Calise et al. 2012; Donati et al. 2013). These results provide increased understanding of the multifaceted roles of S1P in regulating cell fate in normal development, defining specific biological actions exerted by this sphingolipid in skeletal muscle-resident stem cells. The apparent contradictory result of S1P being mitogenic in primary satellite cells and promyogenic in myoblasts could be reconciled by considering that the final biological action of the sphingolipid is strictly dependent on its specific receptor pattern expressed in a given cell type and in view of the multiplicity of signals conveyed by S1PRs that can give raise to distinct and even contrasting cellular effects. Thus, it becomes particularly critical to understand the extracellular cues responsible for the transcriptional regulation of S1PR expression in stem and progenitor cells and the consequent modification of the expression pattern that could act as decisive time switch to accomplish tissue repair.

3 S1P in the Regulation of the Biology and Physiology of Skeletal Muscle

Recently a role of S1P in skeletal muscle fibre regeneration of rat and mouse models has been demonstrated after myotoxic injury induced by bupivacaine. The exogenous addition of S1P resulted in a significant trophic effect in regenerating fibres, whereas reduction of the circulating lipid produced the opposite effect. The use of selective pharmacological S1PR antagonists demonstrated their variant roles during regeneration (Danieli-Betto et al. 2010). Likewise, Sassoli et al. (2011) further showed a role of S1P in skeletal muscle repair/regeneration since exogenous S1P attenuated the excitation–contraction-induced injury of extensor digitorum longus, protecting fibres from apoptosis, preserving satellite cell viability and affecting extracellular matrix remodelling. Moreover, it has been demonstrated that exogenous application of S1P counteracts the reduction of muscle mass caused by denervation, whereas neutralisation of the extracellular lipid with a specific anti-S1P antibody accelerates the atrophy caused by denervation (Zanin et al. 2008). The trophic effect of S1P was found to imply the regulation of myosin heavy chain isoform expression, resulting in the attenuation of the slow-to-fast transformation due to inactivity. Further experimental evidence in favour of a positive role of SK/S1P pathway in skeletal muscle plasticity comes from a report where global gene expression in skeletal muscle of young men following short-term creatine monohydrate supplementation has been examined. SK1 was identified among the most significantly upregulated genes (Safdar et al. 2008), suggesting that enhanced S1P formation is implicated in the increase of fat-free mass and muscle fibre size which accounts for the beneficial effect of creatine monohydrate on muscle (Vandenberghe et al. 1997; Volek et al. 1999).

Muscles that are intensively used show a progressive decline of performance which largely recovers after a period of rest named muscle fatigue. Multiple factors are involved in the onset of fatigue with the relative importance of each dependent on the fibre-type composition and the intensity, type, and duration of the contractile activity. Noticeably, sphingosine and S1P, recognised as modulators of excitation–contraction coupling (Bencini et al. 2003), were found capable of significantly reducing the tension decline during fatigue of extensor digitorum longus muscle (Danieli-Betto et al. 2005). Interestingly, the slowing of fatigue induced by sphingosine was fully prevented in the presence of dimethylsphingosine, suggesting that its mechanism of action implicates its transformation into S1P. Moreover, the beneficial effect of S1P on fatigue development was still detectable in Ca^{2+}-free solutions, whereas sphingosine was inefficacious unless extracellular Ca^{2+} was available, further supporting the view that sphingosine requires to be phosphorylated by calcium-dependent SK in order to attenuate the development of fatigue.

4 Regulation of SK/S1P Axis by Growth Factors, Cytokines and Hormones in Myoblasts

Interestingly, SK/S1P axis in myoblasts appears to be under the control of multiple extracellular cues that exploit this signalling pathway to elicit specific biological responses. In this regard, the first experimental evidence was obtained by examining the mechanism by which the cytokine tumour necrosis factor alpha (TNFα) at low concentration is capable of promoting myogenic differentiation. Indeed, translocation of SK1 isoform to membrane and consequent activation of the enzyme followed by engagement of $S1P_2$ receptor were identified as necessary for the onset of the pro-differentiating effect exerted by TNFα (Donati et al. 2007), further supporting the physiological relevance of the promyogenic effect exerted by the SK/$S1P_2$ signalling pathway (Meacci et al. 2008). Moreover, platelet-derived growth factor (PDGF), which is critically implicated in skeletal muscle repair, cross-communicates with S1P axis in a pathway involving SK1 and $S1P_1$, with important implications in the regulation of cell proliferation as well as cell motility. $S1P_1$ engagement elicited by PDGF via enhancement of SK1 activity attenuates the PDGF-induced cell proliferation since specific gene silencing of SK1 or $S1P_1$ enhances the mitogenic action of the growth factor (Nincheri et al. 2010). Intriguingly, engagement of $S1P_1$ by PDGF also modulates the chemotactic effect exerted by PDGF, significantly increasing the motility of myoblasts in response to the growth factor challenge. Thus, the concomitant reduction of cell proliferation and increase of cell motility mediated by the SK1/$S1P_1$ axis support the occurrence of a dual role of this signalling pathway, likely crucial for the efficient recruitment of myoblasts during muscle repair that requires a temporal shift between proliferation and migration. It is interesting to note that although endogenous S1P production and $S1P_1$ engagement play a significant role in the PDGF signalling, biological response elicited by exogenous S1P appears to be unaffected by $S1P_1$ ablation (Rapizzi et al. 2008), and interestingly, exogenous S1P gives rise to a final outcome distinct from that mediated by S1P endogenously formed in response to PDGF given that it acts as an anti-migratory rather than as a motogenic cue (Becciolini et al. 2006). These observations support the notion that spatial regulation of S1P biosynthesis within the cell, together with tight control of its release in the extracellular environment via ABC transporters or its partition into plasma membrane microdomains, is critical for determining the final biological effect and adds a further layer of complexity to the mechanisms by which this bioactive lipid exerts pleiotropic effects in a given cell type.

Even if the process of skeletal muscle repair is highly regulated, unfortunately, it is not always successful. Indeed, many skeletal muscle diseases are characterised by the onset of fibrosis, a degenerative process that causes replacement of the damaged tissue with extracellular matrix proteins and consequent impairment of the contractile properties of skeletal muscle (Mann et al. 2011). Fibrosis is strongly dependent onto the phenotypic appearance of a new type of cells named myofibroblasts, mainly resulting from transdifferentiation of myoblasts. Strikingly, S1P signalling pathway, which is critically implicated in the transmission of biological responses elicited by growth factors and cytokines with the final goal of favouring successful skeletal muscle regeneration, has been recently identified as key player in the transdifferentiation of myoblasts into myofibroblasts brought about by transforming growth factor beta (TGFβ) and, ultimately, in the appearance of fibrosis. Importantly, treatment with the cytokine upregulates SK1 in C2C12 myoblasts in a Smad-dependent manner and concomitantly modifies the expression of S1P receptors, making $S1P_3$ the dominant S1P receptor subtype (Cencetti et al. 2010). This combined action is responsible for redirecting the signalling of endogenous as well as exogenous S1P from an $S1P_2$-mediated promyogenic action towards a pro-fibrotic effect transmitted by $S1P_3$.

Lastly, it is important to recall that skeletal muscle not only is indispensable for powering the movement of the skeleton but it also plays a key role in the regulation of whole body metabolism. The metabolic requirements of skeletal muscle are highly variable, strictly depending on its working conditions. Moreover, skeletal muscle metabolism is also under hormonal control, being insulin action critical for switching from fatty acid to glucose oxidation. Indeed, after insulin stimulation, skeletal muscle accounts for almost three quarters of glucose utilisation in the body, and, as consequence, reduced insulin-responsive glucose disposal in skeletal muscle is a characteristic feature of metabolic syndrome. Importantly, S1P signalling pathway has been recognised to be crucially involved in the regulation of glucose metabolism in skeletal muscle. Basal and insulin-stimulated glucose uptake in cultured myoblasts is significantly reduced in cells where SK is pharmacologically inhibited or SK1 specifically silenced; in keeping with these results, ectopic expression of SK1 or exogenous S1P enhance basal and insulin-stimulated glucose uptake (Ma et al. 2007). A more recent study has provided the mechanistic explanation of the insulin-mimetic action of SK1 overexpression on glucose uptake, showing that in mouse myoblasts, S1P, acting via $S1P_2$ receptor, transactivates insulin receptor in a ligand-independent manner, via a novel molecular mechanism involving the enhancement of reactive oxygen species and the inhibition of protein tyrosine phoshatase-1B (Rapizzi et al. 2009). This finding suggests that improvement of

S1P signalling via $S1P_2$ could be beneficial for glucose disposal by skeletal muscle and stimulates further investigation to understand whether pathophysiology of insulin resistance and diabetes involves altered S1P production.

On the whole, the here summarised experimental data support a crucial role of SK/S1P axis in skeletal muscle cell biology opening new perspectives for pharmacological interventions aimed at improving skeletal muscle regeneration and attenuating the degenerative processes that characterise chronic myopathies by targeting S1P signalling towards improving the regenerative potential of satellite cells and/or interfering with the progression of skeletal muscle fibrosis.

References

Becciolini L, Meacci E, Donati C, Cencetti F, Rapizzi E, Bruni P (2006) Sphingosine 1-phosphate inhibits cell migration in C2C12 myoblasts. Biochim Biophys Acta 1761:43–51

Bencini C, Squecco R, Piperio C, Formigli L, Meacci E, Nosi D, Tiribilli B, Vassalli M, Quercioli F, Bruni P, Zecchi Orlandini S, Francini F (2003) Effects of sphingosine 1-phosphate on excitation-contraction coupling in mammalian skeletal muscle. J Muscle Res Cell Motil 24:539–554

Bernacchioni C, Cencetti F, Kwon MJ, Gwak HS, Jeong SK, Bruni P, Donati C (2011) The sphingosine kinase activator K6PC-5 stimulates C2C12 myoblast differentiation. Int J Immunopathol Pharmacol 24:55–62

Bondesen BA, Jones KA, Glasgow WC, Pavlath GK (2007) Inhibition of myoblast migration by prostacyclin is associated with enhanced cell fusion. FASEB J 21:3338–3345

Bruni P, Donati C (2008) Pleiotropic effects of sphingolipids in skeletal muscle. Cell Mol Life Sci 65:3725–3736

Calise S, Blescia S, Cencetti F, Bernacchioni C, Donati C, Bruni P (2012) Sphingosine 1-phosphate stimulates proliferation and migration of satellite cells. Role of S1P receptors. Biochem Biophys Acta 1823:439–450

Cencetti F, Bernacchioni C, Nincheri P, Donati C, Bruni P (2010) Transforming growth factor-beta1 induces transdifferentiation of myoblasts into myofibroblasts via up-regulation of sphingosine kinase-1/S1P3 axis. Mol Biol Cell 21:1111–1124

Chargé SBP, Rudnicki MA (2003) Cellular and molecular regulation of muscle regeneration. Physiol Rev 84:209–238

Danieli-Betto D, Germinario E, Esposito A, Megighian A, Midrio M, Ravara B, Damiani E, Libera LD, Sabbadini RA, Betto R (2005) Sphingosine 1-phosphate protects mouse extensor digitorum longus skeletal muscle during fatigue. Am J Physiol Cell Physiol 288:C1367–C1373

Danieli-Betto D, Peron S, Germinario E, Zanin M, Sorci G, Franzoso S, Sandonà D, Betto R (2010) Sphingosine 1-phosphate signaling is involved in skeletal muscle regeneration. Am J Physiol Cell Physiol 298:C550–C558

Donati C, Meacci E, Nuti F, Becciolini L, Farnararo M, Bruni P (2005) Sphingosine 1-phosphate regulates myogenic differentiation: a major role for S1P2 receptor. FASEB J 19:449–451

Donati C, Nincheri P, Cencetti F, Rapizzi E, Farnararo M, Bruni P (2007) Tumor necrosis factor-alpha exerts pro-myogenic action in C2C12 myoblasts via sphingosine kinase/S1P2 signaling. FEBS Lett 581:4384–4388

Donati C, Cencetti F, Bruni P (2013) New insights into the role of sphingosine 1-phosphate and lysophosphatidic acid in the regulation of skeletal muscle cell biology. Biochim Biophys Acta 1831(1):176–184

Formigli L, Meacci E, Sassoli C, Squecco R, Nosi D, Chellini F, Naro F, Francini F, Zecchi-Orlandini S (2007) Cytoskeleton/stretch-activated ion channel interaction regulates myogenic differentiation of skeletal myoblasts. J Cell Physiol 211:296–306

Geeves MA, Holmes KC (1999) Structural mechanisms of muscle contraction. Annu Rev Biochem 68:687–728

Karpati G, Molnar MJ (2008) Regeneration in human skeletal muscle disease. In: Schiaffino S, Partridge T (eds) Skeletal muscle repair and regeneration. Springer, Berlin, pp 199–216

Kim HJ, Lee Y, Chang EJ, Kim HM, Hong SP, Lee ZH, Ryu J, Kim HH (2007) Suppression of osteoclastogenesis by N, N-dimethyl-D-erythro-sphingosine: a sphingosine kinase inhibition-independent action. Mol Pharmacol 72:418–428

Kwon YB, Kim CD, Youm JK, Gwak HS, Park BD, Lee SH, Jeon S, Kim BJ, Seo YJ, Park JK, Lee JH (2007) Novel synthetic ceramide derivatives increase intracellular calcium levels and promote epidermal keratinocyte differentiation. J Lipid Res 48:1936–1943

Lepley D, Paik JH, Hla T, Ferrer F (2005) The G protein-coupled receptor S1P2 regulates Rho/Rho kinase pathway to inhibit tumor cell migration. Cancer Res 65:3788–3795

Ma MM, Chen JL, Wang GG, Wang H, Lu Y, Li JF, Yi J, Yuan YJ, Zhang QW, Mi J, Wang LS, Duan HF, Wu CT (2007) Sphingosine kinase 1 participates in insulin signalling and regulates glucose metabolism and homeostasis in KK/Ay diabetic mice. Diabetologia 50:891–900

Mann CJ, Perdiguero E, Kharraz Y, Aguilar S, Pessina P, Serrano AL, Muñoz-Cánoves P (2011) Aberrant repair and fibrosis development in skeletal muscle. Skelet Muscle 1:21

Meacci E, Vasta V, Donati C, Farnararo M, Bruni P (1999) Receptor-mediated activation of phospholipase D by sphingosine 1-phosphate in skeletal muscle C2C12 cells. A role for protein kinase C. FEBS Lett 457:184–188

Meacci E, Donati C, Cencetti F, Romiti E, Bruni P (2000) Permissive role of protein kinase C alpha but not protein kinase C delta in sphingosine 1-phosphate-induced Rho A activation in C2C12 myoblasts. FEBS Lett 482:97–101

Meacci E, Cencetti F, Formigli L, Squecco R, Donati C, Tiribilli B, Quercioli F, Zecchi Orlandini S, Francini F, Bruni P (2002) Sphingosine 1-phosphate evokes calcium signals in C2C12 myoblasts via Edg3 and Edg5 receptors. Biochem J 362:349–357

Meacci E, Cencetti F, Donati C, Nuti F, Farnararo M, Kohno T, Igarashi Y, Bruni P (2003) Down-regulation of EDG5/S1P2 during myogenic differentiation results in the specific uncoupling of sphingosine 1-phosphate signalling to phospholipase D. Biochim Biophys Acta 1633:133–142

Meacci E, Nuti F, Donati C, Cencetti F, Farnararo M, Bruni P (2008) Sphingosine kinase activity is required for myogenic differentiation of C2C12 myoblasts. J Cell Physiol 214:210–220

Meacci E, Bini F, Sassoli C, Martinesi M, Squecco R, Chellini F, Zecchi-Orlandini S, Francini F, Formigli L (2010) Functional interaction between TRPC1 channel and connexin-43 protein: a novel pathway underlying S1P action on skeletal myogenesis. Cell Mol Life Sci 67:4269–4285

Nagata Y, Kobayashi H, Umeda M, Ohta N, Kawashima S, Zammit PS, Matsuda R (2006a) Sphingomyelin levels in the plasma membrane correlate with the activation state of muscle satellite cells. J Histochem Cytochem 54:375–384

Nagata Y, Partridge TA, Matsuda R, Zammit PS (2006b) Entry of muscle satellite cells into the cell cycle requires sphingolipid signaling. J Cell Biol 174:245–253

Nincheri P, Bernacchioni C, Cencetti F, Donati C, Bruni P (2010) Sphingosine kinase-1/S1P1 signalling axis negatively regulates mitogenic response elicited by PDGF in mouse myoblasts. Cell Signal 22:1688–1699

Rapizzi E, Donati C, Cencetti F, Nincheri P, Bruni P (2008) Sphingosine 1-phosphate differentially regulates proliferation of C2C12 reserve cells and myoblasts. Mol Cell Biochem 314:193–199

Rapizzi E, Taddei ML, Fiaschi T, Donati C, Bruni P, Chiarugi P (2009) Sphingosine 1-phosphate increases glucose uptake through trans-activation of insulin receptor. Cell Mol Life Sci 66:3207–3218

Safdar A, Yardley NJ, Snow R, Melov S, Tarnopolsky MA (2008) Global and targeted gene expression and protein content in skeletal muscle of young men following short-term creatine monohydrate supplementation. Physiol Genomics 32:219–228

Sassoli C, Formigli L, Bini F, Tani A, Squecco R, Battistini C, Zecchi-Orlandini S, Francini F, Meacci E (2011) Effects of S1P on skeletal muscle repair/regeneration during eccentric contraction. J Cell Mol Med 15:2498–2511

Squecco R, Sassoli C, Nuti F, Martinesi M, Chellini F, Nosi D, Zecchi-Orlandini S, Francini F, Formigli L, Meacci E (2006) Sphingosine 1-phosphate induces myoblast differentiation through Cx43 protein expression: a role for a gap junction-dependent and -independent function. Mol Biol Cell 17:4896–4910

Suzuki J, Yamazaki Y, Li G, Kaziro Y, Koide H (2000) Involvement of Ras and Ral in chemotactic migration of skeletal myoblasts. Mol Cell Biol 20:4658–4665

Taha TA, Hannun YA, Obeid LM (2006) Sphingosine kinase: biochemical and cellular regulation and role in disease. J Biochem Mol Biol 39:113–131

Tedesco FS, Dellavalle A, Diaz-Manera J, Messina G, Cossu G (2010) Repairing skeletal muscle: regenerative potential of skeletal muscle stem cells. J Clin Invest 120:11–19

Vandenberghe K, Goris M, Van Hecke P, Van Leemputte M, Vangerven L, Hespel P (1997) Long-term creatine intake is beneficial to muscle performance during resistance training. J Appl Physiol 83:2055–2263

Volek JS, Duncan ND, Mazzetti SA, Staron RS, Putukian M, Gómez AL, Pearson DR, Fink WJ, Kraemer WJ (1999) Performance and muscle fiber adaptations to creatine supplementation and heavy resistance training. Med Sci Sports Exerc 31:1147–1156

Yoshida N, Yoshida S, Koishi K, Masuda K, Nabeshima Y (1998) Cell heterogeneity upon myogenic differentiation: down-regulation of MyoD and Myf-5 generates 'reserve cells'. J Cell Sci 111:769–779

Zanin M, Germinario E, Dalla Libera L, Sandonà D, Sabbadini RA, Betto R, Danieli-Betto D (2008) Trophic action of sphingosine 1-phosphate in denervated rat soleus muscle. Am J Physiol Cell Physiol 294:C36–C46

Index

E. Gulbins and I. Petrache (eds.), *Sphingolipids in Disease*,
Handbook of Experimental Pharmacology 216,
DOI 10.1007/978-3-7091-1511-4,

D

E